THE DESCENT OF MAN,

AND

SELECTION IN RELATION TO SEX

THE DESCENT OF MAN,

AND SELECTION IN RELATION TO SEX

by Charles Darwin

With an Introduction by
John Tyler Bonner and Robert M. May

PRINCETON UNIVERSITY PRESS
PRINCETON, NEW JERSEY

Published by Princeton University Press,
Princeton, New Jersey
In the United Kingdom: Princeton University Press,
Guildford, Surrey

Clothbound editions of Princeton University Press books
are printed on acid-free paper, and binding materials are
chosen for strength and durability

Printed in the United States of America by Princeton
University Press, Princeton, New Jersey

Photoreproduction of the 1871 edition published
by J. Murray, London

This facsimile is made from the copy of the 1871 edition in
Firestone Library, Princeton University

First Princeton Paperback printing, 1981

LIBRARY OF CONGRESS CATALOGING IN PUBLICATION DATA

Darwin, Charles Robert, 1809-1882.
The descent of man, and selection in relation to sex.

Reprint of the 1871 ed. published by J. Murray,
London.
Includes index.
1. Evolution. 2. Sexual selection in animals.
3. Sexual dimorphism (Animals) 4. Sex differences.
5. Man—Origin. I. Title.
QH365.D2 1981 575 80-8679
ISBN 0-691-08278-2
ISBN 0-691-02369-7 (pbk.)

VOLUME II

PART II. SEXUAL SELECTION, *continued*

CONTENTS

INTRODUCTION

By John Tyler Bonner and Robert M. May

THE REASON for reissuing Charles Darwin's *Descent of Man* in 1981, one hundred and ten years after its first appearance, is that it addresses an extraordinary number of problems that are, at this moment, on the minds of many biologists, psychologists, anthropologists, sociologists, and philosophers. It is the genius of Darwin that his ideas, clothed as they are in unhurried Victorian prose, are almost as modern now as they were when they were first published.

In this brief introduction, our aims are to indicate the place the book held in its own time, to point out those areas of present-day inquiry where Darwin's comments and questions are particularly relevant, and to identify where his lack of knowledge (particularly in genetics) imposed limitations on his interpretations. The *Descent* (as we shall henceforth refer to it) is unquestionably second only to the *Origin* in the Darwinian canon, and the ground surveyed here is accordingly covered in much greater depth in other works: Gruber and Barrett have given a careful analysis of the development of Darwin's ideas on this subject, especially as reflected in the so-called M and N notebooks; Ghiselin's *The Triumph of the Darwinian Method* is an insightful account of Darwin's general method of working, with Chapter 9 particularly relevant to the *Descent*; and the collection of essays on *Sexual Selection and the Descent of Man*, edited by Campbell, is also helpful.[1]*

* Notes are given at the end of the Introduction.

A separate Note on the Text at the end of this introduction explains why the first rather than the second edition was chosen for reprinting, and why it was not abridged. We also suggest a way of skipping through the book—a *de facto* abridgment—which will enable a reader to encounter most of the basic thoughts and questions, bypassing many of the catalogues of examples.

The Book in its Own Time

content of *The Descent*

The structure of the *Descent of Man, and Selection in Relation to Sex* is unusual in that, as the title clearly states, it consists of two vast subjects which are cemented together at the end by a discussion of the role of sexual selection in man. The major theme of the first part of the work is simply that man descended from other animals and was not specially created. The book, however, does more than marshal the evidence for the continuity between man and other animals (with all the philosophical materialism thus implied). As observed by Gruber, it also represents Darwin's attempt to "study intelligence as a central feature of adaptive change, and to study it in that organism in which it is most prominent, man."[2] The theme of the second part of the book is that besides natural selection there is sexual selection, so that not only will the general character of a species change over time, but the character of the two sexes may also change, as in, for instance, the peacock and the peahen.

Neither part of the *Descent* can be understood without reference back to the *Origin*, whose three-stranded thread of argument runs as follows. First, there is variation among the individuals in populations of plants and animals, both natural and domesticated, and some of this variation is heritable. Second, organisms in nature tend to produce more offspring than can survive to reproductive age if the population is to remain, on average, steady over any length of time. There is thus a "struggle for existence," in which individuals with certain variations may be favored (or "naturally selected"); these favored individuals will tend to spawn children and grandchildren possessing their traits, their variations. Third, it can happen that over the span of geological time new species eventually evolve by such processes of natural selection. Darwin saw sexual selection as an important variation upon the theme of natural selection, with certain traits in the male (or, less commonly, the female) making him (or her) more successful in mating; the result is a dimorphism between the sexes. Darwin shows that these two kinds of selection can act in concert, or not, and if the latter, then the structure of the sexes reaches some sort of compromise between the pulls of natural and sexual selection.

As we shall discuss further below, one of Darwin's greatest problems was to understand how variations arise, how variability is maintained, and how and to what extent variations are inherited. Without an understanding of genetics this enterprise was doomed. Not the least of Darwin's strengths was the courage and the sense eventually to base his theory on the observed—but unexplained—fact that variability existed. However, in 1838 when he began the M and N notebooks, upon which much of the *Descent* is based, he still hoped to discover the basis of variability and heritability; ideas and information about man's intellect looked like a promising testing ground for exploring hypotheses about heritable variation. Hence the discussions of South American natives, and particularly of the three Tierra del Fuegans transported into English society.

DEVELOPMENT OF IDEAS IN *The Descent*

Working from the notebooks[3] Darwin kept in the years 1837-1839—the B, C, D, and E notebooks dealing with evolution in general, and the M and N notebooks with "Man, Mind and Materialism"—Gruber and others have shown that Darwin's thinking about man was, from the start, an integral part of his exploration of evolutionary questions: "the subject of man and his place in nature was so woven into Darwin's thoughts that it forms an indispensable part of the network of his beliefs."[4] Specifically, the first passage in the M and N notebooks which clearly enunciates the principle of natural selection and applies it to man appears in the N book around 27 November 1838.[5] An earlier passage in the more general C notebook says, "I will never allow that because there is a chasm between man . . . and animals that man has a different origin."[6] And the M notebook of 16 August 1838 contains a triumphal passage with all the combative ring of contemporary sociobiology: "He who understands baboon would do more toward metaphysics than Locke."[7]

In short, although Darwin in his public utterances was for a long time reticent concerning his opinion about man, his private thoughts are clear. The basic intellectual edifice presented in the *Origin* and in the *Descent* was in place by 1838, and his subsequent actions may be seen as a "strategy involving two grand detours: the first, a long delay before publishing his general theory of evolution, and the second, another long delay before revealing his ideas on the evolution of man."[8] The reasons for the first of Darwin's long delays, until Wallace's independent enunciation of the principle of natural selection precipitated the publication of the *Origin* in 1859, have been much discussed. The *Origin* does have a chapter entitled "Instincts," which deals with some psychological issues, but the book avoids discussion of higher mental processes, and (as Darwin admits on the first page of the *Descent*) refers to man only through the evasive phrase that "light will be thrown on the origin of man and his history."

The idea that evolution by natural selection could account for the origin of man was taken up by others as a direct result of Darwin's ideas. The respected T. H. Huxley did this explicitly in 1863 in his *Evidence as to Man's Place in Nature.* So did the flamboyant German biologist Haeckel, who even invented an imaginary missing link between ape and man, *Pithecanthropus alalus*, the speechless apeman. One of the principal dissenting voices was that of Alfred Wallace, who published an essay in 1864 saying that the bodily structure of man could be entirely accounted for on the basis of natural selection, but that the mind of man was created by some "higher intelligence."[9] As a result, Darwin could hardly have expected that his book would create any surprises; rather, he must have wanted to present in detail his own position and publish the great mass of evidence he had accumulated. By 1871 he was a famous person; everyone was eager to know what he had to say of this subject, so central at the time.

A convincing explanation for Darwin's two long hesitations is that he had a lively apprehension of the trouble he was inviting. The *Origin*, doing away with the need for Creation, was bad enough; but (as Wallace and many others showed) its message could be reconciled with a basically religious view of man's place in the world. The frank out-and-out materialism of the *Descent* was worse, leaving no role for the Deity to play. Moreover, as Gruber emphasizes, Darwin knew that he ran real risks, more substantial than the simple storms of public controversy.[10] As an undergraduate at Edinburgh University, he had seen a fellow student's paper formally expunged from the records of the Plinian Society because it argued that "mind is material." Earlier, in 1819, a distinguished surgeon, Lawrence, had published his *Lectures on Physiology, Zoology, and the Natural History of Man*; this work was decried as expounding materialism, and Lawrence withdrew the book and resigned his post as lecturer. When a pirate edition was published in 1822, he lost a suit against the publisher, under a law dating back to the Star Chamber of Charles I whereby an author had no property rights to a "blasphemous, seditious, or immoral" work. Darwin refers to this book in the *Descent*.

The fact that the basic ideas in the *Descent* are contained in the M and N notebooks of 1838 does not imply that fears of fierce controversy alone detained Darwin from producing the book then. Much hard work remained to be done, collecting facts and observations to bind to the framework and testify to its soundness. This work, a lot of which was very original, was done around 1867-1871. The project was initially conceived as having three parts, of which the first two (the descent of man and selection in relation to sex) were realized, while the third (the expression of the emotions in man and animals) grew into a separate and important work, published in 1872.[11]

DARWIN'S SCIENTIFIC METHOD

In this general context, it is interesting to look at the way Darwin went about his work. Gruber gives a particularly fascinating discussion of the disparity between what Darwin actually did, and what he said he did. In his books, Darwin consistently portrays himself as cleaving to the accepted Baconian canons: first marshal the facts, then see what conclusions emerge. Thus on page one of the *Origin*, he claims to have "patiently accumulat[ed] and reflect[ed] on all sorts of facts which could possibly have any bearing on it. After five years' work I allowed myself to speculate on the subject, and drew up some short notes. . . ." Likewise, on page one of *The Expression of the Emotions in Man and Animals*, he explains that "I arrived, however, at these three Principles only at the close of my observations." The actuality of the B, C, D, E, M, and N notebooks tells a very different story, and one which is more familiar to a practicing scientist. Gruber summarizes this beautifully: "The pandemonium of Darwin's notebooks and his actual way of working, in which many different processes tumble over each other in untidy sequences—theorizing, experimenting, casual observing, cagey questioning, reading, etc.—would never have passed muster in a methodological court of inquiry. . . . He gave his work the time and energy necessary to permit this confusion to arise, at the same time persistently sorting it out, finding what order he could. It was an essential part of this 'method' that he worked at all times within the framework of a point of view which gave meaning and coherence to seemingly unrelated facts."[12] In addition to the evidence implicit in the notebooks, Darwin's correspondence shows the divergence between private views and public pieties: "all observation must be for or against some view if it is to be of any service!"; "let theory guide your observations, but till your reputation is well established be sparing in publishing theory. It makes persons doubt your observations."[13]

Naively simple formulations of The Way to do science—be they Baconian, Popperian, or otherwise—are, if anything, more pervasive today than they were in Darwin's time. We believe that Darwin got it right when he wrote, "The only advantage of discovering laws is to foretell what will happen and to see bearing of scattered facts,"[14] and that the scrabbling, nonlinear way he pursued this end is typical of most good science. As Ghiselin puts it: "Viewed from without, science appears to be a body of answers; from within, it is a way of asking questions. . . . The 'predictionist thesis' and 'hypothetico-deductive' model seem a bit trivial as clues to what real scientists are trying to do."[15] This struggle toward understanding and "seeing the bearing of scattered facts" is not entirely concealed in the finished products, and the *Descent* is interesting for the persistent questions that underlie the structure of the book.

A criticism often leveled against the theory of evolution by natural selection is that it is little more than a collection of Just So Stories, in which particular facets of behavior or morphology are argued to be "adaptive" or "optimally designed" to fulfill purposes which are tautologically inferred from the behavioral or morphological feature in question.[16] It is true that individual pieces of the evolutionary puzzle, taken one by one without replicates, controls, or comparisons, are indeed each susceptible to one or more *ad hoc* anecdotal explanations. These are the general grounds for Popper's suggestion that evolut onary thinking is metaphysical, and inherently unfalsifiable.[17] Darwin, however, had a fully modern awareness both of this methodological problem and of one answer to it. By deliberately collecting comparative information for large assemblies of different organisms (geographical races, species, or other taxonomic groupings), he showed that it is possible to document broad trends and patterns among behavior, morphology, or biogeographical features. Moreover, these patterns in such phenomena as geographical distribution or vestigial organs can often be correlated with systematic patterns in the differing environmental circumstances of the organisms under study.

Such comparative studies can then permit predictions to be made about as yet unstudied species. Some instances are given below. For example, the relationship between the degree of sexual dimorphism and the socioeconomic sex ratio (females per reproductive male) recently obtained by Clutton-Brock, Harvey, and Rudder for primate species would enable a rough prediction to be made about either one of these quantities once the other is determined for a species not included in the original study.[18] Whatever their philosophical status, these methods are capable of generating testable predictions that are not qualitatively different from those of the physical sciences. The extreme assertion that evolutionary biology can never aspire to more than "thick description" of individual cases, each unique (as advocated on similar grounds in a different context by some social scientists),[19] fails to comprehend the power of the comparative methods employed by Darwin and refined by later workers.

In one important way, however, many predictions in the biological sciences do differ from those we are familiar with in classical physics. In physics and engineering, most simple predictions or tests of hypotheses are crisply and pleasingly deterministic. But in population biology, ecology, and evolutionary biology, many of the predictions are inherently probabilistic (as, for similar reasons, are the predictions of meteorology or portfolio theory). This point is emphasized by Gruber and Barrett, Monod,[20] and others, and it has implications that range from the grand sweep of evolutionary thinking to important practical aspects of fisheries management.[21]

DARWIN'S STYLE OF PRESENTATION

Finally, it is worth commenting on Darwin's style of presentation. The language is simple and vivid, and to a reader accustomed to the elaborately "objective" circumlocutions of much of the contemporary social sciences it may appear alarmingly anthropomorphic and teleological. Ghiselin analyzes this issue in detail, showing clearly that there is a world of difference between Darwin's "metaphorical use of anthropomorphic expressions and the propositions which he actually asserts." As Ghiselin demonstrates, Darwin "used everyday terminology to convey precise and definite meanings, with elegance and clarity. For instance, Darwin gives two pictures in which he shows the contrasting appearance of cats under conditions involving precisely opposite kinds of behavior. These are entitled 'Cat, Savage, and Prepared to Fight . . .' and 'Cat in an Affectionate Frame of Mind. . . .' Contemporary biologists may regard these captions with amusement, even delight, for the prevailing standards of pedantry are opposed to such expressions, even though nobody ever would take them literally, and even though they could scarcely better express the underlying ideas."[22] Darwin preferred playing with dogs to engaging in ludic activity with canine companions.

Probably the most striking stylistic difference between the book Darwin wrote in 1871 and the one he would write if he lived today lies in his discussion of the races of man and the differences between the human sexes. For his period he is remarkably objective on the matter of race, although he does make it clear that to be civilized means to be like an educated Englishman. Here he was partly influenced by his voyage on the HMS *Beagle* as a young man, where the unspeakable behavior of the Tierra del Fuego Indians impressed him enormously. They had no religion, no care of their hair or their rude clothing, no obvious code of morals, and they ate their grandmothers first when food was short. Yet he sees that those natives who were taken to be educated for a period in England did appear to become civilized; they were capable of improvement. Darwin had the capacity to see beyond the prejudices and ideologies of his own time and culture—more so, indeed, than many who work in these general areas today.

Those readers who are sensitive to sexism may be provoked to considerable outrage by Darwin's short discussion of "the difference in the mental powers of the two sexes." Here he is reflecting no more than the common view of the time, and this unreflective attitude is somewhat mitigated by a curious and somewhat obscure passage on page 329 where he seems to imply that women could become like men in their mental facilities if they were provided with suitable opportunities and training. (It may be remarked that his diligently kept record of backgammon games with his wife show them roughly evenly matched, with Darwin at one time ahead by 2,795 to 2,490.[23] Did he not realize the highly analytical character of this game? Most likely, he never thought about it.)

Some Scientific Problems Confronting Darwin's Work in the Nineteenth Century

Discussion of the public reception of Darwin's work tends to center around the religious and other broadly philosophical issues. It is often not fully appreciated that the theory of natural selection met with very serious scientific objections.

Without a precise understanding of genetics and the laws of Mendelian inheritance, the causes and heritability of variation among individuals in natural and domesticated populations are bound to be a bit mysterious.

Around 1838, Darwin still hoped to solve this mystery and place his theory on a solid base. As we have seen, the M and N notebooks from which the *Descent* sprang were begun partly in the hope of using man's origins to shed light on the sources and mechanisms of heritable variation, and to test hypotheses about habits becoming hereditary. In retreating to the position of taking heritable variation to be an unexplained premise, Darwin was condemned to a lifelong struggle with two large problems which he never satisfactorily resolved.

The first problem stemmed from the received wisdom of Darwin's time, namely, that inheritance worked by a blending of maternal and paternal characters. As emphasized to Darwin by Fleeming Jenkin,[24] under blending inheritance variation simply cannot be maintained! The essentials of the argument can be grasped by considering a trait (such as height or weight) that can be described by a single variable.[25] Suppose the mother departs from the population average in this respect by an amount x, and the father by an amount y. Then, under a scheme of blending inheritance, the progeny will depart from the mean by $\frac{1}{2}(x + y)$. In the parental generation, suppose the statistical scatter of the variable about its mean value is characterized by a variance σ^2; that is, the expectation values of x^2 and y^2 are both σ^2 ($\langle x^2 \rangle = \langle y^2 \rangle = \sigma^2$). In the next generation, the corresponding variance is the expectation value of $(\frac{1}{4})(x + y)^2$ or $(\frac{1}{4})(\langle x^2 \rangle + \langle y^2 \rangle + 2\langle xy \rangle)$, which is equal to $\frac{1}{2}\sigma^2(1 + \rho)$. Here ρ is the correlation coefficient between x and y. If mating is at random, $\rho = 0$, and the variance of the trait among the offspring is $\frac{1}{2}\sigma^2$; the variance is halved in a single generation. Even if there is a tendency for like to seek like in mating, so that $\rho \neq 0$, the variance will still decrease in each generation (except in the unlikely extreme of perfectly sorted mating, with $\rho = 1$). In short, the mid-nineteenth-century conventional wisdom of blending inheritance was flatly inconsistent with the observed propensity for natural populations to exhibit variability. Darwin was troubled by this inconsistency. The answer, of course, lies in the fact that genes are inherited in particulate Mendelian fashion, not by "blending." The basic theorem of Mendelian population genetics—the Hardy-Weinberg theorem, proved in 1908—can be rephrased to state that, in the absence of perturbing factors (such as mutation, selection, drift, migration, or nonrandom mating), variance remains constant from generation to generation.

Mendel's paper on the laws of heredity, providing the key to the puzzle, was published five years before the *Descent* appeared.[26] Although in German, the paper was not in an obscure journal, and was accessible to Darwin and his colleagues. Fisher has made the interesting and plausible suggestion that Mendel's work was overlooked because it was cast in a mathematical idiom, which was truly a foreign language to nineteenth-century British naturalists. De Beer notes Darwin's regret for his ignorance of mathematics;[27] if Fisher's explanation is correct, Darwin had more to regret than he realized! It remained for the early twentieth century to rediscover Mendel.[28]

The second problem associated with Darwin's lack of basic understanding of genetics is that he did not fully comprehend the distinction between what we would call gene inheritance and the inheritance of habits, customs, and behavior. In the *Descent* and elsewhere, Darwin repeatedly says that if an animal behaves in a particular way for a number of generations, this will result in the behavior becoming permanently fixed. Today we know there are some sorts of behaviors that do become genetically fixed, but we understand that the mutations which produce these changes have nothing to do with transmitting the information by teaching and learning. There is a vast difference between behavioral transmission and genetic transmission of information.

In addition to the central difficulty of dealing with variation and heritability in the absence of Mendelian genetics, Darwin had other technical problems when he turned to sexual selection.[29] In general, discussions of sexual selection involve the concepts of sex-linked genes and the role of hormones in producing secondary sexual characteristics. These concepts, whose implications are pursued further below, were not available in Darwin's time.

Another, and quite different, class of difficulties arose because the only fundamental energy sources known to physics in Darwin's day were those associated with the electromagnetic and gravitational forces. Kelvin showed that if the sun's energy source was gravitational, it could not possibly have been burning for more than about 20 million years, and that chemical (electromagnetic) fuels would have given an even shorter life. A different calculation showed that it could not have taken more than around 20-40 million years for the earth to cool from molten rock to its present temperature. These two calculations meant that either the earth was at most a few tens of millions of years old, or that Victorian physics was fundamentally deficient. Faced with Kelvin's arguments, Darwin removed all numerical references to geological time spans in the third and later editions of the *Origin*,[30] and you will look in vain for any explicit chronology in the *Descent*. The discovery of the weak and strong nuclear forces has, of course, shown us that Victorian physics was indeed fundamentally deficient in some respects: the sun has burned nuclear fuel for nearly five billion years; and the heat generated by decay of radioactive elements inside the earth invalidated Kelvin's calculations about cooling rates. Darwin was nearer our modern view in his first purely geological calculations; there is enough time to account for the evolution of man.

The Book in Relation to Research Today

PART ONE: THE DESCENT OF MAN

In Part One of the *Descent*, Darwin argues from comparative analyses of morphology and behavior that man is basically not different from other animals, and that what differences do exist are simply a matter of degree. His arguments about the bodily structure of man, with the exception of the brain, were relatively uncontroversial in his own time, and are entirely acceptable today.

The real controversy surrounding the relation between man and other animals in Darwin's time was essentially a problem of science versus religion. As we have mentioned, even Wallace wrote an essay saying the morphology of man could be entirely accounted for by natural selection, but that man's "intellectual and moral faculties . . . must have had another origin . . . in the unseen universe of Spirit."[31] Except for a disturbing resurgence of anti-intellectual fundamentalism in North America in recent years, this idea that the Creation must have supervened somewhere along the evolutionary line leading to man is no longer the issue. The idea has, however, been replaced by a doctrine held firmly by many social scientists. They see human culture and civilization as being something so special and so unlike anything elsewhere in the animal world that it can only be analyzed in its own terms, and not in terms of the level below (that is, not in terms of biology). The argument is interestingly similar to that of Wallace and other religious people in the nineteenth century: the body of man is indeed a biological structure, clearly descended from the apes, but his culture, which stems from his extraordinary and unique mind, is on a new, higher hierarchical level of its own; evolutionary biology has nothing to tell us about this higher level. This resistance to exploring the possibility that principles of evolutionary biology can shed light on human societies is most recently demonstrated in the reaction[32] of some sociologists, anthropologists and others to E. O. Wilson's *Sociobiology: The New Synthesis* when it was published in 1975.[33] Indeed, many people continue to bring to these issues an unseemly degree of dogmatic certitude.

Setting aside the large questions of culture and social organization, there remain other human attributes—consciousness, morals, language, and other mental qualities—that many people, both in Darwin's time and today, believe clearly separate man from the beasts. In recent years, through the rise of the study of animal behavior, these ancient and entrenched views are again being challenged.

The whole question of consciousness has been reappraised by Griffin in his *The Question of Animal Awareness*, published in 1976.[34] This book, which makes full use of modern work in animal psychology and behavior, comes out strongly for the idea that the difference between man and lesser animals is one of degree, and that there is a continuum. The surprising thing is that when one rereads Darwin's *Descent*, although his facts are not the same, his main points seem completely consistent with the modern views of Griffin and those who have followed him.

More broadly, Darwin's M and N notebooks, and particularly those parts distilled into Chapters 3, 4, and 5 of the *Descent*, establish him as a seminal figure in psychology. As discussed by Gruber, in the M and N notebooks "we can see the wide range of psychological topics Darwin touched upon in the years 1837-39; memory and habit, imagination, language, aesthetic feelings, emotion, motivation and will, animal intelligence, psychopathology, and dreaming."[35] The method, as illustrated in the *Descent* and in *The Expression of the Emotions in Man and Animals* (many of the insights of which are only just being rediscovered by contemporary psychologists),[36] is to search for mental resemblances between man and other animals, and to indicate the line that a more fully developed psychology might take. The *Descent* makes good Darwin's boast at the end of the *Origin*: "Psychology will be based on a new foundation, that of the necessary acquirement of each mental power and capacity by gradation."[37]

The more philosophical question of what evolutionary theory, and natural selection in particular, have to say about human morals and ethics has a long history. Most evolutionary biologists, Darwin included, have succumbed to the temptation to put in a word on this subject. These questions have seen a flurry of activity recently, much of it inspired by new studies in animal behavior. See, for instance, Wilson's *On Human Nature*, Alexander's *Darwinism and Human Affairs*, and the proceedings of a conference on *Morality in Animals*, edited by Stent.[38] Again, the surprising thing is that what Darwin has to say about the subject of these books seems so up to date.

On another philosophical theme, essentially all biologists would agree that Darwin's work, and particularly the *Descent*, disposed of the "mind/brain" dialogue once and for all. As Gruber observes, Darwin once expressed how inseparable he thought mind and brain were by bracketing "intellectual faculties" and "cerebral structure" as equivalent in a passage in the B notebook.[39] It takes much ingenuity and verbal cleverness, plus considerable ignorance of contemporary biological research, to keep this topic alive.

Before going on to take up sexual selection, we return to the general problems of the evolution of culture and social organization in man and other animals, and note some of the directions current research is taking. This survey of contemporary work in sociobiology is necessarily only a very superficial one.

Much of this recent progress stems from a clearer definition of just what it is that is being "naturally selected." Following the rediscovery of Mendelian genetics, the neo-Darwinian revolution of the first half of this century welded together Darwin's basic ideas about natural selection with a rigorous description of the way gene frequencies can change in populations.[40] The upshot was a precise definition of an individual's "Darwinian fitness," which essentially measures the net number of offspring that will, on average, survive to reproductive age. This Darwinian fitness clearly depends on many factors (all of which are therefore susceptible to evolution by natural selection): the probability of successful mating (which can depend in a complicated way on the mating system), the average ratio of males to females among the offspring, along with the more commonly stressed probability that offspring will survive the "struggle for existence," and themselves reproduce. This technical definition of fitness, however, leaves the focus on individuals and their direct progeny. Mainly as a result of the work of Hamilton,[41] it is now realized that the important thing is not how many direct descendants an individual has, but how many of his or her genes get into the next generation. Thus in a diploid sexual system such as possessed by humans, your own offspring share, on average, half your genes, but your brother's children also share one-quarter of your genes, your first cousin's offspring one-sixteenth, and so on. An individual's input of genes into the next generation thus involves not only his or her own net reproductive success or fitness, but also those of all his or her relations, discounted by the degree of relatedness; this quantity has been christened "inclusive fitness" by Hamilton, and it provides a powerful tool for exploring the evolution of some kinds of social behavior.

In particular, the notion of "inclusive fitness" helps explain many apparently altruistic acts observed in nature, such as uttering an alarm call upon noticing a predator, or helping to raise a relative's offspring. Hamilton notes that most of the social insects possess a haplo-diploid sexual structure, whereby females are more closely related to their sisters than to their own offspring, and he suggested that this characteristic may predispose such creatures toward evolving the "eusocial" behavior actually found, in which the sterile female workers help raise more sisters. In the *Origin*, Darwin dwelt at length on the problems posed for his theory by the sterile castes among the social insects, seeing them as "one special difficulty which at first appeared to me insuperable, and actually fatal to my whole theory."[42] With typical insight, he saw that kinship structure within the colony could possibly provide an answer. But, lacking the formal apparatus of modern theory about "inclusive fitness," Darwin could not imagine the fascinating quantitative studies now being done on the social insects as test cases for the evolution of social behavior.[43]

Wilson's *Sociobiology: The New Synthesis* gives a synoptic account of research applying evolutionary thinking to the ecology and behavior of social groups. This research includes field, laboratory, and theoretical studies of territoriality, mating systems, the optimal size of groups, communication within and between groups, social parasitism, energetic and other aspects of foraging, and a host of other things. Some good recent surveys are *Behavioural Ecology* by Krebs and Davies, the collection of papers edited with a commentary by Clutton-Brock and Harvey, and the proceedings of a Dahlem Conference held in 1980.[44]

Many of the studies are characterized by the sort of comparisons so much used by Darwin in the *Descent* and elsewhere. For example, Short has investigated the way various sexual characteristics (size of penis, breasts, etc.) differ among humans, chimpanzees, gorillas, and orangutans.[45] This general subject is touched on briefly in the *Descent*, and less coyly in the M and N notebooks. Short also shows that the patterns of morphological difference correlate well with the mating habits of the species. In a series of interesting papers, Clutton-Brock and Harvey[46] have compiled information about the ecology and social organization of primate societies, and have documented patterns of relationship between quantities such as population density, degree of sexual dimorphism (male weight divided by female weight), "socioeconomic sex ratio" (number of adult females per adult male in breeding groups), the size of feeding and of breeding groups, the size of the home range, and the daily path length of the feeding group. Bertram[47] has made a similar survey of the relation between the behavioral ecology and the social systems of the major vertebrate predators on the Serengeti (lion, leopard, cheetah, hyena, wild dog). Some of the patterns or "evolutionary rules" emerging from these comparative studies are, moreover, what one would deduce by applying optimality arguments to foraging.

As evidenced by the books by Wilson and Alexander referred to above, there is a current surge of interest among biologists in the application to man of the principles of sociobiology. Also, many anthropologists are actively seeking correlations between the social activities of different human societies and the predictions of sociobiology.[48] As Barash observes in reviewing Alexander's book: ". . . research on animals can provide propositions about the manner in which natural selection appears to act upon behaviour, even complex social behaviour. Then these predictions—based ultimately on fitness maximization—can be tested cross-culturally. Thus, the world's human societies constitute a global experiment, with many varying cultures and one constant: our biology."[49]

One important aspect of the enterprise of applying to man the principles derived from the study of social animals has been the examination of the relation of cultural change to evolutionary change. If culture is considered the transmission of information by behavioral means, as Bonner has used the term, then, as he shows, one finds many examples of culture among animals that seem to foreshadow the remarkable ability of man to teach and learn.[50] This was understood by Darwin, and in the *Descent* he gives some good examples of behavioral transmission, or culture among animals. Some modern workers are making a bold attempt to understand the role of simultaneous genetic and cultural transmission in evolutionary change and to put the subject in a constructive theoretical framework. For this it is necessary to extend the technical apparatus of population genetics to embrace the effects of learning and teaching. Essentially, this widens the scope of population genetics to include Lamarckian inheritance of acquired traits. The creation of this body of theory has recently been begun by Cavalli-Sforza and Feldman, Lumsden and Wilson, and others.[51] One of the principal reasons we now seem to have progressed further than Darwin is that, while he appreciated the problems, he lacked one ingredient: genetics.

PART TWO: SELECTION IN RELATION TO SEX

Summarizing passages from the C notebook of 1838, Gruber shows that "in its nascent form the idea of natural selection is suffused with the special notion of sexual selection."[52] This contradicts suggestions that one reason for the extensive treatment of sexual selection in the *Descent* is that Darwin in later life felt impelled to emphasize it to prop up the weakened case for the more general idea of natural selection. Darwin did, however, see "sexual selection" as a mechanism somewhat distinct from "natural selection" (which he often tended to treat as pertaining to survival), and thought it worth pursuing at length in the *Descent*.

A more modern view sees sexual selection as simply one of many particular facets of general questions of natural selection. As we have already pointed out, the current definition of Darwinian fitness deals with an individual's total genetic input into the next generation, and thus includes consideration of mating systems and sex ratios along with simple survival to reproductive age. Thus a species' physical and biological environment is seen as influencing the kind of mating system it adopts (for example, monogamy, polygamy, promiscuity), which in turn influences the evolutionary premium likely to be put on sexual displays and dimorphism. These sexual factors simply have to be weighed along with other factors having to do with foraging, defense, and survival in general; Darwinian fitness is the appropriately weighted sum of all these factors, and it is upon this overall quantity that natural selection acts.

These ideas may be made concrete by considering the evolutionary forces that can bear upon an organism's color.[53] First, color is influenced by questions of getting your own food and avoiding being food for other animals. This can lead to camouflage, both for prey (zebras; arthropods that look like bird droppings or leaves or twigs) and predators (tigers; mantises; angler fish). Alternatively it can lead to brightly colored advertisement of distastefulness or poisonousness; associated with this is mimicry, either Mullerian (where two distasteful species look the same, so that predators have less to learn) or Batesian (in which an edible species cheats by simulating the coloration of a distasteful one). Second, thermodynamic considerations of keeping warm or cool or (for some desert-dwelling burrowers) changing temperature rapidly are influenced by color. Arguably, large herbivores with no significant predators are likely to have their color determined by thermodynamic efficiency; hence the dull gray of elephants, rhinoceroses, and hippopotamuses. Third, sexual selection may give an advantage to brightly colored males or, less commonly, females.

Which of these several factors will assume predominant influence depends on the ecology of the species. Sometimes the explanation remains unclear; there is still debate as to whether polar bears are white for thermodynamic reasons, or for camouflage as they stalk their prey amid the snow and ice. Some recent studies give beautiful examples of the tensions that exist between coloration for sexual selection and for predator avoidance. To mention one, Endler has shown that male guppies living in the rivers of Trinidad have very bright red spots if they are in a section of the stream where there are no predators, but in sections where there is heavy predation from another fish, the male coloring is much subdued.[54]

A reading of the *Descent* reveals that, basically, Darwin understood all this. Moreover, his ideas about sexual selection, as such, are essentially correct (which is just one more reason for the general respect in which he is held by biologists today). But the dichotomy between sexual selection and natural selection (usually taken by Darwin to refer simply to survival) is discordant with contemporary usage.

Many recent studies of sexual selection are in the spirit of Chapter 16 of the *Descent,* in which Darwin uses ecological insights and comparative studies to identify six classes of cases for the color of the plumage of immature birds in relation to that of the adult birds. For example, in *Ecological Adaptations for Breeding in Birds*, Lack[55] compiles information about the percentage of bird species that are monogamous, polygynous, polyandrous, and promiscuous. He relates these mating habits to the diet of the various species, and goes on to discuss the significance of the pair-bond and sexual selection in birds. Orians[56] has widened the discussion to consider the evolution of mating systems in both birds and mammals; he considers the general ecological circumstances favoring the various kinds of mating systems, and shows how this can explain such trends as, for instance, that 92% of all bird species are monogamous, while monogamy is uncommon among mammals. These and other patterns in mating systems and sexual selection are reviewed by Maynard Smith in his recent book, *The Evolution of Sex*.[57]

Male-female dimorphisms are not all necessarily forged by sexual selection. In some cases, the sex differences can be accounted for by ecological factors, usually an expansion of the niche of the species. Thus, on islands, the dimorphism between the beaks of male and female woodpeckers is often significantly more pronounced than for the same species on the mainland;[58] the woodpeckers use this device to broaden their niche, in the absence of many species that constrain them on the mainland. Another example among birds is the remarkable dimorphism in the beaks of the now extinct New Zealand huia. In discussing the huia and other such cases, Darwin considers them primarily the result of natural selection due to ecological factors, but suggests the possibility that the dimorphism might also be initially affected by sexual selection. For further discussion, see Selander's chapter in *Sexual Selection and the Descent of Man.*[59]

As mentioned above, the discussion of sexual selection in the *Descent* is greatly hindered by Darwin's lack of understanding of the specific biological mechanisms producing differences between males and females. As a corollary to nineteenth-century ignorance of the mode of genetic sex determination in different animals, biologists were ignorant of the whole concept of sex-linked genes. Darwin saw that something of this kind was necessary to explain many of the phenomena he discussed, but there is a noticeable gap that we must close as we read his pages. We know now that in most animals, including vertebrates, which have been intensively studied, the genetic constitution and even the chromosome complement of the sexes differ. This means that the sexes may have different genes which affect their appearance on both the sex chromosomes and all the other chromosomes (autosomes). In humans, for instance, women carry two similar *X* chromosomes, while males carry only one *X* which is paired with a quite different *Y* chromosome. The result is that genes on the sex chromosomes may be expressed differently in the opposite sexes. To give an example, in a woman a single recessive gene at one locus on the chromosome will be suppressed by the dominant gene on the other, homologous *X* chromosome, and as a result she will show the dominant trait. However, in a man the recessive gene will be expressed as there is no homologous *X* chromosome to hold a dominant gene. This exactly describes the situation of haemophilia where men suffer the disease if they have the mutant gene, but women will not have it if they have only one such copy. More generally, characters connected with the appearance of males and females are carried on both the sex chromosomes and the autosomes.

By the same token, the role of hormones in the growth and divergence of secondary sexual characters was not known in Darwin's time. He came close to seeing the general idea, pointing out, for instance, that castrated males often appeared like the females or the young. But the detailed mechanism remained a mystery, and he could only talk vaguely about all the sexual characters working together to produce the extreme male coloration or structure. Today we know that sex can be genetically determined by genes on the chromosomes. But the way the differences between the adult male and female characters appear in mammals (which have been studied most intensively) is also by means of hormones. At an early age, usually in the foetus, certain genes induce the production of sex hormones in the male; the female, at least according to current evidence, differentiates directly without significant help from hormones. In the male the chemical messengers stimulate the growth of specific tissues so that both the primary and secondary sexual characters are altered from the female mode to the male. These include the characteristic structure of the genitalia and all the other features of bodily structure: hair distribution, absence of mammary glands, increased size, changes in the skeletal structure, some neuronal patterns in certain regions of the brain, and sexual behavior. Again it will be helpful to the reader to keep all this in mind in the appropriate places.

Another matter which was of concern to Darwin in the

Descent is sex ratios. This subject is again under intensive study because, as discussed by Fisher, and carried further by Hamilton, sex ratios at birth should also be considered in any general discussion of sexual selection and mating systems.[60] This is especially true as one moves from the vertebrates to the larger world of invertebrate species. In an elegant series of papers, Ghiselin, Charnov, Leigh, and others have combined theories of sexual selection and of sex ratios to show that it would benefit some kinds of animals to evolve the ability to change from female to male as they grow older.[61] Charnov has further shown in quantitative detail that some shrimp species conform to these theoretical predictions.[62]

Underlying all this is the question of the evolutionary advantage of sex itself. Why should a female produce offspring carrying only half her genes, when by parthenogenesis or otherwise she could produce clones of herself? The simple answer that the variability produced by sexual recombination makes for greater adaptability, and is therefore "for the good of the species," will not serve. Darwinian natural selection, in both nineteenth century and modern forms, has to do mainly with individuals, and selection for group characteristics has no simple place. The reasons for the evolution of sex remain of absorbing interest today, and are the subject of much argument; see, for example, the work of Ghiselin, Williams, and Maynard Smith.[63]

Conclusion

It is often said that a big difference between science and letters is that in the former the early works provide a foundation of what is to come, but they are no longer quoted. One never sees the name of van Leeuwenhoek or Pasteur in the bibliography of a journal article in microbiology today, yet everyone agrees that their contributions were great milestones. But Shakespeare or, more appropriately, Dickens and Trollope continue to be read with interest and pleasure.

Darwin seems to fall into an intermediate category. He is still read and is cited, we suspect, more than any other nineteenth-century scientist. The *Descent*, for example, runs around forty entries annually in the Science Citation Index in recent years, and the number seems to be increasing. The reason, as we have seen, is that the ideas, the questions, and the methods set out in the *Descent* anticipate much of the work now going on at the frontiers of biology, psychology, sociology, and anthropology, as we try to grasp the evolutionary basis of social organization in animals. Darwin's books, and the *Descent* in particular, make it clear that many of our insights of the last few years were part of his thinking over a century ago.[64]

A Note on the Text

The text reproduced here is the first issue of 1871 of which 2,500 copies were printed (retailing for 24 shillings).* Some readers may wonder why we chose to reprint this first issue rather than a subsequent issue of 1871 or the second edition of 1874. The reason is partly that Darwin had an unfortunate habit, in his revisions, of rewriting some of the freshness out of the initial work.

* This is item no. 937 in the standard handlist, R. B. Freeman, *The Works of Charles Darwin: An Annotated Bibliographical Handlist*, 2nd ed. (Folkestone, England and Hamden, Conn., 1977).

He was very sensitive to criticism, and tried hard to satisfy all his critics by making appropriate alterations and accommodating conflicting points of view. This process is far more evident in the *Origin*, where the first edition nowadays seems much superior to the sixth and last edition. But, to some extent, the problem also arises in the *Descent*. Still, the later revisions do include several new examples which Darwin came across after 1871, some of which are interesting. In the 1874 and later editions he included a table of the principal additions and corrections made since the first printing. We reprint this table, taken from the 1913 edition, following the index in the present volume.

A case can be made for abridging the *Descent*. In the leisurely Victorian tradition, it is an enormously long work. Much of it consists of an overwhelming accumulation of all the evidence Darwin could bring to bear; in particular, Part Two, on sexual selection, includes some 400 pages in which all the instances of sexual dimorphism in the animal kingdom of which he was aware are described, one after the other. A good deal of this could be pruned without losing any of the ideas or the essential flavor of the book. Nonetheless, although such abbreviation could have produced a more crisply readable text, it was decided—for reasons of archival value and service to those who, with Darwin, enjoy the details—to reprint the original in full.

A do-it-yourself abbreviation is, however, easily accomplished. We suggest reading all of Part One on the descent of man (the first seven chapters). Then, after reading the introductory chapter of Part Two on sexual selection (Chapter 8), the reader may skip the next ten chapters (Chapters 9 through 18) which comprise the catalogue of examples of sexual dimorphism. Chapters 19 and 20, on sexual selection in man, should then be read, as should the concluding and summarizing Chapter 21.

Notes and References

1. H. E. Gruber and P. H. Barrett, *Darwin on Man: A Psychological Study of Scientific Creativity* (by H. E. Gruber), *Together with Darwin's Early and Unpublished Notebooks* (transcribed and annotated by P. H. Barrett), Dutton, New York, 1974. (Hereafter cited as GB.) M. T. Ghiselin, *The Triumph of the Darwinian Method*, University of California Press, Berkeley, 1969. B. Campbell (ed.), *Sexual Selection and the Descent of Man, 1871-1971*, Aldine-Atherton, Chicago, 1972.

2. GB, p. 179.

3. Gruber and Barrett (GB) give the chronology of the notebooks as follows: The notebooks on transmutation of species: B, July 1837-February 1838; C, February-July 1838; D, July-October 1838; E, October 1838-July 1839. The M notebook on "Metaphysics—Morals and Speculation on Expression," 15 July 1838-1 October 1838). The N notebook on "Metaphysics and Expression Selected for Species and Theory," 2 October 1838-1 August 1839.

4. GB, p. 10. The designation "Man, Mind and Materialism" is Gruber's, not Darwin's.

5. GB, pp. 338 and 370.

6. Quoted in GB, p. 252.

7. Quoted in GB, p. 243.

8. GB, p. 24.

9. T. H. Huxley, *Evidence as to Man's Place in Nature*, Williams and Norgate, London and Edinburgh, 1864. E. Haeckel, *Die Naturliche Schopfungsgeschichte*, 1868 (referred to on p. 4 of the *Descent*). A. R. Wallace, The Development of Human Races Under the Law of Natural Selection, *Anthropological Review*, May 1864. Reprinted in his book *Natural Selection and Tropical Nature*, Macmillan, London, 1895.

10. GB, pp. 203-208.

11. Charles Darwin, *The Expression of the Emotions in Man and Animals*, John Murray, London, 1872.

12. GB, p. 122.

13. Quoted in GB, p. 123.

14. Quoted in Ghiselin (see note 1 above), pp. 235-236.

15. Ghiselin, p. 236.

16. See, e.g., S. J. Gould and R. C. Lewontin, The spandrels of San Marco and the Panglossian paradigm: a critique of the adaptionist programme, *Proc. Roy. Soc., B205*, 581-598 (1979), and references therein.

17. K. R. Popper, *Unended Quest*, Open Court Publishing Co., LaSalle, Ill., 1976. For an excellent critical analysis of Popper's position see B. Halstead, Popper: good philosophy, bad science?, *New Scientist, 87*, 215-217 (17 July 1980).

18. T. H. Clutton-Brock, P. H. Harvey, and B. Rudder, Sexual dimorphism, socioeconomic sex ratio and body weight in primates, *Nature, 269*, 797-800 (1977).

19. See C. Geertz, *The Interpretation of Culture*, Chapter 1 ("Thick Description, Towards an Interpretive Theory of Culture"), pp. 3-30, Basic Books, New York, 1973.

20. J. Monod, *Chance and Necessity: An Essay on the Natural Philosophy of Modern Biology*, Knopf, New York, 1971.

21. See, e.g., J. A. Gulland, The stability of fish stocks, *J. Cons. Int. Explor. Mer., 37*, 199-204 (1975) and R. M. May, J. R. Beddington, J. W. Horwood, and J. G. Shepherd, Exploiting natural populations in an uncertain world, *Math. Biosci., 42*, 219-252 (1978).

22. Ghiselin, *op. cit.*, pp. 188 and 189.

23. H. Litchfield (ed.), *Emma Darwin: A Century of Family Letters*, Appleton and Co., New York, 1915, Vol. II, p. 221.

24. The history of this problem is discussed, e.g., by G. De Beer in *Charles Darwin: A Scientific Biography*, Doubleday, New York, 1964, pp. 174-175, and more fully in D. Hull, *Darwin and His Critics: The Reception of Darwin's Theory of Evolution by the Scientific Community*, Harvard University Press, Cambridge, Mass., 1973, pp. 302-350. The basic scientific issues are lucidly set out by R. A. Fisher in *The Genetical*

Theory of Natural Selection, Oxford University Press, Oxford, 1930, Chapter 1.

25. The technical details here follow R. A. Fisher, *op. cit.*, pp. 4-5.

26. G. J. Mendel, Versuche über Pflanzen-Hybriden, *Verh. Naturf. Ver. in Brunn*, Vol. X, 1865 (the 1865 issue of this journal actually appeared in 1866).

27. G. De Beer, *op. cit.*, p. 179.

28. See. e.g., E. Mayr and W. B. Provine (eds.), *The Evolutionary Synthesis: Perspectives on the Unification of Biology*, Harvard University Press, Cambridge, Mass., 1981.

29. For an introduction to the quantitative theory of sexual selection, see, e.g., P. O'Donald, *Genetic Models of Sexual Selection*, Cambridge University Press, Cambridge, 1980.

30. See front matter of the third edition (1816) of the *Origin*.

31. Quoted in GB, p. 31, from A. R. Wallace, *Darwinism: An Exposition of the Theory of Natural Selection with Some of Its Applications*, Macmillan, London, 1889.

32. See, e.g., M. Sahlins, *The Use and Abuse of Biology: An Anthropological Critique of Sociobiology*, University of Michigan Press, Ann Arbor, 1976 (a critical review of this book is by N. Humphrey, *New Scientist*, 75, 30 [7 July 1977]); A. Montagu, *Sociobiology Examined*, Oxford University Press, Oxford, 1980; C. Geertz, Sociosexology (review of D. Symons, *The Evolution of Human Sexuality*, Oxford University Press, Oxford, 1979), *New York Review of Books*, *26* (nos. 21 and 22), pp. 3-4 [24 January 1980]); G. W. Barlow and J. Silverberg (eds.), *Sociobiology: Beyond Nature/Nurture?*, Westview Press, Boulder, Colo., 1980; A. L. Caplan (ed.), *The Sociobiology Debate*, Harper and Row, New York, 1978.

33. E. O. Wilson, *Sociobiology: The New Synthesis*, Harvard University Press, Cambridge, Mass., 1975.

34. D. R. Griffin, *The Question of Animal Awareness*, Rockefeller University Press, New York, 1976.

35. GB, p. 221.

36. P. Ekman (ed.), *Darwin and Facial Expression: A Century of Research in Review*, Academic Press, New York, 1973. P. Ekman and H. Oster, Facial expressions of emotion, *Ann. Rev. Psych.*, *30*, 527-554 (1979).

37. Charles Darwin, *On the Origin of Species*, John Murray, London, 1859, p. 488.

38. E. O. Wilson, *On Human Nature*, Harvard University Press, Cambridge, Mass., 1978. R. D. Alexander, *Darwinism and Human Affairs*, University of Washington Press, Seattle, 1979. G. S. Stent (ed.), *Morality in Animals* (Report of a Dahlem Workshop), Verlag Chemie, Weinheim, 1979.

39. Quoted in GB, p. 21, from the B notebook, about September 1837.

40. For an introduction to the revolution wrought by Fisher, Haldane, and Wright, see e.g., J. F. Crow and M. Kimura, *An Introduction to Population Genetics Theory*, Harper and Row, New York, 1970.

41. W. D. Hamilton, The genetical evolution of social behaviour, *J. theor. Biol., 7*, 1-52 (1964).

42. Charles Darwin, *On the Origin of Species* (see note 37 above), p. 290.

43. R. L. Trivers and H. Hare, Haplodiploidy and the evolution of the social insects, *Science, 191*, 249-263 (1976). R. D. Alexander and P. W. Sherman, Local mate competition and parental investment in social insects, *Science, 196*, 494-500 (1977).

44. J. R. Krebs and N. B. Davies (eds.), *Behavioural Ecology: An Evolutionary Approach*, Blackwell, Oxford, 1978. T. H. Clutton-Brock and P. W. Harvey (eds.), *Readings in Sociobiology*, Freeman, San Francisco, 1978. H. Markl (ed.), *Evolution of Social Behaviour* (Report of a Dahlem Workshop), Verlag Chemie, Weinheim, 1980 (there is an account of this conference by R. M. May and M. Robertson, *Nature, 286*, 327-329 [1980]). See also H. S. Horn, Sociobiology, in R. M. May (ed.), *Theoretical Ecology*, 2nd ed., pp. 272-294, Blackwell, 1981.

45. R. V. Short, Sexual selection and its component parts, somatic and genital selection, as illustrated by man and the great apes, *Adv. Stud. Behav.*, *9*, 131-158 (1979).

46. T. H. Clutton-Brock and P. W. Harvey, Primate ecology and social organization, *J. Zool., Lond., 183*, 1-39 (1977). T. H. Clutton-Brock and P. W. Harvey, Evolutionary rules and primate societies, in P.P.G. Bateson and R. A. Hinde (eds.),

Growing Points in Ethology, pp. 195-237, Cambridge University Press, Cambridge, 1976; see also note 18 above. See also J. G. Fleagle, R. F. Kay, and E. L. Simons, Sexual dimorphism in early anthropoids, *Nature, 287*, 328-330 (1980).

47. B.C.R. Bertram, Serengeti predators and their social systems, in A.R.E. Sinclair and M. Norton-Griffiths (eds.), *Serengeti: Dynamics of an Ecosystem*, pp. 221-248, University of Chicago Press, Chicago, 1979.

48. See, e.g., the collection of papers on this subject put together by N. A. Chagnon and W. Irons (eds.), *Evolutionary Biology and Human Social Behavior: An Anthropological Perspective*, Duxbury Press, North Scituate, Mass., 1979.

49. D. P. Barash, Human behaviour: do the genes have it? *Nature, 287*, 173-174 (1980).

50. J. T. Bonner, *The Evolution of Culture in Animals*, Princeton University Press, Princeton, N.J., 1980.

51. L. Cavalli-Sforza and M. W. Feldman, *Cultural Transmission and Evolution: A Quantitative Approach*, Princeton University Press, Princeton, N.J., 1981. C. Lumsden and E. O. Wilson, *Gene-Culture Coevolution*, Harvard University Press, Cambridge, Mass., 1980. R. M. May, Population genetics and cultural inheritance, *Nature, 268*, 11-13 (1977).

52. GB, p. 167.

53. W. J. Hamilton III, *Life's Color Code*, McGraw-Hill, New York, 1973.

54. J. A. Endler, A predator's view of animal color patterns, in M. K. Hecht, W. C. Steere, and B. Wallace (eds.), *Evolutionary Biology*, Vol. 11, pp. 319-364, Plenum, New York, 1978.

55. D. Lack, *Ecological Adaptations for Breeding in Birds*, Methuen, London, 1968.

56. G. H. Orians, On the evolution of mating systems in birds and mammals. *Amer. Natur., 103*, 589-603 (1969). G. H. Orians, *Some Adaptations of Marsh-Nesting Blackbirds*, Princeton University Press, Princeton, N.J., 1980.

57. J. Maynard Smith, *The Evolution of Sex*, Cambridge University Press, Cambridge, 1976.

58. R. K. Selander, Sexual dimorphism and differential niche utilization in birds, *Condor, 68*, 113-151 (1966).

59. R. K. Selander, Sexual selection and dimorphism in birds, in B. Campbell (ed.), *op. cit.*, pp. 180-230.

60. R. A. Fisher, *op. cit.*, Chapter 6. W. D. Hamilton, Extraordinary sex ratios, *Science, 156*, 477-488 (1967). G. C. Williams, The Question of adaptive sex ratio in outcrossed vertebrates, *Proc. Roy. Soc., B205*, 567-580 (1979). E. G. Leigh, *Adaptation and Diversity*, Chapter 3, Freeman, San Francisco, 1971.

61. E. L. Charnov, Simultaneous hermaphroditism and sexual selection, *Proc. Natl. Acad. Sci. USA, 76*, 2480-2484 (1979). E. L. Charnov, J. Maynard Smith, and J. J. Bull, Why be an hermaphrodite? *Nature, 263*, 125-126 (1976). J. H. Werren and E. L. Charnov, Facultative sex ratios and population dynamics, *Nature, 272*, 349-350 (1978). E. G. Leigh, E. L. Charnov, and R. R. Warner, Sex ratio, sex change and natural selection. *Proc. Natl. Acad. Sci. USA, 73*, 3656-3660 (1976). M. T. Ghiselin, The evolution of hermaphroditism among animals, *Q. Rev. Biol., 44*, 189-208 (1969).

62. E. L. Charnov, Natural selection and sex change in Pandalid shrimp: test of a life-history theory, *Amer. Natur., 113*, 715-734 (1979). E. L. Charnov, D. W. Gotshall, and J. G. Robinson, Sex ratio: adaptive response to population fluctuations in Pandalid shrimp, *Science, 200*, 204-206 (1978).

63. M. T. Ghiselin, *The Economy of Nature and the Evolution of Sex*, University of California Press, Berkeley, 1974. G. C. Williams, *Sex and Evolution*, Princeton University Press, Princeton, N.J., 1975. See also note 57 above.

64. We are indebted to G. Geison, M. Ghiselin, and H. Horn for their helpful comments on an earlier version of this introduction.

THE DESCENT OF MAN,

AND

SELECTION IN RELATION TO SEX.

BY CHARLES DARWIN, M.A., F.R.S., &c.

IN TWO VOLUMES.—VOL. I.

WITH ILLUSTRATIONS.

LONDON:
JOHN MURRAY, ALBEMARLE STREET.
1871.

BY THE SAME AUTHOR.

ON THE ORIGIN OF SPECIES BY MEANS OF NATURAL SELECTION; or, THE PRESERVATION OF FAVOURED RACES IN THE STRUGGLE FOR LIFE; Fifth Edition (*Tenth Thousand*), with Additions and Corrections. 1869. MURRAY.

THE VARIATION OF ANIMALS AND PLANTS UNDER DOMESTICATION. In two vols. With Illustrations. 1868. MURRAY.

ON THE VARIOUS CONTRIVANCES BY WHICH BRITISH AND FOREIGN ORCHIDS ARE FERTILISED BY INSECTS; and on the GOOD EFFECTS OF CROSSING. With numerous Woodcuts. MURRAY.

A NATURALIST'S VOYAGE ROUND THE WORLD; or, A JOURNAL OF RESEARCHES INTO THE NATURAL HISTORY AND GEOLOGY OF THE COUNTRIES visited during the voyage of H.M.S. 'Beagle,' under the command of Captain FITZROY, R.N. *Eleventh Thousand.* MURRAY.

ON THE STRUCTURE AND DISTRIBUTION OF CORAL REEFS. SMITH, ELDER, & CO.

GEOLOGICAL OBSERVATIONS ON VOLCANIC ISLANDS. SMITH, ELDER, & CO.

GEOLOGICAL OBSERVATIONS ON SOUTH AMERICA. SMITH, ELDER, & CO.

A MONOGRAPH OF THE CIRRIPEDIA. With numerous Illustrations. 2 vols. 8vo. HARDWICKE.

ON THE MOVEMENTS AND HABITS OF CLIMBING PLANTS. With Woodcuts. WILLIAMS & NORGATE.

LONDON: PRINTED BY WILLIAM CLOWES AND SONS, STAMFORD STREET, AND CHARING CROSS.

CONTENTS.

PART I.

ON THE DESCENT OF MAN.

CHAPTER I.

THE EVIDENCE OF THE DESCENT OF MAN FROM SOME LOWER FORM.

CHAPTER II.

COMPARISON OF THE MENTAL POWERS OF MAN AND THE LOWER ANIMALS.

CHAPTER III.

COMPARISON OF THE MENTAL POWERS OF MAN AND THE LOWER ANIMALS—*continued.*

CHAPTER IV.

On the Manner of Development of Man from some Lower Form.

CHAPTER V.

On the Development of the Intellectual and Moral Faculties during Primeval and Civilised Times.

CHAPTER VI.

On the Affinities and Genealogy of Man.

CHAPTER VII.

On the Races of Man.

PART II.

SEXUAL SELECTION.

CHAPTER VIII.

Principles of Sexual Selection.

CHAPTER IX.

SECONDARY SEXUAL CHARACTERS IN THE LOWER CLASSES OF THE ANIMAL KINGDOM.

CHAPTER X.

SECONDARY SEXUAL CHARACTERS OF INSECTS.

CHAPTER XI.

INSECTS, *continued*.—ORDER LEPIDOPTERA.

THE DESCENT OF MAN;

AND ON

SELECTION IN RELATION TO SEX.

INTRODUCTION.

THE nature of the following work will be best understood by a brief account of how it came to be written. During many years I collected notes on the origin or descent of man, without any intention of publishing on the subject, but rather with the determination not to publish, as I thought that I should thus only add to the prejudices against my views. It seemed to me sufficient to indicate, in the first edition of my 'Origin of Species,' that by this work "light would be thrown on "the origin of man and his history;" and this implies that man must be included with other organic beings in any general conclusion respecting his manner of appearance on this earth. Now the case wears a wholly different aspect. When a naturalist like Carl Vogt ventures to say in his address as President of the National Institution of Geneva (1869), "personne, en Europe "au moins, n'ose plus soutenir la création indépendante "et de toutes pièces, des espèces," it is manifest that at least a large number of naturalists must admit that species are the modified descendants of other species;

and this especially holds good with the younger and rising naturalists. The greater number accept the agency of natural selection; though some urge, whether with justice the future must decide, that I have greatly overrated its importance. Of the older and honoured chiefs in natural science, many unfortunately are still opposed to evolution in every form.

In consequence of the views now adopted by most naturalists, and which will ultimately, as in every other case, be followed by other men, I have been led to put together my notes, so as to see how far the general conclusions arrived at in my former works were applicable to man. This seemed all the more desirable as I had never deliberately applied these views to a species taken singly. When we confine our attention to any one form, we are deprived of the weighty arguments derived from the nature of the affinities which connect together whole groups of organisms—their geographical distribution in past and present times, and their geological succession. The homological structure, embryological development, and rudimentary organs of a species, whether it be man or any other animal, to which our attention may be directed, remain to be considered; but these great classes of facts afford, as it appears to me, ample and conclusive evidence in favour of the principle of gradual evolution. The strong support derived from the other arguments should, however, always be kept before the mind.

The sole object of this work is to consider, firstly, whether man, like every other species, is descended from some pre-existing form; secondly, the manner of his development; and thirdly, the value of the differences between the so-called races of man. As I shall confine myself to these points, it will not be necessary to describe in detail the differences between the several races—an enormous subject which has been fully discussed in many valuable works. The high antiquity of man has recently been demonstrated by the labours of a host of eminent men, beginning with M. Boucher de Perthes; and this is the indispensable basis for understanding his origin. I shall, therefore, take this conclusion for granted, and may refer my readers to the admirable treatises of Sir Charles Lyell, Sir John Lubbock, and others. Nor shall I have occasion to do more than to allude to the amount of difference between man and the anthropomorphous apes; for Prof. Huxley, in the opinion of most competent judges, has conclusively shewn that in every single visible character man differs less from the higher apes than these do from the lower members of the same order of Primates.

This work contains hardly any original facts in regard to man; but as the conclusions at which I arrived, after drawing up a rough draft, appeared to me interesting, I thought that they might interest others. It has often and confidently been asserted, that man's origin can never be known: but ignorance more frequently begets confidence than does knowledge: it is those who know little, and not those who know much, who so positively assert that this or that problem will never be solved by science. The conclusion that man is the co-descendant with other species of some ancient, lower, and extinct form, is not in any degree new. Lamarck long ago came to this conclusion, which has lately been maintained by several eminent naturalists and philosophers; for instance by Wallace, Huxley, Lyell, Vogt, Lubbock, Büchner, Rolle, &c.,[1] and especially by Häckel. This last naturalist, besides his great work, 'Generelle Morphologie' (1866), has recently (1868, with a second edit. in 1870), published his 'Natürliche Schöpfungsgeschichte,' in which he fully discusses the genealogy of man. If this work had appeared before my essay had been written, I should probably never have completed it. Almost all the conclusions at which I have arrived I find confirmed by this naturalist, whose knowledge on many points is much fuller than mine. Wherever I have added any fact or view from Prof. Häckel's writings, I give his authority in the text, other statements I leave as they originally stood in my manuscript, occasionally giving in the foot-notes references to his works, as a confirmation of the more doubtful or interesting points.

During many years it has seemed to me highly probable that sexual selection has played an important part in differentiating the races of man; but in my

[1] As the works of the first-named authors are so well known, I need not give the titles; but as those of the latter are less well known in England, I will give them:—'Sechs Vorlesungen über die Darwin'sche Theorie:' zweite Auflage, 1868, von Dr. L. Büchner; translated into French under the title 'Conférences sur la Théorie Darwinienne,' 1869. 'Der Mensch, im Lichte der Darwin'sche Lehre,' 1865, von Dr. F. Rolle. I will not attempt to give references to all the authors who have taken the same side of the question. Thus G. Canestrini has published ('Annuario della Soc. d. Nat.,' Modena, 1867, p. 81) a very curious paper on rudimentary characters, as bearing on the origin of man. Another work has (1869) been published by Dr. Barrago Francesco, bearing in Italian the title of "Man, made in the image of God, was also made in the image of the ape."

'Origin of Species' (first edition, p. 199) I contented myself by merely alluding to this belief. When I came to apply this view to man, I found it indispensable to treat the whole subject in full detail.[2] Consequently the second part of the present work, treating of sexual selection, has extended to an inordinate length, compared with the first part; but this could not be avoided.

I had intended adding to the present volumes an essay on the expression of the various emotions by man and the lower animals. My attention was called to this subject many years ago by Sir Charles Bell's admirable work. This illustrious anatomist maintains that man is endowed with certain muscles solely for the sake of expressing his emotions. As this view is obviously opposed to the belief that man is descended from some other and lower form, it was necessary for me to consider it. I likewise wished to ascertain how far the emotions are expressed in the same manner by the different races of man. But owing to the length of the present work, I have thought it better to reserve my essay, which is partially completed, for separate publication.

---

[2] Prof. Häckel is the sole author who, since the publication of the 'Origin,' has discussed, in his various works, in a very able manner, the subject of sexual selection, and has seen its full importance.

PART I.

THE DESCENT OR ORIGIN OF MAN.

PART I.—THE DESCENT OF MAN.

CHAPTER I.

THE EVIDENCE OF THE DESCENT OF MAN FROM SOME LOWER FORM.

Nature of the evidence bearing on the origin of man — Homologous structures in man and the lower animals — Miscellaneous points of correspondence — Development — Rudimentary structures, muscles, sense-organs, hair, bones, reproductive organs, &c. — The bearing of these three great classes of facts on the origin of man.

HE who wishes to decide whether man is the modified descendant of some pre-existing form, would probably first enquire whether man varies, however slightly, in bodily structure and in mental faculties; and if so, whether the variations are transmitted to his offspring in accordance with the laws which prevail with the lower animals; such as that of the transmission of characters to the same age or sex. Again, are the variations the result, as far as our ignorance permits us to judge, of the same general causes, and are they governed by the same general laws, as in the case of other organisms; for instance by correlation, the inherited effects of use and disuse, &c.? Is man subject to similar malconformations, the result of arrested development, of reduplication of parts, &c., and does he display in any of his anomalies reversion to some former and ancient type of structure? It might also naturally be enquired whether man, like so many other animals, has given rise to varieties and sub-races, differing but slightly from each other, or to races differing so much that they must be classed as doubtful species? How are such races distributed over the world; and how, when crossed, do they react on each other, both in the first and succeeding generations? And so with many other points.

The enquirer would next come to the important point, whether man tends to increase at so rapid a rate, as to lead to occasional severe struggles for existence, and consequently to beneficial variations, whether in body or mind, being preserved, and injurious ones eliminated. Do the races or species of men, whichever term may be applied, encroach on and replace each other, so that some finally become extinct? We shall see that all these questions, as indeed is obvious in respect to most of them, must be answered in the affirmative, in the same manner as with the lower animals. But the several considerations just referred to may be conveniently deferred for a time; and we will first see how far the bodily structure of man shows traces, more or less plain, of his descent from some lower form. In the two succeeding chapters the mental powers of man, in comparison with those of the lower animals, will be considered.

The Bodily Structure of Man.—It is notorious that man is constructed on the same general type or model with other mammals. All the bones in his skeleton can be compared with corresponding bones in a monkey, bat, or seal. So it is with his muscles, nerves, blood-vessels and internal viscera. The brain, the most important of all the organs, follows the same law, as shewn by Huxley and other anatomists. Bischoff,[1] who is a hostile witness, admits that every chief fissure and fold

[1] 'Grosshirnwindungen des Menschen,' 1868, s. 96.

in the brain of man has its analogy in that of the orang; but he adds that at no period of development do their brains perfectly agree; nor could this be expected, for otherwise their mental powers would have been the same. Vulpian[2] remarks: "Les différences réelles qui existent "entre l'encéphale de l'homme et celui des singes supé-"rieurs, sont bien minimes. Il ne faut pas se faire "d'illusions à cet égard. L'homme est bien plus près "des singes anthropomorphes par les caractères anato-"miques de son cerveau que ceux-ci ne le sont non-"seulement des autres mammifères, mais mêmes de "certains quadrumanes, des guenons et des macaques." But it would be superfluous here to give further details on the correspondence between man and the higher mammals in the structure of the brain and all other parts of the body.

It may, however, be worth while to specify a few points, not directly or obviously connected with structure, by which this correspondence or relationship is well shewn.

Man is liable to receive from the lower animals, and to communicate to them, certain diseases as hydrophobia, variola, the glanders, &c.; and this fact proves the close similarity of their tissues and blood, both in minute structure and composition, far more plainly than does their comparison under the best microscope, or by the aid of the best chemical analysis. Monkeys are liable to many of the same non-contagious diseases as we are; thus Rengger,[3] who carefully observed for a long time the *Cebus Azaræ* in its native land, found it liable to catarrh, with the usual symptoms, and which when

---

[2] 'Leç. sur la Phys.' 1866, p. 890, as quoted by M. Dally, 'L'Ordre des Primates et le Transformisme,' 1868, p. 29.

[3] 'Naturgeschichte der Säugethiere von Paraguay,' 1830, s. 50.

often recurrent led to consumption. These monkeys suffered also from apoplexy, inflammation of the bowels, and cataract in the eye. The younger ones when shedding their milk-teeth often died from fever. Medicines produced the same effect on them as on us. Many kinds of monkeys have a strong taste for tea, coffee, and spirituous liquors: they will also, as I have myself seen, smoke tobacco with pleasure. Brehm asserts that the natives of north-eastern Africa catch the wild baboons by exposing vessels with strong beer, by which they are made drunk. He has seen some of these animals, which he kept in confinement, in this state; and he gives a laughable account of their behaviour and strange grimaces. On the following morning they were very cross and dismal; they held their aching heads with both hands and wore a most pitiable expression: when beer or wine was offered them, they turned away with disgust, but relished the juice of lemons.[4] An American monkey, an Ateles, after getting drunk on brandy, would never touch it again, and thus was wiser than many men. These trifling facts prove how similar the nerves of taste must be in monkeys and man, and how similarly their whole nervous system is affected.

Man is infested with internal parasites, sometimes causing fatal effects, and is plagued by external parasites, all of which belong to the same genera or families with those infesting other mammals. Man is subject like other mammals, birds, and even insects, to that mysterious law, which causes certain normal processes, such as gestation, as well as the maturation and duration of various diseases, to follow lunar periods.[5] His wounds

---

[4] Brehm, 'Thierleben,' B. i. 1864, s. 75, 86. On the Ateles, s. 105. For other analogous statements, see s. 25, 107.

[5] With respect to insects see Dr. Laycock 'On a General Law of Vital Periodicity,' British Association, 1842. Dr. Macculloch, 'Silli-

are repaired by the same process of healing; and the stumps left after the amputation of his limbs occasionally possess, especially during an early embryonic period, some power of regeneration, as in the lowest animals.[6]

The whole process of that most important function, the reproduction of the species, is strikingly the same in all mammals, from the first act of courtship by the male[7] to the birth and nurturing of the young. Monkeys are born in almost as helpless a condition as our own infants; and in certain genera the young differ fully as much in appearance from the adults, as do our children from their full-grown parents.[8] It has been urged by some writers as an important distinction, that with man the young arrive at maturity at a much later age than with any other animal: but if we look to the races of mankind which inhabit tropical countries the difference is not great, for the orang is believed not to be adult till the age of from ten to fifteen years.[9] Man

---

man's North American Journal of Science,' vol. xvii. p. 305, has seen a dog suffering from tertian ague.

[6] I have given the evidence on this head in my 'Variation of Animals and Plants under Domestication,' vol. ii. p. 15.

[7] "Mares e diversis generibus Quadrumanorum sine dubio dignoscunt "feminas humanas a maribus. Primum, credo, odoratu, postea aspectu. "Mr. Youatt, qui diu in Hortis Zoologicis (Bestiariis) medicus animal-"ium erat, vir in rebus observandis cautus et sagax, hoc mihi certissime "probavit, et curatores ejusdem loci et alii e ministris confirmaverunt. "Sir Andrew Smith et Brehm notabant idem in Cynocephalo. Illus-"trissimus Cuvier etiam narrat multa de hac re quâ ut opinor nihil "turpius potest indicari inter omnia hominibus et Quadrumanis com-"munia. Narrat enim Cynocephalum quendam in furorem incidere "aspectu feminarum aliquarum, sed nequaquam accendi tanto furore "ab omnibus. Semper eligebat juniores, et dignoscebat in turba, et "advocabat voce gestuque."

[8] This remark is made with respect to Cynocephalus and the anthropomorphous apes by Geoffroy Saint-Hilaire and F. Cuvier, 'Hist. Nat. des Mammifères,' tom. i. 1824.

[9] Huxley, 'Man's Place in Nature,' 1863, p. 34.

differs from woman in size, bodily strength, hairyness, &c., as well as in mind, in the same manner as do the two sexes of many mammals. It is, in short, scarcely possible to exaggerate the close correspondence in general structure, in the minute structure of the tissues, in chemical composition and in constitution, between man and the higher animals, especially the anthropomorphous apes.

Embryonic Development.—Man is developed from an ovule, about the 125th of an inch in diameter, which differs in no respect from the ovules of other animals. The embryo itself at a very early period can hardly be distinguished from that of other members of the vertebrate kingdom. At this period the arteries run in arch-like branches, as if to carry the blood to branchiæ which are not present in the higher vertebrata, though the slits on the sides of the neck still remain (*f*, *g*, fig. 1), marking their former position. At a somewhat later period, when the extremities are developed, "the feet of "lizards and mammals," as the illustrious Von Baer remarks, "the wings and feet of birds, no less than the "hands and feet of man, all arise from the same fundamental form." It is, says Prof. Huxley,[10] "quite in "the later stages of development that the young human "being presents marked differences from the young "ape, while the latter departs as much from the dog "in its developments, as the man does. Startling as "this last assertion may appear to be, it is demonstrably "true."

As some of my readers may never have seen a drawing of an embryo, I have given one of man and another of a dog, at about the same early stage of development,

[10] 'Man's Place in Nature,' 1863, p. 67.

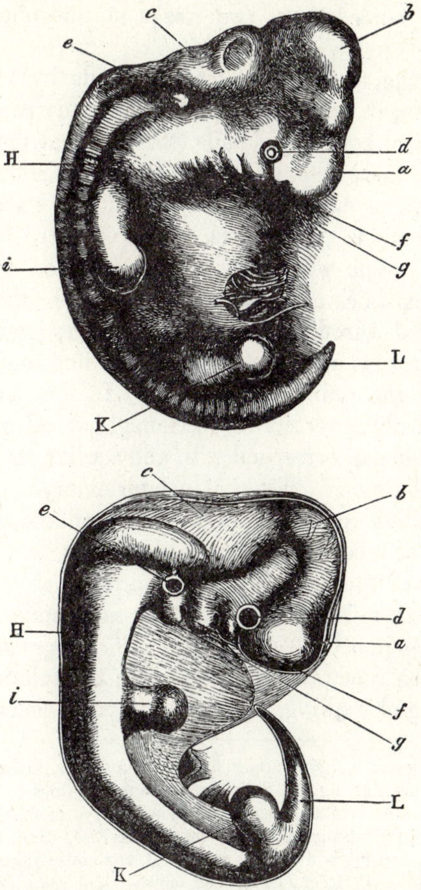

Fig. 1. Upper figure human embryo, from Ecker. Lower figure that of a dog, from Bischoff.

---

a. Fore-brain, cerebral hemispheres, &c.
b. Mid-brain, corpora quadrigemina.
c. Hind-brain, cerebellum, medulla oblongata.
d. Eye.
e. Ear.
f. First visceral arch.
g. Second visceral arch.
H. Vertebral columns and muscles in process of development.
i. Anterior } extremities.
K. Posterior } extremities.
L. Tail or os coccyx.

carefully copied from two works of undoubted accuracy.[11]

After the foregoing statements made by such high authorities, it would be superfluous on my part to give a number of borrowed details, shewing that the embryo of man closely resembles that of other mammals. It may, however, be added that the human embryo likewise resembles in various points of structure certain low forms when adult. For instance, the heart at first exists as a simple pulsating vessel; the excreta are voided through a cloacal passage; and the os coccyx projects like a true tail, "extending considerably "beyond the rudimentary legs."[12] In the embryos of all air-breathing vertebrates, certain glands called the corpora Wolffiana, correspond with and act like the kidneys of mature fishes.[13] Even at a later embryonic period, some striking resemblances between man and the lower animals may be observed. Bischoff says that the convolutions of the brain in a human fœtus at the end of the seventh month reach about the same stage of development as in a baboon when adult.[14] The great toe, as Prof. Owen remarks,[15] "which forms the fulcrum when "standing or walking, is perhaps the most characteristic

[11] The human embryo (upper fig.) is from Ecker, 'Icones Phys.,' 1851-1859, tab. xxx. fig. 2. This embryo was ten lines in length, so that the drawing is much magnified. The embryo of the dog is from Bischoff, 'Entwicklungsgeschichte des Hunde-Eies,' 1845, tab. xi. fig. 42 B. This drawing is five times magnified, the embryo being 25 days old. The internal viscera have been omitted, and the uterine appendages in both drawings removed. I was directed to these figures by Prof. Huxley, from whose work, 'Man's Place in Nature,' the idea of giving them was taken. Häckel has also given analogous drawings in his 'Schöpfungsgeschichte.'

[12] Prof. Wyman in 'Proc. of American Acad. of Sciences,' vol. iv. 1860, p. 17.

[13] Owen, 'Anatomy of Vertebrates,' vol. i. p. 533.

[14] 'Die Grosshirnwindungen des Menschen,' 1868, s. 95.

[15] 'Anatomy of Vertebrates,' vol. ii. p. 553.

"peculiarity in the human structure;" but in an embryo, about an inch in length, Prof. Wyman [16] found "that the great toe was shorter than the others, and, "instead of being parallel to them, projected at an "angle from the side of the foot, thus corresponding "with the permanent condition of this part in the "quadrumana." I will conclude with a quotation from Huxley,[17] who after asking, does man originate in a different way from a dog, bird, frog or fish? says, "the "reply is not doubtful for a moment; without question, "the mode of origin and the early stages of the development of man are identical with those of the animals "immediately below him in the scale: without a doubt "in these respects, he is far nearer to apes, than the apes "are to the dog."

Rudiments.—This subject, though not intrinsically more important than the two last, will for several reasons be here treated with more fullness.[18] Not one of the higher animals can be named which does not bear some part in a rudimentary condition; and man forms no exception to the rule. Rudimentary organs must be distinguished from those that are nascent; though in some cases the distinction is not easy. The former are either absolutely useless, such as the mammæ of male quadrupeds, or the incisor teeth of ruminants which never cut through the gums; or they are of such slight service to their present possessors, that we cannot suppose that they were developed under the conditions

[16] 'Proc. Soc. Nat. Hist.' Boston, 1863, vol. ix. p. 185.

[17] 'Man's Place in Nature,' p. 65.

[18] I had written a rough copy of this chapter before reading a valuable paper, "Caratteri rudimentali in ordine all' origine del uomo" ('Annuario della Soc. d. Nat.,' Modena, 1867, p. 81), by G. Canestrini, to which paper I am considerably indebted. Häckel has given admirable discussions on this whole subject, under the title of Dysteleology, in his 'Generelle Morphologie' and 'Schöpfungsgeschichte.'

which now exist. Organs in this latter state are not strictly rudimentary, but they are tending in this direction. Nascent organs, on the other hand, though not fully developed, are of high service to their possessors, and are capable of further development. Rudimentary organs are eminently variable; and this is partly intelligible, as they are useless or nearly useless, and consequently are no longer subjected to natural selection. They often become wholly suppressed. When this occurs, they are nevertheless liable to occasional reappearance through reversion; and this is a circumstance well worthy of attention.

Disuse at that period of life, when an organ is chiefly used, and this is generally during maturity, together with inheritance at a corresponding period of life, seem to have been the chief agents in causing organs to become rudimentary. The term "disuse" does not relate merely to the lessened action of muscles, but includes a diminished flow of blood to a part or organ, from being subjected to fewer alternations of pressure, or from becoming in any way less habitually active. Rudiments, however, may occur in one sex of parts normally present in the other sex; and such rudiments, as we shall hereafter see, have often originated in a distinct manner. In some cases organs have been reduced by means of natural selection, from having become injurious to the species under changed habits of life. The process of reduction is probably often aided through the two principles of compensation and economy of growth; but the later stages of reduction, after disuse has done all that can fairly be attributed to it, and when the saving to be effected by the economy of growth would be very small,[19] are difficult to understand. The final and com-

[19] Some good criticisms on this subject have been given by Messrs. Murie and Mivart, in 'Transact. Zoolog. Soc.' 1869, vol. vii. p. 92.

plete suppression of a part, already useless and much reduced in size, in which case neither compensation nor economy can come into play, is perhaps intelligible by the aid of the hypothesis of pangenesis, and apparently in no other way. But as the whole subject of rudimentary organs has been fully discussed and illustrated in my former works,[20] I need here say no more on this head.

Rudiments of various muscles have been observed in many parts of the human body;[21] and not a few muscles, which are regularly present in some of the lower animals can occasionally be detected in man in a greatly reduced condition. Every one must have noticed the power which many animals, especially horses, possess of moving or twitching their skin; and this is effected by the panniculus carnosus. Remnants of this muscle in an efficient state are found in various parts of our bodies; for instance, on the forehead, by which the eyebrows are raised. The *platysma myoides*, which is well developed on the neck, belongs to this system, but cannot be voluntarily brought into action. Prof. Turner, of Edinburgh, has occasionally detected, as he informs me, muscular fasciculi in five different situations, namely in the axillæ, near the scapulæ, &c., all of which must be referred to the system of the panniculus. He has also shewn[22] that the *musculus sternalis* or *sternalis brutorum*, which is not an extension of the *rectus abdominalis*, but is closely allied to the panniculus, oc-

[20] 'Variation of Animals and Plants under Domestication,' vol. ii. pp. 317 and 397. See also 'Origin of Species,' 5th edit. p. 535.

[21] For instance M. Richard ('Annales des Sciences Nat.' 3rd series, Zoolog. 1852, tom. xviii. p. 13) describes and figures rudiments of what he calls the "muscle pédieux de la main," which he says is sometimes "infiniment petit." Another muscle, called "le tibial postérieur," is generally quite absent in the hand, but appears from time to time in a more or less rudimentary condition.

[22] Prof. W. Turner, 'Proc. Royal Soc. Edinburgh,' 1866-67, p. 65.

curred in the proportion of about 3 per cent. in upwards of 600 bodies: he adds, that this muscle affords "an "excellent illustration of the statement that occasional "and rudimentary structures are especially liable to "variation in arrangement."

Some few persons have the power of contracting the superficial muscles on their scalps; and these muscles are in a variable and partially rudimentary condition. M. A. de Candolle has communicated to me a curious instance of the long-continued persistence or inheritance of this power, as well as of its unusual development. He knows a family, in which one member, the present head of a family, could, when a youth, pitch several heavy books from his head by the movement of the scalp alone; and he won wagers by performing this feat. His father, uncle, grandfather, and all his three children possess the same power to the same unusual degree. This family became divided eight generations ago into two branches; so that the head of the above-mentioned branch is cousin in the seventh degree to the head of the other branch. This distant cousin resides in another part of France, and on being asked whether he possessed the same faculty, immediately exhibited his power. This case offers a good illustration how persistently an absolutely useless faculty may be transmitted.

The extrinsic muscles which serve to move the whole external ear, and the intrinsic muscles which move the different parts, all of which belong to the system of the panniculus, are in a rudimentary condition in man; they are also variable in development, or at least in function. I have seen one man who could draw his ears forwards, and another who could draw them backwards;[23]

---

[23] Canestrini quotes Hyrt. ('Annuario della Soc. dei Naturalisti,' Modena, 1867, p. 97) to the same effect.

and from what one of these persons told me, it is probable that most of us by often touching our ears and thus directing our attention towards them, could by repeated trials recover some power of movement. The faculty of erecting the ears and of directing them to different points of the compass, is no doubt of the highest service to many animals, as they thus perceive the point of danger; but I have never heard of a man who possessed the least power of erecting his ears,—the one movement which might be of use to him. The whole external shell of the ear may be considered a rudiment, together with the various folds and prominences (helix and anti-helix, tragus and anti-tragus, &c.) which in the lower animals strengthen and support the ear when erect, without adding much to its weight. Some authors, however, suppose that the cartilage of the shell serves to transmit vibrations to the acoustic nerve; but Mr. Toynbee,[24] after collecting all the known evidence on this head, concludes that the external shell is of no distinct use. The ears of the chimpanzee and orang are curiously like those of man, and I am assured by the keepers in the Zoological Gardens that these animals never move or erect them; so that they are in an equally rudimentary condition, as far as function is concerned, as in man. Why these animals, as well as the progenitors of man, should have lost the power of erecting their ears we cannot say. It may be, though I am not quite satisfied with this view, that owing to their arboreal habits and great strength they were but little exposed to danger, and so during a lengthened period moved their ears but little, and thus gradually lost the power of moving them. This would be a parallel case with that of those large and heavy birds,

---

[24] 'The Diseases of the Ear,' by J. Toynbee, F.R.S., 1860, p. 12.

which from inhabiting oceanic islands have not been exposed to the attacks of beasts of prey, and have consequently lost the power of using their wings for flight.

The celebrated sculptor, Mr. Woolner, informs me of one little peculiarity in the external ear, which he has often observed both in men and women, and of which he perceived the full signification. His attention was first called to the subject whilst at work on his figure of Puck, to which he had given pointed ears. He was thus led to examine the ears of various monkeys, and subsequently more carefully those of man. The peculiarity consists in a little blunt point, projecting from the inwardly folded margin, or helix. Mr. Woolner made an exact model of one such case, and has sent me the accompanying drawing. (Fig. 2.) These points not only project inwards, but often a little outwards, so that they are visible when the head is viewed from directly in front or behind. They are variable in size and somewhat in position, standing either a little higher or lower; and they sometimes occur on one ear and not on the other. Now the meaning of these projections is not, I think, doubtful; but it may be thought that they offer too trifling a character to be worth notice. This thought, however, is as false as it is natural. Every character, however slight, must be the result of some definite cause; and if it occurs in many individuals deserves consideration. The helix obviously consists of the extreme margin of the ear folded inwards; and this folding appears to be in some manner connected with the

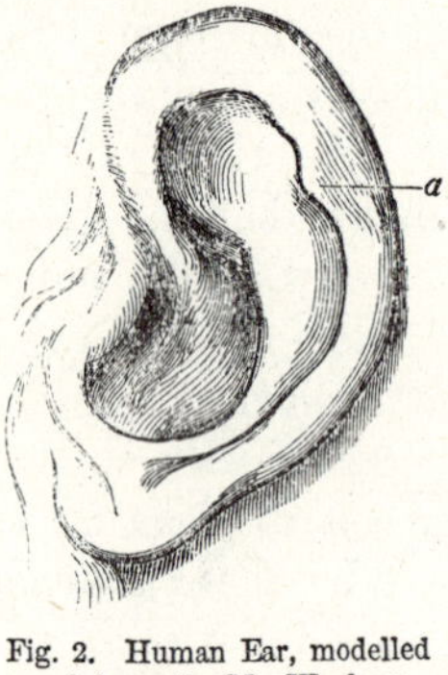

Fig. 2. Human Ear, modelled and drawn by Mr. Woolner.
a. The projecting point.

whole external ear being permanently pressed backwards. In many monkeys, which do not stand high in the order, as baboons and some species of macacus,[25] the upper portion of the ear is slightly pointed, and the margin is not at all folded inwards; but if the margin were to be thus folded, a slight point would necessarily project inwards and probably a little outwards. This could actually be observed in a specimen of the *Ateles beelzebuth* in the Zoological Gardens; and we may safely conclude that it is a similar structure—a vestige of formerly pointed ears—which occasionally reappears in man.

The nictitating membrane, or third eyelid, with its accessory muscles and other structures, is especially well developed in birds, and is of much functional importance to them, as it can be rapidly drawn across the whole eye-ball. It is found in some reptiles and amphibians, and in certain fishes, as in sharks. It is fairly well developed in the two lower divisions of the mammalian series, namely, in the monotremata and marsupials, and in some few of the higher mammals, as in the walrus. But in man, the quadrumana, and most other mammals, it exists, as is admitted by all anatomists, as a mere rudiment, called the semilunar fold.[26]

The sense of smell is of the highest importance to the greater number of mammals—to some, as the ruminants, in warning them of danger; to others, as the

---

[25] See also some remarks, and the drawings of the ears of the Lemuroidea, in Messrs. Murie and Mivart's excellent paper in 'Transact. Zoolog. Soc.' vol. vii. 1869, pp. 6 and 90.

[26] Müller's 'Elements of Physiology,' Eng. translat., 1842, vol. ii. p. 1117. Owen, 'Anatomy of Vertebrates,' vol. iii. p. 260; ibid. on the Walrus, 'Proc. Zoolog. Soc.' November 8th, 1854. See also R. Knox, 'Great Artists and Anatomists,' p. 106. This rudiment apparently is somewhat larger in Negroes and Australians than in Europeans, see Carl Vogt, 'Lectures on Man,' Eng. translat. p. 129.

carnivora, in finding their prey; to others, as the wild boar, for both purposes combined. But the sense of smell is of extremely slight service, if any, even to savages, in whom it is generally more highly developed than in the civilised races. It does not warn them of danger, nor guide them to their food; nor does it prevent the Esquimaux from sleeping in the most fetid atmosphere, nor many savages from eating half-putrid meat. Those who believe in the principle of gradual evolution, will not readily admit that this sense in its present state was originally acquired by man, as he now exists. No doubt he inherits the power in an enfeebled and so far rudimentary condition, from some early progenitor, to whom it was highly serviceable and by whom it was continually used. We can thus perhaps understand how it is, as Dr. Maudsley has truly remarked,[27] that the sense of smell in man "is singu-"larly effective in recalling vividly the ideas and images "of forgotten scenes and places;" for we see in those animals, which have this sense highly developed, such as dogs and horses, that old recollections of persons and places are strongly associated with their odour.

Man differs conspicuously from all the other Primates in being almost naked. But a few short straggling hairs are found over the greater part of the body in the male sex, and fine down on that of the female sex. In individuals belonging to the same race these hairs are highly variable, not only in abundance, but likewise in position: thus the shoulders in some Europeans are quite naked, whilst in others they bear thick tufts of hair.[28] There can be little doubt that the hairs

---

27 'The Physiology and Pathology of Mind,' 2nd edit. 1868, p. 134.

28 Eschricht, Ueber die Richtung der Haare am menschlichen Körper 'Müller's Archiv für Anat. und Phys.' 1837, s. 47. I shall often have to refer to this very curious paper.

thus scattered over the body are the rudiments of the uniform hairy coat of the lower animals. This view is rendered all the more probable, as it is known that fine, short, and pale-coloured hairs on the limbs and other parts of the body occasionally become developed into "thickset, long, and rather coarse dark hairs," when abnormally nourished near old-standing inflamed surfaces.[29]

I am informed by Mr. Paget that persons belonging to the same family often have a few hairs in their eyebrows much longer than the others; so that this slight peculiarity seems to be inherited. These hairs apparently represent the vibrissæ, which are used as organs of touch by many of the lower animals. In a young chimpanzee I observed that a few upright, rather long, hairs, projected above the eyes, where the true eyebrows, if present, would have stood.

The fine wool-like hair, or so-called lanugo, with which the human fœtus during the sixth month is thickly covered, offers a more curious case. It is first developed, during the fifth month, on the eyebrows and face, and especially round the mouth, where it is much longer than that on the head. A moustache of this kind was observed by Eschricht[30] on a female fœtus; but this is not so surprising a circumstance as it may at first appear, for the two sexes generally resemble each other in all external characters during an early period of growth. The direction and arrangement of the hairs on all parts of the fœtal body are the same as in the adult, but are subject to much variability. The whole surface, including even the forehead and ears, is thus thickly clothed; but it is a significant fact that the palms of the hands and

[29] Paget, 'Lectures on Surgical Pathology,' 1853, vol. i. p. 71.
[30] Eschricht, ibid. s. 40, 47.

the soles of the feet are quite naked, like the inferior surfaces of all four extremities in most of the lower animals. As this can hardly be an accidental coincidence, we must consider the woolly covering of the fœtus to be the rudimental representative of the first permanent coat of hair in those mammals which are born hairy. This representation is much more complete, in accordance with the usual law of embryological development, than that afforded by the straggling hairs on the body of the adult.

It appears as if the posterior molar or wisdom-teeth were tending to become rudimentary in the more civilised races of man. These teeth are rather smaller than the other molars, as is likewise the case with the corresponding teeth in the chimpanzee and orang; and they have only two separate fangs. They do not cut through the gums till about the seventeenth year, and I am assured by dentists that they are much more liable to decay, and are earlier lost, than the other teeth. It is also remarkable that they are much more liable to vary both in structure and in the period of their development than the other teeth.[31] In the Melanian races, on the other hand, the wisdom-teeth are usually furnished with three separate fangs, and are generally sound: they also differ from the other molars in size less than in the Caucasian races.[32] Prof. Schaaffhausen accounts for this difference between the races by "the posterior dental portion of the jaw being "always shortened" in those that are civilised,[33] and this shortening may, I presume, be safely attributed to civi-

---

[31] Dr. Webb, 'Teeth in Man and the Anthropoid Apes,' as quoted by Dr. C. Carter Blake in 'Anthropological Review,' July, 1867, p. 299.

[32] Owen, 'Anatomy of Vertebrates,' vol. iii. pp. 320, 321, and 325.

[33] 'On the Primitive Form of the Skull,' Eng. translat. in 'Anthropological Review,' Oct. 1868, p. 426.

lised men habitually feeding on soft, cooked food, and thus using their jaws less. I am informed by Mr. Brace that it is becoming quite a common practice in the United States to remove some of the molar teeth of children, as the jaw does not grow large enough for the perfect development of the normal number.

With respect to the alimentary canal I have met with an account of only a single rudiment, namely the vermiform appendage of the cæcum. The cæcum is a branch or diverticulum of the intestine, ending in a cul-de-sac, and it is extremely long in many of the lower vegetable-feeding mammals. In the marsupial kaola it is actually more than thrice as long as the whole body.[34] It is sometimes produced into a long gradually-tapering point, and is sometimes constricted in parts. It appears as if, in consequence of changed diet or habits, the cæcum had become much shortened in various animals, the vermiform appendage being left as a rudiment of the shortened part. That this appendage is a rudiment, we may infer from its small size, and from the evidence which Prof. Canestrini[35] has collected of its variability in man. It is occasionally quite absent, or again is largely developed. The passage is sometimes completely closed for half or two-thirds of its length, with the terminal part consisting of a flattened solid expansion. In the orang this appendage is long and convoluted: in man it arises from the end of the short cæcum, and is commonly from four to five inches in length, being only about the third of an inch in diameter. Not only is it useless, but it is sometimes the cause of death, of which fact I have lately heard two instances: this is due to small hard bodies,

---

[34] Owen, 'Anatomy of Vertebrates,' vol. iii. pp. 416, 434, 441.
[35] 'Annuario della Soc. d. Nat.' Modena, 1867, p. 94.

such as seeds, entering the passage and causing inflammation.[36]

In the Quadrumana and some other orders of mammals, especially in the Carnivora, there is a passage near the lower end of the humerus, called the supra-condyloid foramen, through which the great nerve of the fore limb passes, and often the great artery. Now in the humerus of man, as Dr. Struthers[37] and others have shewn, there is generally a trace of this passage, and it is sometimes fairly well developed, being formed by a depending hook-like process of bone, completed by a band of ligament. When present the great nerve invariably passes through it, and this clearly indicates that it is the homologue and rudiment of the supra-condyloid foramen of the lower animals. Prof. Turner estimates, as he informs me, that it occurs in about one per cent. of recent skeletons; but during ancient times it appears to have been much more common. Mr. Busk[38] has collected the following evidence on this head: Prof. Broca "noticed the perforation in four and a half per "cent. of the arm-bones collected in the 'Cimetière du "Sud' at Paris; and in the Grotto of Orrony, the con-"tents of which are referred to the Bronze period, as "many as eight humeri out of thirty-two were perfo-"rated; but this extraordinary proportion, he thinks, "might be due to the cavern having been a sort of

---

36 M. C. Martins ("De l'Unité Organique," in 'Revue des Deux Mondes,' June 15, 1862, p. 16), and Häckel ('Generelle Morphologie,' B. ii. s. 278), have both remarked on the singular fact of this rudiment sometimes causing death.

37 'The Lancet,' Jan. 24, 1863, p. 83. Dr. Knox, 'Great Artists and Anatomists,' p. 63. See also an important memoir on this process by Dr. Grube, in the 'Bulletin de l'Acad. Imp. de St. Pétersbourg,' tom. xii. 1867, p. 448.

38 "On the Caves of Gibraltar," 'Transact. Internat. Congress of Prehist. Arch.' Third Session, 1869, p. 54.

" 'family vault.' Again, M. Dupont found 30 per cent. " of perforated bones in the caves of the Valley of the " Lesse, belonging to the Reindeer period; whilst M. " Leguay, in a sort of *dolmen* at Argenteuil, observed " twenty-five per cent. to be perforated; and M. Pruner-" Bey found twenty-six per cent. in the same condition " in bones from Vauréal. Nor should it be left unno-" ticed that M. Pruner-Bey states that this condition is " common in Guanche skeletons." The fact that ancient races, in this and several other cases, more frequently present structures which resemble those of the lower animals than do the modern races, is interesting. One chief cause seems to be that ancient races stand somewhat nearer than modern races in the long line of descent to their remote animal-like progenitors.

The os coccyx in man, though functionless as a tail, plainly represents this part in other vertebrate animals. At an early embryonic period it is free, and, as we have seen, projects beyond the lower extremities. In certain rare and anomalous cases it has been known, according to Isidore Geoffroy St.-Hilaire and others,[39] to form a small external rudiment of a tail. The os coccyx is short, usually including only four vertebræ: and these are in a rudimental condition, for they consist, with the exception of the basal one, of the centrum alone.[40] They are furnished with some small muscles; one of which, as I am informed by Prof. Turner, has been expressly described by Theile as a rudimentary repetition of the extensor of the tail, which is so largely developed in many mammals.

The spinal cord in man extends only as far downwards as the last dorsal or first lumbar vertebra; but a

---

[39] Quatrefages has lately collected the evidence on this subject. 'Revue des Cours Scientifiques,' 1867-1868, p. 625.

[40] Owen, 'On the Nature of Limbs,' 1849, p. 114.

thread-like structure (the *filum terminale*) runs down the axis of the sacral part of the spinal canal, and even along the back of the coccygeal bones. The upper part of this filament, as Prof. Turner informs me, is undoubtedly homologous with the spinal cord; but the lower part apparently consists merely of the *pia mater*, or vascular investing membrane. Even in this case the os coccyx may be said to possess a vestige of so important a structure as the spinal cord, though no longer enclosed within a bony canal. The following fact, for which I am also indebted to Prof. Turner, shews how closely the os coccyx corresponds with the true tail in the lower animals: Luschka has recently discovered at the extremity of the coccygeal bones a very peculiar convoluted body, which is continuous with the middle sacral artery; and this discovery led Krause and Meyer to examine the tail of a monkey (Macacus) and of a cat, in both of which they found, though not at the extremity, a similarly convoluted body.

The reproductive system offers various rudimentary structures; but these differ in one important respect from the foregoing cases. We are not here concerned with a vestige of a part which does not belong to the species in an efficient state; but with a part which is always present and efficient in the one sex, being represented in the other by a mere rudiment. Nevertheless, the occurrence of such rudiments is as difficult to explain on the belief of the separate creation of each species, as in the foregoing cases. Hereafter I shall have to recur to these rudiments, and shall shew that their presence generally depends merely on inheritance; namely, on parts acquired by one sex having been partially transmitted to the other. Here I will only give some instances of such rudiments. It is well known that in the males of all mammals, including man, rudimentary mammæ exist. These in several instances have become well developed, and have yielded a copious supply of milk. Their essential identity in the two sexes is likewise shewn by their occasional sympathetic enlargement in both during an attack of the measles. The *vesicula prostratica,* which has been observed in many male mammals, is now universally acknowledged to be the homologue of the female uterus, together with the connected passage. It is impossible to read Leuckart's able description of this organ, and his reasoning, without admitting the justness of his conclusion. This is especially clear in the case of those mammals in which the true female uterus bifurcates, for in the males of these the vesicula likewise bifurcates.[41] Some additional rudimentary structures belonging to the reproductive system might here have been adduced.[42]

The bearing of the three great classes of facts now given is unmistakeable. But it would be superfluous here fully to recapitulate the line of argument given in detail in my 'Origin of Species.' The homological construction of the whole frame in the members of the same class is intelligible, if we admit their descent from a common progenitor, together with their subsequent adaptation to diversified conditions. On any other view the similarity of pattern between the hand of a man or monkey, the foot of a horse, the flipper of a seal, the wing of a bat, &c., is utterly inexplicable. It is no scientific

---

41 Leuckart, in Todd's 'Cyclop. of Anat.' 1849-52, vol. iv. p. 1415. In man this organ is only from three to six lines in length, but, like so many other rudimentary parts, it is variable in development as well as in other characters.

42 See, on this subject, Owen, 'Anatomy of Vertebrates,' vol. iii. pp. 675, 676, 706.

explanation to assert that they have all been formed on the same ideal plan. With respect to development, we can clearly understand, on the principle of variations supervening at a rather late embryonic period, and being inherited at a corresponding period, how it is that the embryos of wonderfully different forms should still retain, more or less perfectly, the structure of their common progenitor. No other explanation has ever been given of the marvellous fact that the embryo of a man, dog, seal, bat, reptile, &c., can at first hardly be distinguished from each other. In order to understand the existence of rudimentary organs, we have only to suppose that a former progenitor possessed the parts in question in a perfect state, and that under changed habits of life they became greatly reduced, either from simple disuse, or through the natural selection of those individuals which were least encumbered with a superfluous part, aided by the other means previously indicated.

Thus we can understand how it has come to pass that man and all other vertebrate animals have been constructed on the same general model, why they pass through the same early stages of development, and why they retain certain rudiments in common. Consequently we ought frankly to admit their community of descent: to take any other view, is to admit that our own structure and that of all the animals around us, is a mere snare laid to entrap our judgment. This conclusion is greatly strengthened, if we look to the members of the whole animal series, and consider the evidence derived from their affinities or classification, their geographical distribution and geological succession. It is only our natural prejudice, and that arrogance which made our forefathers declare that they were descended from demi-gods, which leads us to demur to this conclusion. But the time will before long come when it will be thought wonderful, that naturalists, who were well acquainted with the comparative structure and development of man and other mammals, should have believed that each was the work of a separate act of creation.

CHAPTER II.

COMPARISON OF THE MENTAL POWERS OF MAN AND THE LOWER ANIMALS.

The difference in mental power between the highest ape and the lowest savage, immense — Certain instincts in common — The emotions — Curiosity — Imitation—Attention — Memory—Imagination — Reason — Progressive improvement — Tools and weapons used by animals — Language — Self-consciousness — Sense of beauty — Belief in God, spiritual agencies, superstitions.

WE have seen in the last chapter that man bears in his bodily structure clear traces of his descent from some lower form; but it may be urged that, as man differs so greatly in his mental power from all other animals, there must be some error in this conclusion. No doubt the difference in this respect is enormous, even if we compare the mind of one of the lowest savages, who has no words to express any number higher than four, and who uses no abstract terms for the commonest objects or affections,[1] with that of the most highly organised ape. The difference would, no doubt, still remain immense, even if one of the higher apes had been improved or civilised as much as a dog has been in comparison with its parent-form, the wolf or jackal. The Fuegians rank amongst the lowest barbarians; but I was continually struck with surprise how closely the three natives on board H.M.S. "Beagle," who had lived some years in England and could talk a little English, resembled us in disposition and in most of our mental faculties. If no

[1] See the evidence on these points, as given by Lubbock, 'Prehistoric Times,' p. 354, &c.

organic being excepting man had possessed any mental power, or if his powers had been of a wholly different nature from those of the lower animals, then we should never have been able to convince ourselves that our high faculties had been gradually developed. But it can be clearly shewn that there is no fundamental difference of this kind. We must also admit that there is a much wider interval in mental power between one of the lowest fishes, as a lamprey or lancelet, and one of the higher apes, than between an ape and man; yet this immense interval is filled up by numberless gradations.

Nor is the difference slight in moral disposition between a barbarian, such as the man described by the old navigator Byron, who dashed his child on the rocks for dropping a basket of sea-urchins, and a Howard or Clarkson; and in intellect, between a savage who does not use any abstract terms, and a Newton or Shakspeare. Differences of this kind between the highest men of the highest races and the lowest savages, are connected by the finest gradations. Therefore it is possible that they might pass and be developed into each other.

My object in this chapter is solely to shew that there is no fundamental difference between man and the higher mammals in their mental faculties. Each division of the subject might have been extended into a separate essay, but must here be treated briefly. As no classification of the mental powers has been universally accepted, I shall arrange my remarks in the order most convenient for my purpose; and will select those facts which have most struck me, with the hope that they may produce some effect on the reader.

With respect to animals very low in the scale, I shall have to give some additional facts under Sexual Selection, shewing that their mental powers are higher than might have been expected. The variability of the faculties in the individuals of the same species is an important point for us, and some few illustrations will here be given. But it would be superfluous to enter into many details on this head, for I have found on frequent enquiry, that it is the unanimous opinion of all those who have long attended to animals of many kinds, including birds, that the individuals differ greatly in every mental characteristic. In what manner the mental powers were first developed in the lowest organisms, is as hopeless an enquiry as how life first originated. These are problems for the distant future, if they are ever to be solved by man.

As man possesses the same senses with the lower animals, his fundamental intuitions must be the same. Man has also some few instincts in common, as that of self-preservation, sexual love, the love of the mother for her new-born offspring, the power possessed by the latter of sucking, and so forth. But man, perhaps, has somewhat fewer instincts than those possessed by the animals which come next to him in the series. The orang in the Eastern islands, and the chimpanzee in Africa, build platforms on which they sleep; and, as both species follow the same habit, it might be argued that this was due to instinct, but we cannot feel sure that it is not the result of both animals having similar wants and possessing similar powers of reasoning. These apes, as we may assume, avoid the many poisonous fruits of the tropics, and man has no such knowledge; but as our domestic animals, when taken to foreign lands and when first turned out in the spring, often eat poisonous herbs, which they afterwards avoid, we cannot feel sure that the apes do not learn from their own experience or from that of their parents what fruits to select. It is however certain, as we shall presently see, that apes have an instinctive dread of serpents, and probably of other dangerous animals.

The fewness and the comparative simplicity of the instincts in the higher animals are remarkable in contrast with those of the lower animals. Cuvier maintained that instinct and intelligence stand in an inverse ratio to each other; and some have thought that the intellectual faculties of the higher animals have been gradually developed from their instincts. But Pouchet, in an interesting essay,[2] has shewn that no such inverse ratio really exists. Those insects which possess the most wonderful instincts are certainly the most intelligent. In the vertebrate series, the least intelligent members, namely fishes and amphibians, do not possess complex instincts; and amongst mammals the animal most remarkable for its instincts, namely the beaver, is highly intelligent, as will be admitted by every one who has read Mr. Morgan's excellent account of this animal.[3]

Although the first dawnings of intelligence, according to Mr. Herbert Spencer,[4] have been developed through the multiplication and co-ordination of reflex actions, and although many of the simpler instincts graduate into actions of this kind and can hardly be distinguished from them, as in the case of young animals sucking, yet the more complex instincts seem to have originated independently of intelligence. I am, however, far from wishing to deny that instinctive actions may lose their fixed and untaught character, and be replaced by others performed by the aid of the free will. On the other hand, some intelligent actions—as when birds on oceanic islands first learn to avoid man—after

[2] 'L'Instinct chez les Insectes.' 'Revue des Deux Mondes,' Feb. 1870, p. 690.

[3] 'The American Beaver and his Works,' 1868.

[4] 'The Principles of Psychology,' 2nd edit. 1870, pp. 418-443

being performed during many generations, become converted into instincts and are inherited. They may then be said to be degraded in character, for they are no longer performed through reason or from experience. But the greater number of the more complex instincts appear to have been gained in a wholly different manner, through the natural selection of variations of simpler instinctive actions. Such variations appear to arise from the same unknown causes acting on the cerebral organisation, which induce slight variations or individual differences in other parts of the body; and these variations, owing to our ignorance, are often said to arise spontaneously. We can, I think, come to no other conclusion with respect to the origin of the more complex instincts, when we reflect on the marvellous instincts of sterile worker-ants and bees, which leave no offspring to inherit the effects of experience and of modified habits.

Although a high degree of intelligence is certainly compatible with the existence of complex instincts, as we see in the insects just named and in the beaver, it is not improbable that they may to a certain extent interfere with each other's development. Little is known about the functions of the brain, but we can perceive that as the intellectual powers become highly developed, the various parts of the brain must be connected by the most intricate channels of intercommunication; and as a consequence each separate part would perhaps tend to become less well fitted to answer in a definite and uniform, that is instinctive, manner to particular sensations or associations.

I have thought this digression worth giving, because we may easily underrate the mental powers of the higher animals, and especially of man, when we compare their actions founded on the memory of past events, on foresight, reason, and imagination, with exactly similar actions instinctively performed by the lower animals; in this latter case the capacity of performing such actions having been gained, step by step, through the variability of the mental organs and natural selection, without any conscious intelligence on the part of the animal during each successive generation. No doubt, as Mr. Wallace has argued,[5] much of the intelligent work done by man is due to imitation and not to reason; but there is this great difference between his actions and many of those performed by the lower animals, namely, that man cannot, on his first trial, make, for instance, a stone hatchet or a canoe, through his power of imitation. He has to learn his work by practice; a beaver, on the other hand, can make its dam or canal, and a bird its nest, as well, or nearly as well, the first time it tries, as when old and experienced.

To return to our immediate subject: the lower animals, like man, manifestly feel pleasure and pain, happiness and misery. Happiness is never better exhibited than by young animals, such as puppies, kittens, lambs, &c., when playing together, like our own children. Even insects play together, as has been described by that excellent observer, P. Huber,[6] who saw ants chasing and pretending to bite each other, like so many puppies.

The fact that the lower animals are excited by the same emotions as ourselves is so well established, that it will not be necessary to weary the reader by many details. Terror acts in the same manner on them as on us, causing the muscles to tremble, the heart to palpitate, the sphincters to be relaxed, and the hair to stand on end. Suspicion, the offspring of fear, is eminently characteristic of most wild animals. Courage

5 'Contributions to the Theory of Natural Selection,' 1870, p. 212.

6 'Recherches sur les Mœurs des Fourmis,' 1810, p. 173.

and timidity are extremely variable qualities in the individuals of the same species, as is plainly seen in our dogs. Some dogs and horses are ill-tempered and easily turn sulky; others are good-tempered; and these qualities are certainly inherited. Every one knows how liable animals are to furious rage, and how plainly they show it. Many anecdotes, probably true, have been published on the long-delayed and artful revenge of various animals. The accurate Rengger and Brehm[7] state that the American and African monkeys which they kept tame, certainly revenged themselves. The love of a dog for his master is notorious; in the agony of death he has been known to caress his master, and every one has heard of the dog suffering under vivisection, who licked the hand of the operator; this man, unless he had a heart of stone, must have felt remorse to the last hour of his life. As Whewell[8] has remarked, "who that reads the touching instances of "maternal affection, related so often of the women of "all nations, and of the females of all animals, can "doubt that the principle of action is the same in the "two cases?"

We see maternal affection exhibited in the most trifling details; thus Rengger observed an American monkey (a Cebus) carefully driving away the flies which plagued her infant; and Duvaucel saw a Hylobates washing the faces of her young ones in a stream. So intense is the grief of female monkeys for the loss of their young, that it invariably caused the death of certain kinds kept under confinement by Brehm in N.

---

7 All the following statements, given on the authority of these two naturalists, are taken from Rengger's 'Naturges. der Säugethiere von Paraguay,' 1830, s. 41-57, and from Brehm's 'Thierleben,' B. i. s. 10-87.

8 'Bridgewater Treatise,' p. 263.

Africa. Orphan-monkeys were always adopted and carefully guarded by the other monkeys, both males and females. One female baboon had so capacious a heart that she not only adopted young monkeys of other species, but stole young dogs and cats, which she continually carried about. Her kindness, however, did not go so far as to share her food with her adopted offspring, at which Brehm was surprised, as his monkeys always divided everything quite fairly with their own young ones. An adopted kitten scratched the above-mentioned affectionate baboon, who certainly had a fine intellect, for she was much astonished at being scratched, and immediately examined the kitten's feet, and without more ado bit off the claws. In the Zoological Gardens, I heard from the keeper that an old baboon (*C. chacma*) had adopted a Rhesus monkey; but when a young drill and mandrill were placed in the cage, she seemed to perceive that these monkeys, though distinct species, were her nearer relatives, for she at once rejected the Rhesus and adopted both of them. The young Rhesus, as I saw, was greatly discontented at being thus rejected, and it would, like a naughty child, annoy and attack the young drill and mandrill whenever it could do so with safety; this conduct exciting great indignation in the old baboon. Monkeys will also, according to Brehm, defend their master when attacked by any one, as well as dogs to whom they are attached, from the attacks of other dogs. But we here trench on the subject of sympathy, to which I shall recur. Some of Brehm's monkeys took much delight in teasing, in various ingenious ways, a certain old dog whom they disliked, as well as other animals.

Most of the more complex emotions are common to the higher animals and ourselves. Every one has seen how jealous a dog is of his master's affection, if lavished on any other creature; and I have observed the same fact with monkeys. This shews that animals not only love, but have the desire to be loved. Animals manifestly feel emulation. They love approbation or praise; and a dog carrying a basket for his master exhibits in a high degree self-complacency or pride. There can, I think, be no doubt that a dog feels shame, as distinct from fear, and something very like modesty when begging too often for food. A great dog scorns the snarling of a little dog, and this may be called magnanimity. Several observers have stated that monkeys certainly dislike being laughed at; and they sometimes invent imaginary offences. In the Zoological Gardens I saw a baboon who always got into a furious rage when his keeper took out a letter or book and read it aloud to him; and his rage was so violent that, as I witnessed on one occasion, he bit his own leg till the blood flowed.

We will now turn to the more intellectual emotions and faculties, which are very important, as forming the basis for the development of the higher mental powers. Animals manifestly enjoy excitement and suffer from ennui, as may be seen with dogs, and, according to Rengger, with monkeys. All animals feel Wonder, and many exhibit Curiosity. They sometimes suffer from this latter quality, as when the hunter plays antics and thus attracts them; I have witnessed this with deer, and so it is with the wary chamois, and with some kinds of wild-ducks. Brehm gives a curious account of the instinctive dread which his monkeys exhibited towards snakes; but their curiosity was so great that they could not desist from occasionally satiating their horror in a most human fashion, by lifting up the lid of the box in which the snakes were kept. I was so much surprised at his account, that I took a stuffed and coiled-up snake into the monkey-house at the Zoological Gardens, and the excitement thus caused was one of the most curious spectacles which I ever beheld. Three species of Cercopithecus were the most alarmed; they dashed about their cages and uttered sharp signal-cries of danger, which were understood by the other monkeys. A few young monkeys and one old Anubis baboon alone took no notice of the snake. I then placed the stuffed specimen on the ground in one of the larger compartments. After a time all the monkeys collected round it in a large circle, and staring intently, presented a most ludicrous appearance. They became extremely nervous; so that when a wooden ball, with which they were familiar as a plaything, was accidently moved in the straw, under which it was partly hidden, they all instantly started away. These monkeys behaved very differently when a dead fish, a mouse, and some other new objects were placed in their cages; for though at first frightened, they soon approached, handled and examined them. I then placed a live snake in a paper bag, with the mouth loosely closed, in one of the larger compartments. One of the monkeys immediately approached, cautiously opened the bag a little, peeped in, and instantly dashed away. Then I witnessed what Brehm has described, for monkey after monkey, with head raised high and turned on one side, could not resist taking momentary peeps into the upright bag, at the dreadful object lying quiet at the bottom. It would almost appear as if monkeys had some notion of zoological affinities, for those kept by Brehm exhibited a strange, though mistaken, instinctive dread of innocent lizards and frogs. An orang, also, has been known to be much alarmed at the first sight of a turtle.[9]

---

9 W. C. L. Martin, 'Nat. Hist. of Mammalia,' 1841, p. 405.

The principle of *Imitation* is strong in man, and especially in man in a barbarous state. Desor[10] has remarked that no animal voluntarily imitates an action performed by man, until in the ascending scale we come to monkeys, which are well-known to be ridiculous mockers. Animals, however, sometimes imitate each others' actions: thus two species of wolves, which had been reared by dogs, learned to bark, as does sometimes the jackal,[11] but whether this can be called voluntary imitation is another question. From one account which I have read, there is reason to believe that puppies nursed by cats sometimes learn to lick their feet and thus to clean their faces: it is at least certain, as I hear from a perfectly trustworthy friend, that some dogs behave in this manner. Birds imitate the songs of their parents, and sometimes those of other birds; and parrots are notorious imitators of any sound which they often hear.

Hardly any faculty is more important for the intellectual progress of man than the power of *Attention*. Animals clearly manifest this power, as when a cat watches by a hole and prepares to spring on its prey. Wild animals sometimes become so absorbed when thus engaged, that they may be easily approached. Mr. Bartlett has given me a curious proof how variable this faculty is in monkeys. A man who trains monkeys to act used to purchase common kinds from the Zoological Society at the price of five pounds for each; but he offered to give double the price, if he might keep three or four of them for a few days, in order to select one. When asked how he could possibly so soon learn whether

---

[10] Quoted by Vogt, 'Mémoire sur les Microcéphales,' 1867, p. 168.

[11] 'The Variation of Animals and Plants under Domestication,' vol. i. p. 27.

a particular monkey would turn out a good actor, he answered that it all depended on their power of attention. If when he was talking and explaining anything to a monkey, its attention was easily distracted, as by a fly on the wall or other trifling object, the case was hopeless. If he tried by punishment to make an inattentive monkey act, it turned sulky. On the other hand, a monkey which carefully attended to him could always be trained.

It is almost superfluous to state that animals have excellent *Memories* for persons and places. A baboon at the Cape of Good Hope, as I have been informed by Sir Andrew Smith, recognised him with joy after an absence of nine months. I had a dog who was savage and averse to all strangers, and I purposely tried his memory after an absence of five years and two days. I went near the stable where he lived, and shouted to him in my old manner; he showed no joy, but instantly followed me out walking and obeyed me, exactly as if I had parted with him only half-an-hour before. A train of old associations, dormant during five years, had thus been instantaneously awakened in his mind. Even ants, as P. Huber[12] has clearly shewn, recognised their fellow-ants belonging to the same community after a separation of four months. Animals can certainly by some means judge of the intervals of time between recurrent events.

The *Imagination* is one of the highest prerogatives of man. By this faculty he unites, independently of the will, former images and ideas, and thus creates brilliant and novel results. A poet, as Jean Paul Richter remarks,[13] "who must reflect whether he shall make a

---

[12] 'Les Mœurs des Fourmis,' 1810, p. 150.

[13] Quoted in Dr. Maudsley's 'Physiology and Pathology of Mind,' 1868, pp. 19, 220.

"character say yes or no—to the devil with him; he is "only a stupid corpse." Dreaming gives us the best notion of this power; as Jean Paul again says, "The "dream is an involuntary art of poetry." The value of the products of our imagination depends of course on the number, accuracy, and clearness of our impressions; on our judgment and taste in selecting or rejecting the involuntary combinations, and to a certain extent on our power of voluntarily combining them. As dogs, cats, horses, and probably all the higher animals, even birds, as is stated on good authority,[14] have vivid dreams, and this is shewn by their movements and voice, we must admit that they possess some power of imagination.

Of all the faculties of the human mind, it will, I presume, be admitted that *Reason* stands at the summit. Few persons any longer dispute that animals possess some power of reasoning. Animals may constantly be seen to pause, deliberate, and resolve. It is a significant fact, that the more the habits of any particular animal are studied by a naturalist, the more he attributes to reason and the less to unlearnt instincts.[15] In future chapters we shall see that some animals extremely low in the scale apparently display a certain amount of reason. No doubt it is often difficult to distinguish between the power of reason and that of instinct. Thus Dr. Hayes, in his work on 'The Open Polar Sea,' repeatedly remarks that his dogs, instead of continuing to draw the sledges in a compact body, diverged and separated when they came to thin ice, so that their weight might be more evenly distributed. This was often the first warn-

[14] Dr. Jerdon, 'Birds of India,' vol. i. 1862, p. xxi.

[15] Mr. L. H. Morgan's work on 'The American Beaver,' 1868, offers a good illustration of this remark. I cannot, however, avoid thinking that he goes too far in underrating the power of Instinct.

ing and notice which the travellers received that the ice was becoming thin and dangerous. Now, did the dogs act thus from the experience of each individual, or from the example of the older and wiser dogs, or from an inherited habit, that is from an instinct? This instinct might possibly have arisen since the time, long ago, when dogs were first employed by the natives in drawing their sledges; or the Arctic wolves, the parent-stock of the Esquimaux dog, may have acquired this instinct, impelling them not to attack their prey in a close pack when on thin ice. Questions of this kind are most difficult to answer.

So many facts have been recorded in various works shewing that animals possess some degree of reason, that I will here give only two or three instances, authenticated by Rengger, and relating to American monkeys, which stand low in their order. He states that when he first gave eggs to his monkeys, they smashed them and thus lost much of their contents; afterwards they gently hit one end against some hard body, and picked off the bits of shell with their fingers. After cutting themselves only once with any sharp tool, they would not touch it again, or would handle it with the greatest care. Lumps of sugar were often given them wrapped up in paper; and Rengger sometimes put a live wasp in the paper, so that in hastily unfolding it they got stung; after this had once happened, they always first held the packet to their ears to detect any movement within. Any one who is not convinced by such facts as these, and by what he may observe with his own dogs, that animals can reason, would not be convinced by anything that I could add. Nevertheless I will give one case with respect to dogs, as it rests on two distinct observers, and can hardly depend on the modification of any instinct.

Mr. Colquhoun[16] winged two wild-ducks, which fell on the opposite side of a stream; his retriever tried to bring over both at once, but could not succeed; she then, though never before known to ruffle a feather, deliberately killed one, brought over the other, and returned for the dead bird. Col. Hutchinson relates that two partridges were shot at once, one being killed, the other wounded; the latter ran away, and was caught by the retriever, who on her return came across the dead bird; "she stopped, evidently greatly puzzled, and "after one or two trials, finding she could not take it up "without permitting the escape of the winged bird, she "considered a moment, then deliberately murdered it "by giving it a severe crunch, and afterwards brought "away both together. This was the only known in-"stance of her ever having wilfully injured any game." Here we have reason, though not quite perfect, for the retriever might have brought the wounded bird first and then returned for the dead one, as in the case of the two wild-ducks.

The muleteers in S. America say, "I will not give "you the mule whose step is easiest, but *la mas racional*, "—the one that reasons best;" and Humboldt[17] adds, "this popular expression, dictated by long experience, "combats the system of animated machines, better per-"haps than all the arguments of speculative philosophy."

It has, I think, now been shewn that man and the higher animals, especially the Primates, have some few instincts in common. All have the same senses, intuitions and sensations—similar passions, affections, and emotions, even the more complex ones; they feel

---

[16] 'The Moor and the Loch,' p. 45. Col. Hutchinson on 'Dog Breaking,' 1850, p. 46.

[17] 'Personal Narrative,' Eng. translat., vol. iii. p. 106.

wonder and curiosity; they possess the same faculties of imitation, attention, memory, imagination, and reason, though in very different degrees. Nevertheless many authors have insisted that man is separated through his mental faculties by an impassable barrier from all the lower animals. I formerly made a collection of above a score of such aphorisms, but they are not worth giving, as their wide difference and number prove the difficulty, if not the impossibility, of the attempt. It has been asserted that man alone is capable of progressive improvement; that he alone makes use of tools or fire, domesticates other animals, possesses property, or employs language; that no other animal is self-conscious, comprehends itself, has the power of abstraction, or possesses general ideas; that man alone has a sense of beauty, is liable to caprice, has the feeling of gratitude, mystery, &c.; believes in God, or is endowed with a conscience. I will hazard a few remarks on the more important and interesting of these points.

Archbishop Sumner formerly maintained [18] that man alone is capable of progressive improvement. With animals, looking first to the individual, every one who has had any experience in setting traps knows that young animals can be caught much more easily than old ones; and they can be much more easily approached by an enemy. Even with respect to old animals, it is impossible to catch many in the same place and in the same kind of trap, or to destroy them by the same kind of poison; yet it is improbable that all should have partaken of the poison, and impossible that all should have been caught in the trap. They must learn caution by seeing their brethren caught or poisoned. In North America, where the fur-bearing animals have long been

---

[18] Quoted by Sir C. Lyell, 'Antiquity of Man,' p. 497.

pursued, they exhibit, according to the unanimous testimony of all observers, an almost incredible amount of sagacity, caution, and cunning; but trapping has been there so long carried on that inheritance may have come into play.

If we look to successive generations, or to the race, there is no doubt that birds and other animals gradually both acquire and lose caution in relation to man or other enemies;[19] and this caution is certainly in chief part an inherited habit or instinct, but in part the result of individual experience. A good observer, Leroy,[20] states that in districts where foxes are much hunted, the young when they first leave their burrows are incontestably much more wary than the old ones in districts where they are not much disturbed.

Our domestic dogs are descended from wolves and jackals,[21] and though they may not have gained in cunning, and may have lost in waryness and suspicion, yet they have progressed in certain moral qualities, such as in affection, trust-worthiness, temper, and probably in general intelligence. The common rat has conquered and beaten several other species throughout Europe, in parts of North America, New Zealand, and recently in Formosa, as well as on the mainland of China. Mr. Swinhoe,[22] who describes these latter cases, attributes the victory of the common rat over the large *Mus coninga* to its superior cunning; and this latter quality may be attributed to the habitual exercise of all its faculties in avoiding extirpation by man, as well

---

19 'Journal of Researches during the Voyage of the "Beagle,"' 1845, p. 398. 'Origin of Species,' 5th edit. p. 260.

20 'Lettres Phil. sur l'Intelligence des Animaux,' nouvelle edit. 1802, p. 86.

21 See the evidence on this head in chap. i. vol. i. 'On the Variation of Animals and Plants under Domestication.'

22 'Proc. Zoolog. Soc.' 1864, p. 186.

as to nearly all the less cunning or weak-minded rats having been successively destroyed by him. To maintain, independently of any direct evidence, that no animal during the course of ages has progressed in intellect or other mental faculties, is to beg the question of the evolution of species. Hereafter we shall see that, according to Lartet, existing mammals belonging to several orders have larger brains than their ancient tertiary prototypes.

It has often been said that no animal uses any tool; but the chimpanzee in a state of nature cracks a native fruit, somewhat like a walnut, with a stone.[23] Rengger[24] easily taught an American monkey thus to break open hard palm-nuts, and afterwards of its own accord it used stones to open other kinds of nuts, as well as boxes. It thus also removed the soft rind of fruit that had a disagreeable flavour. Another monkey was taught to open the lid of a large box with a stick, and afterwards it used the stick as a lever to move heavy bodies; and I have myself seen a young orang put a stick into a crevice, slip his hand to the other end, and use it in the proper manner as a lever. In the cases just mentioned stones and sticks were employed as implements; but they are likewise used as weapons. Brehm[25] states, on the authority of the well-known traveller Schimper, that in Abyssinia when the baboons belonging to one species (*C. gelada*) descend in troops from the mountains to plunder the fields, they sometimes encounter troops of another species (*C. hamadryas*), and then a fight ensues. The Geladas roll down great stones, which the Hamadryas try to avoid, and then both species,

---

[23] Savage and Wyman in 'Boston Journal of Nat. Hist.' vol. iv. 1843-44, p. 383.

[24] 'Säugethiere von Paraguay,' 1830, s. 51-56.

[25] 'Thierleben,' B. i. s. 79, 82.

making a great uproar, rush furiously against each other. Brehm, when accompanying the Duke of Coburg-Gotha, aided in an attack with fire-arms on a troop of baboons in the pass of Mensa in Abyssinia. The baboons in return rolled so many stones down the mountain, some as large as a man's head, that the attackers had to beat a hasty retreat; and the pass was actually for a time closed against the caravan. It deserves notice that these baboons thus acted in concert. Mr. Wallace[26] on three occasions saw female orangs, accompanied by their young, "breaking off branches and "the great spiny fruit of the Durian tree, with every "appearance of rage; causing such a shower of missiles "as effectually kept us from approaching too near the "tree."

In the Zoological Gardens a monkey which had weak teeth used to break open nuts with a stone; and I was assured by the keepers that this animal, after using the stone, hid it in the straw, and would not let any other monkey touch it. Here, then, we have the idea of property; but this idea is common to every dog with a bone, and to most or all birds with their nests.

The Duke of Argyll[27] remarks, that the fashioning of an implement for a special purpose is absolutely peculiar to man; and he considers that this forms an immeasurable gulf between him and the brutes. It is no doubt a very important distinction, but there appears to me much truth in Sir J. Lubbock's suggestion,[28] that when primeval man first used flint-stones for any purpose, he would have accidentally splintered them, and would then have used the sharp fragments. From this step it would be a small one to intentionally break the

---

26 'The Malay Archipelago,' vol. i. 1869, p. 87.
27 'Primeval Man,' 1869, pp. 145, 147.
28 'Prehistoric Times,' 1865, p. 473, &c.

flints, and not a very wide step to rudely fashion them. This latter advance, however, may have taken long ages, if we may judge by the immense interval of time which elapsed before the men of the neolithic period took to grinding and polishing their stone tools. In breaking the flints, as Sir J. Lubbock likewise remarks, sparks would have been emitted, and in grinding them heat would have been evolved: "thus the two usual "methods of obtaining fire may have originated." The nature of fire would have been known in the many volcanic regions where lava occasionally flows through forests. The anthropomorphous apes, guided probably by instinct, build for themselves temporary platforms; but as many instincts are largely controlled by reason, the simpler ones, such as this of building a platform, might readily pass into a voluntary and conscious act. The orang is known to cover itself at night with the leaves of the Pandanus; and Brehm states that one of his baboons used to protect itself from the heat of the sun by throwing a straw-mat over its head. In these latter habits, we probably see the first steps towards some of the simpler arts; namely rude architecture and dress, as they arose amongst the early progenitors of man.

Language.—This faculty has justly been considered as one of the chief distinctions between man and the lower animals. But man, as a highly competent judge, Archbishop Whately remarks, "is not the only animal that "can make use of language to express what is passing in "his mind, and can understand, more or less, what is so "expressed by another."[29] In Paraguay the *Cebus azaræ* when excited utters at least six distinct sounds, which

[29] Quoted in 'Anthropological Review,' 1864, p. 158.

excite in other monkeys similar emotions.[30] The movements of the features and gestures of monkeys are understood by us, and they partly understand ours, as Rengger and others declare. It is a more remarkable fact that the dog, since being domesticated, has learnt to bark[31] in at least four or five distinct tones. Although barking is a new art, no doubt the wild species, the parents of the dog, expressed their feelings by cries of various kinds. With the domesticated dog we have the bark of eagerness, as in the chase; that of anger; the yelping or howling bark of despair, as when shut up; that of joy, as when starting on a walk with his master; and the very distinct one of demand or supplication, as when wishing for a door or window to be opened.

Articulate language is, however, peculiar to man; but he uses in common with the lower animals inarticulate cries to express his meaning, aided by gestures and the movements of the muscles of the face.[32] This especially holds good with the more simple and vivid feelings, which are but little connected with our higher intelligence. Our cries of pain, fear, surprise, anger, together with their appropriate actions, and the murmur of a mother to her beloved child, are more expressive than any words. It is not the mere power of articulation that distinguishes man from other animals, for as every one knows, parrots can talk; but it is his large power of connecting definite sounds with definite ideas; and this obviously depends on the development of the mental faculties.

---

[30] Rengger, ibid. s. 45.

[31] See my 'Variation of Animals and Plants under Domestication,' vol. i. p. 27.

[32] See a discussion on this subject in Mr. E. B. Tylor's very interesting work, 'Researches into the Early History of Mankind,' 1865, chaps. ii. to iv.

As Horne Tooke, one of the founders of the noble science of philology, observes, language is an art, like brewing or baking; but writing would have been a much more appropriate simile. It certainly is not a true instinct, as every language has to be learnt. It differs, however, widely from all ordinary arts, for man has an instinctive tendency to speak, as we see in the babble of our young children; whilst no child has an instinctive tendency to brew, bake, or write. Moreover, no philologist now supposes that any language has been deliberately invented; each has been slowly and unconsciously developed by many steps. The sounds uttered by birds offer in several respects the nearest analogy to language, for all the members of the same species utter the same instinctive cries expressive of their emotions; and all the kinds that have the power of singing exert this power instinctively; but the actual song, and even the call-notes, are learnt from their parents or foster-parents. These sounds, as Daines Barrington[33] has proved, "are no more innate than "language is in man." The first attempts to sing "may be compared to the imperfect endeavour in a "child to babble." The young males continue practising, or, as the bird-catchers say, recording, for ten or eleven months. Their first essays show hardly a rudiment of the future song; but as they grow older we can perceive what they are aiming at; and at last they are said "to sing their song round." Nestlings which have learnt the song of a distinct species, as with the canary-birds educated in the Tyrol, teach and transmit their new song to their offspring. The slight natural differences of song in the same species inha-

---

[33] Hon. Daines Barrington in 'Philosoph. Transactions,' 1773, p. 262. See also Dureau de la Malle, in 'Ann. des Sc. Nat.' 3rd series, Zoolog. tom. x. p. 119.

biting different districts may be appositely compared, as Barrington remarks, "to provincial dialects;" and the songs of allied, though distinct species may be compared with the languages of distinct races of man. I have given the foregoing details to shew that an instinctive tendency to acquire an art is not a peculiarity confined to man.

With respect to the origin of articulate language, after having read on the one side the highly interesting works of Mr. Hensleigh Wedgwood, the Rev. F. Farrar, and Prof. Schleicher,[34] and the celebrated lectures of Prof. Max Müller on the other side, I cannot doubt that language owes its origin to the imitation and modification, aided by signs and gestures, of various natural sounds, the voices of other animals, and man's own instinctive cries. When we treat of sexual selection we shall see that primeval man, or rather some early progenitor of man, probably used his voice largely, as does one of the gibbon-apes at the present day, in producing true musical cadences, that is in singing; we may conclude from a widely-spread analogy that this power would have been especially exerted during the courtship of the sexes, serving to express various emotions, as love, jealousy, triumph, and serving as a challenge to their rivals. The imitation by articulate sounds of musical cries might have given rise to words expressive of various complex emotions. As bearing on the subject of imitation, the strong tendency in our nearest allies, the monkeys, in microcephalous

[34] 'On the Origin of Language,' by H. Wedgwood, 1866. 'Chapters on Language,' by the Rev. F. W. Farrar, 1865. These works are most interesting. See also 'De la Phys. et de Parole,' par Albert Lemoine, 1865, p. 190. The work on this subject, by the late Prof. Aug. Schleicher, has been translated by Dr. Bikkers into English, under the title of 'Darwinism tested by the Science of Language,' 1869.

idiots,[35] and in the barbarous races of mankind, to imitate whatever they hear deserves notice. As monkeys certainly understand much that is said to them by man, and as in a state of nature they utter signal-cries of danger to their fellows,[36] it does not appear altogether incredible, that some unusually wise ape-like animal should have thought of imitating the growl of a beast of prey, so as to indicate to his fellow monkeys the nature of the expected danger. And this would have been a first step in the formation of a language.

As the voice was used more and more, the vocal organs would have been strengthened and perfected through the principle of the inherited effects of use; and this would have reacted on the power of speech. But the relation between the continued use of language and the development of the brain has no doubt been far more important. The mental powers in some early progenitor of man must have been more highly developed than in any existing ape, before even the most imperfect form of speech could have come into use; but we may confidently believe that the continued use and advancement of this power would have reacted on the mind by enabling and encouraging it to carry on long trains of thought. A long and complex train of thought can no more be carried on without the aid of words, whether spoken or silent, than a long calculation without the use of figures or algebra. It appears, also, that even ordinary trains of thought almost require some form of language, for the dumb, deaf, and blind girl, Laura Bridgman, was observed to use her fingers whilst dream-

---

[35] Vogt, 'Mémoire sur les Microcéphales,' 1867, p. 169. With respect to savages, I have given some facts in my 'Journal of Researches,' &c., 1845, p. 206.

[36] See clear evidence on this head in the two works so often quoted, by Brehm and Rengger.

ing.[37] Nevertheless a long succession of vivid and connected ideas, may pass through the mind without the aid of any form of language, as we may infer from the prolonged dreams of dogs. We have, also, seen that retriever-dogs are able to reason to a certain extent; and this they manifestly do without the aid of language. The intimate connection between the brain, as it is now developed in us, and the faculty of speech, is well shewn by those curious cases of brain-disease, in which speech is specially affected, as when the power to remember substantives is lost, whilst other words can be correctly used.[38] There is no more improbability in the effects of the continued use of the vocal and mental organs being inherited, than in the case of hand-writing, which depends partly on the structure of the hand and partly on the disposition of the mind; and hand-writing is certainly inherited.[39]

Why the organs now used for speech should have been originally perfected for this purpose, rather than any other organs, it is not difficult to see. Ants have considerable powers of intercommunication by means of their antennæ, as shewn by Huber, who devotes a whole chapter to their language. We might have used our fingers as efficient instruments, for a person with practice can report to a deaf man every word of a speech rapidly delivered at a public meeting; but the loss of our hands, whilst thus employed, would have been a serious inconvenience. As all the higher mammals possess vocal organs constructed on the same general

---

[37] See remarks on this head by Dr. Maudsley, 'The Physiology and Pathology of Mind,' 2nd edit. 1868, p. 199.

[38] Many curious cases have been recorded. See, for instance, 'Inquiries Concerning the Intellectual Powers,' by Dr. Abercrombie, 1838, p. 150.

[39] 'The Variation of Animals and Plants under Domestication,' vol. ii. p. 6.

plan with ours, and which are used as a means of communication, it was obviously probable, if the power of communication had to be improved, that these same organs would have been still further developed; and this has been effected by the aid of adjoining and well-adapted parts, namely the tongue and lips.[40] The fact of the higher apes not using their vocal organs for speech, no doubt depends on their intelligence not having been sufficiently advanced. The possession by them of organs, which with long-continued practice might have been used for speech, although not thus used, is paralleled by the case of many birds which possess organs fitted for singing, though they never sing. Thus, the nightingale and crow have vocal organs similarly constructed, these being used by the former for diversified song, and by the latter merely for croaking.[41]

The formation of different languages and of distinct species, and the proofs that both have been developed through a gradual process, are curiously the same.[42] But we can trace the origin of many words further back than in the case of species, for we can perceive that they have arisen from the imitation of various sounds, as in alliterative poetry. We find in distinct languages striking homologies due to community of descent, and analogies due to a similar process of

---

[40] See some good remarks to this effect by Dr. Maudsley, 'The Physiology and Pathology of Mind,' 1868, p. 199.

[41] Macgillivray, 'Hist. of British Birds,' vol. ii. 1839, p. 29. An excellent observer, Mr. Blackwall, remarks that the magpie learns to pronounce single words, and even short sentences, more readily than almost any other British bird; yet, as he adds, after long and closely investigating its habits, he has never known it, in a state of nature, display any unusual capacity for imitation. 'Researches in Zoology,' 1834, p. 158.

[42] See the very interesting parallelism between the development of speech and languages, given by Sir C. Lyell in 'The Geolog. Evidences of the Antiquity of Man,' 1863, chap. xxiii.

formation. The manner in which certain letters or sounds change when others change is very like correlated growth. We have in both cases the reduplication of parts, the effects of long-continued use, and so forth. The frequent presence of rudiments, both in languages and in species, is still more remarkable. The letter *m* in the word *am*, means *I;* so that in the expression *I am*, a superfluous and useless rudiment has been retained. In the spelling also of words, letters often remain as the rudiments of ancient forms of pronunciation. Languages, like organic beings, can be classed in groups under groups; and they can be classed either naturally according to descent, or artificially by other characters. Dominant languages and dialects spread widely and lead to the gradual extinction of other tongues. A language, like a species, when once extinct, never, as Sir C. Lyell remarks, reappears. The same language never has two birth-places. Distinct languages may be crossed or blended together.[43] We see variability in every tongue, and new words are continually cropping up; but as there is a limit to the powers of the memory, single words, like whole languages, gradually become extinct. As Max Müller[44] has well remarked:—"A struggle for "life is constantly going on amongst the words and gram-"matical forms in each language. The better, the "shorter, the easier forms are constantly gaining the "upper hand, and they owe their success to their own "inherent virtue." To these more important causes of the survival of certain words, mere novelty may, I think, be added; for there is in the mind of man a strong love for slight changes in all things. The survival or

---

43 See remarks to this effect by the Rev. F. W. Farrar, in an interesting article, entitled "Philology and Darwinism" in 'Nature,' March 24th, 1870, p. 528.

44 'Nature,' Jan. 6th, 1870, p. 257.

preservation of certain favoured words in the struggle for existence is natural selection.

The perfectly regular and wonderfully complex construction of the languages of many barbarous nations has often been advanced as a proof, either of the divine origin of these languages, or of the high art and former civilisation of their founders. Thus F. von Schlegel writes: "In those languages which appear to be at the "lowest grade of intellectual culture, we frequently ob-"serve a very high and elaborate degree of art in their "grammatical structure. This is especially the case with "the Basque and the Lapponian, and many of the Ame-"rican languages."[45] But it is assuredly an error to speak of any language as an art in the sense of its having been elaborately and methodically formed. Philologists now admit that conjugations, declensions, &c., originally existed as distinct words, since joined together; and as such words express the most obvious relations between objects and persons, it is not surprising that they should have been used by the men of most races during the earliest ages. With respect to perfection, the following illustration will best shew how easily we may err: a Crinoid sometimes consists of no less than 150,000 pieces of shell,[46] all arranged with perfect symmetry in radiating lines; but a naturalist does not consider an animal of this kind as more perfect than a bilateral one with comparatively few parts, and with none of these alike, excepting on the opposite sides of the body. He justly considers the differentiation and specialisation of organs as the test of perfection. So with languages, the most symmetrical and complex ought not to be ranked above irregular, abbre-

---

[45] Quoted by C. S. Wake, 'Chapters on Man,' 1868, p. 101.

[46] Buckland, 'Bridgewater Treatise,' p. 411.

viated, and bastardised languages, which have borrowed expressive words and useful forms of construction from various conquering, or conquered, or immigrant races.

From these few and imperfect remarks I conclude that the extremely complex and regular construction of many barbarous languages, is no proof that they owe their origin to a special act of creation.[47] Nor, as we have seen, does the faculty of articulate speech in itself offer any insuperable objection to the belief that man has been developed from some lower form.

Self-consciousness, Individuality, Abstraction, General Ideas, &c.—It would be useless to attempt discussing these high faculties, which, according to several recent writers, make the sole and complete distinction between man and the brutes, for hardly two authors agree in their definitions. Such faculties could not have been fully developed in man until his mental powers had advanced to a high standard, and this implies the use of a perfect language. No one supposes that one of the lower animals reflects whence he comes or whither he goes,—what is death or what is life, and so forth. But can we feel sure that an old dog with an excellent memory and some power of imagination, as shewn by his dreams, never reflects on his past pleasures in the chase? and this would be a form of self-consciousness. On the other hand, as Büchner[48] has remarked, how little can the hard-worked wife of a degraded Australian savage, who uses hardly any abstract words and cannot count above four, exert her self-consciousness, or reflect on the nature of her own existence.

---

[47] See some good remarks on the simplification of languages, by Sir J. Lubbock, 'Origin of Civilisation,' 1870, p. 278.

[48] 'Conférences sur la Théorie Darwinienne,' French translat., 1869, p. 132.

That animals retain their mental individuality is unquestionable. When my voice awakened a train of old associations in the mind of the above-mentioned dog, he must have retained his mental individuality, although every atom of his brain had probably undergone change more than once during the interval of five years. This dog might have brought forward the argument lately advanced to crush all evolutionists, and said, "I abide amid all mental moods and all "material changes. . . . The teaching that atoms leave "their impressions as legacies to other atoms falling "into the places they have vacated is contradictory of "the utterance of consciousness, and is therefore false; "but it is the teaching necessitated by evolutionism, "consequently the hypothesis is a false one."[49]

Sense of Beauty.—This sense has been declared to be peculiar to man. But when we behold male birds elaborately displaying their plumes and splendid colours before the females, whilst other birds not thus decorated make no such display, it is impossible to doubt that the females admire the beauty of their male partners. As women everywhere deck themselves with these plumes, the beauty of such ornaments cannot be disputed. The Bower-birds by tastefully ornamenting their playing-passages with gaily-coloured objects, as do certain humming-birds their nests, offer additional evidence that they possess a sense of beauty. So with the song of birds, the sweet strains poured forth by the males during the season of love are certainly admired by the females, of which fact evidence will hereafter be given. If female birds had been incapable of appreciating the beautiful colours, the ornaments, and voices

---

[49] The Rev. Dr. J. M'Cann, 'Anti-Darwinism,' 1869, p. 13.

of their male partners, all the labour and anxiety exhibited by them in displaying their charms before the females would have been thrown away; and this it is impossible to admit. Why certain bright colours and certain sounds should excite pleasure, when in harmony, cannot, I presume, be explained any more than why certain flavours and scents are agreeable; but assuredly the same colours and the same sounds are admired by us and by many of the lower animals.

The taste for the beautiful, at least as far as female beauty is concerned, is not of a special nature in the human mind; for it differs widely in the different races of man, as will hereafter be shewn, and is not quite the same even in the different nations of the same race. Judging from the hideous ornaments and the equally hideous music admired by most savages, it might be urged that their æsthetic faculty was not so highly developed as in certain animals, for instance, in birds. Obviously no animal would be capable of admiring such scenes as the heavens at night, a beautiful landscape, or refined music; but such high tastes, depending as they do on culture and complex associations, are not enjoyed by barbarians or by uneducated persons.

Many of the faculties, which have been of inestimable service to man for his progressive advancement, such as the powers of the imagination, wonder, curiosity, an undefined sense of beauty, a tendency to imitation, and the love of excitement or novelty, could not fail to have led to the most capricious changes of customs and fashions. I have alluded to this point, because a recent writer[50] has oddly fixed on Caprice "as one of the most remarkable and

[50] 'The Spectator,' Dec. 4th, 1869, p. 1430.

"typical differences between savages and brutes." But not only can we perceive how it is that man is capricious, but the lower animals are, as we shall hereafter see, capricious in their affections, aversions, and sense of beauty. There is also good reason to suspect that they love novelty, for its own sake.

Belief in God—Religion.—There is no evidence that man was aboriginally endowed with the ennobling belief in the existence of an Omnipotent God. On the contrary there is ample evidence, derived not from hasty travellers, but from men who have long resided with savages, that numerous races have existed and still exist, who have no idea of one or more gods, and who have no words in their languages to express such an idea.[51] The question is of course wholly distinct from that higher one, whether there exists a Creator and Ruler of the universe; and this has been answered in the affirmative by the highest intellects that have ever lived.

If, however, we include under the term "religion" the belief in unseen or spiritual agencies, the case is wholly different; for this belief seems to be almost universal with the less civilised races. Nor is it difficult to comprehend how it arose. As soon as the important faculties of the imagination, wonder, and curiosity, together with some power of reasoning, had become partially developed, man would naturally have craved to understand what was passing around him, and have vaguely speculated on his own existence. As

[51] See an excellent article on this subject by the Rev. F. W. Farrar, in the 'Anthropological Review,' Aug. 1864, p. ccxvii. For further facts see Sir J. Lubbock, 'Prehistoric Times,' 2nd edit. 1869, p. 564; and especially the chapters on Religion in his 'Origin of Civilisation,' 1870.

Mr. M'Lennan[52] has remarked, "Some explanation of "the phenomena of life, a man must feign for himself; "and to judge from the universality of it, the simplest "hypothesis, and the first to occur to men, seems to have "been that natural phenomena are ascribable to the pre- "sence in animals, plants, and things, and in the forces "of nature, of such spirits prompting to action as men "are conscious they themselves possess." It is probable, as Mr. Tylor has clearly shewn, that dreams may have first given rise to the notion of spirits; for savages do not readily distinguish between subjective and objective impressions. When a savage dreams, the figures which appear before him are believed to have come from a distance and to stand over him; or "the soul of the "dreamer goes out on its travels, and comes home with "a remembrance of what it has seen."[53] But until the above-named faculties of imagination, curiosity, reason, &c., had been fairly well developed in the mind of man, his dreams would not have led him to believe in spirits, any more than in the case of a dog.

---

[52] The Worship of Animals and Plants, in the 'Fortnightly Review,' Oct. 1, 1869, p. 422.

[53] Tylor, 'Early History of Mankind,' 1865, p. 6. See also the three striking chapters on the Development of Religion, in Lubbock's 'Origin of Civilisation,' 1870. In a like manner Mr. Herbert Spencer, in his ingenious essay in the 'Fortnightly Review' (May 1st, 1870, p. 535), accounts for the earliest forms of religious belief throughout the world, by man being led through dreams, shadows, and other causes, to look at himself as a double essence, corporeal and spiritual. As the spiritual being is supposed to exist after death and to be powerful, it is propitiated by various gifts and ceremonies, and its aid invoked. He then further shews that names or nicknames given from some animal or other object to the early progenitors or founders of a tribe, are supposed after a long interval to represent the real progenitor of the tribe; and such animal or object is then naturally believed still to exist as a spirit, is held sacred, and worshipped as a god. Nevertheless I cannot but suspect that there is a still earlier and ruder stage, when anything which manifests power or movement is thought to be endowed with some form of life, and with mental faculties analogous to our own.

The tendency in savages to imagine that natural objects and agencies are animated by spiritual or living essences, is perhaps illustrated by a little fact which I once noticed: my dog, a full-grown and very sensible animal, was lying on the lawn during a hot and still day; but at a little distance a slight breeze occasionally moved an open parasol, which would have been wholly disregarded by the dog, had any one stood near it. As it was, every time that the parasol slightly moved, the dog growled fiercely and barked. He must, I think, have reasoned to himself in a rapid and unconscious manner, that movement without any apparent cause indicated the presence of some strange living agent, and no stranger had a right to be on his territory.

The belief in spiritual agencies would easily pass into the belief in the existence of one or more gods. For savages would naturally attribute to spirits the same passions, the same love of vengeance or simplest form of justice, and the same affections which they themselves experienced. The Fuegians appear to be in this respect in an intermediate condition, for when the surgeon on board the "Beagle" shot some young ducklings as specimens, York Minster declared in the most solemn manner, "Oh! Mr. Bynoe, much rain, much snow, blow "much;" and this was evidently a retributive punishment for wasting human food. So again he related how, when his brother killed a "wild man," storms long raged, much rain and snow fell. Yet we could never discover that the Fuegians believed in what we should call a God, or practised any religious rites; and Jemmy Button, with justifiable pride, stoutly maintained that there was no devil in his land. This latter assertion is the more remarkable, as with savages the belief in bad spirits is far more common than the belief in good spirits.

The feeling of religious devotion is a highly complex one, consisting of love, complete submission to an exalted and mysterious superior, a strong sense of dependence,[54] fear, reverence, gratitude, hope for the future, and perhaps other elements. No being could experience so complex an emotion until advanced in his intellectual and moral faculties to at least a moderately high level. Nevertheless we see some distant approach to this state of mind, in the deep love of a dog for his master, associated with complete submission, some fear, and perhaps other feelings. The behaviour of a dog when returning to his master after an absence, and, as I may add, of a monkey to his beloved keeper, is widely different from that towards their fellows. In the latter case the transports of joy appear to be somewhat less, and the sense of equality is shewn in every action. Professor Braubach[55] goes so far as to maintain that a dog looks on his master as on a god.

The same high mental faculties which first led man to believe in unseen spiritual agencies, then in fetishism, polytheism, and ultimately in monotheism, would infallibly lead him, as long as his reasoning powers remained poorly developed, to various strange superstitions and customs. Many of these are terrible to think of—such as the sacrifice of human beings to a blood-loving god; the trial of innocent persons by the ordeal of poison or fire; witchcraft, &c.—yet it is well occasionally to reflect on these superstitions, for they shew us what an infinite debt of gratitude we owe to the improvement of our reason, to science, and our

---

[54] See an able article on the Psychical Elements of Religion, by Mr. L. Owen Pike, in 'Anthropolog. Review,' April, 1870, p. lxiii.

[55] 'Religion, Moral, &c., der Darwin'schen Art-Lehre,' 1869, s. 53.

accumulated knowledge.[56] As Sir J. Lubbock has well observed, "it is not too much to say that the horrible "dread of unknown evil hangs like a thick cloud over "savage life, and embitters every pleasure." These miserable and indirect consequences of our highest faculties may be compared with the incidental and occasional mistakes of the instincts of the lower animals.

---

[56] 'Prehistoric Times,' 2nd edit. p. 571. In this work (at p. 553) there will be found an excellent account of the many strange and capricious customs of savages.

CHAPTER III.

COMPARISON OF THE MENTAL POWERS OF MAN AND THE LOWER ANIMALS—*continued.*

The moral sense — Fundamental proposition — The qualities of social animals — Origin of sociability — Struggle between opposed instincts—Man a social animal—The more enduring social instincts conquer other less persistent instincts — The social virtues alone regarded by savages — The self-regarding virtues acquired at a later stage of development — The importance of the judgment of the members of the same community on conduct — Transmission of moral tendencies — Summary.

I FULLY subscribe to the judgment of those writers[1] who maintain that of all the differences between man and the lower animals, the moral sense or conscience is by far the most important. This sense, as Mackintosh[2] remarks, "has a rightful supremacy over every "other principle of human action;" it is summed up in that short but imperious word *ought,* so full of high significance. It is the most noble of all the attributes of man, leading him without a moment's hesitation to risk his life for that of a fellow-creature; or after due deliberation, impelled simply by the deep feeling of right or duty, to sacrifice it in some great cause. Immanuel Kant exclaims, "Duty! Wondrous thought, "that workest neither by fond insinuation, flattery, nor "by any threat, but merely by holding up thy naked "law in the soul, and so extorting for thyself always

[1] See, for instance, on this subject, Quatrefages, 'Unité de l'Espèce Humaine,' 1861, p. 21, &c.

[2] 'Dissertation on Ethical Philosophy,' 1837, p. 231, &c.

"reverence, if not always obedience; before whom all "appetites are dumb, however secretly they rebel; "whence thy original?"[3]

This great question has been discussed by many writers[4] of consummate ability; and my sole excuse for touching on it is the impossibility of here passing it over, and because, as far as I know, no one has approached it exclusively from the side of natural history. The investigation possesses, also, some independent interest, as an attempt to see how far the study of the lower animals can throw light on one of the highest psychical faculties of man.

The following proposition seems to me in a high degree probable—namely, that any animal whatever, endowed with well-marked social instincts,[5] would inevitably acquire a moral sense or conscience, as soon as

---

[3] 'Metaphysics of Ethics,' translated by J. W. Semple, Edinburgh, 1836, p. 136.

[4] Mr. Bain gives a list ('Mental and Moral Science,' 1868, p. 543-725) of twenty-six British authors who have written on this subject, and whose names are familiar to every reader; to these, Mr. Bain's own name, and those of Mr. Lecky, Mr. Shadworth Hodgson, and Sir J. Lubbock, as well as of others, may be added.

[5] Sir B. Brodie, after observing that man is a social animal ('Psychological Enquiries,' 1854, p. 192), asks the pregnant question, "ought not this to settle the disputed question as to the existence of a "moral sense?" Similar ideas have probably occurred to many persons, as they did long ago to Marcus Aurelius. Mr. J. S. Mill speaks, in his celebrated work, 'Utilitarianism,' (1864, p. 46), of the social feelings as a "powerful natural sentiment," and as "the natural basis of "sentiment for utilitarian morality;" but on the previous page he says, "if, as is my own belief, the moral feelings are not innate, but "acquired, they are not for that reason less natural." It is with hesitation that I venture to differ from so profound a thinker, but it can hardly be disputed that the social feelings are instinctive or innate in the lower animals; and why should they not be so in man? Mr. Bain (see, for instance, 'The Emotions and the Will,' 1865, p. 481) and others believe that the moral sense is acquired by each individual during his lifetime. On the general theory of evolution this is at least extremely improbable.

its intellectual powers had become as well developed, or nearly as well developed, as in man. For, *firstly*, the social instincts lead an animal to take pleasure in the society of its fellows, to feel a certain amount of sympathy with them, and to perform various services for them. The services may be of a definite and evidently instinctive nature; or there may be only a wish and readiness, as with most of the higher social animals, to aid their fellows in certain general ways. But these feelings and services are by no means extended to all the individuals of the same species, only to those of the same association. *Secondly*, as soon as the mental faculties had become highly developed, images of all past actions and motives would be incessantly passing through the brain of each individual; and that feeling of dissatisfaction which invariably results, as we shall hereafter see, from any unsatisfied instinct, would arise, as often as it was perceived that the enduring and always present social instinct had yielded to some other instinct, at the time stronger, but neither enduring in its nature, nor leaving behind it a very vivid impression. It is clear that many instinctive desires, such as that of hunger, are in their nature of short duration; and after being satisfied are not readily or vividly recalled. *Thirdly*, after the power of language had been acquired and the wishes of the members of the same community could be distinctly expressed, the common opinion how each member ought to act for the public good, would naturally become to a large extent the guide to action. But the social instincts would still give the impulse to act for the good of the community, this impulse being strengthened, directed, and sometimes even deflected by public opinion, the power of which rests, as we shall presently see, on instinctive sympathy. *Lastly*, habit in the individual would ultimately play a very important part in guiding the conduct of each member; for the social instincts and impulses, like all other instincts, would be greatly strengthened by habit, as would obedience to the wishes and judgment of the community. These several subordinate propositions must now be discussed; and some of them at considerable length.

It may be well first to premise that I do not wish to maintain that any strictly social animal, if its intellectual faculties were to become as active and as highly developed as in man, would acquire exactly the same moral sense as ours. In the same manner as various animals have some sense of beauty, though they admire widely different objects, so they might have a sense of right and wrong, though led by it to follow widely different lines of conduct. If, for instance, to take an extreme case, men were reared under precisely the same conditions as hive-bees, there can hardly be a doubt that our unmarried females would, like the worker-bees, think it a sacred duty to kill their brothers, and mothers would strive to kill their fertile daughters; and no one would think of interfering. Nevertheless the bee, or any other social animal, would in our supposed case gain, as it appears to me, some feeling of right and wrong, or a conscience. For each individual would have an inward sense of possessing certain stronger or more enduring instincts, and others less strong or enduring; so that there would often be a struggle which impulse should be followed; and satisfaction or dissatisfaction would be felt, as past impressions were compared during their incessant passage through the mind. In this case an inward monitor would tell the animal that it would have been better to have followed the one impulse rather than the other. The one course ought to have been followed: the one would have been right and the other wrong; but to these terms I shall have to recur.

Sociability.—Animals of many kinds are social; we find even distinct species living together, as with some American monkeys, and with the united flocks of rooks, jackdaws, and starlings. Man shows the same feeling in his strong love for the dog, which the dog returns with interest. Every one must have noticed how miserable horses, dogs, sheep, &c. are when separated from their companions; and what affection at least the two former kinds show on their reunion. It is curious to speculate on the feelings of a dog, who will rest peacefully for hours in a room with his master or any of the family, without the least notice being taken of him; but if left for a short time by himself, barks or howls dismally. We will confine our attention to the higher social animals, excluding insects, although these aid each other in many important ways. The most common service which the higher animals perform for each other, is the warning each other of danger by means of the united senses of all. Every sportsman knows, as Dr. Jaeger remarks,[6] how difficult it is to approach animals in a herd or troop. Wild horses and cattle do not, I believe, make any danger-signal; but the attitude of any one who first discovers an enemy, warns the others. Rabbits stamp loudly on the ground with their hind-feet as a signal: sheep and chamois do the same, but with their fore-feet, uttering likewise a whistle. Many birds and some mammals post sentinels, which in the case of seals are said[7] generally to be the females. The leader of a troop of monkeys acts as the sentinel, and utters cries expressive both of danger and of safety.[8] Social

---

6 'Die Darwin'sche Theorie,' s. 101.

7 Mr. R. Browne in 'Proc. Zoolog. Soc.' 1868, p. 409.

8 Brehm, 'Thierleben,' B. i. 1864, s. 52, 79. For the case of the

animals perform many little services for each other: horses nibble, and cows lick each other, on any spot which itches: monkeys search for each other's external parasites; and Brehm states that after a troop of the *Cercopithecus griseo-viridis* has rushed through a thorny brake, each monkey stretches itself on a branch, and another monkey sitting by "conscientiously" examines its fur and extracts every thorn or burr.

Animals also render more important services to each other: thus wolves and some other beasts of prey hunt in packs, and aid each other in attacking their victims. Pelicans fish in concert. The Hamadryas baboons turn over stones to find insects, &c.; and when they come to a large one, as many as can stand round, turn it over together and share the booty. Social animals mutually defend each other. The males of some ruminants come to the front when there is danger and defend the herd with their horns. I shall also in a future chapter give cases of two young wild bulls attacking an old one in concert, and of two stallions together trying to drive away a third stallion from a troop of mares. Brehm encountered in Abyssinia a great troop of baboons which were crossing a valley: some had already ascended the opposite mountain, and some were still in the valley: the latter were attacked by the dogs, but the old males immediately hurried down from the rocks, and with mouths widely opened roared so fearfully, that the dogs precipitately retreated. They were again encouraged to the attack; but by this time all the baboons had re-ascended the heights, excepting a young one, about six

---

monkeys extracting thorns from each other, see s. 54. With respect to the Hamadryas turning over stones, the fact is given (s. 76) on the evidence of Alvarez, whose observations Brehm thinks quite trustworthy. For the cases of the old male baboons attacking the dogs, see s. 79; and with respect to the eagle, s. 56.

months old, who, loudly calling for aid, climbed on a block of rock and was surrounded. Now one of the largest males, a true hero, came down again from the mountain, slowly went to the young one, coaxed him, and triumphantly led him away—the dogs being too much astonished to make an attack. I cannot resist giving another scene which was witnessed by this same naturalist; an eagle seized a young Cercopithecus, which, by clinging to a branch, was not at once carried off; it cried loudly for assistance, upon which the other members of the troop with much uproar rushed to the rescue, surrounded the eagle, and pulled out so many feathers, that he no longer thought of his prey, but only how to escape. This eagle, as Brehm remarks, assuredly would never again attack a monkey in a troop.

It is certain that associated animals have a feeling of love for each other which is not felt by adult and non-social animals. How far in most cases they actually sympathise with each other's pains and pleasures is more doubtful, especially with respect to the latter. Mr. Buxton, however, who had excellent means of observation,[9] states that his macaws, which lived free in Norfolk, took "an extravagant interest" in a pair with a nest, and whenever the female left it, she was surrounded by a troop "screaming horrible acclamations in her honour." It is often difficult to judge whether animals have any feeling for each other's sufferings. Who can say what cows feel, when they surround and stare intently on a dying or dead companion? That animals sometimes are far from feeling any sympathy is too certain; for they will expel a wounded animal from the herd, or gore or worry it to death. This is almost the blackest fact in natural

[9] 'Annals and Mag. of Nat. Hist.' November, 1868, p. 382.

history, unless indeed the explanation which has been suggested is true, that their instinct or reason leads them to expel an injured companion, lest beasts of prey, including man, should be tempted to follow the troop. In this case their conduct is not much worse than that of the North American Indians who leave their feeble comrades to perish on the plains, or the Feegeans, who, when their parents get old or fall ill, bury them alive.[10]

Many animals, however, certainly sympathise with each other's distress or danger. This is the case even with birds; Capt. Stansbury[11] found on a salt lake in Utah an old and completely blind pelican, which was very fat, and must have been long and well fed by his companions. Mr. Blyth, as he informs me, saw Indian crows feeding two or three of their companions which were blind; and I have heard of an analogous case with the domestic cock. We may, if we choose, call these actions instinctive; but such cases are much too rare for the development of any special instinct.[12] I have myself seen a dog, who never passed a great friend of his, a cat which lay sick in a basket, without giving her a few licks with his tongue, the surest sign of kind feeling in a dog.

It must be called sympathy that leads a courageous dog to fly at any one who strikes his master, as he certainly will. I saw a person pretending to beat a lady who had a very timid little dog on her lap, and the trial had never before been made. The little crea-

---

10 Sir J. Lubbock, 'Prehistoric Times,' 2nd edit. p. 446.

11 As quoted by Mr. L. H. Morgan, 'The American Beaver,' 1868, p. 272. Capt. Stansbury also gives an interesting account of the manner in which a very young pelican, carried away by a strong stream, was guided and encouraged in its attempts to reach the shore by half a dozen old birds.

12 As Mr. Bain states, "effective aid to a sufferer springs from sympathy proper:" 'Mental and Moral Science,' 1868, p. 245.

ture instantly jumped away, but after the pretended beating was over, it was really pathetic to see how perseveringly he tried to lick his mistress' sface and comfort her. Brehm[13] states that when a baboon in confinement was pursued to be punished, the others tried to protect him. It must have been sympathy in the cases above given which led the baboons and Cercopitheci to defend their young comrades from the dogs and the eagle. I will give only one other instance of sympathetic and heroic conduct in a little American monkey. Several years ago a keeper at the Zoological Gardens, showed me some deep and scarcely healed wounds on the nape of his neck, inflicted on him whilst kneeling on the floor by a fierce baboon. The little American monkey, who was a warm friend of this keeper, lived in the same large compartment, and was dreadfully afraid of the great baboon. Nevertheless, as soon as he saw his friend the keeper in peril, he rushed to the rescue, and by screams and bites so distracted the baboon that the man was able to escape, after running great risk, as the surgeon who attended him thought, of his life.

Besides love and sympathy, animals exhibit other qualities which in us would be called moral; and I agree with Agassiz[14] that dogs possess something very like a conscience. They certainly possess some power of self-command, and this does not appear to be wholly the result of fear. As Braubach[15] remarks, a dog will refrain from stealing food in the absence of his master. Dogs have long been accepted as the very type of fidelity and obedience. All animals living in a body which defend each other or attack their enemies

---

13 'Thierleben,' B. i. s. 85.

14 'De l'Espèce et de la Class.' 1869, p. 97.

15 'Der Darwin'schen Art-Lehre,' 1869, s. 54.

in concert, must be in some degree faithful to each other; and those that follow a leader must be in some degree obedient. When the baboons in Abyssinia[16] plunder a garden, they silently follow their leader; and if an imprudent young animal makes a noise, he receives a slap from the others to teach him silence and obedience; but as soon as they are sure that there is no danger, all show their joy by much clamour.

With respect to the impulse which leads certain animals to associate together, and to aid each other in many ways, we may infer that in most cases they are impelled by the same sense of satisfaction or pleasure which they experience in performing other instinctive actions; or by the same sense of dissatisfaction, as in other cases of prevented instinctive actions. We see this in innumerable instances, and it is illustrated in a striking manner by the acquired instincts of our domesticated animals; thus a young shepherd-dog delights in driving and running round a flock of sheep, but not in worrying them; a young foxhound delights in hunting a fox, whilst some other kinds of dogs as I have witnessed, utterly disregard foxes. What a strong feeling of inward satisfaction must impel a bird, so full of activity, to brood day after day over her eggs. Migratory birds are miserable if prevented from migrating, and perhaps they enjoy starting on their long flight. Some few instincts are determined solely by painful feelings, as by fear, which leads to self-preservation, or is specially directed against certain enemies. No one, I presume, can analyse the sensations of pleasure or pain. In many cases, however, it is probable that instincts are persistently followed from the

[16] Brehm, 'Thierleben,' B. i. s. 76.

mere force of inheritance, without the stimulus of either pleasure or pain. A young pointer, when it first scents game, apparently cannot help pointing. A squirrel in a cage who pats the nuts which it cannot eat, as if to bury them in the ground, can hardly be thought to act thus either from pleasure or pain. Hence the common assumption that men must be impelled to every action by experiencing some pleasure or pain may be erroneous. Although a habit may be blindly and implicitly followed, independently of any pleasure or pain felt at the moment, yet if it be forcibly and abruptly checked, a vague sense of dissatisfaction is generally experienced; and this is especially true in regard to persons of feeble intellect.

It has often been assumed that animals were in the first place rendered social, and that they feel as a consequence uncomfortable when separated from each other, and comfortable whilst together; but it is a more probable view that these sensations were first developed, in order that those animals which would profit by living in society, should be induced to live together. In the same manner as the sense of hunger and the pleasure of eating were, no doubt, first acquired in order to induce animals to eat. The feeling of pleasure from society is probably an extension of the parental or filial affections; and this extension may be in chief part attributed to natural selection, but perhaps in part to mere habit. For with those animals which were benefited by living in close association, the individuals which took the greatest pleasure in society would best escape various dangers; whilst those that cared least for their comrades and lived solitary would perish in greater numbers. With respect to the origin of the parental and filial affections, which apparently lie at the basis of the social affections, it is hopeless to speculate; but we may infer that they have been to a large extent gained through natural selection. So it has almost certainly been with the unusual and opposite feeling of hatred between the nearest relations, as with the worker-bees which kill their brother-drones, and with the queen-bees which kill their daughter-queens; the desire to destroy, instead of loving, their nearest relations having been here of service to the community.

The all-important emotion of sympathy is distinct from that of love. A mother may passionately love her sleeping and passive infant, but she can then hardly be said to feel sympathy for it. The love of a man for his dog is distinct from sympathy, and so is that of a dog for his master. Adam Smith formerly argued, as has Mr. Bain recently, that the basis of sympathy lies in our strong retentiveness of former states of pain or pleasure. Hence, "the sight of another person enduring "hunger, cold, fatigue, revives in us some recollection "of these states, which are painful even in idea." We are thus impelled to relieve the sufferings of another, in order that our own painful feelings may be at the same time relieved. In like manner we are led to participate in the pleasures of others.[17] But I cannot see how this view explains the fact that sympathy is excited in an immeasurably stronger degree by a beloved than by an indifferent person. The mere

[17] See the first and striking chapter in Adam Smith's 'Theory of Moral Sentiments.' Also Mr. Bain's 'Mental and Moral Science,' 1868, p. 244, and 275-282. Mr. Bain states, that "sympathy is, "indirectly, a source of pleasure to the sympathiser;" and he accounts for this through reciprocity. He remarks that "the person benefited, "or others in his stead, may make up, by sympathy and good offices "returned, for all the sacrifice." But if, as appears to be the case, sympathy is strictly an instinct, its exercise would give direct pleasure, in the same manner as the exercise, as before remarked, of almost every other instinct.

sight of suffering, independently of love, would suffice to call up in us vivid recollections and associations. Sympathy may at first have originated in the manner above suggested; but it seems now to have become an instinct, which is especially directed towards beloved objects, in the same manner as fear with animals is especially directed against certain enemies. As sympathy is thus directed, the mutual love of the members of the same community will extend its limits. No doubt a tiger or lion feels sympathy for the sufferings of its own young, but not for any other animal. With strictly social animals the feeling will be more or less extended to all the associated members, as we know to be the case. With mankind selfishness, experience, and imitation probably add, as Mr. Bain has shewn, to the power of sympathy; for we are led by the hope of receiving good in return to perform acts of sympathetic kindness to others; and there can be no doubt that the feeling of sympathy is much strengthened by habit. In however complex a manner this feeling may have originated, as it is one of high importance to all those animals which aid and defend each other, it will have been increased, through natural selection; for those communities, which included the greatest number of the most sympathetic members, would flourish best and rear the greatest number of offspring.

In many cases it is impossible to decide whether certain social instincts have been acquired through natural selection, or are the indirect result of other instincts and faculties, such as sympathy, reason, experience, and a tendency to imitation; or again, whether they are simply the result of long-continued habit. So remarkable an instinct as the placing sentinels to warn the community of danger, can hardly have been the indirect result of any other faculty; it must therefore have been directly acquired. On the other hand, the habit followed by the males of some social animals, of defending the community and of attacking their enemies or their prey in concert, may perhaps have originated from mutual sympathy; but courage, and in most cases strength, must have been previously acquired, probably through natural selection.

Of the various instincts and habits, some are much stronger than others, that is, some either give more pleasure in their performance and more distress in their prevention than others; or, which is probably quite as important, they are more persistently followed through inheritance without exciting any special feeling of pleasure or pain. We are ourselves conscious that some habits are much more difficult to cure or change than others. Hence a struggle may often be observed in animals between different instincts, or between an instinct and some habitual disposition; as when a dog rushes after a hare, is rebuked, pauses, hesitates, pursues again or returns ashamed to his master; or as between the love of a female dog for her young puppies and for her master, for she may be seen to slink away to them, as if half ashamed of not accompanying her master. But the most curious instance known to me of one instinct conquering another, is the migratory instinct conquering the maternal instinct. The former is wonderfully strong; a confined bird will at the proper season beat her breast against the wires of her cage, until it is bare and bloody. It causes young salmon to leap out of the fresh water, where they could still continue to live, and thus unintentionally to commit suicide. Every one knows how strong the maternal instinct is, leading even timid birds to face great danger, though with hesitation and in opposition to the instinct of self-preservation. Nevertheless the migratory instinct is so powerful that late in the autumn swallows and house-martins frequently desert their tender young, leaving them to perish miserably in their nests.[18]

We can perceive that an instinctive impulse, if it be in any way more beneficial to a species than some other or opposed instinct, would be rendered the more potent of the two through natural selection; for the individuals which had it most strongly developed would survive in larger numbers. Whether this is the case with the migratory in comparison with the maternal instinct, may well be doubted. The great persistence or steady action of the former at certain seasons of the year during the whole day, may give it for a time paramount force.

Man a social animal.—Most persons admit that man is a social being. We see this in his dislike of solitude, and in his wish for society beyond that of his own family. Solitary confinement is one of the severest punishments which can be inflicted. Some authors suppose that man primevally lived in single families; but at the present day, though single families, or only two or three together, roam the solitudes of some savage lands, they are always, as far as I can discover, friendly with other families inhabiting the same district. Such families occasionally meet in council, and they unite

---

[18] This fact, the Rev. L. Jenyns states (see his edition of 'White's Nat. Hist. of Selborne,' 1853, p. 204) was first recorded by the illustrious Jenner, in 'Phil. Transact.' 1824, and has since been confirmed by several observers, especially by Mr. Blackwall. This latter careful observer examined, late in the autumn, during two years, thirty-six nests; he found that twelve contained young dead birds, five contained eggs on the point of being hatched, and three eggs not nearly hatched. Many birds not yet old enough for a prolonged flight are likewise deserted and left behind. See Blackwall, 'Researches in Zoology,' 1834, pp. 108, 118. For some additional evidence, although this is not wanted, see Leroy, 'Lettres Phil.' 1802, p. 217.

for their common defence. It is no argument against savage man being a social animal, that the tribes inhabiting adjacent districts are almost always at war with each other; for the social instincts never extend to all the individuals of the same species. Judging from the analogy of the greater number of the Quadrumana, it is probable that the early ape-like progenitors of man were likewise social; but this is not of much importance for us. Although man, as he now exists, has few special instincts, having lost any which his early progenitors may have possessed, this is no reason why he should not have retained from an extremely remote period some degree of instinctive love and sympathy for his fellows. We are indeed all conscious that we do possess such sympathetic feelings;[19] but our consciousness does not tell us whether they are instinctive, having originated long ago in the same manner as with the lower animals, or whether they have been acquired by each of us during our early years. As man is a social animal, it is also probable that he would inherit a tendency to be faithful to his comrades, for this quality is common to most social animals. He would in like manner possess some capacity for self-command, and perhaps of obedience to the leader of the community. He would from an inherited tendency still be willing to defend, in concert with others, his fellow-men, and would be ready to aid them in any way which did not too greatly interfere with his own welfare or his own strong desires.

The social animals which stand at the bottom of the

---

[19] Hume remarks ('An Enquiry Concerning the Principles of Morals,' edit. of 1751, p. 132), "there seems a necessity for confessing that the "happiness and misery of others are not spectacles altogether in-"different to us, but that the view of the former . . . communicates a "secret joy; the appearance of the latter . . . throws a melancholy "damp over the imagination."

scale are guided almost exclusively, and those which stand higher in the scale are largely guided, in the aid which they give to the members of the same community, by special instincts; but they are likewise in part impelled by mutual love and sympathy, assisted apparently by some amount of reason. Although man, as just remarked, has no special instincts to tell him how to aid his fellow-men, he still has the impulse, and with his improved intellectual faculties would naturally be much guided in this respect by reason and experience. Instinctive sympathy would, also, cause him to value highly the approbation of his fellow-men; for, as Mr. Bain has clearly shewn,[20] the love of praise and the strong feeling of glory, and the still stronger horror of scorn and infamy, "are due to the workings of sympathy." Consequently man would be greatly influenced by the wishes, approbation, and blame of his fellow-men, as expressed by their gestures and language. Thus the social instincts, which must have been acquired by man in a very rude state, and probably even by his early ape-like progenitors, still give the impulse to many of his best actions; but his actions are largely determined by the expressed wishes and judgment of his fellow-men, and unfortunately still oftener by his own strong, selfish desires. But as the feelings of love and sympathy and the power of self-command become strengthened by habit, and as the power of reasoning becomes clearer so that man can appreciate the justice of the judgments of his fellow-men, he will feel himself impelled, independently of any pleasure or pain felt at the moment, to certain lines of conduct. He may then say, I am the supreme judge of my own conduct, and in the words of Kant, I will not in my own person violate the dignity of humanity.

---

20 'Mental and Moral Science,' 1868, p. 254.

The more enduring Social Instincts conquer the less Persistent Instincts.—We have, however, not as yet considered the main point, on which the whole question of the moral sense hinges. Why should a man feel that he ought to obey one instinctive desire rather than another? Why does he bitterly regret if he has yielded to the strong sense of self-preservation, and has not risked his life to save that of a fellow-creature; or why does he regret having stolen food from severe hunger?

It is evident in the first place, that with mankind the instinctive impulses have different degrees of strength; a young and timid mother urged by the maternal instinct will, without a moment's hesitation, run the greatest danger for her infant, but not for a mere fellow-creature. Many a man, or even boy, who never before risked his life for another, but in whom courage and sympathy were well developed, has, disregarding the instinct of self-preservation, instantaneously plunged into a torrent to save a drowning fellow-creature. In this case man is impelled by the same instinctive motive, which caused the heroic little American monkey, formerly described, to attack the great and dreaded baboon, to save his keeper. Such actions as the above appear to be the simple result of the greater strength of the social or maternal instincts than of any other instinct or motive; for they are performed too instantaneously for reflection, or for the sensation of pleasure or pain; though if prevented distress would be caused.

I am aware that some persons maintain that actions performed impulsively, as in the above cases, do not come under the dominion of the moral sense, and cannot be called moral. They confine this term to actions done deliberately, after a victory over opposing desires, or to actions prompted by some lofty motive. But it appears scarcely possible to draw any clear line of distinction of this kind; though the distinction may be real. As far as exalted motives are concerned, many instances have been recorded of barbarians, destitute of any feeling of general benevolence towards mankind, and not guided by any religious motive, who have deliberately as prisoners sacrificed their lives,[21] rather than betray their comrades; and surely their conduct ought to be considered as moral. As far as deliberation and the victory over opposing motives are concerned, animals may be seen doubting between opposed instincts, as in rescuing their offspring or comrades from danger; yet their actions, though done for the good of others, are not called moral. Moreover, an action repeatedly performed by us, will at last be done without deliberation or hesitation, and can then hardly be distinguished from an instinct; yet surely no one will pretend that an action thus done ceases to be moral. On the contrary, we all feel that an act cannot be considered as perfect, or as performed in the most noble manner, unless it be done impulsively, without deliberation or effort, in the same manner as by a man in whom the requisite qualities are innate. He who is forced to overcome his fear or want of sympathy before he acts, deserves, however, in one way higher credit than the man whose innate disposition leads him to a good act without effort. As we cannot distinguish between motives, we rank all actions of a certain class as moral, when they are performed by a moral being. A moral being is one who is capable of comparing his past and future actions or motives, and of approving or disapproving of them. We have no reason to suppose that any of the lower animals have

---

21 I have given one such case, namely of three Patagonian Indians who preferred being shot, one after the other, to betraying the plans of their companions in war ('Journal of Researches,' 1845, p. 103).

this capacity; therefore when a monkey faces danger to rescue its comrade, or takes charge of an orphan-monkey, we do not call its conduct moral. But in the case of man, who alone can with certainty be ranked as a moral being, actions of a certain class are called moral, whether performed deliberately after a struggle with opposing motives, or from the effects of slowly-gained habit, or impulsively through instinct.

But to return to our more immediate subject; although some instincts are more powerful than others, thus leading to corresponding actions, yet it cannot be maintained that the social instincts are ordinarily stronger in man, or have become stronger through long-continued habit, than the instincts, for instance, of self-preservation, hunger, lust, vengeance, &c. Why then does man regret, even though he may endeavour to banish any such regret, that he has followed the one natural impulse, rather than the other; and why does he further feel that he ought to regret his conduct? Man in this respect differs profoundly from the lower animals. Nevertheless we can, I think, see with some degree of clearness the reason of this difference.

Man, from the activity of his mental faculties, cannot avoid reflection: past impressions and images are incessantly passing through his mind with distinctness. Now with those animals which live permanently in a body, the social instincts are ever present and persistent. Such animals are always ready to utter the danger-signal, to defend the community, and to give aid to their fellows in accordance with their habits; they feel at all times, without the stimulus of any special passion or desire, some degree of love and sympathy for them; they are unhappy if long separated from them, and always happy to be in their company. So it is with ourselves. A man who possessed no trace of such feelings would be an unnatural monster. On the other hand, the desire to satisfy hunger, or any passion, such as vengeance, is in its nature temporary, and can for a time be fully satisfied. Nor is it easy, perhaps hardly possible, to call up with complete vividness the feeling, for instance, of hunger; nor indeed, as has often been remarked, of any suffering. The instinct of self-preservation is not felt except in the presence of danger; and many a coward has thought himself brave until he has met his enemy face to face. The wish for another man's property is perhaps as persistent a desire as any that can be named; but even in this case the satisfaction of actual possession is generally a weaker feeling than the desire: many a thief, if not an habitual one, after success has wondered why he stole some article.

Thus, as man cannot prevent old impressions continually repassing through his mind, he will be compelled to compare the weaker impressions of, for instance, past hunger, or of vengeance satisfied or danger avoided at the cost of other men, with the instinct of sympathy and good-will to his fellows, which is still present and ever in some degree active in his mind. He will then feel in his imagination that a stronger instinct has yielded to one which now seems comparatively weak; and then that sense of dissatisfaction will inevitably be felt with which man is endowed, like every other animal, in order that his instincts may be obeyed. The case before given, of the swallow, affords an illustration, though of a reversed nature, of a temporary though for the time strongly persistent instinct conquering another instinct which is usually dominant over all others. At the proper season these birds seem all day long to be impressed with the desire to migrate; their habits change; they become restless, are noisy, and congregate in flocks. Whilst the mother-bird is feeding or brooding over her nestlings, the maternal instinct is probably stronger than the migratory; but the instinct which is more persistent gains the victory, and at last, at a moment when her young ones are not in sight, she takes flight and deserts them. When arrived at the end of her long journey, and the migratory instinct ceases to act, what an agony of remorse each bird would feel, if, from being endowed with great mental activity, she could not prevent the image continually passing before her mind of her young ones perishing in the bleak north from cold and hunger.

At the moment of action, man will no doubt be apt to follow the stronger impulse; and though this may occasionally prompt him to the noblest deeds, it will far more commonly lead him to gratify his own desires at the expense of other men. But after their gratification, when past and weaker impressions are contrasted with the ever-enduring social instincts, retribution will surely come. Man will then feel dissatisfied with himself, and will resolve with more or less force to act differently for the future. This is conscience; for conscience looks backwards and judges past actions, inducing that kind of dissatisfaction, which if weak we call regret, and if severe remorse.

These sensations are, no doubt, different from those experienced when other instincts or desires are left unsatisfied; but every unsatisfied instinct has its own proper prompting sensation, as we recognise with hunger, thirst, &c. Man thus prompted, will through long habit acquire such perfect self-command, that his desires and passions will at last instantly yield to his social sympathies, and there will no longer be a struggle between them. The still hungry, or the still revengeful man will not think of stealing food, or of wreaking his vengeance. It is possible, or, as we shall hereafter see, even probable, that the habit of self-command may, like other habits, be inherited. Thus at last man comes to feel, through acquired and perhaps inherited habit, that it is best for him to obey his more persistent instincts. The imperious word *ought* seems merely to imply the consciousness of the existence of a persistent instinct, either innate or partly acquired, serving him as a guide, though liable to be disobeyed. We hardly use the word *ought* in a metaphorical sense, when we say hounds ought to hunt, pointers to point, and retrievers to retrieve their game. If they fail thus to act, they fail in their duty and act wrongly.

If any desire or instinct, leading to an action opposed to the good of others, still appears to a man, when recalled to mind, as strong as, or stronger than, his social instinct, he will feel no keen regret at having followed it; but he will be conscious that if his conduct were known to his fellows, it would meet with their disapprobation; and few are so destitute of sympathy as not to feel discomfort when this is realised. If he has no such sympathy, and if his desires leading to bad actions are at the time strong, and when recalled are not overmastered by the persistent social instincts, then he is essentially a bad man;[22] and the sole restraining motive left is the fear of punishment, and the conviction that in the long run it would be best for his own selfish interests to regard the good of others rather than his own.

It is obvious that every one may with an easy conscience gratify his own desires, if they do not interfere

---

[22] Dr. Prosper Despine, in his 'Psychologie Naturelle,' 1868 (tom. i. p. 243; tom ii. p. 169) gives many curious cases of the worst criminals, who apparently have been entirely destitute of conscience.

with his social instincts, that is with the good of others; but in order to be quite free from self-reproach, or at least of anxiety, it is almost necessary for him to avoid the disapprobation, whether reasonable or not, of his fellow men. Nor must he break through the fixed habits of his life, especially if these are supported by reason; for if he does, he will assuredly feel dissatisfaction. He must likewise avoid the reprobation of the one God or gods, in whom according to his knowledge or superstition he may believe; but in this case the additional fear of divine punishment often supervenes.

The strictly Social Virtues at first alone regarded.—The above view of the first origin and nature of the moral sense, which tells us what we ought to do, and of the conscience which reproves us if we disobey it, accords well with what we see of the early and undeveloped condition of this faculty in mankind. The virtues which must be practised, at least generally, by rude men, so that they may associate in a body, are those which are still recognised as the most important. But they are practised almost exclusively in relation to the men of the same tribe; and their opposites are not regarded as crimes in relation to the men of other tribes. No tribe could hold together if murder, robbery, treachery, &c., were common; consequently such crimes within the limits of the same tribe "are branded "with everlasting infamy;"[23] but excite no such sentiment beyond these limits. A North-American Indian is well pleased with himself, and is honoured by others, when he scalps a man of another tribe; and a Dyak

[23] See an able article in the 'North British Review,' 1867, p. 395. See also. Mr. W. Bagehot's articles on the Importance of Obedience and Coherence to Primitive Man, in the 'Fortnightly Review,' 1867, p. 529, and 1868, p. 457, &c.

cuts off the head of an unoffending person and dries it as a trophy. The murder of infants has prevailed on the largest scale throughout the world,[24] and has met with no reproach; but infanticide, especially of females, has been thought to be good for the tribe, or at least not injurious. Suicide during former times was not generally considered as a crime,[25] but rather from the courage displayed as an honourable act; and it is still largely practised by some semi-civilised nations without reproach, for the loss to a nation of a single individual is not felt: whatever the explanation may be, suicide, as I hear from Sir J. Lubbock, is rarely practised by the lowest barbarians. It has been recorded that an Indian Thug conscientiously regretted that he had not strangled and robbed as many travellers as did his father before him. In a rude state of civilisation the robbery of strangers is, indeed, generally considered as honourable.

The great sin of Slavery has been almost universal, and slaves have often been treated in an infamous manner. As barbarians do not regard the opinion of their women, wives are commonly treated like slaves. Most savages are utterly indifferent to the sufferings of strangers, or even delight in witnessing them. It is well known that the women and children of the North-American Indians aided in torturing their enemies. Some savages take a horrid pleasure in cruelty to animals,[26] and humanity with them is an unknown virtue. Nevertheless, feelings of sympathy and kindness are common, especially

---

[24] The fullest account which I have met with is by Dr. Gerland, in his 'Ueber das Aussterben der Naturvölker,' 1868; but I shall have to recur to the subject of infanticide in a future chapter.

[25] See the very interesting discussion on Suicide in Lecky's 'History of European Morals,' vol. i. 1869, p. 223.

[26] See, for instance, Mr. Hamilton's account of the Kaffirs, 'Anthropological Review,' 1870, p. xv.

during sickness, between the members of the same tribe, and are sometimes extended beyond the limits of the tribe. Mungo Park's touching account of the kindness of the negro women of the interior to him is well known. Many instances could be given of the noble fidelity of savages towards each other, but not to strangers; common experience justifies the maxim of the Spaniard, "Never, never trust an Indian." There cannot be fidelity without truth; and this fundamental virtue is not rare between the members of the same tribe: thus Mungo Park heard the negro women teaching their young children to love the truth. This, again, is one of the virtues which becomes so deeply rooted in the mind that it is sometimes practised by savages even at a high cost, towards strangers; but to lie to your enemy has rarely been thought a sin, as the history of modern diplomacy too plainly shews. As soon as a tribe has a recognised leader, disobedience becomes a crime, and even abject submission is looked at as a sacred virtue.

As during rude times no man can be useful or faithful to his tribe without courage, this quality has universally been placed in the highest rank; and although, in civilised countries, a good, yet timid, man may be far more useful to the community than a brave one, we cannot help instinctively honouring the latter above a coward, however benevolent. Prudence, on the other hand, which does not concern the welfare of others, though a very useful virtue, has never been highly esteemed. As no man can practise the virtues necessary for the welfare of his tribe without self-sacrifice, self-command, and the power of endurance, these qualities have been at all times highly and most justly valued. The American savage voluntarily submits without a groan to the most horrid tortures to prove and strengthen his fortitude and courage; and we cannot help admiring him, or even an Indian Fakir, who, from a foolish religious motive, swings suspended by a hook buried in his flesh.

The other self-regarding virtues, which do not obviously, though they may really, affect the welfare of the tribe, have never been esteemed by savages, though now highly appreciated by civilised nations. The greatest intemperance with savages is no reproach. Their utter licentiousness, not to mention unnatural crimes, is something astounding.[27] As soon, however, as marriage, whether polygamous or monogamous, becomes common, jealousy will lead to the inculcation of female virtue; and this being honoured will tend to spread to the unmarried females. How slowly it spreads to the male sex we see at the present day. Chastity eminently requires self-command; therefore it has been honoured from a very early period in the moral history of civilised man. As a consequence of this, the senseless practice of celibacy has been ranked from a remote period as a virtue.[28] The hatred of indecency, which appears to us so natural as to be thought innate, and which is so valuable an aid to chastity, is a modern virtue, appertaining exclusively, as Sir G. Staunton remarks,[29] to civilised life. This is shewn by the ancient religious rites of various nations, by the drawings on the walls of Pompeii, and by the practices of many savages.

We have now seen that actions are regarded by savages, and were probably so regarded by primeval man, as good or bad, solely as they affect in an obvious manner the welfare of the tribe,—not that of the species, nor that of man as an individual member of the

---

[27] Mr. M'Lennan has given ('Primitive Marriage,' 1865, p. 176) a good collection of facts on this head.

[28] Lecky, 'History of European Morals,' vol. i. 1869, p. 109.

[29] 'Embassy to China,' vol. ii. p. 348.

tribe. This conclusion agrees well with the belief that the so-called moral sense is aboriginally derived from the social instincts, for both relate at first exclusively to the community. The chief causes of the low morality of savages, as judged by our standard, are, firstly, the confinement of sympathy to the same tribe. Secondly, insufficient powers of reasoning, so that the bearing of many virtues, especially of the self-regarding virtues, on the general welfare of the tribe is not recognised. Savages, for instance, fail to trace the multiplied evils consequent on a want of temperance, chastity, &c. And, thirdly, weak power of self-command; for this power has not been strengthened through long-continued, perhaps inherited, habit, instruction and religion.

I have entered into the above details on the immorality of savages,[30] because some authors have recently taken a high view of their moral nature, or have attributed most of their crimes to mistaken benevolence.[31] These authors appear to rest their conclusion on savages possessing, as they undoubtedly do possess, and often in a high degree, those virtues which are serviceable, or even necessary, for the existence of a tribal community.

Concluding Remarks.—Philosophers of the derivative[32] school of morals formerly assumed that the foundation of morality lay in a form of Selfishness; but more recently in the "Greatest Happiness principle." According to the view given above, the moral sense is

---

30 See on this subject copious evidence in Chap. vii. of Sir J. Lubbock, 'Origin of Civilisation,' 1870.

31 For instance Lecky, 'Hist. European Morals,' vol. i. p. 124.

32 This term is used in an able article in the 'Westminister Review,' Oct. 1869, p. 498. For the Greatest Happiness principle, see J. S. Mill, 'Utilitarianism,' p. 17.

fundamentally identical with the social instincts; and in the case of the lower animals it would be absurd to speak of these instincts as having been developed from selfishness, or for the happiness of the community. They have, however, certainly been developed for the general good of the community. The term, general good, may be defined as the means by which the greatest possible number of individuals can be reared in full vigour and health, with all their faculties perfect, under the conditions to which they are exposed. As the social instincts both of man and the lower animals have no doubt been developed by the same steps, it would be advisable, if found practicable, to use the same definition in both cases, and to take as the test of morality, the general good or welfare of the community, rather than the general happiness; but this definition would perhaps require some limitation on account of political ethics.

When a man risks his life to save that of a fellow-creature, it seems more appropriate to say that he acts for the general good or welfare, rather than for the general happiness of mankind. No doubt the welfare and the happiness of the individual usually coincide; and a contented, happy tribe will flourish better than one that is discontented and unhappy. We have seen that at an early period in the history of man, the expressed wishes of the community will have naturally influenced to a large extent the conduct of each member; and as all wish for happiness, the "greatest happiness principle" will have become a most important secondary guide and object; the social instincts, including sympathy, always serving as the primary impulse and guide. Thus the reproach of laying the foundation of the most noble part of our nature in the base principle of selfishness is removed; unless indeed the satisfaction which every animal feels when it follows its proper instincts, and the dissatisfaction felt when prevented, be called selfish.

The expression of the wishes and judgment of the members of the same community, at first by oral and afterwards by written language, serves, as just remarked, as a most important secondary guide of conduct, in aid of the social instincts, but sometimes in opposition to them. This latter fact is well exemplified by the *Law of Honour*, that is the law of the opinion of our equals, and not of all our countrymen. The breach of this law, even when the breach is known to be strictly accordant with true morality, has caused many a man more agony than a real crime. We recognise the same influence in the burning sense of shame which most of us have felt even after the interval of years, when calling to mind some accidental breach of a trifling though fixed rule of etiquette. The judgment of the community will generally be guided by some rude experience of what is best in the long run for all the members; but this judgment will not rarely err from ignorance and from weak powers of reasoning. Hence the strangest customs and superstitions, in complete opposition to the true welfare and happiness of mankind, have become all-powerful throughout the world. We see this in the horror felt by a Hindoo who breaks his caste, in the shame of a Mahometan woman who exposes her face, and in innumerable other instances. It would be difficult to distinguish between the remorse felt by a Hindoo who has eaten unclean food, from that felt after committing a theft; but the former would probably be the more severe.

How so many absurd rules of conduct, as well as so many absurd religious beliefs, have originated we do not know; nor how it is that they have become, in all quarters of the world, so deeply impressed on the mind of men; but it is worthy of remark that a belief constantly inculcated during the early years of life, whilst the brain is impressible, appears to acquire almost the nature of an instinct; and the very essence of an instinct is that it is followed independently of reason. Neither can we say why certain admirable virtues, such as the love of truth, are much more highly appreciated by some savage tribes than by others;[33] nor, again, why similar differences prevail even amongst civilised nations. Knowing how firmly fixed many strange customs and superstitions have become, we need feel no surprise that the self-regarding virtues should now appear to us so natural, supported as they are by reason, as to be thought innate, although they were not valued by man in his early condition.

Notwithstanding many sources of doubt, man can generally and readily distinguish between the higher and lower moral rules. The higher are founded on the social instincts, and relate to the welfare of others. They are supported by the approbation of our fellowmen and by reason. The lower rules, though some of them when implying self-sacrifice hardly deserve to be called lower, relate chiefly to self, and owe their origin to public opinion, when matured by experience and cultivated; for they are not practised by rude tribes.

As man advances in civilisation, and small tribes are united into larger communities, the simplest reason would tell each individual that he ought to extend his social instincts and sympathies to all the members of the same nation, though personally unknown to him. This point being once reached, there is only an arti-

---

[33] Good instances are given by Mr. Wallace in 'Scientific Opinion,' Sept. 15, 1869; and more fully in his 'Contributions to the Theory of Natural Selection,' 1870, p. 353.

ficial barrier to prevent his sympathies extending to the men of all nations and races. If, indeed, such men are separated from him by great differences in appearance or habits, experience unfortunately shews us how long it is before we look at them as our fellow-creatures. Sympathy beyond the confines of man, that is humanity to the lower animals, seems to be one of the latest moral acquisitions. It is apparently unfelt by savages, except towards their pets. How little the old Romans knew of it is shewn by their abhorrent gladiatorial exhibitions. The very idea of humanity, as far as I could observe, was new to most of the Gauchos of the Pampas. This virtue, one of the noblest with which man is endowed, seems to arise incidentally from our sympathies becoming more tender and more widely diffused, until they are extended to all sentient beings. As soon as this virtue is honoured and practised by some few men, it spreads through instruction and example to the young, and eventually through public opinion.

The highest stage in moral culture at which we can arrive, is when we recognise that we ought to control our thoughts, and "not even in inmost thought to think "again the sins that made the past so pleasant to us."[34] Whatever makes any bad action familiar to the mind, renders its performance by so much the easier. As Marcus Aurelius long ago said, "Such as are thy habi-"tual thoughts, such also will be the character of thy "mind; for the soul is dyed by the thoughts."[35]

Our great philosopher, Herbert Spencer, has recently explained his views on the moral sense. He says,[36] "I

---

[34] Tennyson, 'Idylls of the King,' p. 244.

[35] 'The Thoughts of the Emperor M. Aurelius Antoninus,' Eng. translat., 2nd edit., 1869, p. 112. Marcus Aurelius was born A.D. 121.

[36] Letter to Mr. Mill in Bain's 'Mental and Moral Science,' 1868, p. 722.

"believe that the experiences of utility organised and "consolidated through all past generations of the human "race, have been producing corresponding modifications, "which, by continued transmission and accumulation, "have become in us certain faculties of moral intuition— "certain emotions responding to right and wrong con- "duct, which have no apparent basis in the individual "experiences of utility." There is not the least inherent improbability, as it seems to me, in virtuous tendencies being more or less strongly inherited; for, not to mention the various dispositions and habits transmitted by many of our domestic animals, I have heard of cases in which a desire to steal and a tendency to lie appeared to run in families of the upper ranks; and as stealing is so rare a crime in the wealthy classes, we can hardly account by accidental coincidence for the tendency occurring in two or three members of the same family. If bad tendencies are transmitted, it is probable that good ones are likewise transmitted. Excepting through the principle of the transmission of moral tendencies, we cannot understand the differences believed to exist in this respect between the various races of mankind. We have, however, as yet, hardly sufficient evidence on this head.

Even the partial transmission of virtuous tendencies would be an immense assistance to the primary impulse derived directly from the social instincts, and indirectly from the approbation of our fellow-men. Admitting for the moment that virtuous tendencies are inherited, it appears probable, at least in such cases as chastity, temperance, humanity to animals, &c., that they become first impressed on the mental organisation through habit, instruction, and example, continued during several generations in the same family, and in a quite subordinate degree, or not at all, by the individuals possessing such virtues, having succeeded best in the struggle for life. My chief source of doubt with respect to any such inheritance, is that senseless customs, superstitions, and tastes, such as the horror of a Hindoo for unclean food, ought on the same principle to be transmitted. Although this in itself is perhaps not less probable than that animals should acquire inherited tastes for certain kinds of food or fear of certain foes, I have not met with any evidence in support of the transmission of superstitious customs or senseless habits.

Finally, the social instincts which no doubt were acquired by man, as by the lower animals, for the good of the community, will from the first have given to him some wish to aid his fellows, and some feeling of sympathy. Such impulses will have served him at a very early period as a rude rule of right and wrong. But as man gradually advanced in intellectual power and was enabled to trace the more remote consequences of his actions; as he acquired sufficient knowledge to reject baneful customs and superstitions; as he regarded more and more not only the welfare but the happiness of his fellow-men; as from habit, following on beneficial experience, instruction, and example, his sympathies became more tender and widely diffused, so as to extend to the men of all races, to the imbecile, the maimed, and other useless members of society, and finally to the lower animals,—so would the standard of his morality rise higher and higher. And it is admitted by moralists of the derivative school and by some intuitionists, that the standard of morality has risen since an early period in the history of man.[37]

As a struggle may sometimes be seen going on

[37] A writer in the 'North British Review' (July, 1869, p. 531), well capable of forming a sound judgment, expresses himself strongly to this

between the various instincts of the lower animals, it is not surprising that there should be a struggle in man between his social instincts, with their derived virtues, and his lower, though at the moment, stronger impulses or desires. This, as Mr. Galton[38] has remarked, is all the less surprising, as man has emerged from a state of barbarism within a comparatively recent period. After having yielded to some temptation we feel a sense of dissatisfaction, analogous to that felt from other unsatisfied instincts, called in this case conscience; for we cannot prevent past images and impressions continually passing through our minds, and these in their weakened state we compare with the ever-present social instincts, or with habits gained in early youth and strengthened during our whole lives, perhaps inherited, so that they are at last rendered almost as strong as instincts. Looking to future generations, there is no cause to fear that the social instincts will grow weaker, and we may expect that virtuous habits will grow stronger, becoming perhaps fixed by inheritance. In this case the struggle between our higher and lower impulses will be less severe, and virtue will be triumphant.

Summary of the two last Chapters.—There can be no doubt that the difference between the mind of the lowest man and that of the highest animal is immense. An anthropomorphous ape, if he could take a dispassionate view of his own case, would admit that though he could form an artful plan to plunder a garden—though he could use stones for fighting or for breaking open nuts,

effect. Mr. Lecky ('Hist. of Morals,' vol. i. p. 143) seems to a certain extent to coincide.

[38] See his remarkable work on 'Hereditary Genius,' 1869, p. 349. The Duke of Argyll ('Primeval Man,' 1869, p. 188) has some good remarks on the contest in man's nature between right and wrong.

yet that the thought of fashioning a stone into a tool was quite beyond his scope. Still less, as he would admit, could he follow out a train of metaphysical reasoning, or solve a mathematical problem, or reflect on God, or admire a grand natural scene. Some apes, however, would probably declare that they could and did admire the beauty of the coloured skin and fur of their partners in marriage. They would admit, that though they could make other apes understand by cries some of their perceptions and simpler wants, the notion of expressing definite ideas by definite sounds had never crossed their minds. They might insist that they were ready to aid their fellow-apes of the same troop in many ways, to risk their lives for them, and to take charge of their orphans; but they would be forced to acknowledge that disinterested love for all living creatures, the most noble attribute of man, was quite beyond their comprehension.

Nevertheless the difference in mind between man and the higher animals, great as it is, is certainly one of degree and not of kind. We have seen that the senses and intuitions, the various emotions and faculties, such as love, memory, attention, curiosity, imitation, reason, &c., of which man boasts, may be found in an incipient, or even sometimes in a well-developed condition, in the lower animals. They are also capable of some inherited improvement, as we see in the domestic dog compared with the wolf or jackal. If it be maintained that certain powers, such as self-consciousness, abstraction, &c., are peculiar to man, it may well be that these are the incidental results of other highly-advanced intellectual faculties; and these again are mainly the result of the continued use of a highly developed language. At what age does the new-born infant possess the power of abstraction, or become self-conscious and reflect on its own existence? We cannot answer; nor can we answer in regard to the ascending organic scale. The half-art and half-instinct of language still bears the stamp of its gradual evolution. The ennobling belief in God is not universal with man; and the belief in active spiritual agencies naturally follows from his other mental powers. The moral sense perhaps affords the best and highest distinction between man and the lower animals; but I need not say anything on this head, as I have so lately endeavoured to shew that the social instincts,—the prime principle of man's moral constitution[39]—with the aid of active intellectual powers and the effects of habit, naturally lead to the golden rule, "As ye would that men should "do to you, do ye to them likewise;" and this lies at the foundation of morality.

In a future chapter I shall make some few remarks on the probable steps and means by which the several mental and moral faculties of man have been gradually evolved. That this at least is possible ought not to be denied, when we daily see their development in every infant; and when we may trace a perfect gradation from the mind of an utter idiot, lower than that of the lowest animal, to the mind of a Newton.

---

[39] 'The Thoughts of Marcus Aurelius,' &c., p. 139.

CHAPTER IV.

ON THE MANNER OF DEVELOPMENT OF MAN FROM SOME LOWER FORM.

Variability of body and mind in man — Inheritance — Causes of variability — Laws of variation the same in man as in the lower animals — Direct action of the conditions of life — Effects of the increased use and disuse of parts — Arrested development — Reversion — Correlated variation — Rate of increase — Checks to increase — Natural selection — Man the most dominant animal in the world — Importance of his corporeal structure — The causes which have led to his becoming erect—Consequent changes of structure — Decrease in size of the canine teeth — Increased size and altered shape of the skull — Nakedness — Absence of a tail — Defenceless condition of man.

WE have seen in the first chapter that the homological structure of man, his embryological development and the rudiments which he still retains, all declare in the plainest manner that he is descended from some lower form. The possession of exalted mental powers is no insuperable objection to this conclusion. In order that an ape-like creature should have been transformed into man, it is necessary that this early form, as well as many successive links, should all have varied in mind and body. It is impossible to obtain direct evidence on this head; but if it can be shewn that man now varies —that his variations are induced by the same general causes, and obey the same general laws, as in the case of the lower animals—there can be little doubt that the preceding intermediate links varied in a like manner. The variations at each successive stage of descent must, also, have been in some manner accumulated and fixed.

The facts and conclusions to be given in this chapter relate almost exclusively to the probable means by which the transformation of man has been effected, as far as his bodily structure is concerned. The following chapter will be devoted to the development of his intellectual and moral faculties. But the present discussion likewise bears on the origin of the different races or species of mankind, whichever term may be preferred.

It is manifest that man is now subject to much variability. No two individuals of the same race are quite alike. We may compare millions of faces, and each will be distinct. There is an equally great amount of diversity in the proportions and dimensions of the various parts of the body; the length of the legs being one of the most variable points.[1] Although in some quarters of the world an elongated skull, and in other quarters a short skull prevails, yet there is great diversity of shape even within the limits of the same race, as with the aborigines of America and South Australia,—the latter a race "probably as pure and "homogeneous in blood, customs, and language as any "in existence"—and even with the inhabitants of so confined an area as the Sandwich Islands.[2] An eminent dentist assures me that there is nearly as much diversity in the teeth, as in the features. The chief arteries so frequently run in abnormal courses, that it has been found useful for surgical purposes to calculate

---

1 'Investigations in Military and Anthropolog. Statistics of American Soldiers,' by B. A. Gould, 1869, p. 256.

2 With respect to the "Cranial forms of the American aborigines," see Dr. Aitken Meigs in 'Proc. Acad. Nat. Sci.' Philadelphia, May, 1866. On the Australians, see Huxley, in Lyell's 'Antiquity of Man,' 1863, p. 87. On the Sandwich Islanders, Prof. J. Wyman, 'Observations on Crania,' Boston, 1868, p. 18.

from 12,000 corpses how often each course prevails.[3] The muscles are eminently variable: thus those of the foot were found by Prof. Turner[4] not to be strictly alike in any two out of fifty bodies; and in some the deviations were considerable. Prof. Turner adds that the power of performing the appropriate movements must have been modified in accordance with the several deviations. Mr. J. Wood has recorded[5] the occurrence of 295 muscular variations in thirty-six subjects, and in another set of the same number no less than 558 variations, reckoning both sides of the body as one. In the last set, not one body out of the thirty-six was "found "totally wanting in departures from the standard de-"scriptions of the muscular system given in anatomical "text-books." A single body presented the extraordinary number of twenty-five distinct abnormalities. The same muscle sometimes varies in many ways: thus Prof. Macalister describes[6] no less than twenty distinct variations in the *palmaris accessorius.*

The famous old anatomist, Wolff,[7] insists that the internal viscera are more variable than the external parts: *Nulla particula est quæ non aliter et aliter in aliis se habeat hominibus.* He has even written a treatise on the choice of typical examples of the viscera for representation. A discussion on the beau-ideal of the liver, lungs, kidneys, &c., as of the human face divine, sounds strange in our ears.

The variability or diversity of the mental faculties in men of the same race, not to mention the greater

---

[3] 'Anatomy of the Arteries,' by R. Quain.

[4] 'Transact. Royal Soc.' Edinburgh, vol. xxiv. p. 175, 189.

[5] 'Proc. Royal Soc.' 1867, p. 544; also 1868, p. 483, 524. There is a previous paper, 1866, p. 229.

[6] 'Proc. R. Irish Academy,' vol. x. 1868, p. 141.

[7] 'Act. Acad.,' St. Petersburg, 1778, part ii. p. 217.

differences between the men of distinct races, is so notorious that not a word need here be said. So it is with the lower animals, as has been illustrated by a few examples in the last chapter. All who have had charge of menageries admit this fact, and we see it plainly in our dogs and other domestic animals. Brehm especially insists that each individual monkey of those which he kept under confinement in Africa had its own peculiar disposition and temper: he mentions one baboon remarkable for its high intelligence; and the keepers in the Zoological Gardens pointed out to me a monkey, belonging to the New World division, equally remarkable for intelligence. Rengger, also, insists on the diversity in the various mental characters of the monkeys of the same species which he kept in Paraguay; and this diversity, as he adds, is partly innate, and partly the result of the manner in which they have been treated or educated.[8]

I have elsewhere[9] so fully discussed the subject of Inheritance that I need here add hardly anything. A greater number of facts have been collected with respect to the transmission of the most trifling, as well as of the most important characters in man than in any of the lower animals; though the facts are copious enough with respect to the latter. So in regard to mental qualities, their transmission is manifest in our dogs, horses, and other domestic animals. Besides special tastes and habits, general intelligence, courage, bad and good temper, &c., are certainly transmitted. With man we see similar facts in almost every family; and we

---

[8] Brehm, 'Thierleben,' B. i. s. 58, 87. Rengger, 'Säugethiere von Paraguay,' s. 57.

[9] 'Variation of Animals and Plants under Domestication,' vol. ii. chap. xii.

now know through the admirable labours of Mr. Galton[10] that genius, which implies a wonderfully complex combination of high faculties, tends to be inherited; and, on the other hand, it is too certain that insanity and deteriorated mental powers likewise run in the same families.

With respect to the causes of variability we are in all cases very ignorant; but we can see that in man as in the lower animals, they stand in some relation with the conditions to which each species has been exposed during several generations. Domesticated animals vary more than those in a state of nature; and this is apparently due to the diversified and changing nature of their conditions. The different races of man resemble in this respect domesticated animals, and so do the individuals of the same race when inhabiting a very wide area, like that of America. We see the influence of diversified conditions in the more civilised nations, the members of which belong to different grades of rank and follow different occupations, presenting a greater range of character than the members of barbarous nations. But the uniformity of savages has often been exaggerated, and in some cases can hardly be said to exist.[11] It is nevertheless an error to speak of man, even if we look only to the conditions to which he has been subjected, as "far more domesticated"[12] than

[10] 'Hereditary Genius: an Inquiry into its Laws and Consequences,' 1869.

[11] Mr. Bates remarks ('The Naturalist on the Amazons,' 1863, vol. ii. p. 159), with respect to the Indians of the same S. American tribe, "no two of them were at all similar in the shape of the head; one "man had an oval visage with fine features, and another was quite "Mongolian in breadth and prominence of cheek, spread of nostrils, "and obliquity of eyes."

[12] Blumenbach, 'Treatises on Anthropolog.' Eng. translat., 1865, p. 205.

any other animal. Some savage races, such as the Australians, are not exposed to more diversified conditions than are many species which have very wide ranges. In another and much more important respect, man differs widely from any strictly domesticated animal; for his breeding has not been controlled, either through methodical or unconscious selection. No race or body of men has been so completely subjugated by other men, that certain individuals have been preserved and thus unconsciously selected, from being in some way more useful to their masters. Nor have certain male and female individuals been intentionally picked out and matched, except in the well-known case of the Prussian grenadiers; and in this case man obeyed, as might have been expected, the law of methodical selection; for it is asserted that many tall men were reared in the villages inhabited by the grenadiers with their tall wives.

If we consider all the races of man, as forming a single species, his range is enormous; but some separate races, as the Americans and Polynesians, have very wide ranges. It is a well-known law that widely-ranging species are much more variable than species with restricted ranges; and the variability of man may with more truth be compared with that of widely-ranging species, than with that of domesticated animals.

Not only does variability appear to be induced in man and the lower animals by the same general causes, but in both the same characters are affected in a closely analogous manner. This has been proved in such full detail by Godron and Quatrefages, that I need here only refer to their works.[13] Monstrosities, which gra-

[13] Godron, 'De l'Espèce,' 1859, tom. ii. livre 3. Quatrefages, 'Unité de l'Espèce Humaine,' 1861. Also Lectures on Anthropology, given in the 'Revue des Cours Scientifiques,' 1866-1868.

duate into slight variations, are likewise so similar in man and the lower animals, that the same classification and the same terms can be used for both, as may be seen in Isidore Geoffroy St.-Hilaire's great work.[14] This is a necessary consequence of the same laws of change prevailing throughout the animal kingdom. In my work on the variation of domestic animals, I have attempted to arrange in a rude fashion the laws of variation under the following heads:—The direct and definite action of changed conditions, as shewn by all or nearly all the individuals of the same species varying in the same manner under the same circumstances. The effects of the long-continued use or disuse of parts. The cohesion of homologous parts. The variability of multiple parts. Compensation of growth; but of this law I have found no good instances in the case of man. The effects of the mechanical pressure of one part on another; as of the pelvis on the cranium of the infant in the womb. Arrests of development, leading to the diminution or suppression of parts. The reappearance of long-lost characters through reversion. And lastly, correlated variation. All these so-called laws apply equally to man and the lower animals; and most of them even to plants. It would be superfluous here to discuss all of them;[15] but several are so important for us, that they must be treated at considerable length.

The direct and definite action of changed conditions.— This is a most perplexing subject. It cannot be denied

[14] 'Hist. Gen. et Part. des Anomalies de l'Organisation,' in three volumes, tom. i. 1832.

[15] I have fully discussed these laws in my 'Variation of Animals and Plants under Domestication,' vol. ii. chap. xxii. and xxiii. M. J. P. Durand has lately (1868) published a valuable essay 'De l'Influence des Milieux, &c.' He lays much stress on the nature of the soil.

that changed conditions produce some effect, and occasionally a considerable effect, on organisms of all kinds; and it seems at first probable that if sufficient time were allowed this would be the invariable result. But I have failed to obtain clear evidence in favour of this conclusion; and valid reasons may be urged on the other side, at least as far as the innumerable structures are concerned, which are adapted for special ends. There can, however, be no doubt that changed conditions induce an almost indefinite amount of fluctuating variability, by which the whole organisation is rendered in some degree plastic.

In the United States, above 1,000,000 soldiers, who served in the late war, were measured, and the States in which they were born and reared recorded.[16] From this astonishing number of observations it is proved that local influences of some kind act directly on stature; and we further learn that "the State where the physical "growth has in great measure taken place, and the State "of birth, which indicates the ancestry, seem to exert "a marked influence on the stature." For instance it is established, "that residence in the Western States, "during the years of growth, tends to produce increase "of stature." On the other hand, it is certain that with sailors, their manner of life delays growth, as shewn "by "the great difference between the statures of soldiers and "sailors at the ages of 17 and 18 years." Mr. B. A. Gould endeavoured to ascertain the nature of the influences which thus act on stature; but he arrived only at negative results, namely, that they did not relate to climate, the elevation of the land, soil, nor even "in any con- "trolling degree" to the abundance or need of the com-

---

[16] 'Investigations in Military and Anthrop. Statistics,' &c. 1869, by B. A. Gould, p. 93, 107, 126, 131, 134.

forts of life. This latter conclusion is directly opposed to that arrived at by Villermé from the statistics of the height of the conscripts in different parts of France. When we compare the differences in stature between the Polynesian chiefs and the lower orders within the same islands, or between the inhabitants of the fertile volcanic and low barren coral islands of the same ocean,[17] or again between the Fuegians on the eastern and western shores of their country, where the means of subsistence are very different, it is scarcely possible to avoid the conclusion that better food and greater comfort do influence stature. But the preceding statements shew how difficult it is to arrive at any precise result. Dr. Beddoe has lately proved that, with the inhabitants of Britain, residence in towns and certain occupations have a deteriorating influence on height; and he infers that the result is to a certain extent inherited, as is likewise the case in the United States. Dr. Beddoe further believes that wherever a "race attains its maximum of "physical development, it rises highest in energy and "moral vigour."[18]

Whether external conditions produce any other direct effect on man is not known. It might have been expected that differences of climate would have had a marked influence, as the lungs and kidneys are brought into fuller activity under a low temperature, and the liver and skin under a high one.[19] It was formerly thought that the colour of the skin and the character

[17] For the Polynesians, see Prichard's 'Physical Hist. of Mankind,' vol. v. 1847, p. 145, 283. Also Godron, 'De l'Espèce,' tom. ii. p. 289. There is also a remarkable difference in appearance between the closely-allied Hindoos inhabiting the Upper Ganges and Bengal; see Elphinstone's 'History of India,' vol. i. p. 324.

[18] 'Memoirs, Anthropolog. Soc.' vol. iii. 1867-69, p. 561, 565, 567.

[19] Dr. Brakenridge, 'Theory of Diathesis,' 'Medical Times,' June 19 and July 17, 1869.

of the hair were determined by light or heat; and although it can hardly be denied that some effect is thus produced, almost all observers now agree that the effect has been very small, even after exposure during many ages. But this subject will be more properly discussed when we treat of the different races of mankind. With our domestic animals there are grounds for believing that cold and damp directly affect the growth of the hair; but I have not met with any evidence on this head in the case of man.

Effects of the increased Use and Disuse of Parts.—It is well known that use strengthens the muscles in the individual, and complete disuse, or the destruction of the proper nerve, weakens them. When the eye is destroyed the optic nerve often becomes atrophied. When an artery is tied, the lateral channels increase not only in diameter, but in the thickness and strength of their coats. When one kidney ceases acting from disease, the other increases in size and does double work. Bones increase not only in thickness, but in length, from carrying a greater weight.[20] Different occupations habitually followed lead to changed proportions in various parts of the body. Thus it was clearly ascertained by the United States Commission[21] that the legs of the sailors employed in the late war were longer by 0·217 of an inch than those of the soldiers, though the sailors were on an average shorter men; whilst their arms were shorter by 1·09 of an inch, and therefore out of proportion shorter in relation to

[20] I have given authorities for these several statements in my 'Variation of Animals under Domestication,' vol. ii. p. 297-300. Dr. Jaeger, "Ueber das Längenwachsthum der Knochen," 'Jenaischen Zeitschrift,' B. v. Heft i.

[21] 'Investigations,' &c. By B. A. Gould, 1869, p. 288.

their lesser height. This shortness of the arms is apparently due to their greater use, and is an unexpected result; but sailors chiefly use their arms in pulling and not in supporting weights. The girth of the neck and the depth of the instep are greater, whilst the circumference of the chest, waist, and hips is less in sailors than in soldiers.

Whether the several foregoing modifications would become hereditary, if the same habits of life were followed during many generations, is not known, but is probable. Rengger[22] attributes the thin legs and thick arms of the Payaguas Indians to successive generations having passed nearly their whole lives in canoes, with their lower extremities motionless. Other writers have come to a similar conclusion in other analogous cases. According to Cranz,[23] who lived for a long time with the Esquimaux, "the natives believe that ingenuity and "dexterity in seal-catching (their highest art and virtue) "is hereditary; there is really something in it, for the "son of a celebrated seal-catcher will distinguish him-"self though he lost his father in childhood." But in this case it is mental aptitude, quite as much as bodily structure, which appears to be inherited. It is asserted that the hands of English labourers are at birth larger than those of the gentry.[24] From the correlation which exists, at least in some cases,[25] between the development of the extremities and of the jaws, it is possible that in those classes which do not labour much with their hands and feet, the jaws would be reduced in size from this cause. That they are generally smaller in refined and civilised men than in hard-working men or savages,

---

22 'Säugethiere von Paraguay,' 1830, s. 4.

23 'History of Greenland,' Eng. translat. 1767, vol. i. p. 230.

24 'Intermarriage.' By Alex. Walker, 1838, p. 377.

25 'The Variation of Animals under Domestication,' vol. i. p. 173.

is certain. But with savages, as Mr. Herbert Spencer[26] has remarked, the greater use of the jaws in chewing coarse, uncooked food, would act in a direct manner on the masticatory muscles and on the bones to which they are attached. In infants long before birth, the skin on the soles of the feet is thicker than on any other part of the body;[27] and it can hardly be doubted that this is due to the inherited effects of pressure during a long series of generations.

It is familiar to every one that watchmakers and engravers are liable to become short-sighted, whilst sailors and especially savages are generally long-sighted. Short-sight and long-sight certainly tend to be inherited.[28] The inferiority of Europeans, in comparison with savages, in eye-sight and in the other senses, is no doubt the accumulated and transmitted effect of lessened use during many generations; for Rengger[29] states that he has repeatedly observed Europeans, who had been brought up and spent their whole lives with the wild Indians, who nevertheless did not equal them in the sharpness of their senses. The same naturalist observes that the cavities in the skull for the reception of the several sense-organs are larger in the American aborigines than in Europeans; and this no doubt indicates a corresponding difference in the dimensions of the organs themselves. Blumenbach has also remarked on the large size of the nasal cavities

---

[26] 'Principles of Biology,' vol. i. p. 455.

[27] Paget, 'Lectures on Surgical Pathology,' vol. i. 1853, p. 209.

[28] 'The Variation of Animals under Domestication,' vol. i. p. 8.

[29] 'Säugethiere von Paraguay,' s. 8, 10. I have had good opportunities for observing the extraordinary power of eyesight in the Fuegians. See also Lawrence ('Lectures on Physiology,' &c., 1822, p. 404) on this same subject. M. Giraud-Teulon has recently collected ('Revue des Cours Scientifiques,' 1870, p. 625) a large and valuable body of evidence proving that the cause of short-sight, "*C'est le travail* "*assidu, de près.*"

in the skulls of the American aborigines, and connects this fact with their remarkably acute power of smell. The Mongolians of the plains of Northern Asia, according to Pallas, have wonderfully perfect senses; and Prichard believes that the great breadth of their skulls across the zygomas follows from their highly-developed sense-organs.[30]

The Quechua Indians inhabit the lofty plateaux of Peru, and Alcide d'Orbigny states[31] that from continually breathing a highly rarefied atmosphere they have acquired chests and lungs of extraordinary dimensions. The cells, also, of the lungs are larger and more numerous than in Europeans. These observations have been doubted; but Mr. D. Forbes carefully measured many Aymaras, an allied race, living at the height of between ten and fifteen thousand feet; and he informs me[32] that they differ conspicuously from the men of all other races seen by him, in the circumference and length of their bodies. In his table of measurements, the stature of each man is taken at 1000, and the other measurements are reduced to this standard. It is here seen that the extended arms of the Aymaras are shorter than those of Europeans, and much shorter than those of Negroes. The legs are likewise shorter, and they present this remarkable peculiarity, that in every Aymara measured the femur is actually shorter than the tibia. On an average the length of the femur to that of the tibia is as 211 to 252; whilst in two Europeans measured at the same

---

[30] Prichard, 'Phys. Hist. of Mankind,' on the authority of Blumenbach, vol. i. 1851, p. 311; for the statement by Pallas, vol. iv. 1844, p. 407.

[31] Quoted by Prichard, 'Researches into the Phys. Hist. of Mankind,' vol. v. p. 463.

[32] Mr. Forbes' valuable paper is now published in the 'Journal of the Ethnological Soc. of London,' new series, vol. ii. 1870, p. 193.

time, the femora to the tibiæ were as 244 to 230; and in three Negroes as 258 to 241. The humerus is likewise shorter relatively to the forearm. This shortening of that part of the limb which is nearest to the body, appears to be, as suggested to me by Mr. Forbes, a case of compensation in relation with the greatly increased length of the trunk. The Aymaras present some other singular points of structure, for instance, the very small projection of the heel.

These men are so thoroughly acclimatised to their cold and lofty abode, that when formerly carried down by the Spaniards to the low Eastern plains, and when now tempted down by high wages to the gold-washings, they suffer a frightful rate of mortality. Nevertheless Mr. Forbes found a few pure families which had survived during two generations; and he observed that they still inherited their characteristic peculiarities. But it was manifest, even without measurement, that these peculiarities had all decreased; and on measurement their bodies were found not to be so much elongated as those of the men on the high plateau; whilst their femora had become somewhat lengthened, as had their tibiæ but in a less degree. The actual measurements may be seen by consulting Mr. Forbes' memoir. From these valuable observations, there can, I think, be no doubt that residence during many generations at a great elevation tends, both directly and indirectly, to induce inherited modifications in the proportions of the body.[33]

Although man may not have been much modified during the latter stages of his existence through the

[33] Dr. Wilckens ('Landwirthschaft. Wochenblatt,' No. 10, 1869) has lately published an interesting essay shewing how domestic animals, which live in mountainous regions, have their frames modified.

increased or decreased use of parts, the facts now given shew that his liability in this respect has not been lost; and we positively know that the same law holds good with the lower animals. Consequently we may infer, that when at a remote epoch the progenitors of man were in a transitional state, and were changing from quadrupeds into bipeds, natural selection would probably have been greatly aided by the inherited effects of the increased or diminished use of the different parts of the body.

Arrests of Development.—Arrested development differs from arrested growth, as parts in the former state continue to grow whilst still retaining their early condition. Various monstrosities come under this head, and some are known to be occasionally inherited, as a cleft-palate. It will suffice for our purpose to refer to the arrested brain-development of microcephalous idiots, as described in Vogt's great memoir.[34] Their skulls are smaller, and the convolutions of the brain are less complex than in normal men. The frontal sinus, or the projection over the eye-brows, is largely developed, and the jaws are prognathous to an "*effrayant*" degree; so that these idiots somewhat resemble the lower types of mankind. Their intelligence and most of their mental faculties are extremely feeble. They cannot acquire the power of speech, and are wholly incapable of prolonged attention, but are much given to imitation. They are strong and remarkably active, continually gamboling and jumping about, and making grimaces. They often ascend stairs on all-fours; and are curiously fond of climbing up furniture or trees. We are thus reminded of the delight

[34] 'Mémoire sur les Microcéphales,' 1867, p. 50, 125, 169, 171, 184-198.

shewn by almost all boys in climbing trees; and this again reminds us how lambs and kids, originally alpine animals, delight to frisk on any hillock, however small.

Reversion.—Many of the cases to be here given might have been introduced under the last heading. Whenever a structure is arrested in its development, but still continues growing until it closely resembles a corresponding structure in some lower and adult member of the same group, we may in one sense consider it as a case of reversion. The lower members in a group give us some idea how the common progenitor of the group was probably constructed; and it is hardly credible that a part arrested at an early phase of embryonic development should be enabled to continue growing so as ultimately to perform its proper function, unless it had acquired this power of continued growth during some earlier state of existence, when the present exceptional or arrested structure was normal. The simple brain of a microcephalous idiot, in as far as it resembles that of an ape, may in this sense be said to offer a case of reversion. There are other cases which come more strictly under our present heading of reversion. Certain structures, regularly occurring in the lower members of the group to which man belongs, occasionally make their appearance in him, though not found in the normal human embryo; or, if present in the normal human embryo, they become developed in an abnormal manner, though this manner of development is proper to the lower members of the same group. These remarks will be rendered clearer by the following illustrations.

In various mammals the uterus graduates from a double organ with two distinct orifices and two passages, as in the marsupials, into a single organ, showing no signs of doubleness except a slight internal fold, as in the higher apes and man. The rodents exhibit a perfect series of gradations between these two extreme states. In all mammals the uterus is developed from two simple primitive tubes, the inferior portions of which form the cornua; and it is in the words of Dr. Farre "by the coalescence of the two cornua at "their lower extremities that the body of the uterus is "formed in man; while in those animals in which no "middle portion or body exists, the cornua remain un-"united. As the development of the uterus proceeds, "the two cornua become gradually shorter, until at "length they are lost, or, as it were, absorbed into the "body of the uterus." The angles of the uterus are still produced into cornua, even so high in the scale as in the lower apes, and their allies the lemurs.

Now in women anomalous cases are not very infrequent, in which the mature uterus is furnished with cornua, or is partially divided into two organs; and such cases, according to Owen, repeat "the grade of con-"centrative development," attained by certain rodents. Here perhaps we have an instance of a simple arrest of embryonic development, with subsequent growth and perfect functional development, for either side of the partially double uterus is capable of performing the proper office of gestation. In other and rarer cases, two distinct uterine cavities are formed, each having its proper orifice and passage.[35] No such stage is passed through during the ordinary development of the embryo, and it is difficult to believe, though perhaps not impossible, that the two simple, minute, primitive tubes could know how (if such an expression may be used) to

[35] See Dr. A. Farre's well-known article in the 'Cyclop. of Anat. and Phys.' vol. v. 1859, p. 642. Owen 'Anatomy of Vertebrates,' vol. iii. 1868, p. 687. Prof. Turner in 'Edinburgh Medical Journal,' Feb. 1865.

grow into two distinct uteri, each with a well-constructed orifice and passage, and each furnished with numerous muscles, nerves, glands and vessels, if they had not formerly passed through a similar course of development, as in the case of existing marsupials. No one will pretend that so perfect a structure as the abnormal double uterus in woman could be the result of mere chance. But the principle of reversion, by which long-lost dormant structures are called back into existence, might serve as the guide for the full development of the organ, even after the lapse of an enormous interval of time.

Professor Canestrini,[36] after discussing the foregoing and various analogous cases, arrives at the same conclusion as that just given. He adduces, as another instance, the malar bone, which, in some of the Quadrumana and other mammals, normally consists of two portions. This is its condition in the two-months-old human fœtus; and thus it sometimes remains, through arrested development, in man when adult, more especially in the lower prognathous races. Hence Canestrini concludes that some ancient progenitor of man must have possessed this bone normally divided into two portions, which subsequently became fused together. In man the frontal bone consists of a single piece, but in the embryo and in children, and in almost all the lower mammals, it consists of two pieces separated by a distinct suture. This suture occasionally persists, more or less distinctly, in man after maturity, and more fre-

---

[36] 'Annuario della Soc. dei Naturalisti in Modena,' 1867, p. 83. Prof. Canestrini gives extracts on this subject from various authorities. Laurillard remarks, that as he has found a complete similarity in the form, proportions, and connexion of the two malar bones in several human subjects and in certain apes, he cannot consider this disposition of the parts as simply accidental.

quently in ancient than in recent crania, especially as Canestrini has observed in those exhumed from the Drift and belonging to the brachycephalic type. Here again he comes to the same conclusion as in the analogous case of the malar bones. In this and other instances presently to be given, the cause of ancient races approaching the lower animals in certain characters more frequently than do the modern races, appears to be that the latter stand at a somewhat greater distance in the long line of descent from their early semi-human progenitors.

Various other anomalies in man, more or less analogous with the foregoing, have been advanced by different authors[37] as cases of reversion; but these seem not a little doubtful, for we have to descend extremely low in the mammalian series before we find such structures normally present.[38]

---

37 A whole series of cases is given by Isid. Geoffroy St.-Hilaire, 'Hist. des Anomalies,' tom. iii. p. 437.

38 In my 'Variation of Animals under Domestication' (vol. ii. p. 57) I attributed the not very rare cases of supernumerary mammæ in women to reversion. I was led to this as a *probable* conclusion, by the additional mammæ being generally placed symmetrically on the breast, and more especially from one case, in which a single efficient mamma occurred in the inguinal region of a woman, the daughter of another woman with supernumerary mammæ. But Prof. Preyer ('Der Kampf um das Dasein,' 1869, s. 45) states that *mammæ erraticæ* have been known to occur in other situations, even on the back; so that the force of my argument is greatly weakened or perhaps quite destroyed.

With much hesitation I, in the same work (vol. ii. p. 12), attributed the frequent cases of polydactylism in men to reversion. I was partly led to this through Prof. Owen's statement, that some of the Ichthyopterygia possess more than five digits, and therefore, as I supposed, had retained a primordial condition; but after reading Prof. Gegenbaur's paper ('Jenaischen Zeitschrift,' B. v. Heft 3, s. 341), who is the highest authority in Europe on such a point, and who disputes Owen's conclusion, I see that it is extremely doubtful whether supernumerary digits can thus be accounted for. It was the fact that such digits not only frequently occur and are strongly inherited, but have the power of regrowth after amputation, like the normal digits of the lower verte-

In man the canine teeth are perfectly efficient instruments for mastication. But their true canine character, as Owen[39] remarks, "is indicated by the conical form "of the crown, which terminates in an obtuse point, is "convex outward and flat or sub-concave within, at the "base of which surface there is a feeble prominence. "The conical form is best expressed in the Melanian "races, especially the Australian. The canine is more "deeply implanted, and by a stronger fang than the "incisors." Nevertheless this tooth no longer serves man as a special weapon for tearing his enemies or prey; it may, therefore, as far as its proper function is concerned, be considered as rudimentary. In every large collection of human skulls some may be found, as Häckel[40] observes, with the canine teeth projecting considerably beyond the others in the same manner, but in a less degree, as in the anthropomorphous apes. In these cases, open spaces between the teeth in the one jaw are left for the reception of the canines belonging to the opposite jaw. An interspace of this kind in a Kaffir skull, figured by Wagner, is surprisingly wide.[41] Considering how few ancient skulls have been examined in comparison with recent skulls, it is an interesting fact that in at least three cases the canines project largely; and in the Naulette jaw they are spoken of as enormous.[42]

---

brata, that chiefly led me to the above conclusion. This extraordinary fact of their regrowth remains inexplicable, if the belief in reversion to some extremely remote progenitor must be rejected. I cannot, however, follow Prof. Gegenbaur in supposing that additional digits could not reappear through reversion, without at the same time other parts of the skeleton being simultaneously and similarly modified; for single characters often reappear through reversion.

[39] 'Anatomy of Vertebrates,' vol. iii. 1868, p. 323.

[40] 'Generelle Morphologie,' 1866, B. ii. s. clv.

[41] Carl Vogt's 'Lectures on Man,' Eng. translat. 1864, p. 151.

[42] C. Carter Blake, on a jaw from La Naulette, 'Anthropolog. Review,' 1867, p. 295. Schaaffhausen, ibid. 1868, p. 426.

The males alone of the anthropomorphous apes have their canines fully developed; but in the female gorilla, and in a less degree in the female orang, these teeth project considerably beyond the others; therefore the fact that women sometimes have, as I have been assured, considerably projecting canines, is no serious objection to the belief that their occasional great development in man is a case of reversion to an ape-like progenitor. He who rejects with scorn the belief that the shape of his own canines, and their occasional great development in other men, are due to our early progenitors having been provided with these formidable weapons, will probably reveal by sneering the line of his descent. For though he no longer intends, nor has the power, to use these teeth as weapons, he will unconsciously retract his "snarling muscles" (thus named by Sir C. Bell)[43] so as to expose them ready for action, like a dog prepared to fight.

Many muscles are occasionally developed in man, which are proper to the Quadrumana or other mammals. Professor Vlacovich[44] examined forty male subjects, and found a muscle, called by him the ischio-pubic, in nineteen of them; in three others there was a ligament which represented this muscle; and in the remaining eighteen no trace of it. Out of thirty female subjects this muscle was developed on both sides in only two, but in three others the rudimentary ligament was present. This muscle, therefore, appears to be much more common in the male than in the female sex; and on the principle of the descent of man from some lower form, its presence can be understood; for it has been detected in several of the lower animals, and in all of

---

43 'The Anatomy of Expression,' 1844, p. 110, 131.

44 Quoted by Prof. Canestrini in the 'Annuario,' &c., 1867, p. 90.

these it serves exclusively to aid the male in the act of reproduction.

Mr. J. Wood, in his valuable series of papers,[45] has minutely described a vast number of muscular variations in man, which resemble normal structures in the lower animals. Looking only to the muscles which closely resemble those regularly present in our nearest allies, the Quadrumana, they are too numerous to be here even specified. In a single male subject, having a strong bodily frame and well-formed skull, no less than seven muscular variations were observed, all of which plainly represented muscles proper to various kinds of apes. This man, for instance, had on both sides of his neck a true and powerful "*levator claviculæ*," such as is found in all kinds of apes, and which is said to occur in about one out of sixty human subjects.[46] Again, this man had "a special abductor of "the metatarsal bone of the fifth digit, such as Pro-"fessor Huxley and Mr. Flower have shewn to exist "uniformly in the higher and lower apes." The hands and arms of man are eminently characteristic structures, but their muscles are extremely liable to vary, so as to resemble the corresponding muscles in the lower animals.[47] Such resemblances are either complete and per-

[45] These papers deserve careful study by any one who desires to learn how frequently our muscles vary, and in varying come to resemble those of the Quadrumana. The following references relate to the few points touched on in my text: vol. xiv. 1865, p. 379-384; vol. xv. 1866, p. 241, 242; vol. xv. 1867, p. 544; vol. xvi. 1868, p. 524. I may here add that Dr. Murie and Mr. St. George Mivart have shewn in their Memoir on the Lemuroidea ('Transact. Zoolog. Soc.' vol. vii. 1869, p. 96), how extraordinarily variable some of the muscles are in these animals, the lowest members of the Primates. Gradations, also, in the muscles leading to structures found in animals still lower in the scale, are numerous in the Lemuroidea.

[46] Prof. Macalister in 'Proc. R. Irish Academy,' vol. x. 1868, p. 124.

[47] Prof. Macalister (ibid. p. 121) has tabulated his observations, and finds that muscular abnormalities are most frequent in the forearms, secondly in the face, thirdly in the foot, &c.

fect or imperfect, yet in this latter case manifestly of a transitional nature. Certain variations are more common in man, and others in woman, without our being able to assign any reason. Mr. Wood, after describing numerous cases, makes the following pregnant remark: "Notable departures from the ordinary type of the "muscular structures run in grooves or directions, which "must be taken to indicate some unknown factor, of "much importance to a comprehensive knowledge of "general and scientific anatomy."[48]

That this unknown factor is reversion to a former state of existence may be admitted as in the highest degree probable. It is quite incredible that a man should through mere accident abnormally resemble, in no less than seven of his muscles, certain apes, if there had been no genetic connection between them. On the other hand, if man is descended from some ape-like creature, no valid reason can be assigned why certain muscles should not suddenly reappear after an interval of many thousand generations, in the same manner as with horses, asses, and mules, dark-coloured stripes suddenly reappear on the legs and shoulders, after an interval of hundreds, or more probably thousands, of generations.

These various cases of reversion are so closely related

---

[48] The Rev. Dr. Haughton, after giving ('Proc. R. Irish Academy,' June 27, 1864, p. 715) a remarkable case of variation in the human *flexor pollicis longus*, adds, "This remarkable example shews that man "may sometimes possess the arrangement of tendons of thumb and "fingers characteristic of the macaque; but whether such a case should "be regarded as a macaque passing upwards into a man, or a man "passing downwards into a macaque, or as a congenital freak of "nature, I cannot undertake to say." It is satisfactory to hear so capable an anatomist, and so embittered an opponent of evolutionism, admitting even the possibility of either of his first propositions. Prof. Macalister has also described ('Proc. R. Irish Acad.' vol. x. 1864, p. 138) variations in the *flexor pollicis longus*, remarkable from their relations to the same muscle in the Quadrumana.

to those of rudimentary organs given in the first chapter, that many of them might have been indifferently introduced in either chapter. Thus a human uterus furnished with cornua may be said to represent in a rudimentary condition the same organ in its normal state in certain mammals. Some parts which are rudimental in man, as the os coccyx in both sexes and the mammæ in the male sex, are always present; whilst others, such as the supracondyloid foramen, only occasionally appear, and therefore might have been introduced under the head of reversion. These several reversionary, as well as the strictly rudimentary, structures reveal the descent of man from some lower form in an unmistakeable manner.

Correlated Variation.—In man, as in the lower animals, many structures are so intimately related, that when one part varies so does another, without our being able, in most cases, to assign any reason. We cannot say whether the one part governs the other, or whether both are governed by some earlier developed part. Various monstrosities, as I. Geoffroy repeatedly insists, are thus intimately connected. Homologous structures are particularly liable to change together, as we see on the opposite sides of the body, and in the upper and lower extremities. Meckel long ago remarked that when the muscles of the arm depart from their proper type, they almost always imitate those of the leg; and so conversely with the muscles of the legs. The organs of sight and hearing, the teeth and hair, the colour of the skin and hair, colour and constitution, are more or less correlated.[49] Professor Schaaffhausen first drew attention to the rela-

[49] The authorities for these several statements are given in my 'Variation of Animals under Domestication,' vol. ii. p. 320-335.

tion apparently existing between a muscular frame and strongly-pronounced supra-orbital ridges, which are so characteristic of the lower races of man.

Besides the variations which can be grouped with more or less probability under the foregoing heads, there is a large class of variations which may be provisionally called spontaneous, for they appear, owing to our ignorance, to arise without any exciting cause. It can, however, be shewn that such variations, whether consisting of slight individual differences, or of strongly-marked and abrupt deviations of structure, depend much more on the constitution of the organism than on the nature of the conditions to which it has been subjected.[50]

Rate of Increase.—Civilised populations have been known under favourable conditions, as in the United States, to double their number in twenty-five years; and according to a calculation by Euler, this might occur in a little over twelve years.[51] At the former rate the present population of the United States, namely, thirty millions, would in 657 years cover the whole terraqueous globe so thickly, that four men would have to stand on each square yard of surface. The primary or fundamental check to the continued increase of man is the difficulty of gaining subsistence and of living in comfort. We may infer that this is the case from what we see, for instance, in the United States, where subsistence is easy and there is plenty of room. If such means were suddenly doubled in Great Britain, our number would be quickly doubled. With civilised nations the

[50] This whole subject has been discussed in chap. xxiii. vol. ii. of my 'Variation of Animals and Plants under Domestication.'

[51] See the ever memorable 'Essay on the Principle of Population,' by the Rev. T. Malthus, vol. i. 1826, p. 6, 517.

above primary check acts chiefly by restraining marriages. The greater death-rate of infants in the poorest classes is also very important; as well as the greater mortality at all ages, and from various diseases, of the inhabitants of crowded and miserable houses. The effects of severe epidemics and wars are soon counterbalanced, and more than counterbalanced, in nations placed under favourable conditions. Emigration also comes in aid as a temporary check, but not to any great extent with the extremely poor classes.

There is reason to suspect, as Malthus has remarked, that the reproductive power is actually less in barbarous than in civilised races. We know nothing positively on this head, for with savages no census has been taken; but from the concurrent testimony of missionaries, and of others who have long resided with such people, it appears that their families are usually small, and large ones rare. This may be partly accounted for, as it is believed, by the women suckling their infants for a prolonged period; but it is highly probable that savages, who often suffer much hardship, and who do not obtain so much nutritious food as civilised men, would be actually less prolific. I have shewn in a former work,[52] that all our domesticated quadrupeds and birds, and all our cultivated plants, are more fertile than the corresponding species in a state of nature. It is no valid objection to this conclusion that animals suddenly supplied with an excess of food, or when rendered very fat, and that most plants when suddenly removed from very poor to very rich soil, are rendered more or less sterile. We might, therefore, expect that civilised men, who in one sense are highly domesticated, would

---

52 'Variation of Animals and Plants under Domestication,' vol. ii. p. 111-113, 163.

be more prolific than wild men. It is also probable that the increased fertility of civilised nations would become, as with our domestic animals, an inherited character: it is at least known that with mankind a tendency to produce twins runs in families.[53]

Notwithstanding that savages appear to be less prolific than civilised people, they would no doubt rapidly increase if their numbers were not by some means rigidly kept down. The Santali, or hill-tribes of India, have recently afforded a good illustration of this fact; for they have increased, as shewn by Mr. Hunter,[54] at an extraordinary rate since vaccination has been introduced, other pestilences mitigated, and war sternly repressed. This increase, however, would not have been possible had not these rude people spread into the adjoining districts and worked for hire. Savages almost always marry; yet there is some prudential restraint, for they do not commonly marry at the earliest possible age. The young men are often required to show that they can support a wife, and they generally have first to earn the price with which to purchase her from her parents. With savages the difficulty of obtaining subsistence occasionally limits their number in a much more direct manner than with civilised people, for all tribes periodically suffer from severe famines. At such times savages are forced to devour much bad food, and their health can hardly fail to be injured. Many accounts have been published of their protruding stomachs and emaciated limbs after and during famines. They are then, also, compelled to wander much about, and their infants, as I was assured in Australia, perish

---

53 Mr. Sedgwick, 'British and Foreign Medico-Chirurg. Review,' July, 1863, p. 170.

54 'The Annals of Rural Bengal,' by W. W. Hunter, 1868, p. 259.

in large numbers. As famines are periodical, depending chiefly on extreme seasons, all tribes must fluctuate in number. They cannot steadily and regularly increase, as there is no artificial increase in the supply of food. Savages when hardly pressed encroach on each other's territories, and war is the result; but they are indeed almost always at war with their neighbours. They are liable to many accidents on land and water in their search for food; and in some countries they must suffer much from the larger beasts of prey. Even in India, districts have been depopulated by the ravages of tigers.

Malthus has discussed these several checks, but he does not lay stress enough on what is probably the most important of all, namely infanticide, especially of female infants, and the habit of procuring abortion. These practices now prevail in many quarters of the world, and infanticide seems formerly to have prevailed, as Mr. M'Lennan[55] has shewn, on a still more extensive scale. These practices appear to have originated in savages recognising the difficulty, or rather the impossibility of supporting all the infants that are born. Licentiousness may also be added to the foregoing checks; but this does not follow from failing means of subsistence; though there is reason to believe that in some cases (as in Japan) it has been intentionally encouraged as a means of keeping down the population.

If we look back to an extremely remote epoch, before man had arrived at the dignity of manhood, he would have been guided more by instinct and less by reason than are savages at the present time. Our early semi-human progenitors would not have practised infanticide, for the instincts of the lower animals are never so perverted as to lead them regularly to destroy their own

---

55 'Primitive Marriage,' 1865.

offspring. There would have been no prudential restraint from marriage, and the sexes would have freely united at an early age. Hence the progenitors of man would have tended to increase rapidly, but checks of some kind, either periodical or constant, must have kept down their numbers, even more severely than with existing savages. What the precise nature of these checks may have been, we cannot say, any more than with most other animals. We know that horses and cattle, which are not highly prolific animals, when first turned loose in South America, increased at an enormous rate. The slowest breeder of all known animals, namely the elephant, would in a few thousand years stock the whole world. The increase of every species of monkey must be checked by some means; but not, as Brehm remarks, by the attacks of beasts of prey. No one will assume that the actual power of reproduction in the wild horses and cattle of America, was at first in any sensible degree increased; or that, as each district became fully stocked, this same power was diminished. No doubt in this case and in all others, many checks concur, and different checks under different circumstances; periodical dearths, depending on unfavourable seasons, being probably the most important of all. So it will have been with the early progenitors of man.

Natural Selection.—We have now seen that man is variable in body and mind; and that the variations are induced, either directly or indirectly, by the same general causes, and obey the same general laws, as with the lower animals. Man has spread widely over the face of the earth, and must have been exposed, during his incessant migrations,[56] to the most diversified con-

---

[56] See some good remarks to this effect by W. Stanley Jevons, "A Deduction from Darwin's Theory," 'Nature,' 1869, p. 231.

ditions. The inhabitants of Tierra del Fuego, the Cape of Good Hope, and Tasmania in the one hemisphere, and of the Arctic regions in the other, must have passed through many climates and changed their habits many times, before they reached their present homes.[57] The early progenitors of man must also have tended, like all other animals, to have increased beyond their means of subsistence; they must therefore occasionally have been exposed to a struggle for existence, and consequently to the rigid law of natural selection. Beneficial variations of all kinds will thus, either occasionally or habitually, have been preserved, and injurious ones eliminated. I do not refer to strongly-marked deviations of structure, which occur only at long intervals of time, but to mere individual differences. We know, for instance, that the muscles of our hands and feet, which determine our powers of movement, are liable, like those of the lower animals,[58] to incessant variability. If then the ape-like progenitors of man which inhabited any district, especially one undergoing some change in its conditions, were divided into two equal bodies, the one half which included all the individuals best adapted by their powers of movement for gaining subsistence or for defending themselves, would on an average survive in greater number and procreate more offspring than the other and less well endowed half.

Man in the rudest state in which he now exists is the most dominant animal that has ever appeared on the earth. He has spread more widely than any

---

[57] Latham, 'Man and his Migrations,' 1851, p. 135.

[58] Messrs. Murie and Mivart in their "Anatomy of the Lemuroidea" ('Transact. Zoolog. Soc.' vol. vii. 1869, p. 96-98) say, "some muscles "are so irregular in their distribution that they cannot be well classed "in any of the above groups." These muscles differ even on the opposite sides of the same individual.

other highly organised form; and all others have yielded before him. He manifestly owes this immense superiority to his intellectual faculties, his social habits, which lead him to aid and defend his fellows, and to his corporeal structure. The supreme importance of these characters has been proved by the final arbitrament of the battle for life. Through his powers of intellect, articulate language has been evolved; and on this his wonderful advancement has mainly depended. He has invented and is able to use various weapons, tools, traps, &c., with which he defends himself, kills or catches prey, and otherwise obtains food. He has made rafts or canoes on which to fish or cross over to neighbouring fertile islands. He has discovered the art of making fire, by which hard and stringy roots can be rendered digestible, and poisonous roots or herbs innocuous. This last discovery, probably the greatest, excepting language, ever made by man, dates from before the dawn of history. These several inventions, by which man in the rudest state has become so preeminent, are the direct result of the development of his powers of observation, memory, curiosity, imagination, and reason. I cannot, therefore, understand how it is that Mr. Wallace[59] maintains, that "natural selec-

---

[59] 'Quarterly Review,' April, 1869, p. 392. This subject is more fully discussed in Mr. Wallace's 'Contributions to the Theory of Natural Selection,' 1870, in which all the essays referred to in this work are republished. The 'Essay on Man' has been ably criticised by Prof. Claparède, one of the most distinguished zoologists in Europe, in an article published in the 'Bibliothèque Universelle,' June, 1870. The remark quoted in my text will surprise every one who has read Mr. Wallace's celebrated paper on 'The Origin of Human Races deduced from the Theory of Natural Selection,' originally published in the 'Anthropological Review,' May, 1864, p. clviii. I cannot here resist quoting a most just remark by Sir J. Lubbock ('Prehistoric Times,' 1865, p. 479) in reference to this paper, namely, that Mr. Wallace, "with characteristic unselfishness, ascribes it (*i. e.* the idea of

" tion could only have endowed the savage with a brain " a little superior to that of an ape."

Although the intellectual powers and social habits of man are of paramount importance to him, we must not underrate the importance of his bodily structure, to which subject the remainder of this chapter will be devoted. The development of the intellectual and social or moral faculties will be discussed in the following chapter.

Even to hammer with precision is no easy matter, as every one who has tried to learn carpentry will admit. To throw a stone with as true an aim as can a Fuegian in defending himself, or in killing birds, requires the most consummate perfection in the correlated action of the muscles of the hand, arm, and shoulder, not to mention a fine sense of touch. In throwing a stone or spear, and in many other actions, a man must stand firmly on his feet; and this again demands the perfect coadaptation of numerous muscles. To chip a flint into the rudest tool, or to form a barbed spear or hook from a bone, demands the use of a perfect hand; for, as a most capable judge, Mr. Schoolcraft,[60] remarks, the shaping fragments of stone into knives, lances, or arrow-heads, shews " extra-" ordinary ability and long practice." We have evidence of this in primeval men having practised a division of labour; each man did not manufacture his own flint tools or rude pottery; but certain individuals appear to have devoted themselves to such work, no doubt receiving in exchange the produce of the chase. Archæologists are convinced that an enormous interval of time

---

" natural selection) unreservedly to Mr. Darwin, although, as is well " known, he struck out the idea independently, and published it, " though not with the same elaboration, at the same time."

[60] Quoted by Mr. Lawson Tait in his " Law of Natural Selection," —'Dublin Quarterly Journal of Medical Science,' Feb. 1869. Dr. Keller is likewise quoted to the same effect.

elapsed before our ancestors thought of grinding chipped flints into smooth tools. A man-like animal who possessed a hand and arm sufficiently perfect to throw a stone with precision or to form a flint into a rude tool, could, it can hardly be doubted, with sufficient practice make almost anything, as far as mechanical skill alone is concerned, which a civilised man can make. The structure of the hand in this respect may be compared with that of the vocal organs, which in the apes are used for uttering various signal-cries, or, as in one species, musical cadences; but in man closely similar vocal organs have become adapted through the inherited effects of use for the utterance of articulate language.

Turning now to the nearest allies of man, and therefore to the best representatives of our early progenitors, we find that the hands in the Quadrumana are constructed on the same general pattern as in us, but are far less perfectly adapted for diversified uses. Their hands do not serve so well as the feet of a dog for locomotion; as may be seen in those monkeys which walk on the outer margins of the palms, or on the backs of their bent fingers, as in the chimpanzee and orang.[16] Their hands, however, are admirably adapted for climbing trees. Monkeys seize thin branches or ropes, with the thumb on one side and the fingers and palm on the other side, in the same manner as we do. They can thus also carry rather large objects, such as the neck of a bottle, to their mouths. Baboons turn over stones and scratch up roots with their hands. They seize nuts, insects, or other small objects with the thumb in opposition to the fingers, and no doubt they thus extract eggs and the young from the nests of birds. American monkeys beat the wild oranges on the

---

[61] Owen, 'Anatomy of Vertebrates,' vol. iii. p. 71.

branches until the rind is cracked, and then tear it off with the fingers of the two hands. Other monkeys open mussel-shells with the two thumbs. With their fingers they pull out thorns and burs, and hunt for each other's parasites. In a state of nature they break open hard fruits with the aid of stones. They roll down stones or throw them at their enemies; nevertheless, they perform these various actions clumsily, and they are quite unable, as I have myself seen, to throw a stone with precision.

It seems to me far from true that because "objects "are grasped clumsily" by monkeys, "a much less "specialised organ of prehension" would have served them[62] as well as their present hands. On the contrary, I see no reason to doubt that a more perfectly constructed hand would have been an advantage to them, provided, and it is important to note this, that their hands had not thus been rendered less well adapted for climbing trees. We may suspect that a perfect hand would have been disadvantageous for climbing; as the most arboreal monkeys in the world, namely Ateles in America and Hylobates in Asia, either have their thumbs much reduced in size and even rudimentary, or their fingers partially coherent, so that their hands are converted into mere grasping-hooks.[63]

As soon as some ancient member in the great series of the Primates came, owing to a change in its manner of procuring subsistence, or to a change in the conditions of its native country, to live somewhat less on trees and more on the ground, its manner of progression would have been modified; and in this case it

[62] 'Quarterly Review,' April, 1869, p. 392.

[63] In *Hylobates syndactylus*, as the name expresses, two of the digits regularly cohere; and this, as Mr. Blyth informs me, is occasionally the case with the digits of *H. agilis*, *lar*, and *leuciscus*.

would have had to become either more strictly quadrupedal or bipedal. Baboons frequent hilly and rocky districts, and only from necessity climb up high trees;[64] and they have acquired almost the gait of a dog. Man alone has become a biped; and we can, I think, partly see how he has come to assume his erect attitude, which forms one of the most conspicuous differences between him and his nearest allies. Man could not have attained his present dominant position in the world without the use of his hands which are so admirably adapted to act in obedience to his will. As Sir C. Bell[65] insists "the hand supplies all instruments, and by its "correspondence with the intellect gives him universal "dominion." But the hands and arms could hardly have become perfect enough to have manufactured weapons, or to have hurled stones and spears with a true aim, as long as they were habitually used for locomotion and for supporting the whole weight of the body, or as long as they were especially well adapted, as previously remarked, for climbing trees. Such rough treatment would also have blunted the sense of touch, on which their delicate use largely depends. From these causes alone it would have been an advantage to man to have become a biped; but for many actions it is almost necessary that both arms and the whole upper part of the body should be free; and he must for this end stand firmly on his feet. To gain this great advantage, the feet have been rendered flat, and the great toe peculiarly modified, though this has entailed the loss of the power of prehension. It accords with the principle of the division of physiological labour, which prevails throughout the animal kingdom, that

---

[64] Brehm, 'Thierleben,' B. i. s. 80.

[65] "The Hand, its mechanism," &c. 'Bridgewater Treatise,' 1833, p. 38.

as the hands became perfected for prehension, the feet should have become perfected for support and locomotion. With some savages, however, the foot has not altogether lost its prehensile power, as shewn by their manner of climbing trees and of using them in other ways.[66]

If it be an advantage to man to have his hands and arms free and to stand firmly on his feet, of which there can be no doubt from his pre-eminent success in the battle of life, then I can see no reason why it should not have been advantageous to the progenitors of man to have become more and more erect or bipedal. They would thus have been better able to have defended themselves with stones or clubs, or to have attacked their prey, or otherwise obtained food. The best constructed individuals would in the long run have succeeded best, and have survived in larger numbers. If the gorilla and a few allied forms had become extinct, it might have been argued with great force and apparent truth, that an animal could not have been gradually converted from a quadruped into a biped; as all the individuals in an intermediate condition would have been miserably ill-fitted for progression. But we know (and this is well worthy of reflection) that several kinds of apes are now actually in this intermediate condition; and no one doubts that they are on the whole well adapted for their conditions of life. Thus the gorilla runs with a sidelong shambling gait, but more commonly

---

[66] Häckel has an excellent discussion on the steps by which man became a biped: 'Natürliche Schöpfungsgeschichte,' 1868, s. 507. Dr. Büchner ('Conférences sur la Théorie Darwinienne,' 1869, p. 135) has given good cases of the use of the foot as a prehensile organ by man; also on the manner of progression of the higher apes to which I allude in the following paragraph: see also Owen ('Anatomy of Vertebrates,' vol. iii. p. 71) on this latter subject.

progresses by resting on its bent hands. The long-armed apes occasionally use their arms like crutches, swinging their bodies forward between them, and some kinds of Hylobates, without having been taught, can walk or run upright with tolerable quickness; yet they move awkwardly, and much less securely than man. We see, in short, with existing monkeys various gradations between a form of progression strictly like that of a quadruped and that of a biped or man.

As the progenitors of man became more and more erect, with their hands and arms more and more modified for prehension and other purposes, with their feet and legs at the same time modified for firm support and progression, endless other changes of structure would have been necessary. The pelvis would have had to be made broader, the spine peculiarly curved and the head fixed in an altered position, and all these changes have been attained by man. Prof. Schaaffhausen[67] maintains that "the powerful mastoid processes of the human skull are the result of his erect position;" and these processes are absent in the orang, chimpanzee, &c., and are smaller in the gorilla than in man. Various other structures might here have been specified, which appear connected with man's erect position. It is very difficult to decide how far all these correlated modifications are the result of natural selection, and how far of the inherited effects of the increased use of certain parts, or of the action of one part on another. No doubt these means of change act and react on each other: thus when certain muscles, and the crests of bone to which they are attached, become enlarged by

---

67 "On the Primitive Form of the Skull," translated in 'Anthropological Review,' Oct. 1868, p. 428. Owen ('Anatomy of Vertebrates,' vol. ii. 1866, p. 551) on the mastoid processes in the higher apes.

habitual use, this shews that certain actions are habitually performed and must be serviceable. Hence the individuals which performed them best, would tend to survive in greater numbers.

The free use of the arms and hands, partly the cause and partly the result of man's erect position, appears to have led in an indirect manner to other modifications of structure. The early male progenitors of man were, as previously stated, probably furnished with great canine teeth; but as they gradually acquired the habit of using stones, clubs, or other weapons, for fighting with their enemies, they would have used their jaws and teeth less and less. In this case, the jaws, together with the teeth, would have become reduced in size, as we may feel sure from innumerable analogous cases. In a future chapter we shall meet with a closely-parallel case, in the reduction or complete disappearance of the canine teeth in male ruminants, apparently in relation with the development of their horns; and in horses, in relation with their habit of fighting with their incisor teeth and hoofs.

In the adult male anthropomorphous apes, as Rütimeyer,[68] and others have insisted, it is precisely the effect which the jaw-muscles by their great development have produced on the skull, that causes it to differ so greatly in many respects from that of man, and has given to it "a truly frightful physiognomy." Therefore as the jaws and teeth in the progenitors of man gradually become reduced in size, the adult skull would have presented nearly the same characters which it offers in the young of the anthropomorphous apes, and would thus have come to resemble more nearly that of existing

---

68 'Die Grenzen der Thierwelt, eine Betrachtung zu Darwin's Lehre,' 1868, s. 51.

man. A great reduction of the canine teeth in the males would almost certainly, as we shall hereafter see, have affected through inheritance the teeth of the females.

As the various mental faculties were gradually developed, the brain would almost certainly have become larger. No one, I presume, doubts that the large size of the brain in man, relatively to his body, in comparison with that of the gorilla or orang, is closely connected with his higher mental powers. We meet with closely analogous facts with insects, in which the cerebral ganglia are of extraordinary dimensions in ants; these ganglia in all the Hymenoptera being many times larger than in the less intelligent orders, such as beetles.[69] On the other hand, no one supposes that the intellect of any two animals or of any two men can be accurately gauged by the cubic contents of their skulls. It is certain that there may be extraordinary mental activity with an extremely small absolute mass of nervous matter: thus the wonderfully diversified instincts, mental powers, and affections of ants are generally known, yet their cerebral ganglia are not so large as the quarter of a small pin's head. Under this latter point of view, the brain of an ant is one of the most marvellous atoms of matter in the world, perhaps more marvellous than the brain of man.

The belief that there exists in man some close relation between the size of the brain and the development of the intellectual faculties is supported by the comparison of the skulls of savage and civilised races, of ancient and modern people, and by the analogy of the whole verte-

---

[69] Dujardin, 'Annales des Sc. Nat.' 3rd series, Zoolog. tom. xiv. 1850, p. 203. See also Mr. Lowne, 'Anatomy and Phys. of the *Musca vomitoria*,' 1870, p. 14. My son, Mr. F. Darwin, dissected for me the cerebral ganglia of the *Formica rufa*.

brate series. Dr. J. Barnard Davis has proved[70] by many careful measurements, that the mean internal capacity of the skull in Europeans is 92·3 cubic inches; in Americans 87·5; in Asiatics 87·1; and in Australians only 81·9 inches. Professor Broca[71] found that skulls from graves in Paris of the nineteenth century, were larger than those from vaults of the twelfth century, in the proportion of 1484 to 1426; and Prichard is persuaded that the present inhabitants of Britain have "much more capacious brain-cases" than the ancient inhabitants. Nevertheless it must be admitted that some skulls of very high antiquity, such as the famous one of Neanderthal, are well developed and capacious. With respect to the lower animals, M. E. Lartet,[72] by comparing the crania of tertiary and recent mammals, belonging to the same groups, has come to the remarkable conclusion that the brain is generally larger and the convolutions more complex in the more recent form. On the other hand I have shewn[73] that the brains of domestic rabbits are considerably reduced in bulk, in comparison with those of the wild rabbit or hare; and this may be attributed to their having been closely confined during many generations, so that they have exerted but little their intellect, instincts, senses, and voluntary movements.

The gradually increasing weight of the brain and skull in man must have influenced the development of the supporting spinal column, more especially whilst he was becoming erect. As this change of position was

---

70 'Philosophical Transactions,' 1869, p. 513.

71 Quoted in C. Vogt's 'Lectures on Man,' Eng. translat. 1864, p. 88, 90. Prichard, 'Phys. Hist. of Mankind,' vol. i. 1838, p. 305.

72 'Comptes Rendus des Séances,' &c. June 1, 1868.

73 'The Variation of Animals and Plants under Domestication,' vol. i. p. 124-129.

being brought about, the internal pressure of the brain, will, also, have influenced the form of the skull; for many facts shew how easily the skull is thus affected. Ethnologists believe that it is modified by the kind of cradle in which infants sleep. Habitual spasms of the muscles and a cicatrix from a severe burn have permanently modified the facial bones. In young persons whose heads from disease have become fixed either sideways or backwards, one of the eyes has changed its position, and the bones of the skull have been modified; and this apparently results from the brain pressing in a new direction.[74] I have shewn that with long-eared rabbits, even so trifling a cause as the lopping forward of one ear drags forward on that side almost every bone of the skull; so that the bones on the opposite sides no longer strictly correspond. Lastly, if any animal were to increase or diminish much in general size, without any change in its mental powers; or if the mental powers were to be much increased or diminished without any great change in the size of the body; the shape of the skull would almost certainly be altered. I infer this from my observations on domestic rabbits, some kinds of which have become very much larger than the wild animal, whilst others have retained nearly the same size, but in both cases the brain has been much reduced relatively to the size of the body. Now I was at first much surprised by finding that in all these rabbits the skull had become elongated or dolicho-

[74] Schaaffhausen gives from Blumenbach and Busch, the cases of the spasms and cicatrix, in 'Anthropolog. Review,' Oct. 1868, p. 420. Dr. Jarrold ('Anthropologia,' 1808, p. 115, 116) adduces from Camper and from his own observations, cases of the modification of the skull from the head being fixed in an unnatural position. He believes that certain trades, such as that of a shoemaker, by causing the head to be habitually held forward, makes the forehead more rounded and prominent.

cephalic; for instance, of two skulls of nearly equal breadth, the one from a wild rabbit and the other from a large domestic kind, the former was only 3·15 and the latter 4·3 inches in length.[75] One of the most marked distinctions in different races of man is that the skull in some is elongated, and in others rounded; and here the explanation suggested by the case of the rabbits may partially hold good; for Welcker finds that short "men incline more to brachycephaly, and tall men to dolichocephaly;"[76] and tall men may be compared with the larger and longer-bodied rabbits, all of which have elongated skulls, or are dolichocephalic.

From these several facts we can to a certain extent understand the means through which the great size and more or less rounded form of the skull has been acquired by man; and these are characters eminently distinctive of him in comparison with the lower animals.

Another most conspicuous difference between man and the lower animals is the nakedness of his skin. Whales and dolphins (Cetacea), dugongs (Sirenia) and the hippopotamus are naked; and this may be advantageous to them for gliding through the water; nor would it be injurious to them from the loss of warmth, as the species which inhabit the colder regions are protected by a thick layer of blubber, serving the same purpose as the fur of seals and otters. Elephants and rhinoceroses are almost hairless; and as certain extinct species which formerly lived under an arctic climate were covered with long wool or hair, it would almost appear as if the existing species of both genera had lost

---

[75] 'Variation of Animals,' &c., vol. i. p. 117 on the elongation of the skull; p. 119, on the effect of the lopping of one ear.

[76] Quoted by Schaaffhausen, in 'Anthropolog. Review,' Oct. 1868, p. 419.

their hairy covering from exposure to heat. This appears the more probable, as the elephants in India which live on elevated and cool districts are more hairy[77] than those on the lowlands. May we then infer that man became divested of hair from having aboriginally inhabited some tropical land? The fact of the hair being chiefly retained in the male sex on the chest and face, and in both sexes at the junction of all four limbs with the trunk, favours this inference, assuming that the hair was lost before man became erect; for the parts which now retain most hair would then have been most protected from the heat of the sun. The crown of the head, however, offers a curious exception, for at all times it must have been one of the most exposed parts, yet it is thickly clothed with hair. In this respect man agrees with the great majority of quadrupeds, which generally have their upper and exposed surfaces more thickly clothed than the lower surface. Nevertheless, the fact that the other members of the order of Primates, to which man belongs, although inhabiting various hot regions, are well clothed with hair, generally thickest on the upper surface,[78] is strongly opposed to the supposition that man became naked through the action of the sun. I am inclined to believe, as we shall see under sexual selection, that man, or rather primarily woman, became divested of hair for ornamental purposes; and according to this belief it is not

---

77 Owen, 'Anatomy of Vertebrates,' vol. iii. p. 619.

78 Isidore Geoffroy St.-Hilaire remarks ('Hist. Nat. Générale,' tom. ii. 1859, p. 215-217) on the head of man being covered with long hair; also on the upper surfaces of monkeys and of other mammals being more thickly clothed than the lower surfaces. This has likewise been observed by various authors. Prof. P. Gervais ('Hist. Nat. des Mammifères,' tom. i. 1854, p. 28), however, states that in the Gorilla the hair is thinner on the back, where it is partly rubbed off, than on the lower surface.

surprising that man should differ so greatly in hairiness from all his lower brethren, for characters gained through sexual selection often differ in closely-related forms to an extraordinary degree.

According to a popular impression, the absence of a tail is eminently distinctive of man; but as those apes which come nearest to man are destitute of this organ, its disappearance does not especially concern us. Nevertheless it may be well to own that no explanation, as far as I am aware, has ever been given of the loss of the tail by certain apes and man. Its loss, however, is not surprising, for it sometimes differs remarkably in length in species of the same genera: thus in some species of Macacus the tail is longer than the whole body, consisting of twenty-four vertebræ; in others it consists of a scarcely visible stump, containing only three or four vertebræ. In some kinds of baboons there are twenty-five, whilst in the mandrill there are ten very small stunted caudal vertebræ, or, according to Cuvier,[79] sometimes only five. This great diversity in the structure and length of the tail in animals belonging to the same genera, and following nearly the same habits of life, renders it probable that the tail is not of much importance to them; and if so, we might have expected that it would sometimes have become more or less rudimentary, in accordance with what we incessantly see with other structures. The tail almost always tapers towards the end whether it be long or short; and this, I presume, results from the atrophy, through disuse, of the terminal muscles together with their arteries and nerves, leading to the atrophy of the terminal bones. With respect

[79] Mr. St. George Mivart, 'Proc. Zoolog. Soc.' 1865, p. 562, 583. Dr. J. E. Gray, 'Cat. Brit. Mus.: Skeletons.' Owen, 'Anatomy of Vertebrates,' vol. ii. p. 517. Isidore Geoffroy, 'Hist. Nat. Gén.' tom. ii. p. 244.

to the os coccyx, which in man and the higher apes manifestly consists of the few basal and tapering segments of an ordinary tail, I have heard it asked how could these have become completely embedded within the body; but there is no difficulty in this respect, for in many monkeys the basal segments of the true tail are thus embedded. For instance, Mr. Murie informs me that in the skeleton of a not full-grown *Macacus inornatus*, he counted nine or ten caudal vertebræ, which altogether were only 1·8 inch in length. Of these the three basal ones appeared to have been embedded; the remainder forming the free part of the tail, which was only one inch in length, and half an inch in diameter. Here, then, the three embedded caudal vertebræ plainly correspond with the four coalesced vertebræ of the human os coccyx.

I have now endeavoured to shew that some of the most distinctive characters of man have in all probability been acquired, either directly, or more commonly indirectly, through natural selection. We should bear in mind that modifications in structure or constitution, which are of no service to an organism in adapting it to its habits of life, to the food which it consumes, or passively to the surrounding conditions, cannot have been thus acquired. We must not, however, be too confident in deciding what modifications are of service to each being: we should remember how little we know about the use of many parts, or what changes in the blood or tissues may serve to fit an organism for a new climate or some new kind of food. Nor must we forget the principle of correlation, by which, as Isidore Geoffroy has shewn in the case of man, many strange deviations of structure are tied together. Independently of correlation, a change in one part often leads through the increased or decreased use of other parts, to other changes of a quite unexpected nature. It is also well to reflect on such facts, as the wonderful growth of galls on plants caused by the poison of an insect, and on the remarkable changes of colour in the plumage of parrots when fed on certain fishes, or inoculated with the poison of toads;[80] for we can thus see that the fluids of the system, if altered for some special purpose, might induce other strange changes. We should especially bear in mind that modifications acquired and continually used during past ages for some useful purpose would probably become firmly fixed and might be long inherited.

Thus a very large yet undefined extension may safely be given to the direct and indirect results of natural selection; but I now admit, after reading the essay by Nägeli on plants, and the remarks by various authors with respect to animals, more especially those recently made by Professor Broca, that in the earlier editions of my 'Origin of Species' I probably attributed too much to the action of natural selection or the survival of the fittest. I have altered the fifth edition of the Origin so as to confine my remarks to adaptive changes of structure. I had not formerly sufficiently considered the existence of many structures which appear to be, as far as we can judge, neither beneficial nor injurious; and this I believe to be one of the greatest oversights as yet detected in my work. I may be permitted to say as some excuse, that I had two distinct objects in view, firstly, to shew that species had not been separately created, and secondly, that natural selection had been the chief agent of change, though largely aided by the

---

[80] 'The Variation of Animals and Plants under Domestication,' vol. ii. p. 280, 282.

inherited effects of habit, and slightly by the direct action of the surrounding conditions. Nevertheless I was not able to annul the influence of my former belief, then widely prevalent, that each species had been purposely created; and this led to my tacitly assuming that every detail of structure, excepting rudiments, was of some special, though unrecognised, service. Any one with this assumption in his mind would naturally extend the action of natural selection, either during past or present times, too far. Some of those who admit the principle of evolution, but reject natural selection, seem to forget, when criticising my book, that I had the above two objects in view; hence if I have erred in giving to natural selection great power, which I am far from admitting, or in having exaggerated its power, which is in itself probable, I have at least, as I hope, done good service in aiding to overthrow the dogma of separate creations.

That all organic beings, including man, present many modifications of structure which are of no service to them at present, nor have been formerly, is, as I can now see, probable. We know not what produces the numberless slight differences between the individuals of each species, for reversion only carries the problem a few steps backwards; but each peculiarity must have had its own efficient cause. If these causes, whatever they may be, were to act more uniformly and energetically during a lengthened period (and no reason can be assigned why this should not sometimes occur), the result would probably be not mere slight individual differences, but well-marked, constant modifications. Modifications which are in no way beneficial cannot have been kept uniform through natural selection, though any which were injurious would have been thus eliminated. Uniformity of character would, however, naturally follow from the assumed uniformity of the exciting causes, and likewise from the free intercrossing of many individuals. The same organism might acquire in this manner during successive periods successive modifications, and these would be transmitted in a nearly uniform state as long as the exciting causes remained the same and there was free intercrossing. With respect to the exciting causes we can only say, as when speaking of so-called spontaneous variations, that they relate much more closely to the constitution of the varying organism, than to the nature of the conditions to which it has been subjected.

Conclusion.—In this chapter we have seen that as man at the present day is liable, like every other animal, to multiform individual differences or slight variations, so no doubt were the early progenitors of man; the variations being then as now induced by the same general causes, and governed by the same general and complex laws. As all animals tend to multiply beyond their means of subsistence, so it must have been with the progenitors of man; and this will inevitably have led to a struggle for existence and to natural selection. This latter process will have been greatly aided by the inherited effects of the increased use of parts; these two processes incessantly reacting on each other. It appears, also, as we shall hereafter see, that various unimportant characters have been acquired by man through sexual selection. An unexplained residuum of change, perhaps a large one, must be left to the assumed uniform action of those unknown agencies, which occasionally induce strongly-marked and abrupt deviations of structure in our domestic productions.

Judging from the habits of savages and of the greater number of the Quadrumana, primeval men, and even the ape-like progenitors of man, probably lived in society. With strictly social animals, natural selection sometimes acts indirectly on the individual, through the preservation of variations which are beneficial only to the community. A community including a large number of well-endowed individuals increases in number and is victorious over other and less well-endowed communities; although each separate member may gain no advantage over the other members of the same community. With associated insects many remarkable structures, which are of little or no service to the individual or its own offspring, such as the pollen-collecting apparatus, or the sting of the worker-bee, or the great jaws of soldier-ants, have been thus acquired. With the higher social animals, I am not aware that any structure has been modified solely for the good of the community, though some are of secondary service to it. For instance, the horns of ruminants and the great canine teeth of baboons appear to have been acquired by the males as weapons for sexual strife, but they are used in defence of the herd or troop. In regard to certain mental faculties the case, as we shall see in the following chapter, is wholly different; for these faculties have been chiefly, or even exclusively, gained for the benefit of the community; the individuals composing the community being at the same time indirectly benefited.

It has often been objected to such views as the foregoing, that man is one of the most helpless and defenceless creatures in the world; and that during his early and less well-developed condition he would have been still more helpless. The Duke of Argyll, for instance, insists[81] that "the human frame has diverged from

[81] 'Primeval Man,' 1869, p. 66.

"the structure of brutes, in the direction of greater "physical helplessness and weakness. That is to say, "it is a divergence which of all others it is most "impossible to ascribe to mere natural selection." He adduces the naked and unprotected state of the body, the absence of great teeth or claws for defence, the little strength of man, his small speed in running, and his slight power of smell, by which to discover food or to avoid danger. To these deficiencies there might have been added the still more serious loss of the power of quickly climbing trees, so as to escape from enemies. Seeing that the unclothed Fuegians can exist under their wretched climate, the loss of hair would not have been a great injury to primeval man, if he inhabited a warm country. When we compare defenceless man with the apes, many of which are provided with formidable canine teeth, we must remember that these in their fully-developed condition are possessed by the males alone, being chiefly used by them for fighting with their rivals; yet the females which are not thus provided, are able to survive.

In regard to bodily size or strength, we do not know whether man is descended from some comparatively small species, like the chimpanzee, or from one as powerful as the gorilla; and, therefore, we cannot say whether man has become larger and stronger, or smaller and weaker, in comparison with his progenitors. We should, however, bear in mind that an animal possessing great size, strength, and ferocity, and which, like the gorilla, could defend itself from all enemies, would probably, though not necessarily, have failed to become social; and this would most effectually have checked the acquirement by man of his higher mental qualities, such as sympathy and the love of his fellow-creatures. Hence it might have been an immense advantage to man to have sprung from some comparatively weak creature.

The slight corporeal strength of man, his little speed, his want of natural weapons, &c., are more than counterbalanced, firstly by his intellectual powers, through which he has, whilst still remaining in a barbarous state, formed for himself weapons, tools, &c., and secondly by his social qualities which lead him to give aid to his fellow-men and to receive it in return. No country in the world abounds in a greater degree with dangerous beasts than Southern Africa; no country presents more fearful physical hardships than the Arctic regions; yet one of the puniest races, namely, the Bushmen, maintain themselves in Southern Africa, as do the dwarfed Esquimaux in the Arctic regions. The early progenitors of man were, no doubt, inferior in intellect, and probably in social disposition, to the lowest existing savages; but it is quite conceivable that they might have existed, or even flourished, if, whilst they gradually lost their brute-like powers, such as climbing trees, &c., they at the same time advanced in intellect. But granting that the progenitors of man were far more helpless and defenceless than any existing savages, if they had inhabited some warm continent or large island, such as Australia or New Guinea, or Borneo (the latter island being now tenanted by the orang), they would not have been exposed to any special danger. In an area as large as one of these islands, the competition between tribe and tribe would have been sufficient, under favourable conditions, to have raised man, through the survival of the fittest, combined with the inherited effects of habit, to his present high position in the organic scale.

CHAPTER V.

ON THE DEVELOPMENT OF THE INTELLECTUAL AND MORAL FACULTIES DURING PRIMEVAL AND CIVILISED TIMES.

The advancement of the intellectual powers through natural selection — Importance of imitation — Social and moral faculties — Their development within the limits of the same tribe—Natural selection as affecting civilised nations—Evidence that civilised nations were once barbarous.

THE subjects to be discussed in this chapter are of the highest interest, but are treated by me in a most imperfect and fragmentary manner. Mr. Wallace, in an admirable paper before referred to,[1] argues that man after he had partially acquired those intellectual and moral faculties which distinguish him from the lower animals, would have been but little liable to have had his bodily structure modified through natural selection or any other means. For man is enabled through his mental faculties "to keep with an un- "changed body in harmony with the changing universe." He has great power of adapting his habits to new conditions of life. He invents weapons, tools and various stratagems, by which he procures food and defends himself. When he migrates into a colder climate he uses clothes, builds sheds, and makes fires; and, by the aid of fire, cooks food otherwise indigestible. He aids his fellow-men in many ways, and anticipates future events. Even at a remote period he practised some subdivision of labour.

---

[1] 'Anthropological Review,' May, 1864, p. clviii.

The lower animals, on the other hand, must have their bodily structure modified in order to survive under greatly changed conditions. They must be rendered stronger, or acquire more effective teeth or claws, in order to defend themselves from new enemies; or they must be reduced in size so as to escape detection and danger. When they migrate into a colder climate they must become clothed with thicker fur, or have their constitutions altered. If they fail to be thus modified, they will cease to exist.

The case, however, is widely different, as Mr. Wallace has with justice insisted, in relation to the intellectual and moral faculties of man. These faculties are variable; and we have every reason to believe that the variations tend to be inherited. Therefore, if they were formerly of high importance to primeval man and to his ape-like progenitors, they would have been perfected or advanced through natural selection. Of the high importance of the intellectual faculties there can be no doubt, for man mainly owes to them his pre-eminent position in the world. We can see that, in the rudest state of society, the individuals who were the most sagacious, who invented and used the best weapons or traps, and who were best able to defend themselves, would rear the greatest number of offspring. The tribes which included the largest number of men thus endowed would increase in number and supplant other tribes. Numbers depend primarily on the means of subsistence, and this, partly on the physical nature of the country, but in a much higher degree on the arts which are there practised. As a tribe increases and is victorious, it is often still further increased by the absorption of other tribes.[2] The stature and strength of the men of a tribe

[2] After a time the members or tribes which are absorbed into another tribe assume, as Mr. Maine remarks ('Ancient Law,' 1861, p. 131), that they are the co-descendants of the same ancestors.

are likewise of some importance for its success, and these depend in part on the nature and amount of the food which can be obtained. In Europe the men of the Bronze period were supplanted by a more powerful and, judging from their sword-handles, larger-handed race;[3] but their success was probably due in a much higher degree to their superiority in the arts.

All that we know about savages, or may infer from their traditions and from old monuments, the history of which is quite forgotten by the present inhabitants, shew that from the remotest times successful tribes have supplanted other tribes. Relics of extinct or forgotten tribes have been discovered throughout the civilised regions of the earth, on the wild plains of America, and on the isolated islands in the Pacific Ocean. At the present day civilised nations are everywhere supplanting barbarous nations, excepting where the climate opposes a deadly barrier; and they succeed mainly, though not exclusively, through their arts, which are the products of the intellect. It is, therefore, highly probable that with mankind the intellectual faculties have been gradually perfected through natural selection; and this conclusion is sufficient for our purpose. Undoubtedly it would have been very interesting to have traced the development of each separate faculty from the state in which it exists in the lower animals to that in which it exists in man; but neither my ability nor knowledge permit the attempt.

It deserves notice that as soon as the progenitors of man became social (and this probably occurred at a very early period), the advancement of the intellectual faculties will have been aided and modified in an important manner, of which we see only traces in

---

[3] Morlot, 'Soc. Vaud. Sc. Nat.' 1860, p. 294.

the lower animals, namely, through the principle of imitation, together with reason and experience. Apes are much given to imitation, as are the lowest savages; and the simple fact previously referred to, that after a time no animal can be caught in the same place by the same sort of trap, shews that animals learn by experience, and imitate each others' caution. Now, if some one man in a tribe, more sagacious than the others, invented a new snare or weapon, or other means of attack or defence, the plainest self-interest, without the assistance of much reasoning power, would prompt the other members to imitate him; and all would thus profit. The habitual practice of each new art must likewise in some slight degree strengthen the intellect. If the new invention were an important one, the tribe would increase in number, spread, and supplant other tribes. In a tribe thus rendered more numerous there would always be a rather better chance of the birth of other superior and inventive members. If such men left children to inherit their mental superiority, the chance of the birth of still more ingenious members would be somewhat better, and in a very small tribe decidedly better. Even if they left no children, the tribe would still include their blood-relations; and it has been ascertained by agriculturists[4] that by preserving and breeding from the family of an animal, which when slaughtered was found to be valuable, the desired character has been obtained.

Turning now to the social and moral faculties. In order that primeval men, or the ape-like progenitors of man, should have become social, they must have

---

[4] I have given instances in my 'Variation of Animals under Domestication,' vol. ii. p. 196.

acquired the same instinctive feelings which impel other animals to live in a body; and they no doubt exhibited the same general disposition. They would have felt uneasy when separated from their comrades, for whom they would have felt some degree of love; they would have warned each other of danger, and have given mutual aid in attack or defence. All this implies some degree of sympathy, fidelity, and courage. Such social qualities, the paramount importance of which to the lower animals is disputed by no one, were no doubt acquired by the progenitors of man in a similar manner, namely, through natural selection, aided by inherited habit. When two tribes of primeval man, living in the same country, came into competition, if the one tribe included (other circumstances being equal) a greater number of courageous, sympathetic, and faithful members, who were always ready to warn each other of danger, to aid and defend each other, this tribe would without doubt succeed best and conquer the other. Let it be borne in mind how all-important, in the never-ceasing wars of savages, fidelity and courage must be. The advantage which disciplined soldiers have over undisciplined hordes follows chiefly from the confidence which each man feels in his comrades. Obedience, as Mr. Bagehot has well shewn,[5] is of the highest value, for any form of government is better than none. Selfish and contentious people will not cohere, and without coherence nothing can be effected. A tribe possessing the above qualities in a high degree would spread and be victorious over other tribes; but in the course of time it would, judging from all past history, be in its turn overcome by some other

---

[5] See a remarkable series of articles on Physics and Politics in the 'Fortnightly Review,' Nov. 1867; April 1, 1868; July 1, 1869.

and still more highly endowed tribe. Thus the social and moral qualities would tend slowly to advance and be diffused throughout the world.

But it may be asked, how within the limits of the same tribe did a large number of members first become endowed with these social and moral qualities, and how was the standard of excellence raised? It is extremely doubtful whether the offspring of the more sympathetic and benevolent parents, or of those which were the most faithful to their comrades, would be reared in greater number than the children of selfish and treacherous parents of the same tribe. He who was ready to sacrifice his life, as many a savage has been, rather than betray his comrades, would often leave no offspring to inherit his noble nature. The bravest men, who were always willing to come to the front in war, and who freely risked their lives for others, would on an average perish in larger number than other men. Therefore it seems scarcely possible (bearing in mind that we are not here speaking of one tribe being victorious over another) that the number of men gifted with such virtues, or that the standard of their excellence, could be increased through natural selection, that is, by the survival of the fittest.

Although the circumstances which lead to an increase in the number of men thus endowed within the same tribe are too complex to be clearly followed out, we can trace some of the probable steps. In the first place, as the reasoning powers and foresight of the members became improved, each man would soon learn from experience that if he aided his fellow-men, he would commonly receive aid in return. From this low motive he might acquire the habit of aiding his fellows; and the habit of performing benevolent actions certainly strengthens the feeling of sympathy, which gives the first impulse to benevolent actions. Habits, moreover, followed during many generations probably tend to be inherited.

But there is another and much more powerful stimulus to the development of the social virtues, namely, the praise and the blame of our fellow-men. The love of approbation and the dread of infamy, as well as the bestowal of praise or blame, are primarily due, as we have seen in the third chapter, to the instinct of sympathy; and this instinct no doubt was originally acquired, like all the other social instincts, through natural selection. At how early a period the progenitors of man, in the course of their development, became capable of feeling and being impelled by the praise or blame of their fellow-creatures, we cannot, of course, say. But it appears that even dogs appreciate encouragement, praise, and blame. The rudest savages feel the sentiment of glory, as they clearly show by preserving the trophies of their prowess, by their habit of excessive boasting, and even by the extreme care which they take of their personal appearance and decorations; for unless they regarded the opinion of their comrades, such habits would be senseless.

They certainly feel shame at the breach of some of their lesser rules; but how far they experience remorse is doubtful. I was at first surprised that I could not recollect any recorded instances of this feeling in savages; and Sir J. Lubbock[6] states that he knows of none. But if we banish from our minds all cases given in novels and plays and in death-bed confessions made to priests, I doubt whether many of us have actually witnessed remorse; though we may have often seen shame and contrition for smaller offences. Remorse is

---

[6] 'Origin of Civilisation,' 1870, p. 265.

a deeply hidden feeling. It is incredible that a savage, who will sacrifice his life rather than betray his tribe, or one who will deliver himself up as a prisoner rather than break his parole,[7] would not feel remorse in his inmost soul, though he might conceal it, if he had failed in a duty which he held sacred.

We may therefore conclude that primeval man, at a very remote period, would have been influenced by the praise and blame of his fellows. It is obvious, that the members of the same tribe would approve of conduct which appeared to them to be for the general good, and would reprobate that which appeared evil. To do good unto others—to do unto others as ye would they should do unto you,—is the foundation-stone of morality. It is, therefore, hardly possible to exaggerate the importance during rude times of the love of praise and the dread of blame. A man who was not impelled by any deep, instinctive feeling, to sacrifice his life for the good of others, yet was roused to such actions by a sense of glory, would by his example excite the same wish for glory in other men, and would strengthen by exercise the noble feeling of admiration. He might thus do far more good to his tribe than by begetting offspring with a tendency to inherit his own high character.

With increased experience and reason, man perceives the more remote consequences of his actions, and the self-regarding virtues, such as temperance, chastity, &c., which during early times are, as we have before seen, utterly disregarded, come to be highly esteemed or even held sacred. I need not, however, repeat what I have said on this head in the third chapter. Ultimately a highly complex sentiment, having its first origin in the

---

[7] Mr. Wallace gives cases in his 'Contributions to the Theory of Natural Selection,' 1870, p. 354.

social instincts, largely guided by the approbation of our fellow-men, ruled by reason, self-interest, and in later times by deep religious feelings, confirmed by instruction and habit, all combined, constitute our moral sense or conscience.

It must not be forgotten that although a high standard of morality gives but a slight or no advantage to each individual man and his children over the other men of the same tribe, yet that an advancement in the standard of morality and an increase in the number of well-endowed men will certainly give an immense advantage to one tribe over another. There can be no doubt that a tribe including many members who, from possessing in a high degree the spirit of patriotism, fidelity, obedience, courage, and sympathy, were always ready to give aid to each other and to sacrifice themselves for the common good, would be victorious over most other tribes; and this would be natural selection. At all times throughout the world tribes have supplanted other tribes; and as morality is one element in their success, the standard of morality and the number of well-endowed men will thus everywhere tend to rise and increase.

It is, however, very difficult to form any judgment why one particular tribe and not another has been successful and has risen in the scale of civilisation. Many savages are in the same condition as when first discovered several centuries ago. As Mr. Bagehot has remarked, we are apt to look at progress as the normal rule in human society; but history refutes this. The ancients did not even entertain the idea; nor do the oriental nations at the present day. According to another high authority, Mr. Maine,[8] "the greatest part of mankind has never

[8] 'Ancient Law,' 1861, p. 22. For Mr. Bagehot's remarks, 'Fortnightly Review,' April 1, 1868, p. 452.

"shewn a particle of desire that its civil institutions "should be improved." Progress seems to depend on many concurrent favourable conditions, far too complex to be followed out. But it has often been remarked, that a cool climate from leading to industry and the various arts has been highly favourable, or even indispensable for this end. The Esquimaux, pressed by hard necessity, have succeeded in many ingenious inventions, but their climate has been too severe for continued progress. Nomadic habits, whether over wide plains, or through the dense forests of the tropics, or along the shores of the sea, have in every case been highly detrimental. Whilst observing the barbarous inhabitants of Tierra del Fuego, it struck me that the possession of some property, a fixed abode, and the union of many families under a chief, were the indispensable requisites for civilisation. Such habits almost necessitate the cultivation of the ground; and the first steps in cultivation would probably result, as I have elsewhere shewn,[9] from some such accident as the seeds of a fruit-tree falling on a heap of refuse and producing an unusually fine variety. The problem, however, of the first advance of savages towards civilisation is at present much too difficult to be solved.

Natural Selection as affecting Civilised Nations.—In the last and present chapters I have considered the advancement of man from a former semi-human condition to his present state as a barbarian. But some remarks on the agency of natural selection on civilised nations may be here worth adding. This subject has been ably discussed by Mr. W. R. Greg,[10] and previously

[9] 'The Variation of Animals and Plants under Domestication,' vol. i. p. 309.

[10] 'Fraser's Magazine,' Sept. 1868, p. 353. This article seems to have struck many persons, and has given rise to two remarkable essays

by Mr. Wallace and Mr. Galton.[11] Most of my remarks are taken from these three authors. With savages, the weak in body or mind are soon eliminated; and those that survive commonly exhibit a vigorous state of health. We civilised men, on the other hand, do our utmost to check the process of elimination; we build asylums for the imbecile, the maimed, and the sick; we institute poor-laws; and our medical men exert their utmost skill to save the life of every one to the last moment. There is reason to believe that vaccination has preserved thousands, who from a weak constitution would formerly have succumbed to small-pox. Thus the weak members of civilised societies propagate their kind. No one who has attended to the breeding of domestic animals will doubt that this must be highly injurious to the race of man. It is surprising how soon a want of care, or care wrongly directed, leads to the degeneration of a domestic race; but excepting in the case of man himself, hardly any one is so ignorant as to allow his worst animals to breed.

The aid which we feel impelled to give to the helpless is mainly an incidental result of the instinct of sympathy, which was originally acquired as part of the social instincts, but subsequently rendered, in the manner previously indicated, more tender and more widely diffused. Nor could we check our sympathy, if so urged by hard reason, without deterioration in the

---

and a rejoinder in the 'Spectator,' Oct. 3rd and 17th 1868. It has also been discussed in the 'Q. Journal of Science,' 1869, p. 152, and by Mr. Lawson Tait in the 'Dublin Q. Journal of Medical Science,' Feb. 1869, and by Mr. E. Ray Lankester in his 'Comparative Longevity,' 1870, p. 128. Similar views appeared previously in the 'Australasian,' July 13, 1867. I have borrowed ideas from several of these writers.

[11] For Mr. Wallace, see 'Anthropolog. Review,' as before cited. Mr. Galton in 'Macmillan's Magazine,' Aug. 1865, p. 318; also his great work, 'Hereditary Genius,' 1870.

noblest part of our nature. The surgeon may harden himself whilst performing an operation, for he knows that he is acting for the good of his patient; but if we were intentionally to neglect the weak and helpless, it could only be for a contingent benefit, with a certain and great present evil. Hence we must bear without complaining the undoubtedly bad effects of the weak surviving and propagating their kind; but there appears to be at least one check in steady action, namely the weaker and inferior members of society not marrying so freely as the sound; and this check might be indefinitely increased, though this is more to be hoped for than expected, by the weak in body or mind refraining from marriage.

In all civilised countries man accumulates property and bequeaths it to his children. So that the children in the same country do not by any means start fair in the race for success. But this is far from an unmixed evil; for without the accumulation of capital the arts could not progress; and it is chiefly through their power that the civilised races have extended, and are now everywhere extending, their range, so as to take the place of the lower races. Nor does the moderate accumulation of wealth interfere with the process of selection. When a poor man becomes rich, his children enter trades or professions in which there is struggle enough, so that the able in body and mind succeed best. The presence of a body of well-instructed men, who have not to labour for their daily bread, is important to a degree which cannot be over-estimated; as all high intellectual work is carried on by them, and on such work material progress of all kinds mainly depends, not to mention other and higher advantages. No doubt wealth when very great tends to convert men into useless drones, but their number is never large; and some degree of elimination here occurs, as we daily see rich men, who happen to be fools or profligate, squandering away all their wealth.

Primogeniture with entailed estates is a more direct evil, though it may formerly have been a great advantage by the creation of a dominant class, and any government is better than anarchy. The eldest sons, though they may be weak in body or mind, generally marry, whilst the younger sons, however superior in these respects, do not so generally marry. Nor can worthless eldest sons with entailed estates squander their wealth. But here, as elsewhere, the relations of civilised life are so complex that some compensatory checks intervene. The men who are rich through primogeniture are able to select generation after generation the more beautiful and charming women; and these must generally be healthy in body and active in mind. The evil consequences, such as they may be, of the continued preservation of the same line of descent, without any selection, are checked by men of rank always wishing to increase their wealth and power; and this they effect by marrying heiresses. But the daughters of parents who have produced single children, are themselves, as Mr. Galton has shewn,[12] apt to be sterile; and thus noble families are continually cut off in the direct line, and their wealth flows into some side channel; but unfortunately this channel is not determined by superiority of any kind.

Although civilisation thus checks in many ways the action of natural selection, it apparently favours, by means of improved food and the freedom from occasional hardships, the better development of the body. This may be inferred from civilised men having been

---

[12] 'Hereditary Genius,' 1870, p. 132-140.

found, wherever compared, to be physically stronger than savages. They appear also to have equal powers of endurance, as has been proved in many adventurous expeditions. Even the great luxury of the rich can be but little detrimental; for the expectation of life of our aristocracy, at all ages and of both sexes, is very little inferior to that of healthy English lives in the lower classes.[13]

We will now look to the intellectual faculties alone. If in each grade of society the members were divided into two equal bodies, the one including the intellectually superior and the other the inferior, there can be little doubt that the former would succeed best in all occupations and rear a greater number of children. Even in the lowest walks of life, skill and ability must be of some advantage, though in many occupations, owing to the great division of labour, a very small one. Hence in civilised nations there will be some tendency to an increase both in the number and in the standard of the intellectually able. But I do not wish to assert that this tendency may not be more than counterbalanced in other ways, as by the multiplication of the reckless and improvident; but even to such as these, ability must be some advantage.

It has often been objected to views like the foregoing, that the most eminent men who have ever lived have left no offspring to inherit their great intellect. Mr. Galton says,[14] "I regret I am unable to solve the " simple question whether, and how far, men and women " who are prodigies of genius are infertile. I have, how- " ever, shewn that men of eminence are by no means so."

[13] See the fifth and sixth columns, compiled from good authorities, in the table given in Mr. E. R. Lankester's 'Comparative Longevity,' 1870, p. 115.

[14] 'Hereditary Genius,' 1870, p. 330.

Great lawgivers, the founders of beneficent religions, great philosophers and discoverers in science, aid the progress of mankind in a far higher degree by their works than by leaving a numerous progeny. In the case of corporeal structures, it is the selection of the slightly better-endowed and the elimination of the slightly less well-endowed individuals, and not the preservation of strongly-marked and rare anomalies, that leads to the advancement of a species.[15] So it will be with the intellectual faculties, namely from the somewhat more able men in each grade of society succeeding rather better than the less able, and consequently increasing in number, if not otherwise prevented. When in any nation the standard of intellect and the number of intellectual men have increased, we may expect from the law of the deviation from an average, as shewn by Mr. Galton, that prodigies of genius will appear somewhat more frequently than before.

In regard to the moral qualities, some elimination of the worst dispositions is always in progress even in the most civilised nations. Malefactors are executed, or imprisoned for long periods, so that they cannot freely transmit their bad qualities. Melancholic and insane persons are confined, or commit suicide. Violent and quarrelsome men often come to a bloody end. Restless men who will not follow any steady occupation—and this relic of barbarism is a great check to civilisation[16]—emigrate to newly-settled countries, where they prove useful pioneers. Intemperance is so highly destructive, that the expectation of life of the intemperate, at the age, for instance, of thirty, is only 13·8 years; whilst for the rural labourers of England at the same age it is

---

[15] 'Origin of Species' (fifth edition, 1869), p. 104.
[16] 'Hereditary Genius,' 1870, p. 347.

40·59 years.[17] Profligate women bear few children, and profligate men rarely marry; both suffer from disease. In the breeding of domestic animals, the elimination of those individuals, though few in number, which are in any marked manner inferior, is by no means an unimportant element towards success. This especially holds good with injurious characters which tend to reappear through reversion, such as blackness in sheep; and with mankind some of the worst dispositions, which occasionally without any assignable cause make their appearance in families, may perhaps be reversions to a savage state, from which we are not removed by very many generations. This view seems indeed recognised in the common expression that such men are the black sheep of the family.

With civilised nations, as far as an advanced standard of morality, and an increased number of fairly well-endowed men are concerned, natural selection apparently effects but little; though the fundamental social instincts were originally thus gained. But I have already said enough, whilst treating of the lower races, on the causes which lead to the advance of morality, namely, the approbation of our fellow-men—the strengthening of our sympathies by habit—example and imitation—reason—experience and even self-interest—instruction during youth, and religious feelings.

A most important obstacle in civilised countries to an increase in the number of men of a superior class has been strongly urged by Mr. Greg and Mr. Galton,[18]

---

[17] E. Ray Lankester, 'Comparative Longevity,' 1870, p. 115. The table of the intemperate is from Neison's 'Vital Statistics.' In regard to profligacy, see Dr. Farr, "Influence of Marriage on Mortality," 'Nat. Assoc. for the Promotion of Social Science,' 1858.

[18] 'Fraser's Magazine,' Sept. 1868, p. 353. 'Macmillan's Magazine,' Aug. 1865, p. 318. The Rev. F. W. Farrar ('Fraser's Mag.,' Aug. 1870, p. 264) takes a different view.

namely, the fact that the very poor and reckless, who are often degraded by vice, almost invariably marry early, whilst the careful and frugal, who are generally otherwise virtuous, marry late in life, so that they may be able to support themselves and their children in comfort. Those who marry early produce within a given period not only a greater number of generations, but, as shewn by Dr. Duncan,[19] they produce many more children. The children, moreover, that are born by mothers during the prime of life are heavier and larger, and therefore probably more vigorous, than those born at other periods. Thus the reckless, degraded, and often vicious members of society, tend to increase at a quicker rate than the provident and generally virtuous members. Or as Mr. Greg puts the case: "The care- "less, squalid, unaspiring Irishman multiplies like "rabbits: the frugal, foreseeing, self-respecting, am- "bitious Scot, stern in his morality, spiritual in his "faith, sagacious and disciplined in his intelligence, "passes his best years in struggle and in celibacy, "marries late, and leaves few behind him. Given a "land originally peopled by a thousand Saxons and a "thousand Celts—and in a dozen generations five-sixths "of the population would be Celts, but five-sixths of "the property, of the power, of the intellect, would "belong to the one-sixth of Saxons that remained. "In the eternal 'struggle for existence,' it would be "the inferior and *less* favoured race that had prevailed "—and prevailed by virtue not of its good qualities "but of its faults."

There are, however, some checks to this downward tendency. We have seen that the intemperate suffer

[19] "On the Laws of the Fertility of Women," in 'Transact. Royal Soc.' Edinburgh, vol. xxiv. p. 287. See, also, Mr. Galton, 'Hereditary Genius,' p. 352-357, for observations to the above effect.

from a high rate of mortality, and the extremely profligate leave few offspring. The poorest classes crowd into towns, and it has been proved by Dr. Stark from the statistics of ten years in Scotland,[20] that at all ages the death-rate is higher in towns than in rural districts, "and during the first five years of life the town death-"rate is almost exactly double that of the rural districts." As these returns include both the rich and the poor, no doubt more than double the number of births would be requisite to keep up the number of the very poor inhabitants in the towns, relatively to those in the country. With women, marriage at too early an age is highly injurious; for it has been found in France that, "twice "as many wives under twenty die in the year, as died out "of the same number of the unmarried." The mortality, also, of husbands under twenty is "excessively high,"[21] but what the cause of this may be seems doubtful. Lastly, if the men who prudently delay marrying until they can bring up their families in comfort, were to select, as they often do, women in the prime of life, the rate of increase in the better class would be only slightly lessened.

It was established from an enormous body of statistics, taken during 1853, that the unmarried men throughout France, between the ages of twenty and eighty, die in a much larger proportion than the married: for instance, out of every 1000 unmarried men, between the ages of twenty and thirty, 11·3 annually died, whilst of the married only 6·5 died.[22] A similar law was proved to

---

[20] 'Tenth Annual Report of Births, Deaths, &c., in Scotland,' 1867, p. xxix.

[21] These quotations are taken from our highest authority on such questions, namely, Dr. Farr, in his paper "On the Influence of Marriage on the Mortality of the French People," read before the Nat. Assoc. for the Promotion of Social Science, 1858.

[22] Dr. Farr, ibid. The quotations given below are extracted from the same striking paper.

hold good, during the years 1863 and 1864, with the entire population above the age of twenty in Scotland: for instance, out of every 1000 unmarried men, between the ages of twenty and thirty, 14·97 annually died, whilst of the married only 7·24 died, that is less than half.[23] Dr. Stark remarks on this, "Bachelorhood is "more destructive to life than the most unwholesome "trades, or than residence in an unwholesome house or "district where there has never been the most distant "attempt at sanitary improvement." He considers that the lessened mortality is the direct result of "marriage, "and the more regular domestic habits which attend that "state." He admits, however, that the intemperate, profligate, and criminal classes, whose duration of life is low, do not commonly marry; and it must likewise be admitted that men with a weak constitution, ill health, or any great infirmity in body or mind, will often not wish to marry, or will be rejected. Dr. Stark seems to have come to the conclusion that marriage in itself is a main cause of prolonged life, from finding that aged married men still have a considerable advantage in this respect over the unmarried of the same advanced age; but every one must have known instances of men, who with weak health during youth did not marry, and yet have survived to old age, though remaining weak and therefore always with a lessened chance of life. There is another remarkable circumstance which seems to support Dr. Stark's conclusion, namely, that widows and widowers in France suffer in comparison with the married a very heavy rate of mortality; but Dr. Farr attributes this to the poverty and

---

[23] I have taken the mean of the quinquennial means, given in 'The Tenth Annual Report of Births, Deaths, &c., in Scotland,' 1867. The quotation from Dr. Stark is copied from an article in the 'Daily News,' Oct. 17th, 1868, which Dr. Farr considers very carefully written.

evil habits consequent on the disruption of the family, and to grief. On the whole we may conclude with Dr. Farr that the lesser mortality of married than of unmarried men, which seems to be a general law, "is mainly "due to the constant elimination of imperfect types, and "to the skilful selection of the finest individuals out of "each successive generation;" the selection relating only to the marriage state, and acting on all corporeal, intellectual, and moral qualities. We may, therefore, infer that sound and good men who out of prudence remain for a time unmarried do not suffer a high rate of mortality.

If the various checks specified in the two last paragraphs, and perhaps others as yet unknown, do not prevent the reckless, the vicious and otherwise inferior members of society from increasing at a quicker rate than the better class of men, the nation will retrograde, as has occurred too often in the history of the world. We must remember that progress is no invariable rule. It is most difficult to say why one civilised nation rises, becomes more powerful, and spreads more widely, than another; or why the same nation progresses more at one time than at another. We can only say that it depends on an increase in the actual number of the population, on the number of the men endowed with high intellectual and moral faculties, as well as on their standard of excellence. Corporeal structure, except so far as vigour of body leads to vigour of mind, appears to have little influence.

It has been urged by several writers that as high intellectual powers are advantageous to a nation, the old Greeks, who stood some grades higher in intellect than any race that has ever existed,[24] ought to have

[24] See the ingenious and original argument on this subject by Mr. Galton, 'Hereditary Genius,' p. 340-342.

risen, if the power of natural selection were real, still higher in the scale, increased in number, and stocked the whole of Europe. Here we have the tacit assumption, so often made with respect to corporeal structures, that there is some innate tendency towards continued development in mind and body. But development of all kinds depends on many concurrent favourable circumstances. Natural selection acts only in a tentative manner. Individuals and races may have acquired certain indisputable advantages, and yet have perished from failing in other characters. The Greeks may have retrograded from a want of coherence between the many small states, from the small size of their whole country, from the practice of slavery, or from extreme sensuality; for they did not succumb until "they were enervated "and corrupt to the very core."[25] The western nations of Europe, who now so immeasurably surpass their former savage progenitors and stand at the summit of civilisation, owe little or none of their superiority to direct inheritance from the old Greeks; though they owe much to the written works of this wonderful people.

Who can positively say why the Spanish nation, so dominant at one time, has been distanced in the race. The awakening of the nations of Europe from the dark ages is a still more perplexing problem. At this early period, as Mr. Galton[26] has remarked, almost all the men of a gentle nature, those given to meditation or culture of the mind, had no refuge except in the bosom of the Church which demanded celibacy;

---

[25] Mr. Greg, 'Fraser's Magazine,' Sept. 1868, p. 357.

[26] 'Hereditary Genius,' 1870, p. 357-359. The Rev. F. H. Farrar ('Fraser's Mag.', Aug. 1870, p. 257) advances arguments on the other side. Sir C. Lyell had already ('Principles of Geology,' vol. ii. 1868, p. 489) called attention, in a striking passage, to the evil influence of the Holy Inquisition in having lowered, through selection, the general standard of intelligence in Europe.

and this could hardly fail to have had a deteriorating influence on each successive generation. During this same period the Holy Inquisition selected with extreme care the freest and boldest men in order to burn or imprison them. In Spain alone some of the best men—those who doubted and questioned, and without doubting there can be no progress—were eliminated during three centuries at the rate of a thousand a year. The evil which the Catholic Church has thus effected, though no doubt counterbalanced to a certain, perhaps large extent in other ways, is incalculable; nevertheless, Europe has progressed at an unparalleled rate.

The remarkable success of the English as colonists over other European nations, which is well illustrated by comparing the progress of the Canadians of English and French extraction, has been ascribed to their "daring "and persistent energy;" but who can say how the English gained their energy. There is apparently much truth in the belief that the wonderful progress of the United States, as well as the character of the people, are the results of natural selection; the more energetic, restless, and courageous men from all parts of Europe having emigrated during the last ten or twelve generations to that great country, and having there succeeded best.[27] Looking to the distant future, I do not think that the Rev. Mr. Zincke takes an exaggerated view when he says:[28] "All other series of events—as "that which resulted in the culture of mind in Greece, "and that which resulted in the empire of Rome—only "appear to have purpose and value when viewed in "connection with, or rather as subsidiary to . . . . the "great stream of Anglo-Saxon emigration to the west."

---

[27] Mr. Galton, 'Macmillan's Magazine,' August, 1865, p. 325. See, also, 'Nature,' "On Darwinism and National Life," Dec. 1869, p. 184.

[28] 'Last Winter in the United States,' 1868, p. 29.

Obscure as is the problem of the advance of civilisation, we can at least see that a nation which produced during a lengthened period the greatest number of highly intellectual, energetic, brave, patriotic, and benevolent men, would generally prevail over less favoured nations.

Natural selection follows from the struggle for existence; and this from a rapid rate of increase. It is impossible not bitterly to regret, but whether wisely is another question, the rate at which man tends to increase; for this leads in barbarous tribes to infanticide and many other evils, and in civilised nations to abject poverty, celibacy, and to the late marriages of the prudent. But as man suffers from the same physical evils with the lower animals, he has no right to expect an immunity from the evils consequent on the struggle for existence. Had he not been subjected to natural selection, assuredly he would never have attained to the rank of manhood. When we see in many parts of the world enormous areas of the most fertile land peopled by a few wandering savages, but which are capable of supporting numerous happy homes, it might be argued that the struggle for existence had not been sufficiently severe to force man upwards to his highest standard. Judging from all that we know of man and the lower animals, there has always been sufficient variability in the intellectual and moral faculties, for their steady advancement through natural selection. No doubt such advancement demands many favourable concurrent circumstances; but it may well be doubted whether the most favourable would have sufficed, had not the rate of increase been rapid, and the consequent struggle for existence severe to an extreme degree.

On the evidence that all civilised nations were once barbarous.—As we have had to consider the steps by which some semi-human creature has been gradually raised to the rank of man in his most perfect state, the present subject cannot be quite passed over. But it has been treated in so full and admirable a manner by Sir J. Lubbock,[29] Mr. Tylor, Mr. M'Lennan, and others, that I need here give only the briefest summary of their results. The arguments recently advanced by the Duke of Argyll[30] and formerly by Archbishop Whately, in favour of the belief that man came into the world as a civilised being and that all savages have since undergone degradation, seem to me weak in comparison with those advanced on the other side. Many nations, no doubt, have fallen away in civilisation, and some may have lapsed into utter barbarism, though on this latter head I have not met with any evidence. The Fuegians were probably compelled by other conquering hordes to settle in their inhospitable country, and they may have become in consequence somewhat more degraded; but it would be difficult to prove that they have fallen much below the Botocudos who inhabit the finest parts of Brazil.

The evidence that all civilised nations are the descendants of barbarians, consists, on the one side, of clear traces of their former low condition in still-existing customs, beliefs, language, &c.; and on the other side, of proofs that savages are independently able to raise themselves a few steps in the scale of civilisation, and have actually thus risen. The evidence on the first head is extremely curious, but cannot be here given: I refer to such cases as that, for instance, of the art of enumeration, which, as Mr. Tylor clearly shews by the words still used in some places, originated in counting

---

29 'On the Origin of Civilisation,' 'Proc. Ethnological Soc.' Nov. 26, 1867.

30 'Primeval Man,' 1869.

the fingers, first of one hand and then of the other, and lastly of the toes. We have traces of this in our own decimal system, and in the Roman numerals, which after reaching to the number V., change into VI., &c., when the other hand no doubt was used. So again, "when we speak of three-score and ten, we are counting by the vigesimal system, each score thus ideally "made, standing for 20—for 'one man' as a Mexican "or Carib would put it."[31] According to a large and increasing school of philologists, every language bears the marks of its slow and gradual evolution. So it is with the art of writing, as letters are rudiments of pictorial representations. It is hardly possible to read Mr. M'Lennan's work[32] and not admit that almost all civilised nations still retain some traces of such rude habits as the forcible capture of wives. What ancient nation, as the same author asks, can be named that was originally monogamous? The primitive idea of justice, as shewn by the law of battle and other customs of which traces still remain, was likewise most rude. Many existing superstitions are the remnants of former false religious beliefs. The highest form of religion—the grand idea of God hating sin and loving righteousness—was unknown during primeval times.

Turning to the other kind of evidence: Sir J. Lubbock has shewn that some savages have recently improved a little in some of their simpler arts. From the

---

[31] 'Royal Institution of Great Britain,' March 15, 1867. Also, 'Researches into the Early History of Mankind,' 1865.

[32] 'Primitive Marriage,' 1865. See, likewise, an excellent article, evidently by the same author, in the 'North British Review,' July, 1869. Also, Mr. L. H. Morgan, "A Conjectural Solution of the Origin of the Class. System of Relationship," in 'Proc. American Acad. of Sciences,' vol. vii. Feb. 1868. Prof. Schaaffhausen ('Anthropolog. Review,' Oct. 1869, p. 373) remarks on "the vestiges of human sacrifices found both in Homer and the Old Testament."

extremely curious account which he gives of the weapons, tools, and arts, used or practised by savages in various parts of the world, it cannot be doubted that these have nearly all been independent discoveries, excepting perhaps the art of making fire.[33] The Australian boomerang is a good instance of one such independent discovery. The Tahitians when first visited had advanced in many respects beyond the inhabitants of most of the other Polynesian islands. There are no just grounds for the belief that the high culture of the native Peruvians and Mexicans was derived from any foreign source;[34] many native plants were there cultivated, and a few native animals domesticated. We should bear in mind that a wandering crew from some semi-civilised land, if washed to the shores of America, would not, judging from the small influence of most missionaries, have produced any marked effect on the natives, unless they had already become somewhat advanced. Looking to a very remote period in the history of the world, we find, to use Sir J. Lubbock's well-known terms, a paleolithic and neolithic period; and no one will pretend that the art of grinding rough flint tools was a borrowed one. In all parts of Europe, as far east as Greece, in Palestine, India, Japan, New Zealand, and Africa, including Egypt, flint tools have been discovered in abundance; and of their use the existing inhabitants retain no tradition. There is also indirect evidence of their former use by the Chinese and ancient Jews. Hence there can hardly be a doubt that the inhabitants of these many countries, which include nearly the whole civilised world, were once in a barbarous condition. To believe that man was abori-

---

[33] Sir J. Lubbock, 'Prehistoric Times,' 2nd edit. 1869, chap. xv. and xvi. *et passim.*

[34] Dr. F. Müller has made some good remarks to this effect in the 'Reise der Novara: Anthropolog. Theil,' Abtheil. iii. 1868, s. 127.

ginally civilised and then suffered utter degradation in so many regions, is to take a pitiably low view of human nature. It is apparently a truer and more cheerful view that progress has been much more general than retrogression; that man has risen, though by slow and interrupted steps, from a lowly condition to the highest standard as yet attained by him in knowledge, morals, and religion.

CHAPTER VI.

ON THE AFFINITIES AND GENEALOGY OF MAN.

Position of man in the animal series — The natural system genealogical — Adaptive characters of slight value — Various small points of resemblance between man and the Quadrumana — Rank of man in the natural system — Birthplace and antiquity of man — Absence of fossil connecting-links — Lower stages in the genealogy of man, as inferred, firstly from his affinities and secondly from his structure — Early androgynous condition of the Vertebrata — Conclusion.

EVEN if it be granted that the difference between man and his nearest allies is as great in corporeal structure as some naturalists maintain, and although we must grant that the difference between them is immense in mental power, yet the facts given in the previous chapters declare, as it appears to me, in the plainest manner, that man is descended from some lower form, notwithstanding that connecting-links have not hitherto been discovered.

Man is liable to numerous, slight, and diversified variations, which are induced by the same general causes, are governed and transmitted in accordance with the same general laws, as in the lower animals. Man tends to multiply at so rapid a rate that his offspring are necessarily exposed to a struggle for existence, and consequently to natural selection. He has given rise to many races, some of which are so different that they have often been ranked by naturalists as distinct species. His body is constructed on the same homological plan as that of other mammals, independently of the uses to which the several parts may be put. He passes through the same phases of embryological development. He retains many rudimentary and useless structures, which no doubt were once serviceable. Characters occasionally make their re-appearance in him, which we have every reason to believe were possessed by his early progenitors. If the origin of man had been wholly different from that of all other animals, these various appearances would be mere empty deceptions; but such an admission is incredible. These appearances, on the other hand, are intelligible, at least to a large extent, if man is the co-descendant with other mammals of some unknown and lower form.

Some naturalists, from being deeply impressed with the mental and spiritual powers of man, have divided the whole organic world into three kingdoms, the Human, the Animal, and the Vegetable, thus giving to man a separate kingdom.[1] Spiritual powers cannot be compared or classed by the naturalist; but he may endeavour to shew, as I have done, that the mental faculties of man and the lower animals do not differ in kind, although immensely in degree. A difference in degree, however great, does not justify us in placing man in a distinct kingdom, as will perhaps be best illustrated by comparing the mental powers of two insects, namely, a coccus or scale-insect and an ant, which undoubtedly belong to the same class. The difference is here greater, though of a somewhat different kind, than that between man and the highest mammal. The female coccus, whilst young, attaches itself by its proboscis to a plant; sucks the sap but never moves again; is fertilised and lays eggs; and this is its whole history. On the other hand, to describe the habits and mental

[1] Isidore Geoffroy St.-Hilaire gives a detailed account of the position assigned to man by various naturalists in their classifications: 'Hist. Nat. Gén.' tom. ii. 1859, p. 170-189.

powers of a female ant, would require, as Pierre Huber has shewn, a large volume; I may, however, briefly specify a few points. Ants communicate information to each other, and several unite for the same work, or games of play. They recognise their fellow-ants after months of absence. They build great edifices, keep them clean, close the doors in the evening, and post sentries. They make roads, and even tunnels under rivers. They collect food for the community, and when an object, too large for entrance, is brought to the nest, they enlarge the door, and afterwards build it up again.[2] They go out to battle in regular bands, and freely sacrifice their lives for the common weal. They emigrate in accordance with a preconcerted plan. They capture slaves. They keep Aphides as milch-cows. They move the eggs of their aphides, as well as their own eggs and cocoons, into warm parts of the nest, in order that they may be quickly hatched; and endless similar facts could be given. On the whole, the difference in mental power between an ant and a coccus is immense; yet no one has ever dreamed of placing them in distinct classes, much less in distinct kingdoms. No doubt this interval is bridged over by the intermediate mental powers of many other insects; and this is not the case with man and the higher apes. But we have every reason to believe that breaks in the series are simply the result of many forms having become extinct.

Professor Owen, relying chiefly on the structure of the brain, has divided the mammalian series into four sub-classes. One of these he devotes to man; in another he places both the marsupials and the monotremata; so that he makes man as distinct from all other mam-

[2] See the very interesting article, "L'Instinct chez les Insectes," by M. George Pouchet, 'Revue des Deux Mondes,' Feb. 1870, p. 682.

mals as are these two latter groups conjoined. This view has not been accepted, as far as I am aware, by any naturalist capable of forming an independent judgment, and therefore need not here be further considered.

We can understand why a classification founded on any single character or organ—even an organ so wonderfully complex and important as the brain—or on the high development of the mental faculties, is almost sure to prove unsatisfactory. This principle has indeed been tried with hymenopterous insects; but when thus classed by their habits or instincts, the arrangement proved thoroughly artificial.[3] Classifications may, of course, be based on any character whatever, as on size, colour, or the element inhabited; but naturalists have long felt a profound conviction that there is a natural system. This system, it is now generally admitted, must be, as far as possible, genealogical in arrangement,—that is, the co-descendants of the same form must be kept together in one group, separate from the co-descendants of any other form; but if the parent-forms are related, so will be their descendants, and the two groups together will form a larger group. The amount of difference between the several groups—that is the amount of modification which each has undergone—will be expressed by such terms as genera, families, orders, and classes. As we have no record of the lines of descent, these lines can be discovered only by observing the degrees of resemblance between the beings which are to be classed. For this object numerous points of resemblance are of much more importance than the amount of similarity or dissimilarity in a few points. If two languages were found to resemble each other in a multitude of

---

[3] Westwood, 'Modern Class. of Insects,' vol. ii. 1840, p. 87.

words and points of construction, they would be universally recognised as having sprung from a common source, notwithstanding that they differed greatly in some few words or points of construction. But with organic beings the points of resemblance must not consist of adaptations to similar habits of life: two animals may, for instance, have had their whole frames modified for living in the water, and yet they will not be brought any nearer to each other in the natural system. Hence we can see how it is that resemblances in unimportant structures, in useless and rudimentary organs, and in parts not as yet fully developed or functionally active, are by far the most serviceable for classification; for they can hardly be due to adaptations within a late period; and thus they reveal the old lines of descent or of true affinity.

We can further see why a great amount of modification in some one character ought not to lead us to separate widely any two organisms. A part which already differs much from the same part in other allied forms has already, according to the theory of evolution, varied much; consequently it would (as long as the organism remained exposed to the same exciting conditions) be liable to further variations of the same kind; and these, if beneficial, would be preserved, and thus continually augmented. In many cases the continued development of a part, for instance, of the beak of a bird, or of the teeth of a mammal, would not be advantageous to the species for gaining its food, or for any other object; but with man we can see no definite limit, as far as advantage is concerned, to the continued development of the brain and mental faculties. Therefore in determining the position of man in the natural or genealogical system, the extreme development of his brain ought not to outweigh a multitude of resemblances in other less important or quite unimportant points.

The greater number of naturalists who have taken into consideration the whole structure of man, including his mental faculties, have followed Blumenbach and Cuvier, and have placed man in a separate Order, under the title of the Bimana, and therefore on an equality with the Orders of the Quadrumana, Carnivora, &c. Recently many of our best naturalists have recurred to the view first propounded by Linnæus, so remarkable for his sagacity, and have placed man in the same Order with the Quadrumana, under the title of the Primates. The justice of this conclusion will be admitted if, in the first place, we bear in mind the remarks just made on the comparatively small importance for classification of the great development of the brain in man; bearing, also, in mind that the strongly-marked differences between the skulls of man and the Quadrumana (lately insisted upon by Bischoff, Aeby, and others) apparently follow from their differently developed brains. In the second place, we must remember that nearly all the other and more important differences between man and the Quadrumana are manifestly adaptive in their nature, and relate chiefly to the erect position of man; such as the structure of his hand, foot, and pelvis, the curvature of his spine, and the position of his head. The family of seals offers a good illustration of the small importance of adaptive characters for classification. These animals differ from all other Carnivora in the form of their bodies and in the structure of their limbs, far more than does man from the higher apes; yet in every system, from that of Cuvier to the most recent one by Mr. Flower,[4] seals are ranked as a mere family

[4] 'Proc. Zoolog. Soc.' 1869, p. 4.

in the Order of the Carnivora. If man had not been his own classifier, he would never have thought of founding a separate order for his own reception.

It would be beyond my limits, and quite beyond my knowledge, even to name the innumerable points of structure in which man agrees with the other Primates. Our great anatomist and philosopher, Prof. Huxley, has fully discussed this subject,[5] and has come to the conclusion that man in all parts of his organisation differs less from the higher apes, than these do from the lower members of the same group. Consequently there "is "no justification for placing man in a distinct order."

In an early part of this volume I brought forward various facts, shewing how closely man agrees in constitution with the higher mammals; and this agreement, no doubt, depends on our close similarity in minute structure and chemical composition. I gave, as instances, our liability to the same diseases, and to the attacks of allied parasites; our tastes in common for the same stimulants, and the similar effects thus produced, as well as by various drugs; and other such facts.

As small unimportant points of resemblance between man and the higher apes are not commonly noticed in systematic works, and as, when numerous, they clearly reveal our relationship, I will specify a few such points. The relative position of the features are manifestly the same in man and the Quadrumana; and the various emotions are displayed by nearly similar movements of the muscles and skin, chiefly above the eyebrows and round the mouth. Some few expressions are, indeed, almost the same, as in the weeping of certain kinds of monkeys, and in the laughing noise made by others, during which the corners of the mouth are drawn back-

[5] 'Evidence as to Man's Place in Nature,' 1863, p. 70, *et passim*.

wards, and the lower eyelids wrinkled. The external ears are curiously alike. In man the nose is much more prominent than in most monkeys; but we may trace the commencement of an aquiline curvature in the nose of the Hoolock Gibbon; and this in the *Semnopithecus nasica* is carried to a ridiculous extreme.

The faces of many monkeys are ornamented with beards, whiskers, or moustaches. The hair on the head grows to a great length in some species of Semnopithecus;[6] and in the Bonnet monkey (*Macacus radiatus*) it radiates from a point on the crown, with a parting down the middle, as in man. It is commonly said that the forehead gives to man his noble and intellectual appearance; but the thick hair on the head of the Bonnet monkey terminates abruptly downwards, and is succeeded by such short and fine hair, or down, that at a little distance the forehead, with the exception of the eyebrows, appears quite naked. It has been erroneously asserted that eyebrows are not present in any monkey. In the species just named the degree of nakedness of the forehead differs in different individuals; and Eschricht states[7] that in our children the limit between the hairy scalp and the naked forehead is sometimes not well defined; so that here we seem to have a trifling case of reversion to a progenitor, in whom the forehead had not as yet become quite naked.

It is well known that the hair on our arms tends to converge from above and below to a point at the elbow. This curious arrangement, so unlike that in most of the lower mammals, is common to the gorilla, chimpanzee, orang, some species of Hylobates, and even to some few American monkeys. But in *Hylobates agilis* the hair

[6] Isid. Geoffroy, 'Hist. Nat. Gén.' tom. ii. 1859, p. 217.

[7] "Ueber die Richtung der Haare," &c., Müller's 'Archiv für Anat. und Phys.' 1837, s. 51.

on the fore-arm is directed downwards or towards the wrist in the ordinary manner; and in *H. lar* it is nearly erect, with only a very slight forward inclination; so that in this latter species it is in a transitional state. It can hardly be doubted that with most mammals the thickness of the hair and its direction on the back is adapted to throw off the rain; even the transverse hairs on the fore-legs of a dog may serve for this end when he is coiled up asleep. Mr. Wallace remarks that the convergence of the hair towards the elbow on the arms of the orang (whose habits he has so carefully studied) serves to throw off the rain, when, as is the custom of this animal, the arms are bent, with the hands clasped round a branch or over its own head. We should, however, bear in mind that the attitude of an animal may perhaps be in part determined by the direction of the hair; and not the direction of the hair by the attitude. If the above explanation is correct in the case of the orang, the hair on our fore-arms offers a curious record of our former state; for no one supposes that it is now of any use in throwing off the rain, nor in our present erect condition is it properly directed for this purpose.

It would, however, be rash to trust too much to the principle of adaptation in regard to the direction of the hair in man or his early progenitors; for it is impossible to study the figures given by Eschricht of the arrangement of the hair on the human fœtus (this being the same as in the adult) and not agree with this excellent observer that other and more complex causes have intervened. The points of convergence seem to stand in some relation to those points in the embryo which are last closed in during development. There appears, also, to exist some relation between the arrangement of the hair on the limbs, and the course of the medullary arteries.[8]

It must not be supposed that the resemblances between man and certain apes in the above and many other points—such as in having a naked forehead, long tresses on the head, &c.—are all necessarily the result of unbroken inheritance from a common progenitor thus characterised, or of subsequent reversion. Many of these resemblances are more probably due to analogous variation, which follows, as I have elsewhere attempted to shew,[9] from co-descended organisms having a similar constitution and having been acted on by similar causes inducing variability. With respect to the similar direction of the hair on the forearms of man and certain monkeys, as this character is common to almost all the anthropomorphous apes, it may probably be attributed to inheritance; but not certainly so, as some very distinct American monkeys are thus characterised. The same remark is applicable to the tailless condition of man; for the tail is absent in all the anthropomorphous apes. Nevertheless this character cannot with certainty be attributed to inheritance, as the tail, though not absent, is rudimentary in several other Old World and in some New World species, and is quite absent in several species belonging to the allied group of Lemurs.

Although, as we have now seen, man has no just right to form a separate Order for his own reception, he may

---

[8] On the hair in Hylobates, see 'Nat. Hist. of Mammals,' by C. L. Martin, 1841, p. 415. Also, Isid. Geoffroy on the American monkeys and other kinds, 'Hist. Nat. Gén.' vol. ii. 1859, p. 216, 243. Eschricht, ibid, s. 46, 55, 61. Owen, 'Anat. of Vertebrates,' vol. iii. p. 619. Wallace, 'Contributions to the Theory of Natural Selection,' 1870, p. 344.

[9] 'Origin of Species,' 5th edit. 1869, p. 194. 'The Variation of Animals and Plants under Domestication,' vol. ii. 1868, p. 348.

perhaps claim a distinct Sub-order or Family. Prof. Huxley, in his last work,[10] divides the Primates into three Sub-orders; namely, the Anthropidæ with man alone, the Simiadæ including monkeys of all kinds, and the Lemuridæ with the diversified genera of lemurs. As far as differences in certain important points of structure are concerned, man may no doubt rightly claim the rank of a Sub-order; and this rank is too low, if we look chiefly to his mental faculties. Nevertheless, under a genealogical point of view it appears that this rank is too high, and that man ought to form merely a Family, or possibly even only a Sub-family. If we imagine three lines of descent proceeding from a common source, it is quite conceivable that two of them might after the lapse of ages be so slightly changed as still to remain as species of the same genus; whilst the third line might become so greatly modified as to deserve to rank as a distinct Sub-family, Family, or even Order. But in this case it is almost certain that the third line would still retain through inheritance numerous small points of resemblance with the other two lines. Here then would occur the difficulty, at present insoluble, how much weight we ought to assign in our classifications to strongly-marked differences in some few points,—that is to the amount of modification undergone; and how much to close resemblance in numerous unimportant points, as indicating the lines of descent or genealogy. The former alternative is the most obvious, and perhaps the safest, though the latter appears the most correct as giving a truly natural classification.

To form a judgment on this head, with reference to man we must glance at the classification of the

---

[10] 'An Introduction to the Classification of Animals,' 1869, p. 99.

Simiadæ. This family is divided by almost all naturalists into the Catarhine group, or Old World monkeys, all of which are characterised (as their name expresses) by the peculiar structure of their nostrils and by having four premolars in each jaw; and into the Platyrhine group or New World monkeys (including two very distinct sub-groups), all of which are characterised by differently-constructed nostrils and by having six premolars in each jaw. Some other small differences might be mentioned. Now man unquestionably belongs in his dentition, in the structure of his nostrils, and some other respects, to the Catarhine or Old World division; nor does he resemble the Platyrhines more closely than the Catarhines in any characters, excepting in a few of not much importance and apparently of an adaptive nature. Therefore it would be against all probability to suppose that some ancient New World species had varied, and had thus produced a man-like creature with all the distinctive characters proper to the Old World division; losing at the same time all its own distinctive characters. There can consequently hardly be a doubt that man is an offshoot from the Old World Simian stem; and that under a genealogical point of view, he must be classed with the Catarhine division.[11]

The anthropomorphous apes, namely the gorilla, chimpanzee, orang, and hylobates, are separated as a distinct sub-group from the other Old World monkeys by most naturalists. I am aware that Gratiolet, relying on the structure of the brain, does not admit the exist-

[11] This is nearly the same classification as that provisionally adopted by Mr. St. George Mivart ('Transact. Philosoph. Soc.' 1867, p. 300), who, after separating the Lemuridæ, divides the remainder of the Primates into the Hominidæ, the Simiadæ answering to the Catarhines, the Cebidæ, and the Hapalidæ,—these two latter groups answering to the Platyrhines.

ence of this sub-group, and no doubt it is a broken one; thus the orang, as Mr. St. G. Mivart remarks,[12] "is one of the most peculiar and aberrant forms to be "found in the Order." The remaining, non-anthropomorphous, Old World monkeys, are again divided by some naturalists into two or three smaller sub-groups; the genus Semnopithecus, with its peculiar sacculated stomach, being the type of one such sub-group. But it appears from M. Gaudry's wonderful discoveries in Attica, that during the Miocene period a form existed there, which connected Semnopithecus and Macacus; and this probably illustrates the manner in which the other and higher groups were once blended together.

If the anthropomorphous apes be admitted to form a natural sub-group, then as man agrees with them, not only in all those characters which he possesses in common with the whole Catarhine group, but in other peculiar characters, such as the absence of a tail and of callosities and in general appearance, we may infer that some ancient member of the anthropomorphous sub-group gave birth to man. It is not probable that a member of one of the other lower sub-groups should, through the law of analogous variation, have given rise to a man-like creature, resembling the higher anthropomorphous apes in so many respects. No doubt man, in comparison with most of his allies, has undergone an extraordinary amount of modification, chiefly in consequence of his greatly developed brain and erect position; nevertheless we should bear in mind that he "is but one of several exceptional forms "of Primates."[13]

Every naturalist, who believes in the principle of

---

12 'Transact. Zoolog. Soc.' vol. vi. 1867, p. 214.

13 Mr. St. G. Mivart, 'Transact. Phil. Soc.' 1867, p. 410.

evolution, will grant that the two main divisions of the Simiadæ, namely the Catarhine and Platyrhine monkeys, with their sub-groups, have all proceeded from some one extremely ancient progenitor. The early descendants of this progenitor, before they had diverged to any considerable extent from each other, would still have formed a single natural group; but some of the species or incipient genera would have already begun to indicate by their diverging characters the future distinctive marks of the Catarhine and Platyrhine divisions. Hence the members of this supposed ancient group would not have been so uniform in their dentition or in the structure of their nostrils, as are the existing Catarhine monkeys in one way and the Platyrhines in another way, but would have resembled in this respect the allied Lemuridæ which differ greatly from each other in the form of their muzzles,[14] and to an extraordinary degree in their dentition.

The Catarhine and Platyrhine monkeys agree in a multitude of characters, as is shewn by their unquestionably belonging to one and the same Order. The many characters which they possess in common can hardly have been independently acquired by so many distinct species; so that these characters must have been inherited. But an ancient form which possessed many characters common to the Catarhine and Platyrhine monkeys, and others in an intermediate condition, and some few perhaps distinct from those now present in either group, would undoubtedly have been ranked, if seen by a naturalist, as an ape or monkey. And as man under a genealogical point of view belongs to the Catarhine or Old World stock, we must conclude, how-

---

14 Messrs. Murie and Mivart on the Lemuroidea, 'Transact. Zoolog. Soc.' vol. vii. 1869, p. 5.

ever much the conclusion may revolt our pride, that our early progenitors would have been properly thus designated.[15] But we must not fall into the error of supposing that the early progenitor of the whole Simian stock, including man, was identical with, or even closely resembled, any existing ape or monkey.

On the Birthplace and Antiquity of Man.—We are naturally led to enquire where was the birthplace of man at that stage of descent when our progenitors diverged from the Catarhine stock. The fact that they belonged to this stock clearly shews that they inhabited the Old World; but not Australia nor any oceanic island, as we may infer from the laws of geographical distribution. In each great region of the world the living mammals are closely related to the extinct species of the same region. It is therefore probable that Africa was formerly inhabited by extinct apes closely allied to the gorilla and chimpanzee; and as these two species are now man's nearest allies, it is somewhat more probable that our early progenitors lived on the African continent than elsewhere. But it is useless to speculate on this subject, for an ape nearly as large as a man, namely the Dryopithecus of Lartet, which was closely allied to the anthropomorphous Hylobates, existed in Europe during the Upper Miocene period; and since so remote a period the earth has certainly undergone many great revolutions, and there has been ample time for migration on the largest scale.

---

[15] Häckel has come to this same conclusion. See 'Ueber die Entstehung des Menschengeschlechts,' in Virchow's 'Sammlung. gemein. wissen. Vorträge,' 1868, s. 61. Also his 'Natürliche Schöpfungsgeschichte,' 1868, in which he gives in detail his views on the genealogy of man.

At the period and place, whenever and wherever it may have been, when man first lost his hairy covering, he probably inhabited a hot country; and this would have been favourable for a frugiferous diet, on which, judging from analogy, he subsisted. We are far from knowing how long ago it was when man first diverged from the Catarhine stock; but this may have occurred at an epoch as remote as the Eocene period; for the higher apes had diverged from the lower apes as early as the Upper Miocene period, as shewn by the existence of the Dryopithecus. We are also quite ignorant at how rapid a rate organisms, whether high or low in the scale, may under favourable circumstances be modified: we know, however, that some have retained the same form during an enormous lapse of time. From what we see going on under domestication, we learn that within the same period some of the co-descendants of the same species may be not at all changed, some a little, and some greatly changed. Thus it may have been with man, who has undergone a great amount of modification in certain characters in comparison with the higher apes.

The great break in the organic chain between man and his nearest allies, which cannot be bridged over by any extinct or living species, has often been advanced as a grave objection to the belief that man is descended from some lower form; but this objection will not appear of much weight to those who, convinced by general reasons, believe in the general principle of evolution. Breaks incessantly occur in all parts of the series, some being wide, sharp and defined, others less so in various degrees; as between the orang and its nearest allies—between the Tarsius and the other Lemuridæ—between the elephant and in a more striking manner between the Ornithorhynchus or

Echidna, and other mammals. But all these breaks depend merely on the number of related forms which have become extinct. At some future period, not very distant as measured by centuries, the civilised races of man will almost certainly exterminate and replace throughout the world the savage races. At the same time the anthropomorphous apes, as Professor Schaaffhausen has remarked,[16] will no doubt be exterminated. The break will then be rendered wider, for it will intervene between man in a more civilised state, as we may hope, than the Caucasian, and some ape as low as a baboon, instead of as at present between the negro or Australian and the gorilla.

With respect to the absence of fossil remains, serving to connect man with his ape-like progenitors, no one will lay much stress on this fact, who will read Sir C. Lyell's discussion,[17] in which he shews that in all the vertebrate classes the discovery of fossil remains has been an extremely slow and fortuitous process. Nor should it be forgotten that those regions which are the most likely to afford remains connecting man with some extinct ape-like creature, have not as yet been searched by geologists.

Lower Stages in the Genealogy of Man.—We have seen that man appears to have diverged from the Catarhine or Old World division of the Simiadæ, after these had diverged from the New World division. We will now endeavour to follow the more remote traces of his genealogy, trusting in the first place to the mutual affinities between the various classes and orders, with some slight aid from the periods, as far as ascertained,

---

16 'Anthropological Review,' April, 1867, p. 236.

17 'Elements of Geology,' 1865, p. 583-585. 'Antiquity of Man, 1863, p. 145.

of their successive appearance on the earth. The Lemuridæ stand below and close to the Simiadæ, constituting a very distinct family of the Primates, or, according to Häckel, a distinct Order. This group is diversified and broken to an extraordinary degree, and includes many aberrant forms. It has, therefore, probably suffered much extinction. Most of the remnants survive on islands, namely in Madagascar and in the islands of the Malayan archipelago, where they have not been exposed to such severe competition as they would have been on well-stocked continents. This group likewise presents many gradations, leading, as Huxley remarks,[18] "insensibly from the crown and "summit of the animal creation down to creatures "from which there is but a step, as it seems, to the "lowest, smallest, and least intelligent of the placental "mammalia." From these various considerations it is probable that the Simiadæ were originally developed from the progenitors of the existing Lemuridæ; and these in their turn from forms standing very low in the mammalian series.

The Marsupials stand in many important characters below the placental mammals. They appeared at an earlier geological period, and their range was formerly much more extensive than what it now is. Hence the Placentata are generally supposed to have been derived from the Implacentata or Marsupials; not, however, from forms closely like the existing Marsupials, but from their early progenitors. The Monotremata are plainly allied to the Marsupials; forming a third and still lower division in the great mammalian series. They are represented at the present day solely by the Ornithorhynchus and Echidna; and these two forms may

---

[18] 'Man's Place in Nature,' p. 105.

be safely considered as relics of a much larger group which have been preserved in Australia through some favourable concurrence of circumstances. The Monotremata are eminently interesting, as in several important points of structure they lead towards the class of reptiles.

In attempting to trace the genealogy of the Mammalia, and therefore of man, lower down in the series, we become involved in greater and greater obscurity. He who wishes to see what ingenuity and knowledge can effect, may consult Prof. Häckel's works.[19] I will content myself with a few general remarks. Every evolutionist will admit that the five great vertebrate classes, namely, mammals, birds, reptiles, amphibians, and fishes, are all descended from some one prototype; for they have much in common, especially during their embryonic state. As the class of fishes is the most lowly organised and appeared before the others, we may conclude that all the members of the vertebrate kingdom are derived from some fish-like animal, less highly organised than any as yet found in the lowest known formations. The belief that animals so distinct as a monkey or elephant and a humming-bird, a snake, frog, and fish, &c., could all have sprung from the same parents, will appear monstrous to those who have not attended to the recent progress of natural history. For this belief implies the former existence of links closely binding together all these forms, now so utterly unlike.

---

[19] Elaborate tables are given in his 'Generelle Morphologie' (B. ii. s. cliii. and s. 425); and with more especial reference to man in his 'Natürliche Schöpfungsgeschichte,' 1868. Prof. Huxley, in reviewing this latter work ('The Academy,' 1869, p. 42) says, that he considers the phylum or lines of descent of the Vertebrata to be admirably discussed by Häckel, although he differs on some points. He expresses, also, his high estimate of the value of the general tenor and spirit of the whole work.

Nevertheless it is certain that groups of animals have existed, or do now exist, which serve to connect more or less closely the several great vertebrate classes. We have seen that the Ornithorhynchus graduates towards reptiles; and Prof. Huxley has made the remarkable discovery, confirmed by Mr. Cope and others, that the old Dinosaurians are intermediate in many important respects between certain reptiles and certain birds—the latter consisting of the ostrich-tribe (itself evidently a widely-diffused remnant of a larger group) and of the Archeopteryx, that strange Secondary bird having a long tail like that of the lizard. Again, according to Prof. Owen,[20] the Ichthyosaurians—great sea-lizards furnished with paddles—present many affinities with fishes, or rather, according to Huxley, with amphibians. This latter class (including in its highest division frogs and toads) is plainly allied to the Ganoid fishes. These latter fishes swarmed during the earlier geological periods, and were constructed on what is called a highly generalised type, that is they presented diversified affinities with other groups of organisms. The amphibians and fishes are also so closely united by the Lepidosiren, that naturalists long disputed in which of these two classes it ought to be placed. The Lepidosiren and some few Ganoid fishes have been preserved from utter extinction by inhabiting our rivers, which are harbours of refuge, bearing the same relation to the great waters of the ocean that islands bear to continents.

Lastly, one single member of the immense and diversified class of fishes, namely the lancelet or amphioxus, is so different from all other fishes, that Häckel maintains that it ought to form a distinct class in the vertebrate kingdom. This fish is remarkable for its

---

[20] 'Palæontology,' 1860, p. 199.

negative characters; it can hardly be said to possess a brain, vertebral column, or heart, &c.; so that it was classed by the older naturalists amongst the worms. Many years ago Prof. Goodsir perceived that the lancelet presented some affinities with the Ascidians, which are invertebrate, hermaphrodite, marine creatures permanently attached to a support. They hardly appear like animals, and consist of a simple, tough, leathery sack, with two small projecting orifices. They belong to the Molluscoida of Huxley—a lower division of the great kingdom of the Mollusca; but they have recently been placed by some naturalists amongst the Vermes or worms. Their larvæ somewhat resemble tadpoles in shape,[21] and have the power of swimming freely about. Some observations lately made by M. Kowalevsky,[22] since confirmed by Prof. Kuppfer, will form a discovery of extraordinary interest, if still further extended, as I hear from M. Kowalevsky in Naples he has now effected. The discovery is that the larvæ of Ascidians are related to the Vertebrata, in their manner of development, in the relative position of the nervous system, and in possessing a structure closely like the *chorda dorsalis* of vertebrate animals. It thus appears, if we may rely on embryology, which has always proved the safest guide in classification, that we have at last gained a clue to the source whence the Vertebrata have

---

[21] I had the satisfaction of seeing, at the Falkland Islands, in April, 1833, and therefore some years before any other naturalist, the locomotive larvæ of a compound Ascidian, closely allied to, but apparently generically distinct from, Synoicum. The tail was about five times as long as the oblong head, and terminated in a very fine filament. It was plainly divided, as sketched by me under a simple microscope, by transverse opaque partitions, which I presume represent the great cells figured by Kowalevsky. At an early stage of development the tail was closely coiled round the head of the larva.

[22] 'Mémoires de l'Acad. des Sciences de St. Pétersbourg,' tom. x. No. 15, 1866.

been derived. We should thus be justified in believing that at an extremely remote period a group of animals existed, resembling in many respects the larvæ of our present Ascidians, which diverged into two great branches—the one retrograding in development and producing the present class of Ascidians, the other rising to the crown and summit of the animal kingdom by giving birth to the Vertebrata.

We have thus far endeavoured rudely to trace the genealogy of the Vertebrata by the aid of their mutual affinities. We will now look to man as he exists; and we shall, I think, be able partially to restore during successive periods, but not in due order of time, the structure of our early progenitors. This can be effected by means of the rudiments which man still retains, by the characters which occasionally make their appearance in him through reversion, and by the aid of the principles of morphology and embryology. The various facts, to which I shall here allude, have been given in the previous chapters. The early progenitors of man were no doubt once covered with hair, both sexes having beards; their ears were pointed and capable of movement; and their bodies were provided with a tail, having the proper muscles. Their limbs and bodies were also acted on by many muscles which now only occasionally reappear, but are normally present in the Quadrumana. The great artery and nerve of the humerus ran through a supra-condyloid foramen. At this or some earlier period, the intestine gave forth a much larger diverticulum or cæcum than that now existing. The foot, judging from the condition of the great toe in the fœtus, was then prehensile; and our progenitors, no doubt, were arboreal in their habits, frequenting some warm, forest-clad and. The males were provided with great canine teeth, which served them as formidable weapons.

At a much earlier period the uterus was double; the excreta were voided through a cloaca; and the eye was protected by a third eyelid or nictitating membrane. At a still earlier period the progenitors of man must have been aquatic in their habits; for morphology plainly tells us that our lungs consist of a modified swim-bladder, which once served as a float. The clefts on the neck in the embryo of man show where the branchiæ once existed. At about this period the true kidneys were replaced by the corpora wolffiana. The heart existed as a simple pulsating vessel; and the chorda dorsalis took the place of a vertebral column. These early predecessors of man, thus seen in the dim recesses of time, must have been as lowly organised as the lancelet or amphioxus, or even still more lowly organised.

There is one other point deserving a fuller notice. It has long been known that in the vertebrate kingdom one sex bears rudiments of various accessory parts, appertaining to the reproductive system, which properly belong to the opposite sex; and it has now been ascertained that at a very early embryonic period both sexes possess true male and female glands. Hence some extremely remote progenitor of the whole vertebrate kingdom appears to have been hermaphrodite or androgynous.[23] But here we encounter a singular

---

[23] This is the conclusion of one of the highest authorities in comparative anatomy, namely, Prof. Gegenbaur: 'Grundzüge der vergleich. Anat.' 1870, s. 876. The result has been arrived at chiefly from the study of the Amphibia; but it appears from the researches of Waldeyer (as quoted in Humphry's 'Journal of Anat. and Phys.' 1869, p. 161), that the sexual organs of even "the higher vertebrata are, in their early "condition, hermaphrodite." Similar views have long been held by some authors, though until recently not well based.

difficulty. In the mammalian class the males possess in their vesiculæ prostraticæ rudiments of a uterus with the adjacent passage; they bear also rudiments of mammæ, and some male marsupials have rudiments of a marsupial sack.[24] Other analogous facts could be added. Are we, then, to suppose that some extremely ancient mammal possessed organs proper to both sexes, that is, continued androgynous after it had acquired the chief distinctions of its proper class, and therefore after it had diverged from the lower classes of the vertebrate kingdom? This seems improbable in the highest degree; for had this been the case, we might have expected that some few members of the two lower classes, namely fishes[25] and amphibians, would still have remained androgynous. We must, on the contrary, believe that when the five vertebrate classes diverged from their common progenitor the sexes had already become separated. To account, however, for male mammals possessing rudiments of the accessory female organs, and for female mammals possessing rudiments of the masculine organs, we need not suppose that their early progenitors were still androgynous after they had assumed their chief mammalian characters. It is quite possible that as the one sex gradually acquired the accessory organs proper to it, some of the successive steps or modifications were transmitted to the opposite sex. When we treat of sexual selection, we shall meet with innumerable instances of this form of transmission,—as in the case of the spurs, plumes,

---

[24] The male Thylacinus offers the best instance. Owen, 'Anatomy of Vertebrates,' vol. iii. p. 771.

[25] Serranus is well known often to be in an hermaphrodite condition; but Dr. Günther informs me that he is convinced that this is not its normal state. Descent from an ancient androgynous prototype would, however, naturally favour and explain, to a certain extent, the recurrence of this condition in these fishes.

and brilliant colours, acquired by male birds for battle or ornament, and transferred to the females in an imperfect or rudimentary condition.

The possession by male mammals of functionally imperfect mammary organs is, in some respects, especially curious. The Monotremata have the proper milk-secreting glands with orifices, but no nipples; and as these animals stand at the very base of the mammalian series, it is probable that the progenitors of the class possessed, in like manner, the milk-secreting glands, but no nipples. This conclusion is supported by what is known of their manner of development; for Professor Turner informs me, on the authority of Kölliker and Lauger, that in the embryo the mammary glands can be distinctly traced before the nipples are in the least visible; and it should be borne in mind that the development of successive parts in the individual generally seems to represent and accord with the development of successive beings in the same line of descent. The Marsupials differ from the Monotremata by possessing nipples; so that these organs were probably first acquired by the Marsupials after they had diverged from, and risen above, the Monotremata, and were then transmitted to the placental mammals. No one will suppose that after the Marsupials had approximately acquired their present structure, and therefore at a rather late period in the development of the mammalian series, any of its members still remained androgynous. We seem, therefore, compelled to recur to the foregoing view, and to conclude that the nipples were first developed in the females of some very early marsupial form, and were then, in accordance with a common law of inheritance, transferred in a functionally imperfect condition to the males.

Nevertheless a suspicion has sometimes crossed my

mind that long after the progenitors of the whole mammalian class had ceased to be androgynous, both sexes might have yielded milk and thus nourished their young; and in the case of the Marsupials, that both sexes might have carried their young in marsupial sacks. This will not appear utterly incredible, if we reflect that the males of syngnathous fishes receive the eggs of the females in their abdominal pouches, hatch them, and afterwards, as some believe, nourish the young;[26]—that certain other male fishes hatch the eggs within their mouths or branchial cavities;—that certain male toads take the chaplets of eggs from the females and wind them round their own thighs, keeping them there until the tadpoles are born;—that certain male birds undertake the whole duty of incubation, and that male pigeons, as well as the females, feed their nestlings with a secretion from their crops. But the above suspicion first occurred to me from the mammary glands in male mammals being developed so much more perfectly than the rudiments of those other accessory reproductive parts, which are found in the one sex though proper to the other. The mammary glands and nipples, as they exist in male mammals, can indeed hardly be called rudimentary; they are simply not fully developed and not functionally active. They are sympathetically affected under the influence of certain diseases, like the same organs in the female. At birth they often secrete a few drops of milk; and they have

---

[26] Mr. Lockwood believes (as quoted in 'Quart. Journal of Science,' April, 1868, p. 269), from what he has observed of the development of Hippocampus, that the walls of the abdominal pouch of the male in some way afford nourishment. On male fishes hatching the ova in their mouths, see a very interesting paper by Prof. Wyman, in 'Proc. Boston Soc. of Nat. Hist.' Sept. 15, 1857; also Prof. Turner, in 'Journal of Anat. and Phys.' Nov. 1, 1866, p. 78. Dr. Günther has likewise described similar cases.

been known occasionally in man and other mammals to become well developed, and to yield a fair supply of milk. Now if we suppose that during a former prolonged period male mammals aided the females in nursing their offspring, and that afterwards from some cause, as from a smaller number of young being produced, the males ceased giving this aid, disuse of the organs during maturity would lead to their becoming inactive; and from two well-known principles of inheritance this state of inactivity would probably be transmitted to the males at the corresponding age of maturity. But at all earlier ages these organs would be left unaffected, so that they would be equally well developed in the young of both sexes.

Conclusion.—The best definition of advancement or progress in the organic scale ever given, is that by Von Baer; and this rests on the amount of differentiation and specialisation of the several parts of the same being, when arrived, as I should be inclined to add, at maturity. Now as organisms have become slowly adapted by means of natural selection for diversified lines of life, their parts will have become, from the advantage gained by the division of physiological labour, more and more differentiated and specialised for various functions. The same part appears often to have been modified first for one purpose, and then long afterwards for some other and quite distinct purpose; and thus all the parts are rendered more and more complex. But each organism will still retain the general type of structure of the progenitor from which it was aboriginally derived. In accordance with this view it seems, if we turn to geological evidence, that organisation on the whole has advanced throughout the world by slow and interrupted steps. In the great

P 2

kingdom of the Vertebrata it has culminated in man. It must not, however, be supposed that groups of organic beings are always supplanted and disappear as soon as they have given birth to other and more perfect groups. The latter, though victorious over their predecessors, may not have become better adapted for all places in the economy of nature. Some old forms appear to have survived from inhabiting protected sites, where they have not been exposed to very severe competition; and these often aid us in constructing our genealogies, by giving us a fair idea of former and lost populations. But we must not fall into the error of looking at the existing members of any lowly-organised group as perfect representatives of their ancient predecessors.

The most ancient progenitors in the kingdom of the Vertebrata, at which we are able to obtain an obscure glance, apparently consisted of a group of marine animals,[27] resembling the larvæ of existing Ascidians. These animals probably gave rise to a group of fishes, as lowly organised as the lancelet; and from these the Ganoids, and other fishes like the Lepidosiren, must have been developed. From such fish a very small advance would

---

[27] All vital functions tend to run their course in fixed and recurrent periods, and with tidal animals the periods would probably be lunar; for such animals must have been left dry or covered deep with water,—supplied with copious food or stinted,—during endless generations, at regular lunar intervals. If then the Vertebrata are descended from an animal allied to the existing tidal Ascidians, the mysterious fact, that with the higher and now terrestrial Vertebrata, not to mention other classes, many normal and abnormal vital processes run their course according to lunar periods, is rendered intelligible. A recurrent period, if approximately of the right duration, when once gained, would not, as far as we can judge, be liable to be changed; consequently it might be thus transmitted during almost any number of generations. This conclusion, if it could be proved sound, would be curious; for we should then see that the period of gestation in each mammal, and the hatching of each bird's eggs, and many other vital processes, still betrayed the primordial birthplace of these animals.

carry us on to the amphibians. We have seen that birds and reptiles were once intimately connected together; and the Monotremata now, in a slight degree, connect mammals with reptiles. But no one can at present say by what line of descent the three higher and related classes, namely, mammals, birds, and reptiles, were derived from either of the two lower vertebrate classes, namely amphibians and fishes. In the class of mammals the steps are not difficult to conceive which led from the ancient Monotremata to the ancient Marsupials; and from these to the early progenitors of the placental mammals. We may thus ascend to the Lemuridæ; and the interval is not wide from these to the Simiadæ. The Simiadæ then branched off into two great stems, the New World and Old World monkeys; and from the latter, at a remote period, Man, the wonder and glory of the Universe, proceeded.

Thus we have given to man a pedigree of prodigious length, but not, it may be said, of noble quality. The world, it has often been remarked, appears as if it had long been preparing for the advent of man; and this, in one sense is strictly true, for he owes his birth to a long line of progenitors. If any single link in this chain had never existed, man would not have been exactly what he now is. Unless we wilfully close our eyes, we may, with our present knowledge, approximately recognise our parentage; nor need we feel ashamed of it. The most humble organism is something much higher than the inorganic dust under our feet; and no one with an unbiassed mind can study any living creature, however humble, without being struck with enthusiasm at its marvellous structure and properties.

CHAPTER VII.

ON THE RACES OF MAN.

The nature and value of specific characters—Application to the races of man—Arguments in favour of, and opposed to, ranking the so-called races of man as distinct species — Sub-species —Monogenists and polygenists — Convergence of character — Numerous points of resemblance in body and mind between the most distinct races of man — The state of man when he first spread over the earth — Each race not descended from a single pair — The extinction of races — The formation of races — The effects of crossing — Slight influence of the direct action of the conditions of life — Slight or no influence of natural selection — Sexual selection.

IT is not my intention here to describe the several so-called races of men; but to inquire what is the value of the differences between them under a classificatory point of view, and how they have originated. In determining whether two or more allied forms ought to be ranked as species or varieties, naturalists are practically guided by the following considerations; namely, the amount of difference between them, and whether such differences relate to few or many points of structure, and whether they are of physiological importance; but more especially whether they are constant. Constancy of character is what is chiefly valued and sought for by naturalists. Whenever it can be shewn, or rendered probable, that the forms in question have remained distinct for a long period, this becomes an argument of much weight in favour of treating them as species. Even a slight degree of sterility between any two forms when first crossed, or in their offspring, is generally considered as a decisive test of their specific distinctness; and their continued persistence without blending within the same area, is usually accepted as sufficient evidence, either of some degree of mutual sterility, or in the case of animals of some repugnance to mutual pairing.

Independently of blending from intercrossing, the complete absence, in a well-investigated region, of varieties linking together any two closely-allied forms, is probably the most important of all the criterions of their specific distinctness; and this is a somewhat different consideration from mere constancy of character, for two forms may be highly variable and yet not yield intermediate varieties. Geographical distribution is often unconsciously and sometimes consciously brought into play; so that forms living in two widely separated areas, in which most of the other inhabitants are specifically distinct, are themselves usually looked at as distinct; but in truth this affords no aid in distinguishing geographical races from so-called good or true species.

Now let us apply these generally-admitted principles to the races of man, viewing him in the same spirit as a naturalist would any other animal. In regard to the amount of difference between the races, we must make some allowance for our nice powers of discrimination gained by the long habit of observing ourselves. In India, as Elphinstone remarks,[1] although a newly-arrived European cannot at first distinguish the various native races, yet they soon appear to him extremely dissimilar; and the Hindoo cannot at first perceive any difference between the several European nations. Even the most distinct races of man, with the exception of certain negro tribes, are much more like each other in form

---

1 'History of India,' 1841, vol. i. p. 323. Father Ripa makes exactly the same remark with respect to the Chinese.

than would at first be supposed. This is well shewn by the French photographs in the Collection Anthropologique du Muséum of the men belonging to various races, the greater number of which, as many persons to whom I have shown them have remarked, might pass for Europeans. Nevertheless, these men if seen alive would undoubtedly appear very distinct, so that we are clearly much influenced in our judgment by the mere colour of the skin and hair, by slight differences in the features, and by expression.

There is, however, no doubt that the various races, when carefully compared and measured, differ much from each other,—as in the texture of the hair, the relative proportions of all parts of the body,[2] the capacity of the lungs, the form and capacity of the skull, and even in the convolutions of the brain.[3] But it would be an endless task to specify the numerous points of structural difference. The races differ also in constitution, in acclimatisation, and in liability to certain diseases. Their mental characteristics are likewise very distinct; chiefly as it would appear in their emotional, but partly in their intellectual, faculties. Every one who has had the opportunity of comparison, must have been struck with the contrast between the taciturn, even morose, aborigines of S. America and the light-hearted, talkative negroes. There is a nearly similar contrast between the Malays and the Papuans,[4] who live

[2] A vast number of measurements of Whites, Blacks, and Indians, are given in the 'Investigations in the Military and Anthropolog. Statistics of American Soldiers,' by B. A. Gould, 1869, p. 298-358; on the capacity of the lungs, p. 471. See also the numerous and valuable tables, by Dr. Weisbach, from the observations of Dr. Scherzer and Dr. Schwarz, in the 'Reise der Novara: Anthropolog. Theil,' 1867.

[3] See, for instance, Mr. Marshall's account of the brain of a Bushwoman, in 'Phil. Transact.' 1864, p. 519.

[4] Wallace, 'The Malay Archipelago,' vol. ii. 1869, p. 178.

under the same physical conditions, and are separated from each other only by a narrow space of sea.

We will first consider the arguments which may be advanced in favour of classing the races of man as distinct species, and then those on the other side. If a naturalist, who had never before seen such beings, were to compare a Negro, Hottentot, Australian, or Mongolian, he would at once perceive that they differed in a multitude of characters, some of slight and some of considerable importance. On inquiry he would find that they were adapted to live under widely different climates, and that they differed somewhat in bodily constitution and mental disposition. If he were then told that hundreds of similar specimens could be brought from the same countries, he would assuredly declare that they were as good species as many to which he had been in the habit of affixing specific names. This conclusion would be greatly strengthened as soon as he had ascertained that these forms had all retained the same character for many centuries; and that negroes, apparently identical with existing negroes, had lived at least 4000 years ago.[5] He would also hear from an excellent observer,

[5] With respect to the figures in the famous Egyptian caves of Abou-Simbel, M. Pouchet says ('The Plurality of the Human Races,' Eng. translat. 1864, p. 50), that he was far from finding recognisable representations of the dozen or more nations which some authors believe that they can recognise. Even some of the most strongly-marked races cannot be identified with that degree of unanimity which might have been expected from what has been written on the subject. Thus Messrs. Nott and Gliddon ('Types of Mankind,' p. 148) state that Rameses II., or the Great, has features superbly European; whereas Knox, another firm believer in the specific distinction of the races of man ('Races of Man,' 1850, p. 201), speaking of young Memnon (the same person with Rameses II., as I am informed by Mr. Birch) insists in the strongest manner that he is identical in character with the Jews of Antwerp. Again, whilst looking in the British Museum with two competent judges, officers of the establishment, at the statue of Amunoph III., we agreed that he had a strongly negro cast of features;

Dr. Lund,[6] that the human skulls found in the caves of Brazil, entombed with many extinct mammals, belonged to the same type as that now prevailing throughout the American Continent.

Our naturalist would then perhaps turn to geographical distribution, and he would probably declare that forms differing not only in appearance, but fitted for the hottest and dampest or driest countries, as well as for the arctic regions, must be distinct species. He might appeal to the fact that no one species in the group next to man, namely the Quadrumana, can resist a low temperature or any considerable change of climate; and that those species which come nearest to man have never been reared to maturity, even under the temperate climate of Europe. He would be deeply impressed with the fact, first noticed by Agassiz,[7] that the different races of man are distributed over the world in the same zoological provinces, as those inhabited by undoubtedly distinct species and genera of mammals. This is manifestly the case with the Australian, Mongolian, and Negro races of man; in a less well-marked manner with the Hottentots; but plainly with the Papuans and Malays, who are separated, as Mr. Wallace has shewn, by nearly the same line which divides the great Malayan and Australian zoological provinces. The aborigines of America range throughout the Continent; and this at first appears opposed to the above rule, for most of the productions of the Southern and Northern halves differ widely; yet some few living forms, as the

---

but Messrs. Nott and Gliddon (ibid. p. 146, fig. 53) describe him as "a hybrid, but not of negro intermixture."

[6] As quoted by Nott and Gliddon, 'Types of Mankind,' 1854, p. 439. They give also corroborative evidence; but C. Vogt thinks that the subject requires further investigation.

[7] "Diversity of Origin of the Human Races," in the 'Christian Examiner,' July, 1850.

opossum, range from the one into the other, as did formerly some of the gigantic Edentata. The Esquimaux, like other Arctic animals, extend round the whole polar regions. It should be observed that the mammalian forms which inhabit the several zoological provinces, do not differ from each other in the same degree; so that it can hardly be considered as an anomaly that the Negro differs more, and the American much less, from the other races of man than do the mammals of the same continents from those of the other provinces. Man, it may be added, does not appear to have aboriginally inhabited any oceanic island; and in this respect he resembles the other members of his class.

In determining whether the varieties of the same kind of domestic animal should be ranked as specifically distinct, that is, whether any of them are descended from distinct wild species, every naturalist would lay much stress on the fact, if established, of their external parasites being specifically distinct. All the more stress would be laid on this fact, as it would be an exceptional one, for I am informed by Mr. Denny that the most different kinds of dogs, fowls, and pigeons, in England, are infested by the same species of Pediculi or lice. Now Mr. A. Murray has carefully examined the Pediculi collected in different countries from the different races of man;[8] and he finds that they differ, not only in colour, but in the structure of their claws and limbs. In every case in which numerous specimens were obtained the differences were constant. The surgeon of a whaling ship in the Pacific assured me that when the Pediculi, with which some Sandwich Islanders on board swarmed, strayed on to the bodies of the English sailors, they died in the course of three or four days. These Pediculi

[8] 'Transact. R. Soc. of Edinburgh,' vol. xxii. 1861, p. 567.

were darker coloured and appeared different from those proper to the natives of Chiloe in South America, of which he gave me specimens. These, again, appeared larger and much softer than European lice. Mr. Murray procured four kinds from Africa, namely from the Negroes of the Eastern and Western coasts, from the Hottentots and Caffres; two kinds from the natives of Australia; two from North, and two from South America. In these latter cases it may be presumed that the Pediculi came from natives inhabiting different districts. With insects slight structural differences, if constant, are generally esteemed of specific value: and the fact of the races of man being infested by parasites, which appear to be specifically distinct, might fairly be urged as an argument that the races themselves ought to be classed as distinct species.

Our supposed naturalist having proceeded thus far in his investigation, would next inquire whether the races of men, when crossed, were in any degree sterile. He might consult the work[9] of a cautious and philosophical observer, Professor Broca; and in this he would find good evidence that some races were quite fertile together; but evidence of an opposite nature in regard to other races. Thus it has been asserted that the native women of Australia and Tasmania rarely produce children to European men; the evidence, however, on this head has now been shewn to be almost valueless. The half-castes are killed by the pure blacks; and an account has lately been published of eleven half-caste youths murdered and burnt at the same time, whose remains were found by the police.[10] Again, it has often

[9] 'On the Phenomena of Hybridity in the Genus Homo,' Eng. translat. 1864.

[10] See the interesting letter by Mr. T. A. Murray, in the 'Anthropolog. Review,' April, 1868, p. liii. In this letter Count Strzelecki's

been said that when mulattoes intermarry they produce few children; on the other hand, Dr. Bachman of Charlestown[11] positively asserts that he has known mulatto families which have intermarried for several generations, and have continued on an average as fertile as either pure whites or pure blacks. Inquiries formerly made by Sir C. Lyell on this subject led him, as he informs me, to the same conclusion. In the United States the census for the year 1854 included, according to Dr. Bachman, 405,751 mulattoes; and this number, considering all the circumstances of the case, seems small; but it may partly be accounted for by the degraded and anomalous position of the class, and by the profligacy of the women. A certain amount of absorption of mulattoes into negroes must always be in progress; and this would lead to an apparent diminution of the former. The inferior vitality of mulattoes is spoken of in a trustworthy work[12] as a well-known phenomenon; but this is a different consideration from their lessened fertility; and can hardly be advanced as a proof of the specific distinctness of the parent races. No doubt both animal and vegetable hybrids, when produced from extremely distinct species, are liable to premature death; but the parents of mulattoes cannot be put under the category of extremely distinct species. The common Mule, so notorious for long life and vigour, and yet so sterile, shews how little necessary connection

statement, that Australian women who have borne children to a white man are afterwards sterile with their own race, is disproved. M. A. de Quatrefages has also collected ('Revue des Cours Scientifiques,' March, 1869, p. 239) much evidence that Australians and Europeans are not sterile when crossed.

11 'An Examination of Prof. Agassiz's Sketch of the Nat. Provinces of the Animal World,' Charleston, 1855, p. 44.

12 'Military and Anthropolog. Statistics of American Soldiers,' by B. A. Gould, 1869, p. 319.

there is in hybrids between lessened fertility and vitality: other analogous cases could be added.

Even if it should hereafter be proved that all the races of men were perfectly fertile together, he who was inclined from other reasons to rank them as distinct species, might with justice argue that fertility and sterility are not safe criterions of specific distinctness. We know that these qualities are easily affected by changed conditions of life or by close inter-breeding, and that they are governed by highly complex laws, for instance that of the unequal fertility of reciprocal crosses between the same two species. With forms which must be ranked as undoubted species, a perfect series exists from those which are absolutely sterile when crossed, to those which are almost or quite fertile. The degrees of sterility do not coincide strictly with the degrees of difference in external structure or habits of life. Man in many respects may be compared with those animals which have long been domesticated, and a large body of evidence can be advanced in favour of the Pallasian doctrine[13] that domestication tends to eliminate the

---

[13] 'The Variation of Animals and Plants under Domestication,' vol. ii. p. 109. I may here remind the reader that the sterility of species when crossed is not a specially-acquired quality; but, like the incapacity of certain trees to be grafted together, is incidental on other acquired differences. The nature of these differences is unknown, but they relate more especially to the reproductive system, and much less to external structure or to ordinary differences in constitution. One important element in the sterility of crossed species apparently lies in one or both having been long habituated to fixed conditions; for we know that changed conditions have a special influence on the reproductive system, and we have good reason to believe (as before remarked) that the fluctuating conditions of domestication tend to eliminate that sterility which is so general with species in a natural state when crossed. It has elsewhere been shewn by me (ibid. vol. ii. p. 185, and 'Origin of Species,' 5th edit. p. 317) that the sterility of crossed species has not been acquired through natural selection: we can see that when two forms have already been rendered very sterile, it is scarcely

sterility which is so general a result of the crossing of species in a state of nature. From these several considerations, it may be justly urged that the perfect fertility of the intercrossed races of man, if established, would not absolutely preclude us from ranking them as distinct species.

Independently of fertility, the character of the offspring from a cross has sometimes been thought to afford evidence whether the parent-forms ought to be ranked as species or varieties; but after carefully studying the evidence, I have come to the conclusion that no general rules of this kind can be trusted. Thus with mankind the offspring of distinct races resemble in all respects the offspring of true species and of varieties. This is shewn, for instance, by the manner in which the characters of both parents are blended, and by one form absorbing another through repeated crosses. In this latter case the progeny both of crossed species and varieties retain for a long period a tendency to revert to their ancestors, especially to that one which is prepotent in transmission. When any character has suddenly appeared in a race or species as the result of a

possible that their sterility should be augmented by the preservation or survival of the more and more sterile individuals; for as the sterility increases fewer and fewer offspring will be produced from which to breed, and at last only single individuals will be produced, at the rarest intervals. But there is even a higher grade of sterility than this. Both Gärtner and Kölreuter have proved that in genera of plants including numerous species, a series can be formed from species which when crossed yield fewer and fewer seeds, to species which never produce a single seed, but yet are affected by the pollen of the other species, for the germen swells. It is here manifestly impossible to select the more sterile individuals, which have already ceased to yield seeds; so that the acme of sterility, when the germen alone is affected, cannot be gained through selection. This acme, and no doubt the other grades of sterility, are the incidental results of certain unknown differences in the constitution of the reproductive system of the species which are crossed.

single act of variation, as is general with monstrosities,[14] and this race is crossed with another not thus characterised, the characters in question do not commonly appear in a blended condition in the young, but are transmitted to them either perfectly developed or not at all. As with the crossed races of man cases of this kind rarely or never occur, this may be used as an argument against the view suggested by some ethnologists, namely that certain characters, for instance the blackness of the negro, first appeared as a sudden variation or sport. Had this occurred, it is probable that mulattoes would often have been born, either completely black or completely white.

We have now seen that a naturalist might feel himself fully justified in ranking the races of man as distinct species; for he has found that they are distinguished by many differences in structure and constitution, some being of importance. These differences have, also, remained nearly constant for very long periods of time. He will have been in some degree influenced by the enormous range of man, which is a great anomaly in the class of mammals, if mankind be viewed as a single species. He will have been struck with the distribution of the several so-called races, in accordance with that of other undoubtedly distinct species of mammals. Finally he might urge that the mutual fertility of all the races has not as yet been fully proved; and even if proved would not be an absolute proof of their specific identity.

On the other side of the question, if our supposed naturalist were to enquire whether the forms of man kept distinct like ordinary species, when mingled to-

[14] 'The Variation of Animals,' &c., vol. ii. p. 92.

gether in large numbers in the same country, he would immediately discover that this was by no means the case. In Brazil he would behold an immense mongrel population of Negroes and Portuguese; in Chiloe and other parts of South America, he would behold the whole population consisting of Indians and Spaniards blended in various degrees.[15] In many parts of the same continent he would meet with the most complex crosses between Negroes, Indians, and Europeans; and such triple crosses afford the severest test, judging from the vegetable kingdom, of the mutual fertility of the parent-forms. In one island of the Pacific he would find a small population of mingled Polynesian and English blood; and in the Viti Archipelago a population of Polynesians and Negritos crossed in all degrees. Many analogous cases could be added, for instance, in South Africa. Hence the races of man are not sufficiently distinct to co-exist without fusion; and this it is, which in all ordinary cases affords the usual test of specific distinctness.

Our naturalist would likewise be much disturbed as soon as he perceived that the distinctive characters of every race of man were highly variable. This strikes every one when he first beholds the negro-slaves in Brazil, who have been imported from all parts of Africa. The same remark holds good with the Polynesians, and with many other races. It may be doubted whether any character can be named which is distinctive of a race and is constant. Savages, even within the limits of the same tribe, are not nearly so uniform in character, as has often been said. Hottentot women offer certain

[15] M. de Quatrefages has given ('Anthropolog. Review,' Jan. 1869, p. 22) an interesting account of the success and energy of the Paulistas in Brazil, who are a much crossed race of Portuguese and Indians, with a mixture of the blood of other races.

peculiarities, more strongly marked than those occurring in any other race, but these are known not to be of constant occurrence. In the several American tribes, colour and hairyness differ considerably; as does colour to a certain degree, and the shape of the features greatly, in the Negroes of Africa. The shape of the skull varies much in some races;[16] and so it is with every other character. Now all naturalists have learnt by dearly-bought experience, how rash it is to attempt to define species by the aid of inconstant characters.

But the most weighty of all the arguments against treating the races of man as distinct species, is that they graduate into each other, independently in many cases, as far as we can judge, of their having intercrossed. Man has been studied more carefully than any other organic being, and yet there is the greatest possible diversity amongst capable judges whether he should be classed as a single species or race, or as two (Virey), as three (Jacquinot), as four (Kant), five (Blumenbach), six (Buffon), seven (Hunter), eight (Agassiz), eleven (Pickering), fifteen (Bory St. Vincent), sixteen (Desmoulins), twenty-two (Morton), sixty (Crawfurd), or as sixty-three, according to Burke.[17] This diversity of judgment does not prove that the races ought not to be ranked as species, but it shews that they graduate into each other, and that it is hardly possible to discover clear distinctive characters between them.

Every naturalist who has had the misfortune to under-

---

[16] For instance with the aborigines of America and Australia. Prof. Huxley says ('Transact. Internat. Congress of Prehist. Arch.' 1868, p. 105) that the skulls of many South Germans and Swiss are "as short "and as broad as those of the Tartars," &c.

[17] See a good discussion on this subject in Waitz, 'Introduct. to Anthropology,' Eng. translat. 1863, p. 198-208, 227. I have taken some of the above statements from H. Tuttle's 'Origin and Antiquity of Physical Man,' Boston, 1866, p. 35.

take the description of a group of highly varying organisms, has encountered cases (I speak after experience) precisely like that of man; and if of a cautious disposition, he will end by uniting all the forms which graduate into each other as a single species; for he will say to himself that he has no right to give names to objects which he cannot define. Cases of this kind occur in the Order which includes man, namely in certain genera of monkeys; whilst in other genera, as in Cercopithecus, most of the species can be determined with certainty. In the American genus Cebus, the various forms are ranked by some naturalists as species, by others as mere geographical races. Now if numerous specimens of Cebus were collected from all parts of South America, and those forms which at present appear to be specifically distinct, were found to graduate into each other by close steps, they would be ranked by most naturalists as mere varieties or races; and thus the greater number of naturalists have acted with respect to the races of man. Nevertheless it must be confessed that there are forms, at least in the vegetable kingdom,[18] which we cannot avoid naming as species, but which are connected together, independently of intercrossing, by numberless gradations.

Some naturalists have lately employed the term "sub-species" to designate forms which possess many of the characteristics of true species, but which hardly deserve so high a rank. Now if we reflect on the weighty arguments, above given, for raising the races of man to the dignity of species, and the insuperable difficulties on the other side in defining them, the term "sub-

[18] Prof. Nägeli has carefully described several striking cases in his 'Botanische Mittheilungen,' B. ii. 1866, s. 294-369. Prof. Asa Gray has made analogous remarks on some intermediate forms in the Compositæ of N. America.

species" might here be used with much propriety. But from long habit the term "race" will perhaps always be employed. The choice of terms is only so far important as it is highly desirable to use, as far as that may be possible, the same terms for the same degrees of difference. Unfortunately this is rarely possible; for within the same family the larger genera generally include closely-allied forms, which can be distinguished only with much difficulty, whilst the smaller genera include forms that are perfectly distinct; yet all must equally be ranked as species. So again the species within the same large genus by no means resemble each other to the same degree: on the contrary, in most cases some of them can be arranged in little groups round other species, like satellites round planets.[19]

The question whether mankind consists of one or several species has of late years been much agitated by anthropologists, who are divided into two schools of monogenists and polygenists. Those who do not admit the principle of evolution, must look at species either as separate creations or as in some manner distinct entities; and they must decide what forms to rank as species by the analogy of other organic beings which are commonly thus received. But it is a hopeless endeavour to decide this point on sound grounds, until some definition of the term "species" is generally accepted; and the definition must not include an element which cannot possibly be ascertained, such as an act of creation. We might as well attempt without any definition to decide whether a certain number of houses should be called a village, or town, or city. We have a practical illustration of the difficulty in the never-

[19] 'Origin of Species,' 5th edit. p. 68.

ending doubts whether many closely-allied mammals, birds, insects, and plants, which represent each other in North America and Europe, should be ranked species or geographical races; and so it is with the productions of many islands situated at some little distance from the nearest continent.

Those naturalists, on the other hand, who admit the principle of evolution, and this is now admitted by the greater number of rising men, will feel no doubt that all the races of man are descended from a single primitive stock; whether or not they think fit to designate them as distinct species, for the sake of expressing their amount of difference.[20] With our domestic animals the question whether the various races have arisen from one or more species is different. Although all such races, as well as all the natural species within the same genus, have undoubtedly sprung from the same primitive stock, yet it is a fit subject for discussion, whether, for instance, all the domestic races of the dog have acquired their present differences since some one species was first domesticated and bred by man; or whether they owe some of their characters to inheritance from distinct species, which had already been modified in a state of nature. With mankind no such question can arise, for he cannot be said to have been domesticated at any particular period.

When the races of man diverged at an extremely remote epoch from their common progenitor, they will have differed but little from each other, and been few in number; consequently they will then, as far as their distinguishing characters are concerned, have had less claim to rank as distinct species, than the existing so-

[20] See Prof. Huxley to this effect in the 'Fortnightly Review,' 1865, p. 275.

called races. Nevertheless such early races would perhaps have been ranked by some naturalists as distinct species, so arbitrary is the term, if their differences, although extremely slight, had been more constant than at present, and had not graduated into each other.

It is, however, possible, though far from probable, that the early progenitors of man might at first have diverged much in character, until they became more unlike each other than are any existing races; but that subsequently, as suggested by Vogt,[21] they converged in character. When man selects for the same object the offspring of two distinct species, he sometimes induces, as far as general appearance is concerned, a considerable amount of convergence. This is the case, as shewn by Von Nathusius,[22] with the improved breeds of pigs, which are descended from two distinct species; and in a less well-marked manner with the improved breeds of cattle. A great anatomist, Gratiolet, maintains that the anthropomorphous apes do not form a natural sub-group; but that the orang is a highly developed gibbon or semnopithecus; the chimpanzee a highly developed macacus; and the gorilla a highly developed mandrill. If this conclusion, which rests almost exclusively on brain-characters, be admitted, we should have a case of convergence at least in external characters, for the anthropomorphous apes are certainly more like each other in many points than they are to other apes. All analogical resemblances, as of a whale to a fish, may indeed be said to be cases of convergence; but this term has never been applied to superficial and adaptive resemblances. It

---

[21] 'Lectures on Man,' Eng. translat. 1864, p. 468.

[22] 'Die Racen des Schweines,' 1860, s. 46. 'Vorstudien für Geschichte, &c., Schweineschädel,' 1864, s. 104. With respect to cattle, see M. de Quatrefages, 'Unité de l'Espèce Humaine,' 1861, p. 119.

would be extremely rash in most cases to attribute to convergence close similarity in many points of structure in beings which had once been widely different. The form of a crystal is determined solely by the molecular forces, and it is not surprising that dissimilar substances should sometimes assume the same form; but with organic beings we should bear in mind that the form of each depends on an infinitude of complex relations, namely on the variations which have arisen, these being due to causes far too intricate to be followed out,—on the nature of the variations which have been preserved, and this depends on the surrounding physical conditions, and in a still higher degree on the surrounding organisms with which each has come into competition,—and lastly, on inheritance (in itself a fluctuating element) from innumerable progenitors, all of which have had their forms determined through equally complex relations. It appears utterly incredible that two organisms, if differing in a marked manner, should ever afterwards converge so closely as to lead to a near approach to identity throughout their whole organisation. In the case of the convergent pigs above referred to, evidence of their descent from two primitive stocks is still plainly retained, according to Von Nathusius, in certain bones of their skulls. If the races of man were descended, as supposed by some naturalists, from two or more distinct species, which had differed as much, or nearly as much, from each other, as the orang differs from the gorilla, it can hardly be doubted that marked differences in the structure of certain bones would still have been discoverable in man as he now exists.

Although the existing races of man differ in many respects, as in colour, hair, shape of skull, proportions of the body, &c., yet if their whole organisation be taken into consideration they are found to resemble each other closely in a multitude of points. Many of these points are of so unimportant or of so singular a nature, that it is extremely improbable that they should have been independently acquired by aboriginally distinct species or races. The same remark holds good with equal or greater force with respect to the numerous points of mental similarity between the most distinct races of man. The American aborigines, Negroes and Europeans differ as much from each other in mind as any three races that can be named; yet I was incessantly struck, whilst living with the Fuegians on board the "Beagle," with the many little traits of character, shewing how similar their minds were to ours; and so it was with a full-blooded negro with whom I happened once to be intimate.

He who will carefully read Mr. Tylor's and Sir J. Lubbock's interesting works[23] can hardly fail to be deeply impressed with the close similarity between the men of all races in tastes, dispositions and habits. This is shewn by the pleasure which they all take in dancing, rude music, acting, painting, tattooing, and otherwise decorating themselves,—in their mutual comprehension of gesture-language—and, as I shall be able to shew in a future essay, by the same expression in their features, and by the same inarticulate cries, when they are excited by various emotions. This similarity, or rather identity, is striking, when contrasted with the different expressions which may be observed in distinct species of monkeys. There is good evidence that the art of shooting with bows and arrows has not been handed down from any common progenitor of

---

[23] Tylor's 'Early History of Mankind,' 1865; for the evidence with respect to gesture-language, see p. 54. Lubbock's 'Prehistoric Times,' 2nd edit. 1869.

mankind, yet the stone arrow-heads, brought from the most distant parts of the world and manufactured at the most remote periods, are, as Nilsson has shewn,[24] almost identical; and this fact can only be accounted for by the various races having similar inventive or mental powers. The same observation has been made by archæologists[25] with respect to certain widely-prevalent ornaments, such as zigzags, &c.; and with respect to various simple beliefs and customs, such as the burying of the dead under megalithic structures. I remember observing in South America,[26] that there, as in so many other parts of the world, man has generally chosen the summits of lofty hills, on which to throw up piles of stones, either for the sake of recording some remarkable event, or for burying his dead.

Now when naturalists observe a close agreement in numerous small details of habits, tastes and dispositions between two or more domestic races, or between nearly-allied natural forms, they use this fact as an argument that all are descended from a common progenitor who was thus endowed; and consequently that all should be classed under the same species. The same argument may be applied with much force to the races of man.

As it is improbable that the numerous and unimportant points of resemblance between the several races of man in bodily structure and mental faculties (I do not here refer to similar customs) should all have been independently acquired, they must have been inherited from progenitors who were thus characterised. We thus gain some insight into the early state of man,

---

[24] 'The Primitive Inhabitants of Scandinavia,' Eng. translat. edited by Sir J. Lubbock, 1868, p. 104.

[25] Hodder M. Westropp, on Cromlechs, &c., 'Journal of Ethnological Soc.' as given in 'Scientific Opinion,' June 2nd, 1869, p. 3.

[26] 'Journal of Researches: Voyage of the "Beagle,"' p. 46.

before he had spread step by step over the face of the earth. The spreading of man to regions widely separated by the sea, no doubt, preceded any considerable amount of divergence of character in the several races; for otherwise we should sometimes meet with the same race in distinct continents; and this is never the case. Sir J. Lubbock, after comparing the arts now practised by savages in all parts of the world, specifies those which man could not have known, when he first wandered from his original birth-place; for if once learnt they would never have been forgotten.[27] He thus shews that "the spear, which is but a development of the "knife-point, and the club, which is but a long hammer, "are the only things left." He admits, however, that the art of making fire probably had already been discovered, for it is common to all the races now existing, and was known to the ancient cave-inhabitants of Europe. Perhaps the art of making rude canoes or rafts was likewise known; but as man existed at a remote epoch, when the land in many places stood at a very different level, he would have been able, without the aid of canoes, to have spread widely. Sir J. Lubbock further remarks how improbable it is that our earliest ancestors could have "counted as high as ten, considering that so many races now in existence cannot get "beyond four." Nevertheless, at this early period, the intellectual and social faculties of man could hardly have been inferior in any extreme degree to those now possessed by the lowest savages; otherwise primeval man could not have been so eminently successful in the struggle for life, as proved by his early and wide diffusion.

From the fundamental differences between certain

---

27 'Prehistoric Times,' 1869, p. 574.

languages, some philologists have inferred that when man first became widely diffused he was not a speaking animal; but it may be suspected that languages, far less perfect than any now spoken, aided by gestures, might have been used, and yet have left no traces on subsequent and more highly-developed tongues. Without the use of some language, however imperfect, it appears doubtful whether man's intellect could have risen to the standard implied by his dominant position at an early period.

Whether primeval man, when he possessed very few arts of the rudest kind, and when his power of language was extremely imperfect, would have deserved to be called man, must depend on the definition which we employ. In a series of forms graduating insensibly from some ape-like creature to man as he now exists, it would be impossible to fix on any definite point when the term "man" ought to be used. But this is a matter of very little importance. So again it is almost a matter of indifference whether the so-called races of man are thus designated, or are ranked as species or sub-species; but the latter term appears the most appropriate. Finally, we may conclude that when the principles of evolution are generally accepted, as they surely will be before long, the dispute between the monogenists and the polygenists will die a silent and unobserved death.

One other question ought not to be passed over without notice, namely, whether, as is sometimes assumed, each sub-species or race of man has sprung from a single pair of progenitors. With our domestic animals a new race can readily be formed from a single pair possessing some new character, or even from a single individual thus characterised, by carefully matching the varying offspring; but most of our races have been formed, not intentionally from a selected pair, but unconsciously by the preservation of many individuals which have varied, however slightly, in some useful or desired manner. If in one country stronger and heavier horses, and in another country lighter and fleeter horses, were habitually preferred, we may feel sure that two distinct sub-breeds would, in the course of time, be produced, without any particular pairs or individuals having been separated and bred from in either country. Many races have been thus formed, and their manner of formation is closely analogous with that of natural species. We know, also, that the horses which have been brought to the Falkland Islands have become, during successive generations, smaller and weaker, whilst those which have run wild on the Pampas have acquired larger and coarser heads; and such changes are manifestly due, not to any one pair, but to all the individuals having been subjected to the same conditions, aided, perhaps, by the principle of reversion. The new sub-breeds in none of these cases are descended from any single pair, but from many individuals which have varied in different degrees, but in the same general manner; and we may conclude that the races of man have been similarly produced, the modifications being either the direct result of exposure to different conditions, or the indirect result of some form of selection. But to this latter subject we shall presently return.

On the Extinction of the Races of Man.—The partial and complete extinction of many races and sub-races of man are historically known events. Humboldt saw in South America a parrot which was the sole living creature that could speak the language of a lost tribe.

Ancient monuments and stone implements found in all parts of the world, of which no tradition is preserved by the present inhabitants, indicate much extinction. Some small and broken tribes, remnants of former races, still survive in isolated and generally mountainous districts. In Europe the ancient races were all, according to Schaaffhausen,[28] "lower in "the scale than the rudest living savages;" they must therefore have differed, to a certain extent, from any existing race. The remains described by Professor Broca[29] from Les Eyzies, though they unfortunately appear to have belonged to a single family, indicate a race with a most singular combination of low or simious and high characteristics, and is "entirely different "from any other race, ancient or modern, that we have "ever heard of." It differed, therefore, from the quaternary race of the caverns of Belgium.

Unfavourable physical conditions appear to have had but little effect in the extinction of races.[30] Man has long lived in the extreme regions of the North, with no wood wherewith to make his canoes or other implements, and with blubber alone for burning and giving him warmth, but more especially for melting the snow. In the Southern extremity of America the Fuegians survive without the protection of clothes, or of any building worthy to be called a hovel. In South Africa the aborigines wander over the most arid plains, where dangerous beasts abound. Man can withstand the deadly influence of the Terai at the foot of the Himalaya, and the pestilential shores of tropical Africa.

---

[28] Translation in 'Anthropological Review,' Oct. 1868, p. 431.

[29] 'Transact. Internat. Congress of Prehistoric Arch.' 1868, p. 172-175. See also Broca (translation) in 'Anthropological Review,' Oct. 1868, p. 410.

[30] Dr. Gerland, 'Ueber das Aussterben der Naturvölker,' 1868, s. 82.

Extinction follows chiefly from the competition of tribe with tribe, and race with race. Various checks are always in action, as specified in a former chapter, which serve to keep down the numbers of each savage tribe,—such as periodical famines, the wandering of the parents and the consequent deaths of infants, prolonged suckling, the stealing of women, wars, accidents, sickness, licentiousness, especially infanticide, and, perhaps, lessened fertility from less nutritious food, and many hardships. If from any cause any one of these checks is lessened, even in a slight degree, the tribe thus favoured will tend to increase; and when one of two adjoining tribes becomes more numerous and powerful than the other, the contest is soon settled by war, slaughter, cannibalism, slavery, and absorption. Even when a weaker tribe is not thus abruptly swept away, if it once begins to decrease, it generally goes on decreasing until it is extinct.[31]

When civilised nations come into contact with barbarians the struggle is short, except where a deadly climate gives its aid to the native race. Of the causes which lead to the victory of civilised nations, some are plain and some very obscure. We can see that the cultivation of the land will be fatal in many ways to savages, for they cannot, or will not, change their habits. New diseases and vices are highly destructive; and it appears that in every nation a new disease causes much death, until those who are most susceptible to its destructive influence are gradually weeded out;[32] and so it may be with the evil effects from spirituous liquors, as well as with the unconquerably strong taste for them shewn by so many savages. It further appears, mysterious as is

---

31 Gerland (ibid. s. 12) gives facts in support of this statement.

32 See remarks to this effect in Sir H. Holland's 'Medical Notes and Reflections,' 1839, p. 390.

the fact, that the first meeting of distinct and separated people generates disease.[33] Mr. Sproat, who in Vancouver Island closely attended to the subject of extinction, believes that changed habits of life, which always follow from the advent of Europeans, induces much ill-health. He lays, also, great stress on so trifling a cause as that the natives become "bewildered and dull by the "new life around them; they lose the motives for exer-"tion, and get no new ones in their place."[34]

The grade of civilisation seems a most important element in the success of nations which come in competition. A few centuries ago Europe feared the inroads of Eastern barbarians; now, any such fear would be ridiculous. It is a more curious fact, that savages did not formerly waste away, as Mr. Bagehot has remarked, before the classical nations, as they now do before modern civilised nations; had they done so, the old moralists would have mused over the event; but there is no lament in any writer of that period over the perishing barbarians.[35]

Although the gradual decrease and final extinction of the races of man is an obscure problem, we can see that it depends on many causes, differing in different places and at different times. It is the same difficult problem as that presented by the extinction of one of the higher animals—of the fossil horse, for instance, which disappeared from South America, soon afterwards to be replaced, within the same districts, by countless troops of the Spanish horse. The New Zealander seems

[33] I have collected ('Journal of Researches, Voyage of the "Beagle,"' p. 435) a good many cases bearing on this subject: see also Gerland, ibid. s. 8. Poeppig speaks of the "breath of civilisation as poisonous "to savages."

[34] Sproat, 'Scenes and Studies of Savage Life,' 1868, p. 284.

[35] Bagehot, "Physics and Politics," 'Fortnightly Review,' April 1, 1868, p. 455.

conscious of this parallelism, for he compares his future fate with that of the native rat almost exterminated by the European rat. The difficulty, though great to our imagination, and really great if we wish to ascertain the precise causes, ought not to be so to our reason, as long as we keep steadily in mind that the increase of each species and each race is constantly hindered by various checks; so that if any new check, or cause of destruction, even a slight one, be superadded, the race will surely decrease in number; and as it has everywhere been observed that savages are much opposed to any change of habits, by which means injurious checks could be counterbalanced, decreasing numbers will sooner or later lead to extinction; the end, in most cases, being promptly determined by the inroads of increasing and conquering tribes.

On the Formation of the Races of Man.—It may be premised that when we find the same race, though broken up into distinct tribes, ranging over a great area, as over America, we may attribute their general resemblance to descent from a common stock. In some cases the crossing of races already distinct has led to the formation of new races. The singular fact that Europeans and Hindoos, who belong to the same Aryan stock and speak a language fundamentally the same, differ widely in appearance, whilst Europeans differ but little from Jews, who belong to the Semitic stock and speak quite another language, has been accounted for by Broca[36] through the Aryan branches having been largely crossed during their wide diffusion by various indigenous tribes. When two races in close contact

---

[36] "On Anthropology," translation, 'Anthropolog. Review,' Jan. 1868, p. 38.

cross, the first result is a heterogeneous mixture: thus Mr. Hunter, in describing the Santali or hill-tribes of India, says that hundreds of imperceptible gradations may be traced "from the black, squat tribes "of the mountains to the tall olive-coloured Brahman, "with his intellectual brow, calm eyes, and high but "narrow head;" so that it is necessary in courts of justice to ask the witnesses whether they are Santalis or Hindoos.[37] Whether a heterogeneous people, such as the inhabitants of some of the Polynesian islands, formed by the crossing of two distinct races, with few or no pure members left, would ever become homogeneous, is not known from direct evidence. But as with our domesticated animals, a crossed breed can certainly, in the course of a few generations, be fixed and made uniform by careful selection,[38] we may infer that the free and prolonged intercrossing during many generations of a heterogeneous mixture would supply the place of selection, and overcome any tendency to reversion, so that a crossed race would ultimately become homogeneous, though it might not partake in an equal degree of the characters of the two parent-races.

Of all the differences between the races of man, the colour of the skin is the most conspicuous and one of the best marked. Differences of this kind, it was formerly thought, could be accounted for by long exposure under different climates; but Pallas first shewed that this view is not tenable, and he has been followed by almost all anthropologists.[39] The view has been

[37] 'The Annals of Rural Bengal,' 1868, p. 134.

[38] 'The Variation of Animals and Plants under Domestication,' vol. ii. p. 95.

[39] Pallas, 'Act. Acad. St. Petersburgh,' 1780, part ii. p. 69. He was followed by Rudolphi, in his 'Beyträge zur Anthropologie,' 1812. An excellent summary of the evidence is given by Godron, 'De l'Espèce,' 1859, vol. ii. p. 246, &c.

rejected chiefly because the distribution of the variously coloured races, most of whom must have long inhabited their present homes, does not coincide with corresponding differences of climate. Weight must also be given to such cases as that of the Dutch families, who, as we hear on excellent authority,[40] have not undergone the least change of colour, after residing for three centuries in South Africa. The uniform appearance in various parts of the world of gypsies and Jews, though the uniformity of the latter has been somewhat exaggerated,[41] is likewise an argument on the same side. A very damp or a very dry atmosphere has been supposed to be more influential in modifying the colour of the skin than mere heat; but as D'Orbigny in South America, and Livingstone in Africa, arrived at diametrically opposite conclusions with respect to dampness and dryness, any conclusion on this head must be considered as very doubtful.[42]

Various facts, which I have elsewhere given, prove that the colour of the skin and hair is sometimes correlated in a surprising manner with a complete immunity from the action of certain vegetable poisons and from the attacks of certain parasites. Hence it occurred to me, that negroes and other dark races might have acquired their dark tints by the darker individuals escaping during a long series of generations from the deadly influence of the miasmas of their native countries.

I afterwards found that the same idea had long ago

---

40 Sir Andrew Smith, as quoted by Knox, 'Races of Man,' 1850, p. 473.

41 See De Quatrefages on this head, 'Revue des Cours Scientifiques,' Oct. 17, 1868, p. 731.

42 Livingstone's 'Travels and Researches in S. Africa,' 1857, p. 338, 329. D'Orbigny, as quoted by Godron, 'De l'Espèce,' vol. ii. p. 266.

occurred to Dr. Wells.[43] That negroes, and even mulattoes, are almost completely exempt from the yellow-fever, which is so destructive in tropical America, has long been known.[44] They likewise escape to a large extent the fatal intermittent fevers that prevail along, at least, 2600 miles of the shores of Africa, and which annually cause one-fifth of the white settlers to die, and another fifth to return home invalided.[45] This immunity in the negro seems to be partly inherent, depending on some unknown peculiarity of constitution, and partly the result of acclimatisation. Pouchet[46] states that the negro regiments, borrowed from the Viceroy of Egypt for the Mexican war, which had been recruited near the Soudan, escaped the yellow-fever almost equally well with the negroes originally brought from various parts of Africa, and accustomed to the climate of the West Indies. That acclimatisation plays a part is shewn by the many cases in which negroes, after having resided for some time in a colder climate, have become to a certain extent liable to tropical fevers.[47] The nature of the climate under which the white races have long resided, likewise has some influence on them; for during the fearful epidemic of yellow-fever in Demerara during 1837, Dr. Blair found that the death-rate of the immigrants was proportional

[43] See a paper read before the Royal Soc. in 1813, and published in his Essays in 1818. I have given an account of Dr. Wells' views in the Historical Sketch (p. xvi) to my 'Origin of Species.' Various cases of colour correlated with constitutional peculiarities are given in my 'Variation of Animals under Domestication,' vol. ii. p. 227, 335.

[44] See, for instance, Nott and Gliddon, 'Types of Mankind,' p. 68.

[45] Major Tulloch, in a paper read before the Statistical Society, April 20th, 1840, and given in the 'Athenæum,' 1840, p. 353.

[46] 'The Plurality of the Human Race' (translat.), 1864, p. 60.

[47] Quatrefages, 'Unité de l'Espèce Humaine,' 1861, p. 205. Waitz, 'Introduct. to Anthropology,' translat. vol. i. 1863, p. 124. Livingstone gives analogous cases in his 'Travels.'

to the latitude of the country whence they had come. With the negro the immunity, as far as it is the result of acclimatisation, implies exposure during a prodigious length of time; for the aborigines of tropical America, who have resided there from time immemorial, are not exempt from yellow-fever; and the Rev. B. Tristram states, that there are districts in Northern Africa which the native inhabitants are compelled annually to leave, though the negroes can remain with safety.

That the immunity of the negro is in any degree correlated with the colour of his skin is a mere conjecture: it may be correlated with some difference in his blood, nervous system, or other tissues. Nevertheless, from the facts above alluded to, and from some connection apparently existing between complexion and a tendency to consumption, the conjecture seemed to me not improbable. Consequently I endeavoured, with but little success,[48] to ascertain how far it held good. The

---

[48] In the spring of 1862 I obtained permission from the Director-General of the Medical department of the Army, to transmit to the surgeons of the various regiments on foreign service a blank table, with the following appended remarks, but I have received no returns. "As several well-marked cases have been recorded with our domestic "animals of a relation between the colour of the dermal appendages "and the constitution; and it being notorious that there is some limited "degree of relation between the colour of the races of man and the "climate inhabited by them; the following investigation seems worth "consideration. Namely, whether there is any relation in Europeans "between the colour of their hair, and their liability to the diseases of "tropical countries. If the surgeons of the several regiments, when "stationed in unhealthy tropical districts, would be so good as first to "count, as a standard of comparison, how many men, in the force "whence the sick are drawn, have dark and light-coloured hair, and "hair of intermediate or doubtful tints; and if a similar account were "kept by the same medical gentlemen, of all the men who suffered "from malarious and yellow fevers, or from dysentery, it would soon "be apparent, after some thousand cases had been tabulated, whether "there exists any relation between the colour of the hair and consti- "tutional liability to tropical diseases. Perhaps no such relation would

late Dr. Daniell, who had long lived on the West Coast of Africa, told me that he did not believe in any such relation. He was himself unusually fair, and had withstood the climate in a wonderful manner. When he first arrived as a boy on the coast, an old and experienced negro chief predicted from his appearance that this would prove the case. Dr. Nicholson, of Antigua, after having attended to this subject, wrote to me that he did not think that dark-coloured Europeans escaped the yellow-fever better than those that were light-coloured. Mr. J. M. Harris altogether denies[49] that Europeans with dark hair withstand a hot climate better than other men; on the contrary, experience has taught him in making a selection of men for service on the coast of Africa, to choose those with red hair. As far, therefore, as these slight indications serve, there seems no foundation for the hypothesis, which has been accepted by several writers, that the colour of the black races may have resulted from darker and darker individuals having survived in greater numbers, during their exposure to the fever-generating miasmas of their native countries.

Although with our present knowledge we cannot account for the strongly-marked differences in colour between the races of man, either through correlation with constitutional peculiarities, or through the direct action of climate; yet we must not quite ignore the

"be discovered, but the investigation is well worth making. In case "any positive result were obtained, it might be of some practical use "in selecting men for any particular service. Theoretically the result "would be of high interest, as indicating one means by which a race "of men inhabiting from a remote period an unhealthy tropical climate, "might have become dark-coloured by the better preservation of dark-"haired or dark-complexioned individuals during a long succession of "generations."

49 'Anthropological Review,' Jan. 1866, p. xxi.

latter agency, for there is good reason to believe that some inherited effect is thus produced.[50]

We have seen in our third chapter that the conditions of life, such as abundant food and general comfort, affect in a direct manner the development of the bodily frame, the effects being transmitted. Through the combined influences of climate and changed habits of life, European settlers in the United States undergo, as is generally admitted, a slight but extraordinarily rapid change of appearance. There is, also, a considerable body of evidence shewing that in the Southern States the house-slaves of the third generation present a markedly different appearance from the field-slaves.[51]

If, however, we look to the races of man, as distributed over the world, we must infer that their characteristic differences cannot be accounted for by the direct action of different conditions of life, even after exposure to them for an enormous period of time. The Esquimaux live exclusively on animal food; they are clothed in thick fur, and are exposed to intense cold and to prolonged darkness; yet they do not differ in any extreme degree from the inhabitants of Southern China, who live entirely on vegetable food and are exposed almost naked to a hot, glaring climate. The unclothed Fuegians live on the marine productions of their inhospitable shores; the Botocudos of Brazil wander

---

50 See, for instance, Quatrefages ('Revue des Cours Scientifiques,' Oct. 10, 1868, p. 724) on the effects of residence in Abyssinia and Arabia, and other analogous cases. Dr. Rolle ('Der Mensch, seine Abstammung,' &c., 1865, s. 99) states, on the authority of Khanikof, that the greater number of German families settled in Georgia, have acquired in the course of two generations dark hair and eyes. Mr. D. Forbes informs me that the Quichuas in the Andes vary greatly in colour, according to the position of the valleys inhabited by them.

51 Harlan, 'Medical Researches,' p. 532. Quatrefages ('Unité de l'Espèce Humaine,' 1861, p. 128) has collected much evidence on this head.

about the hot forests of the interior and live chiefly on vegetable productions; yet these tribes resemble each other so closely that the Fuegians on board the "Beagle" were mistaken by some Brazilians for Botocudos. The Botocudos again, as well as the other inhabitants of tropical America, are wholly different from the Negroes who inhabit the opposite shores of the Atlantic, are exposed to a nearly similar climate, and follow nearly the same habits of life.

Nor can the differences between the races of man be accounted for, except to a quite insignificant degree, by the inherited effects of the increased or decreased use of parts. Men who habitually live in canoes, may have their legs somewhat stunted; those who inhabit lofty regions have their chests enlarged; and those who constantly use certain sense-organs have the cavities in which they are lodged somewhat increased in size, and their features consequently a little modified. With civilised nations, the reduced size of the jaws from lessened use, the habitual play of different muscles serving to express different emotions, and the increased size of the brain from greater intellectual activity, have together produced a considerable effect on their general appearance in comparison with savages.[52] It is also possible that increased bodily stature, with no corresponding increase in the size of the brain, may have given to some races (judging from the previously adduced cases of the rabbits) an elongated skull of the dolichocephalic type.

Lastly, the little-understood principle of correlation will almost certainly have come into action, as in the case of great muscular development and strongly pro-

---

[52] See Prof. Schaaffhausen, translat. in 'Anthropological Review, Oct. 1868, p. 429.

jecting supra-orbital ridges. It is not improbable that the texture of the hair, which differs much in the different races, may stand in some kind of correlation with the structure of the skin; for the colour of the hair and skin are certainly correlated, as is its colour and texture with the Mandans.[53] The colour of the skin and the odour emitted by it are likewise in some manner connected. With the breeds of sheep the number of hairs within a given space and the number of the excretory pores stand in some relation to each other.[54] If we may judge from the analogy of our domesticated animals, many modifications of structure in man probably come under this principle of correlated growth.

We have now seen that the characteristic differences between the races of man cannot be accounted for in a satisfactory manner by the direct action of the conditions of life, nor by the effects of the continued use of parts, nor through the principle of correlation. We are therefore led to inquire whether slight individual differences, to which man is eminently liable, may not have been preserved and augmented during a long series of generations through natural selection. But here we are at once met by the objection that beneficial variations alone can be thus preserved; and as far as we are enabled to judge (although always liable to error on this head) not one of the external differences between the races of man are of any direct or

---

[53] Mr. Catlin states ('N. American Indians,' 3rd edit. 1842, vol. i. p. 49) that in the whole tribe of the Mandans, about one in ten or twelve of the members of all ages and both sexes have bright silvery grey hair, which is hereditary. Now this hair is as coarse and harsh as that of a horse's mane, whilst the hair of other colours is fine and soft.

[54] On the odour of the skin, Godron, 'Sur l'Espèce,' tom. ii. p. 217. On the pores in the skin, Dr. Wilckens, 'Die Aufgaben der landwirth. Zootechnik,' 1869, s. 7.

special service to him. The intellectual and moral or social faculties must of course be excepted from this remark; but differences in these faculties can have had little or no influence on external characters. The variability of all the characteristic differences between the races, before referred to, likewise indicates that these differences cannot be of much importance; for, had they been important, they would long ago have been either fixed and preserved, or eliminated. In this respect man resembles those forms, called by naturalists protean or polymorphic, which have remained extremely variable, owing, as it seems, to their variations being of an indifferent nature, and consequently to their having escaped the action of natural selection.

We have thus far been baffled in all our attempts to account for the differences between the races of man; but there remains one important agency, namely Sexual Selection, which appears to have acted as powerfully on man, as on many other animals. I do not intend to assert that sexual selection will account for all the differences between the races. An unexplained residuum is left, about which we can in our ignorance only say, that as individuals are continually born with, for instance, heads a little rounder or narrower, and with noses a little longer or shorter, such slight differences might become fixed and uniform, if the unknown agencies which induced them were to act in a more constant manner, aided by long-continued intercrossing. Such modifications come under the provisional class, alluded to in our fourth chapter, which for the want of a better term have been called spontaneous variations. Nor do I pretend that the effects of sexual selection can be indicated with scientific precision; but it can be shewn that it would be an inexplicable fact if man had not been modified by this agency, which has acted so powerfully on innumerable animals, both high and low in the scale. It can further be shewn that the differences between the races of man, as in colour, hairyness, form of features, &c., are of the nature which it might have been expected would have been acted on by sexual selection. But in order to treat this subject in a fitting manner, I have found it necessary to pass the whole animal kingdom in review; I have therefore devoted to it the Second Part of this work. At the close I shall return to man, and, after attempting to shew how far he has been modified through sexual selection, will give a brief summary of the chapters in this First Part.

Part II.

SEXUAL SELECTION.

PART II.—SEXUAL SELECTION.

CHAPTER VIII.

PRINCIPLES OF SEXUAL SELECTION.

Secondary sexual characters — Sexual selection — Manner of action — Excess of males — Polygamy — The male alone generally modified through sexual selection — Eagerness of the male — Variability of the male — Choice exerted by the female — Sexual compared with natural selection — Inheritance, at corresponding periods of life, at corresponding seasons of the year, and as limited by sex — Relations between the several forms of inheritance — Causes why one sex and the young are not modified through sexual selection — Supplement on the proportional numbers of the two sexes throughout the animal kingdom — On the limitation of the numbers of the two sexes through natural selection.

WITH animals which have their sexes separated, the males necessarily differ from the females in their organs of reproduction; and these afford the primary sexual characters. But the sexes often differ in what Hunter has called secondary sexual characters, which are not directly connected with the act of reproduction; for instance, in the male possessing certain organs of sense or locomotion, of which the female is quite destitute, or in having them more highly-developed, in order that he may readily find or reach her; or again, in the male having special organs of prehension so as to hold her securely. These latter organs of infinitely diversified kinds graduate into, and in some cases can hardly be distinguished from, those which are commonly ranked as primary, such as the complex appendages at the apex of the abdomen in male insects. Unless indeed we confine the term "primary" to the reproductive glands, it is scarcely possible to decide, as far as the organs of prehension are concerned, which ought to be called primary and which secondary.

The female often differs from the male in having organs for the nourishment or protection of her young, as the mammary glands of mammals, and the abdominal sacks of the marsupials. The male, also, in some few cases differs from the female in possessing analogous organs, as the receptacles for the ova possessed by the males of certain fishes, and those temporarily developed in certain male frogs. Female bees have a special apparatus for collecting and carrying pollen, and their ovipositor is modified into a sting for the defence of their larvæ and the community. In the females of many insects the ovipositor is modified in the most complex manner for the safe placing of the eggs. Numerous similar cases could be given, but they do not here concern us. There are, however, other sexual differences quite disconnected with the primary organs with which we are more especially concerned—such as the greater size, strength, and pugnacity of the male, his weapons of offence or means of defence against rivals, his gaudy colouring and various ornaments, his power of song, and other such characters.

Besides the foregoing primary and secondary sexual differences, the male and female sometimes differ in structures connected with different habits of life, and not at all, or only indirectly, related to the reproductive functions. Thus the females of certain flies (Culicidæ and Tabanidæ) are blood-suckers, whilst the males live on flowers and have their mouths destitute of mandibles.[1] The males alone of certain moths and of some

---

[1] Westwood, 'Modern Class. of Insects,' vol. ii. 1840, p. 541. In

crustaceans (*e.g.* Tanais) have imperfect, closed mouths, and cannot feed. The Complemental males of certain cirripedes live like epiphytic plants either on the female or hermaphrodite form, and are destitute of a mouth and prehensile limbs. In these cases it is the male which has been modified and has lost certain important organs, which the other members of the same group possess. In other cases it is the female which has lost such parts; for instance, the female glowworm is destitute of wings, as are many female moths, some of which never leave their cocoons. Many female parasitic crustaceans have lost their natatory legs. In some weevil-beetles (Curculionidæ) there is a great difference between the male and female in the length of the rostrum or snout;[2] but the meaning of this and of many analogous differences, is not at all understood. Differences of structure between the two sexes in relation to different habits of life are generally confined to the lower animals; but with some few birds the beak of the male differs from that of the female. No doubt in most, but apparently not in all these cases, the differences are indirectly connected with the propagation of the species: thus a female which has to nourish a multitude of ova will require more food than the male, and consequently will require special means for procuring it. A male animal which lived for a very short time might without detriment lose through disuse its organs for procuring food; but he would retain his locomotive organs in a perfect state, so that he might reach the female. The female, on the other hand, might safely lose her organs for flying, swimming,

---

regard to the statement about Tanais, mentioned below, I am indebted to Fritz Müller.

[2] Kirby and Spence, 'Introduction to Entomology,' vol. iii. 1826, p. 309.

or walking, if she gradually acquired habits which rendered such powers useless.

We are, however, here concerned only with that kind of selection, which I have called sexual selection. This depends on the advantage which certain individuals have over other individuals of the same sex and species, in exclusive relation to reproduction. When the two sexes differ in structure in relation to different habits of life, as in the cases above mentioned, they have no doubt been modified through natural selection, accompanied by inheritance limited to one and the same sex. So again the primary sexual organs, and those for nourishing or protecting the young, come under this same head; for those individuals which generated or nourished their offspring best, would leave, *cæteris paribus*, the greatest number to inherit their superiority; whilst those which generated or nourished their offspring badly, would leave but few to inherit their weaker powers. As the male has to search for the female, he requires for this purpose organs of sense and locomotion, but if these organs are necessary for the other purposes of life, as is generally the case, they will have been developed through natural selection. When the male has found the female he sometimes absolutely requires prehensile organs to hold her; thus Dr. Wallace informs me that the males of certain moths cannot unite with the females if their tarsi or feet are broken. The males of many oceanic crustaceans have their legs and antennæ modified in an extraordinary manner for the prehension of the female; hence we may suspect that owing to these animals being washed about by the waves of the open sea, they absolutely require these organs in order to propagate their kind, and if so their development will have been the result of ordinary or natural selection.

When the two sexes follow exactly the same habits

of life, and the male has more highly developed sense or locomotive organs than the female, it may be that these in their perfected state are indispensable to the male for finding the female; but in the vast majority of cases, they serve only to give one male an advantage over another, for the less well-endowed males, if time were allowed them, would succeed in pairing with the females; and they would in all other respects, judging from the structure of the female, be equally well adapted for their ordinary habits of life. In such cases sexual selection must have come into action, for the males have acquired their present structure, not from being better fitted to survive in the struggle for existence, but from having gained an advantage over other males, and from having transmitted this advantage to their male offspring alone. It was the importance of this distinction which led me to designate this form of selection as sexual selection. So again, if the chief service rendered to the male by his prehensile organs is to prevent the escape of the female before the arrival of other males, or when assaulted by them, these organs will have been perfected through sexual selection, that is by the advantage acquired by certain males over their rivals. But in most cases it is scarcely possible to distinguish between the effects of natural and sexual selection. Whole chapters could easily be filled with details on the differences between the sexes in their sensory, locomotive, and prehensile organs. As, however, these structures are not more interesting than others adapted for the ordinary purposes of life, I shall almost pass them over, giving only a few instances under each class.

There are many other structures and instincts which must have been developed through sexual selection—such as the weapons of offence and the means of defence possessed by the males for fighting with and driving away their rivals—their courage and pugnacity—their ornaments of many kinds—their organs for producing vocal or instrumental music—and their glands for emitting odours; most of these latter structures serving only to allure or excite the female. That these characters are the result of sexual and not of ordinary selection is clear, as unarmed, unornamented, or unattractive males would succeed equally well in the battle for life and in leaving a numerous progeny, if better endowed males were not present. We may infer that this would be the case, for the females, which are unarmed and unornamented, are able to survive and procreate their kind. Secondary sexual characters of the kind just referred to, will be fully discussed in the following chapters, as they are in many respects interesting, but more especially as they depend on the will, choice, and rivalry of the individuals of either sex. When we behold two males fighting for the possession of the female, or several male birds displaying their gorgeous plumage, and performing the strangest antics before an assembled body of females, we cannot doubt that, though led by instinct, they know what they are about, and consciously exert their mental and bodily powers.

In the same manner as man can improve the breed of his game-cocks by the selection of those birds which are victorious in the cockpit, so it appears that the strongest and most vigorous males, or those provided with the best weapons, have prevailed under nature, and have led to the improvement of the natural breed or species. Through repeated deadly contests, a slight degree of variability, if it led to some advantage, however slight, would suffice for the work of sexual selection; and it is certain that secondary sexual characters are eminently variable. In the same manner as man can give beauty, according to his standard of taste, to his male poultry—can give to the Sebright bantam a new and elegant plumage, an erect and peculiar carriage—so it appears that in a state of nature female birds, by having long selected the more attractive males, have added to their beauty. No doubt this implies powers of discrimination and taste on the part of the female which will at first appear extremely improbable; but I hope hereafter to shew that this is not the case.

From our ignorance on several points, the precise manner in which sexual selection acts is to a certain extent uncertain. Nevertheless if those naturalists who already believe in the mutability of species, will read the following chapters, they will, I think, agree with me that sexual selection has played an important part in the history of the organic world. It is certain that with almost all animals there is a struggle between the males for the possession of the female. This fact is so notorious that it would be superfluous to give instances. Hence the females, supposing that their mental capacity sufficed for the exertion of a choice, could select one out of several males. But in numerous cases it appears as if it had been specially arranged that there should be a struggle between many males. Thus with migratory birds, the males generally arrive before the females at their place of breeding, so that many males are ready to contend for each female. The bird-catchers assert that this is invariably the case with the nightingale and blackcap, as I am informed by Mr. Jenner Weir, who confirms the statement with respect to the latter species.

Mr. Swaysland of Brighton, who has been in the habit, during the last forty years, of catching our migratory birds on their first arrival, writes to me that he has never known the females of any species to arrive before their males. During one spring he shot thirty-nine males of Ray's wagtail (*Budytes Raii*) before he saw a single female. Mr. Gould has ascertained by dissection, as he informs me, that male snipes arrive in this country before the females. In the case of fish, at the period when the salmon ascend our rivers, the males in large numbers are ready to breed before the females. So it apparently is with frogs and toads. Throughout the great class of insects the males almost always emerge from the pupal state before the other sex, so that they generally swarm for a time before any females can be seen.[3] The cause of this difference between the males and females in their periods of arrival and maturity is sufficiently obvious. Those males which annually first migrated into any country, or which in the spring were first ready to breed, or were the most eager, would leave the largest number of offspring; and these would tend to inherit similar instincts and constitutions. On the whole there can be no doubt that with almost all animals, in which the sexes are separate, there is a constantly recurrent struggle between the males for the possession of the females.

Our difficulty in regard to sexual selection lies in understanding how it is that the males which conquer other males, or those which prove the most attractive to the females, leave a greater number of offspring to inherit their superiority than the beaten and less

---

[3] Even with those of plants in which the sexes are separate, the male flowers are generally mature before the female. Many hermaphrodite plants are, as first shewn by C. K. Sprengel, dichogamous; that is, their male and female organs are not ready at the same time, so that they cannot be self-fertilised. Now with such plants the pollen is generally mature in the same flower before the stigma, though there are some exceptional species in which the female organs are mature before the male.

attractive males. Unless this result followed, the characters which gave to certain males an advantage over others, could not be perfected and augmented through sexual selection. When the sexes exist in exactly equal numbers, the worst-endowed males will ultimately find females (excepting where polygamy prevails), and leave as many offspring, equally well fitted for their general habits of life, as the best-endowed males. From various facts and considerations, I formerly inferred that with most animals, in which secondary sexual characters were well developed, the males considerably exceeded the females in number; and this does hold good in some few cases. If the males were to the females as two to one, or as three to two, or even in a somewhat lower ratio, the whole affair would be simple; for the better-armed or more attractive males would leave the largest number of offspring. But after investigating, as far as possible, the numerical proportions of the sexes, I do not believe that any great inequality in number commonly exists. In most cases sexual selection appears to have been effective in the following manner.

Let us take any species, a bird for instance, and divide the females inhabiting a district into two equal bodies: the one consisting of the more vigorous and better-nourished individuals, and the other of the less vigorous and healthy. The former, there can be little doubt, would be ready to breed in the spring before the others; and this is the opinion of Mr. Jenner Weir, who has during many years carefully attended to the habits of birds. There can also be no doubt that the most vigorous, healthy, and best-nourished females would on an average succeed in rearing the largest number of offspring. The males, as we have seen, are generally ready to breed before the females; of the males the strongest, and with some species the best armed, drive away the weaker males; and the former would then unite with the more vigorous and best-nourished females, as these are the first to breed. Such vigorous pairs would surely rear a larger number of offspring than the retarded females, which would be compelled, supposing the sexes to be numerically equal, to unite with the conquered and less powerful males; and this is all that is wanted to add, in the course of successive generations, to the size, strength and courage of the males, or to improve their weapons.

But in a multitude of cases the males which conquer other males, do not obtain possession of the females, independently of choice on the part of the latter. The courtship of animals is by no means so simple and short an affair as might be thought. The females are most excited by, or prefer pairing with, the more ornamented males, or those which are the best songsters, or play the best antics; but it is obviously probable, as has been actually observed in some cases, that they would at the same time prefer the more vigorous and lively males.[4] Thus the more vigorous females, which are the first to breed, will have the choice of many males; and though they may not always select the strongest or best armed, they will select those which are vigorous and well armed, and in other respects the most attractive. Such early pairs would have the same advantage in rearing offspring on the female side as above explained, and nearly the same advantage on the male side. And this apparently has sufficed during a long course of generations to add not only to the strength and fighting-powers of

---

[4] I have received information, hereafter to be given, to this effect with respect to poultry. Even with birds, such as pigeons, which pair for life, the female, as I hear from Mr. Jenner Weir, will desert her mate if he is injured or grows weak.

the males, but likewise to their various ornaments or other attractions.

In the converse and much rarer case of the males selecting particular females, it is plain that those which were the most vigorous and had conquered others, would have the freest choice; and it is almost certain that they would select vigorous as well as attractive females. Such pairs would have an advantage in rearing offspring, more especially if the male had the power to defend the female during the pairing-season, as occurs with some of the higher animals, or aided in providing for the young. The same principles would apply if both sexes mutually preferred and selected certain individuals of the opposite sex; supposing that they selected not only the more attractive, but likewise the more vigorous individuals.

Numerical Proportion of the Two Sexes.—I have remarked that sexual selection would be a simple affair if the males considerably exceeded in number the females. Hence I was led to investigate, as far as I could, the proportions between the two sexes of as many animals as possible; but the materials are scanty. I will here give only a brief abstract of the results, retaining the details for a supplementary discussion, so as not to interfere with the course of my argument. Domesticated animals alone afford the opportunity of ascertaining the proportional numbers at birth; but no records have been specially kept for this purpose. By indirect means, however, I have collected a considerable body of statistical data, from which it appears that with most of our domestic animals the sexes are nearly equal at birth. Thus with race-horses, 25,560 births have been recorded during twenty-one years, and the male births have been to the female births as 99·7 to 100. With greyhounds the inequality is greater than with any other animal, for during twelve years, out of 6878 births, the male births have been as 110·1 to 100 female births. It is, however, in some degree doubtful whether it is safe to infer that the same proportional numbers would hold good under natural conditions as under domestication; for slight and unknown differences in the conditions affect to a certain extent the proportion of the sexes. Thus with mankind, the male births in England are as 104·5, in Russia as 108·9, and with the Jews of Livornia as 120 to 100 females. The proportion is also mysteriously affected by the circumstance of the births being legitimate or illegitimate.

For our present purpose we are concerned with the proportion of the sexes, not at birth, but at maturity, and this adds another element of doubt; for it is a well ascertained fact that with man a considerably larger proportion of males than of females die before or during birth, and during the first few years of infancy. So it almost certainly is with male lambs, and so it may be with the males of other animals. The males of some animals kill each other by fighting; or they drive each other about until they become greatly emaciated. They must, also, whilst wandering about in eager search for the females, be often exposed to various dangers. With many kinds of fish the males are much smaller than the females, and they are believed often to be devoured by the latter or by other fishes. With some birds the females appear to die in larger proportion than the males: they are also liable to be destroyed on their nests, or whilst in charge of their young. With insects the female larvæ are often larger than those of the males, and would consequently be more likely to be devoured: in some cases the mature females are less active and less rapid in their movements than the males, and would not be so well able to escape from danger. Hence, with animals in a state of nature, in order to judge of the proportions of the sexes at maturity, we must rely on mere estimation; and this, except perhaps when the inequality is strongly marked, is but little trustworthy. Nevertheless, as far as a judgment can be formed, we may conclude from the facts given in the supplement, that the males of some few mammals, of many birds, of some fish and insects, considerably exceed in number the females.

The proportion between the sexes fluctuates slightly during successive years: thus with race-horses, for every 100 females born, the males varied from 107·1 in one year to 92·6 in another year, and with greyhounds from 116·3 to 95·3. But had larger numbers been tabulated throughout a more extensive area than England, these fluctuations would probably have disappeared; and such as they are, they would hardly suffice to lead under a state of nature to the effective action of sexual selection. Nevertheless with some few wild animals, the proportions seem, as shewn in the supplement, to fluctuate either during different seasons or in different localities in a sufficient degree to lead to such action. For it should be observed that any advantage gained during certain years or in certain localities by those males which were able to conquer other males, or were the most attractive to the females, would probably be transmitted to the offspring and would not subsequently be eliminated. During the succeeding seasons, when from the equality of the sexes every male was everywhere able to procure a female, the stronger or more attractive males previously produced would still have at least as good a chance of leaving offspring as the less strong or less attractive.

Polygamy.—The practice of polygamy leads to the

same results as would follow from an actual inequality in the number of the sexes; for if each male secures two or more females, many males will not be able to pair; and the latter assuredly will be the weaker or less attractive individuals. Many mammals and some few birds are polygamous, but with animals belonging to the lower classes I have found no evidence of this habit. The intellectual powers of such animals are, perhaps, not sufficient to lead them to collect and guard a harem of females. That some relation exists between polygamy and the development of secondary sexual characters, appears nearly certain; and this supports the view that a numerical preponderance of males would be eminently favourable to the action of sexual selection. Nevertheless many animals, especially birds, which are strictly monogamous, display strongly-marked secondary sexual characters; whilst some few animals, which are polygamous, are not thus characterised.

We will first briefly run through the class of mammals, and then turn to birds. The gorilla seems to be a polygamist, and the male differs considerably from the female; so it is with some baboons which live in herds containing twice as many adult females as males. In South America the *Mycetes caraya* presents well-marked sexual differences in colour, beard, and vocal organs, and the male generally lives with two or three wives: the male of the *Cebus capucinus* differs somewhat from the female, and appears to be polygamous.[5] Little is known on this head with respect to most other monkeys, but some species are strictly monogamous. The ruminants are eminently polygamous, and they

[5] On the Gorilla, Savage and Wyman, 'Boston Journal of Nat. Hist.' vol. v. 1845-47, p. 423. On Cynocephalus, Brehm, 'Illust. Thierleben,' B. i. 1864, s. 77. On Mycetes, Rengger, 'Naturgesch.: Säugethiere von Paraguay,' 1830, s. 14, 20. On Cebus, Brehm, ibid. s. 108.

more frequently present sexual differences than almost any other group of mammals, especially in their weapons, but likewise in other characters. Most deer, cattle, and sheep are polygamous; as are most antelopes, though some of the latter are monogamous. Sir Andrew Smith, in speaking of the antelopes of South Africa, says that in herds of about a dozen there was rarely more than one mature male. The Asiatic *Antilope saiga* appears to be the most inordinate polygamist in the world; for Pallas[6] states that the male drives away all rivals, and collects a herd of about a hundred, consisting of females and kids: the female is hornless and has softer hair, but does not otherwise differ much from the male. The horse is polygamous, but, except in his greater size and in the proportions of his body, differs but little from the mare. The wild boar, in his great tusks and some other characters, presents well-marked sexual characters; in Europe and in India he leads a solitary life, except during the breeding-season; but at this season he consorts in India with several females, as Sir W. Elliot, who has had large experience in observing this animal, believes: whether this holds good in Europe is doubtful, but is supported by some statements. The adult male Indian elephant, like the boar, passes much of his time in solitude; but when associating with others, "it is rare to find," as Dr. Campbell states, "more than one male with a whole "herd of females." The larger males expel or kill the smaller and weaker ones. The male differs from the female by his immense tusks and greater size, strength, and endurance; so great is the difference in these latter

[6] Pallas, 'Spicilegia Zoolog.' Fasc. xii. 1777, p. 29. Sir Andrew Smith, 'Illustrations of the Zoology of S. Africa,' 1849, pl. 29, on the Kobus. Owen, in his 'Anatomy of Vertebrates' (vol. iii. 1868, p. 633) gives a table incidentally showing which species of Antelopes pair and which are gregarious.

respects, that the males when caught are valued at twenty per cent. above the females.[7] With other pachydermatous animals the sexes differ very little or not at all, and they are not, as far as known, polygamists. Hardly a single species amongst the Cheiroptera and Edentata, or in the great Orders of the Rodents and Insectivora, presents well-developed secondary sexual differences; and I can find no account of any species being polygamous, excepting, perhaps, the common rat, the males of which, as some rat-catchers affirm, live with several females.

The lion in South Africa, as I hear from Sir Andrew Smith, sometimes lives with a single female, but generally with more than one, and, in one case, was found with as many as five females, so that he is polygamous. He is, as far as I can discover, the sole polygamist in the whole group of the terrestrial Carnivora, and he alone presents well-marked sexual characters. If, however, we turn to the marine Carnivora, the case is widely different; for many species of seals offer, as we shall hereafter see, extraordinary sexual differences, and they are eminently polygamous. Thus the male sea-elephant of the Southern Ocean, always possesses, according to Péron, several females, and the sea-lion of Forster is said to be surrounded by from twenty to thirty females. In the North, the male sea-bear of Steller is accompanied by even a greater number of females.

With respect to birds, many species, the sexes of which differ greatly from each other, are certainly monogamous. In Great Britain we see well-marked sexual differences in, for instance, the wild-duck which pairs with a single female, with the common blackbird,

---

7 Dr. Campbell, in 'Proc. Zoolog. Soc.' 1869, p. 138. See also an interesting paper, by Lieut. Johnstone, in 'Proc. Asiatic Soc. of Bengal,' May, 1868.

and with the bullfinch which is said to pair for life. So it is, as I am informed by Mr. Wallace, with the Chatterers or Cotingidæ of South America, and numerous other birds. In several groups I have not been able to discover whether the species are polygamous or monogamous. Lesson says that birds of paradise, so remarkable for their sexual differences, are polygamous, but Mr. Wallace doubts whether he had sufficient evidence. Mr. Salvin informs me that he has been led to believe that humming-birds are polygamous. The male widow-bird, remarkable for his caudal plumes, certainly seems to be a polygamist.[8] I have been assured by Mr. Jenner Weir and by others, that three starlings not rarely frequent the same nest; but whether this is a case of polygamy or polyandry has not been ascertained.

The Gallinaceæ present almost as strongly marked sexual differences as birds of paradise or humming-birds, and many of the species are, as is well known, polygamous; others being strictly monogamous. What a contrast is presented between the sexes by the polygamous peacock or pheasant, and the monogamous guinea-fowl or partridge! Many similar cases could be given, as in the grouse tribe, in which the males of the polygamous capercailzie and black-cock differ greatly from the females; whilst the sexes of the monogamous red grouse and ptarmigan differ very little. Amongst the Cursores, no great number of species offer strongly-marked sexual differences, except the bustards, and the great bustard (*Otis tarda*), is said to

---

[8] 'The Ibis,' vol. iii. 1861, p. 133, on the Progne Widow-bird. See also on the Vidua axillaris, ibid. vol. ii. 1860, p. 211. On the polygamy of the Capercailzie and Great Bustard, see L. Lloyd, 'Game Birds of Sweden,' 1867, p. 19, and 182. Montagu and Selby speak of the Black Grouse as polygamous and of the Red Grouse as monogamous.

be polygamous. With the Grallatores, extremely few species differ sexually, but the ruff (*Machetes pugnax*) affords a strong exception, and this species is believed by Montagu to be a polygamist. Hence it appears that with birds there often exists a close relation between polygamy and the development of strongly-marked sexual differences. On asking Mr. Bartlett, at the Zoological Gardens, who has had such large experience with birds, whether the male tragopan (one of the Gallinaceæ) was polygamous, I was struck by his answering, "I do not know, but should think so from "his splendid colours."

It deserves notice that the instinct of pairing with a single female is easily lost under domestication. The wild-duck is strictly monogamous, the domestic-duck highly polygamous. The Rev. W. D. Fox informs me that with some half-tamed wild-ducks, kept on a large pond in his neighbourhood, so many mallards were shot by the gamekeeper that only one was left for every seven or eight females; yet unusually large broods were reared. The guinea-fowl is strictly monogamous; but Mr. Fox finds that his birds succeed best when he keeps one cock to two or three hens.[9] Canary-birds pair in a state of nature, but the breeders in England successfully put one male to four or five females; nevertheless the first female, as Mr. Fox has been assured, is alone treated as the wife, she and her young ones being fed by him; the others are treated as concubines. I have noticed these cases, as it renders it in some degree probable that monogamous species, in a state of nature, might readily become either temporarily or permanently polygamous.

9 The Rev. E. S. Dixon, however, speaks positively ('Ornamental Poultry,' 1848, p. 76) about the eggs of the guinea-fowl being infertile when more than one female is kept with the same male.

With respect to reptiles and fishes, too little is known of their habits to enable us to speak of their marriage arrangements. The stickle-back Gasterosteus), however, is said to be a polygamist;[10] and the male during the breeding-season differs conspicuously from the female.

To sum up on the means through which, as far as we can judge, sexual selection has led to the development of secondary sexual characters. It has been shewn that the largest number of vigorous offspring will be reared from the pairing of the strongest and best-armed males, which have conquered other males, with the most vigorous and best-nourished females, which are the first to breed in the spring. Such females, if they select the more attractive, and at the same time vigorous, males, will rear a larger number of offspring than the retarded females, which must pair with the less vigorous and less attractive males. So it will be if the more vigorous males select the more attractive and at the same time healthy and vigorous females; and this will especially hold good if the male defends the female, and aids in providing food for the young. The advantage thus gained by the more vigorous pairs in rearing a larger number of offspring has apparently sufficed to render sexual selection efficient. But a large preponderance in number of the males over the females would be still more efficient; whether the preponderance was only occasional and local, or permanent; whether it occurred at birth, or subsequently from the greater destruction of the females; or whether it indirectly followed from the practice of polygamy.

The Male generally more modified than the Female.— Throughout the animal kingdom, when the sexes differ

---

[10] Noel Humphreys, 'River Gardens,' 1857.

from each other in external appearance, it is the male which, with rare exceptions, has been chiefly modified; for the female still remains more like the young of her own species, and more like the other members of the same group. The cause of this seems to lie in the males of almost all animals having stronger passions than the females. Hence it is the males that fight together and sedulously display their charms before the females; and those which are victorious transmit their superiority to their male offspring. Why the males do not transmit their characters to both sexes will hereafter be considered. That the males of all mammals eagerly pursue the females is notorious to every one. So it is with birds; but many male birds do not so much pursue the female, as display their plumage, perform strange antics, and pour forth their song, in her presence. With the few fish which have been observed, the male seems much more eager than the female; and so it is with alligators, and apparently with Batrachians. Throughout the enormous class of insects, as Kirby remarks,[11] "the law is, that the male "shall seek the female." With spiders and crustaceans, as I hear from two great authorities, Mr. Blackwall and Mr. C. Spence Bate, the males are more active and more erratic in their habits than the females. With insects and crustaceans, when the organs of sense or locomotion are present in the one sex and absent in the other, or when, as is frequently the case, they are more highly developed in the one than the other, it is almost invariably the male, as far as I can discover, which retains such organs, or has them most developed; and this shews that the male is the more active member in the courtship of the sexes.[12]

11 Kirby and Spence, 'Introduction to Entomology,' vol. iii. 1826, p. 342.

12 One parasitic Hymenopterous insect (Westwood, 'Modern Class. of Insects,' vol. ii. p. 160) forms an exception to the rule, as the male

The female, on the other hand, with the rarest exception, is less eager than the male. As the illustrious Hunter[13] long ago observed, she generally "requires to "be courted;" she is coy, and may often be seen endeavouring for a long time to escape from the male. Every one who has attended to the habits of animals will be able to call to mind instances of this kind. Judging from various facts, hereafter to be given, and from the results which may fairly be attributed to sexual selection, the female, though comparatively passive, generally exerts some choice and accepts one male in preference to others. Or she may accept, as appearances would sometimes lead us to believe, not the male which is the most attractive to her, but the one which is the least distasteful. The exertion of some choice on the part of the female seems almost as general a law as the eagerness of the male.

We are naturally led to enquire why the male in so many and such widely distinct classes has been rendered more eager than the female, so that he searches for her and plays the more active part in courtship. It would be no advantage and some loss of power if both sexes were mutually to search for each other; but why should the male almost always be the seeker? With plants, the ovules after fertilisation have to be nourished for a time; hence the pollen is necessarily brought to the female organs—being placed on the stigma, through the agency of insects or of the wind,

has rudimentary wings, and never quits the cell in which it is born, whilst the female has well-developed wings. Audouin believes that the females are impregnated by the males which are born in the same cells with them; but it is much more probable that the females visit other cells, and thus avoid close interbreeding. We shall hereafter meet with a few exceptional cases, in various classes, in which the female, instead of the male, is the seeker and wooer.

13 'Essays and Observations,' edited by Owen, vol. i. 1861, p. 194.

or by the spontaneous movements of the stamens; and with the Algæ, &c., by the locomotive power of the antherozooids. With lowly-organised animals permanently affixed to the same spot and having their sexes separate, the male element is invariably brought to the female; and we can see the reason; for the ova, even if detached before being fertilised and not requiring subsequent nourishment or protection, would be, from their larger relative size, less easily transported than the male element. Hence plants [14] and many of the lower animals are, in this respect, analogous. In the case of animals not affixed to the same spot, but enclosed within a shell with no power of protruding any part of their bodies, and in the case of animals having little power of locomotion, the males must trust the fertilising element to the risk of at least a short transit through the waters of the sea. It would, therefore, be a great advantage to such animals, as their organisation became perfected, if the males when ready to emit the fertilising element, were to acquire the habit of approaching the female as closely as possible. The males of various lowly-organised animals having thus aboriginally acquired the habit of approaching and seeking the females, the same habit would naturally be transmitted to their more highly developed male descendants; and in order that they should become efficient seekers, they would have to be endowed with strong passions. The acquirement of such passions would naturally follow from the more eager males leaving a larger number of offspring than the less eager.

The great eagerness of the male has thus indirectly

---

[14] Prof. Sachs ('Lehrbuch der Botanik,' 1870, s. 633) in speaking of the male and female reproductive cells, remarks, "verhält sich die eine "bei der Vereinigung activ, . . . die andere erscheint bei der Vereinigung passiv."

led to the much more frequent development of secondary sexual characters in the male than in the female. But the development of such characters will have been much aided, if the conclusion at which I arrived after studying domesticated animals, can be trusted, namely, that the male is more liable to vary than the female. I am aware how difficult it is to verify a conclusion of this kind. Some slight evidence, however, can be gained by comparing the two sexes in mankind, as man has been more carefully observed than any other animal. During the Novara Expedition[15] a vast number of measurements of various parts of the body in different races were made, and the men were found in almost every case to present a greater range of variation than the women; but I shall have to recur to this subject in a future chapter. Mr. J. Wood,[16] who has carefully attended to the variation of the muscles in man, puts in italics the conclusion that "the greatest number of "abnormalities in each subject is found in the males." He had previously remarked that "altogether in 102 "subjects the varieties of redundancy were found to "be half as many again as in females, contrasting "widely with the greater frequency of deficiency in "females before described." Professor Macalister like wise remarks[17] that variations in the muscles "are "probably more common in males than females." Certain muscles which are not normally present in mankind are also more frequently developed in the male than in the female sex, although exceptions to this rule

---

[15] 'Reise der Novara: Anthropolog. Theil,' 1867, s. 216-269. The results were calculated by Dr. Weisbach from measurements made by Drs. K. Scherzer and Schwarz. On the greater variability of the males of domesticated animals, see my 'Variation of Animals and Plants under Domestication,' vol. ii. 1868, p. 75.

[16] 'Proceedings Royal Soc.' vol. xvi. July, 1868, p. 519 and 524.

[17] 'Proc. Royal Irish Academy,' vol. x. 1868, p. 123.

are said to occur. Dr. Burt Wilder[18] has tabulated the cases of 152 individuals with supernumerary digits, of which 86 were males, and 39, or less than half, females; the remaining 27 being of unknown sex. It should not, however, be overlooked that women would more frequently endeavour to conceal a deformity of this kind than men. Whether the large proportional number of deaths of the male offspring of man and apparently of sheep, compared with the female offspring, before, during, and shortly after birth (see supplement), has any relation to a stronger tendency in the organs of the male to vary and thus to become abnormal in structure or function, I will not pretend to conjecture.

In various classes of animals a few exceptional cases occur, in which the female instead of the male has acquired well pronounced secondary sexual characters, such as brighter colours, greater size, strength, or pugnacity. With birds, as we shall hereafter see, there has sometimes been a complete transposition of the ordinary characters proper to each sex; the females having become the more eager in courtship, the males remaining comparatively passive, but apparently selecting, as we may infer from the results, the more attractive females. Certain female birds have thus been rendered more highly coloured or otherwise ornamented, as well as more powerful and pugnacious than the males, these characters being transmitted to the female offspring alone.

It may be suggested that in some cases a double process of selection has been carried on; the males having selected the more attractive females, and the latter the more attractive males. This process however, though it might lead to the modification of both sexes,

---

18 'Massachusetts Medical Soc.' vol. ii. No. 3, 1868, p. 9.

would not make the one sex different from the other, unless indeed their taste for the beautiful differed; but this is a supposition too improbable in the case of any animal, excepting man, to be worth considering. There are, however, many animals, in which the sexes resemble each other, both being furnished with the same ornaments, which analogy would lead us to attribute to the agency of sexual selection. In such cases it may be suggested with more plausibility, that there has been a double or mutual process of sexual selection; the more vigorous and precocious females having selected the more attractive and vigorous males, the latter having rejected all except the more attractive females. But from what we know of the habits of animals, this view is hardly probable, the male being generally eager to pair with any female. It is more probable that the ornaments common to both sexes were acquired by one sex, generally the male, and then transmitted to the offspring of both sexes. If, indeed, during a lengthened period the males of any species were greatly to exceed the females in number, and then during another lengthened period under different conditions the reverse were to occur, a double, but not simultaneous, process of sexual selection might easily be carried on, by which the two sexes might be rendered widely different.

We shall hereafter see that many animals exist, of which neither sex is brilliantly coloured or provided with special ornaments, and yet the members of both sexes or of one alone have probably been modified through sexual selection. The absence of bright tints or other ornaments may be the result of variations of the right kind never having occurred, or of the animals themselves preferring simple colours, such as plain black or white. Obscure colours have often been acquired through natural selection for the sake of protection, and the acquirement through sexual selection of conspicuous colours, may have been checked from the danger thus incurred. But in other cases the males have probably struggled together during long ages, through brute force, or by the display of their charms, or by both means combined, and yet no effect will have been produced unless a larger number of offspring were left by the more successful males to inherit their superiority, than by the less successful males; and this, as previously shewn, depends on various complex contingencies.

Sexual selection acts in a less rigorous manner than natural selection. The latter produces its effects by the life or death at all ages of the more or less successful individuals. Death, indeed, not rarely ensues from the conflicts of rival males. But generally the less successful male merely fails to obtain a female, or obtains later in the season a retarded and less vigorous female, or, if polygamous, obtains fewer females; so that they leave fewer, or less vigorous, or no offspring. In regard to structures acquired through ordinary or natural selection, there is in most cases, as long as the conditions of life remain the same, a limit to the amount of advantageous modification in relation to certain special ends; but in regard to structures adapted to make one male victorious over another, either in fighting or in charming the female, there is no definite limit to the amount of advantageous modification; so that as long as the proper variations arise the work of sexual selection will go on. This circumstance may partly account for the frequent and extraordinary amount of variability presented by secondary sexual characters. Nevertheless, natural selection will determine that characters of this kind shall not be acquired by the victorious males, which would be injurious to them in any high degree, either by expending too much of their vital powers, or by exposing them to any great danger. The development, however, of certain structures—of the horns, for instance, in certain stags—has been carried to a wonderful extreme; and in some instances to an extreme which, as far as the general conditions of life are concerned, must be slightly injurious to the male. From this fact we learn that the advantages which favoured males have derived from conquering other males in battle or courtship, and thus leaving a numerous progeny, have been in the long run greater than those derived from rather more perfect adaptation to the external conditions of life. We shall further see, and this could never have been anticipated, that the power to charm the female has been in some few instances more important than the power to conquer other males in battle.

LAWS OF INHERITANCE.

In order to understand how sexual selection has acted, and in the course of ages has produced conspicuous results with many animals of many classes, it is necessary to bear in mind the laws of inheritance, as far as they are known. Two distinct elements are included under the term "inheritance," namely the transmission and the development of characters; but as these generally go together, the distinction is often overlooked. We see this distinction in those characters which are transmitted through the early years of life, but are developed only at maturity or during old age. We see the same distinction more clearly with secondary sexual characters, for these are transmitted through both sexes, though developed in one alone. That they are present in both sexes, is manifest when two species, having strongly-marked sexual characters, are crossed, for each transmits the characters proper to its own male and female sex to the hybrid offspring of both sexes. The same fact is likewise manifest, when characters proper to the male are occasionally developed in the female when she grows old or becomes diseased; and so conversely with the male. Again, characters occasionally appear, as if transferred from the male to the female, as when, in certain breeds of the fowl, spurs regularly appear in the young and healthy females; but in truth they are simply developed in the female; for in every breed each detail in the structure of the spur is transmitted through the female to her male offspring. In all cases of reversion, characters are transmitted through two, three, or many generations, and are then under certain unknown favourable conditions developed. This important distinction between transmission and development will be easiest kept in mind by the aid of the hypothesis of pangenesis, whether or not it be accepted as true. According to this hypothesis, every unit or cell of the body throws off gemmules or undeveloped atoms, which are transmitted to the offspring of both sexes, and are multiplied by self-division. They may remain undeveloped during the early years of life or during successive generations; their development into units or cells, like those from which they were derived, depending on their affinity for, and union with, other units or cells previously developed in the due order of growth.

Inheritance at Corresponding Periods of Life.—This tendency is well established. If a new character appears in an animal whilst young, whether it endures throughout life or lasts only for a time, it will reappear, as a general rule, at the same age and in the same manner in the offspring. If, on the other hand, a new character appears at maturity, or even during old age, it tends to reappear in the offspring at the same advanced age. When deviations from this rule occur, the transmitted characters much oftener appear before than after the corresponding age. As I have discussed this subject at sufficient length in another work,[19] I will here merely give two or three instances, for the sake of recalling the subject to the reader's mind. In several breeds of the Fowl, the chickens whilst covered with down, in their first true plumage, and in their adult plumage, differ greatly from each other, as well as from their common parent-form, the *Gallus bankiva;* and these characters are faithfully transmitted by each breed to their offspring at the corresponding period of life. For instance, the chickens of spangled Hamburghs, whilst covered with down, have a few dark spots on the head and rump, but are not longitudinally striped, as in many other breeds; in their first true plumage, "they "are beautifully pencilled," that is each feather is transversely marked by numerous dark bars; but in their second plumage the feathers all become spangled or tipped with a dark round spot.[20] Hence in this breed variations have occurred and have been transmitted at three distinct periods of life. The Pigeon offers a more remarkable case, because the aboriginal parent-species does not undergo with advancing age any change of plumage, excepting that at maturity the breast becomes more iridescent; yet there are breeds which do not acquire their characteristic colours until they

---

[19] 'The Variation of Animals and Plants under Domestication,' vol. ii. 1868, p. 75. In the last chapter but one, the provisional hypothesis of pangenesis, above alluded to, is fully explained.

[20] These facts are given on the high authority of a great breeder, Mr. Teebay, in Tegetmeier's 'Poultry Book,' 1868, p. 158. On the characters of chickens of different breeds, and on the breeds of the pigeon, alluded to in the above paragraph, see 'Variation of Animals,' &c., vol. i. p. 160, 249; vol. ii. p. 77.

have moulted two, three, or four times; and these modifications of plumage are regularly transmitted.

Inheritance at Corresponding Seasons of the Year.—With animals in a state of nature innumerable instances occur of characters periodically appearing at different seasons. We see this with the horns of the stag, and with the fur of arctic animals which becomes thick and white during the winter. Numerous birds acquire bright colours and other decorations during the breeding-season alone. I can throw but little light on this form of inheritance from facts observed under domestication. Pallas states,[21] that in Siberia domestic cattle and horses periodically become lighter-coloured during the winter; and I have observed a similar marked change of colour in certain ponies in England. Although I do not know that this tendency to assume a differently coloured coat during different seasons of the year is transmitted, yet it probably is so, as all shades of colour are strongly inherited by the horse. Nor is this form of inheritance, as limited by season, more remarkable than inheritance as limited by age or sex.

Inheritance as Limited by Sex.—The equal transmission of characters to both sexes is the commonest form of inheritance, at least with those animals which do not present strongly-marked sexual differences, and indeed with many of these. But characters are not rarely transferred exclusively to that sex, in which they first appeared. Ample evidence on this head has been advanced in my work on Variation under Domestica-

---

[21] 'Novæ species Quadrupedum e Glirium ordine,' 1778, p. 7. On the transmission of colour by the horse, see 'Variation of Animals, &c. under Domestication,' vol. i. p. 21. Also vol. ii. p. 71, for a general discussion on Inheritance as limited by Sex.

tion; but a few instances may here be given. There are breeds of the sheep and goat, in which the horns of the male differ greatly in shape from those of the female; and these differences, acquired under domestication, are regularly transmitted to the same sex. With tortoise-shell cats the females alone, as a general rule, are thus coloured, the males being rusty-red. With most breeds of the fowl, the characters proper to each sex are transmitted to the same sex alone. So general is this form of transmission that it is an anomaly when we see in certain breeds variations transmitted equally to both sexes. There are also certain sub-breeds of the fowl in which the males can hardly be distinguished from each other, whilst the females differ considerably in colour. With the pigeon the sexes of the parent-species do not differ in any external character; nevertheless in certain domesticated breeds the male is differently coloured from the female.[22] The wattle in the English Carrier pigeon and the crop in the Pouter are more highly developed in the male than in the female; and although these characters have been gained through long-continued selection by man, the difference between the two sexes is wholly due to the form of inheritance which has prevailed; for it has arisen, not from, but rather in opposition to, the wishes of the breeder.

Most of our domestic races have been formed by the accumulation of many slight variations; and as some of the successive steps have been transmitted to one sex alone, and some to both sexes, we find in the different breeds of the same species all gradations between great sexual dissimilarity and complete similarity. In-

---

[22] Dr. Chapuis, 'Le Pigeon Voyageur Belge,' 1865, p. 87. Boitard et Corbié, 'Les Pigeons de Volière,' &c., 1824, p. 173.

stances have already been given with the breeds of the fowl and pigeon; and under nature analogous cases are of frequent occurrence. With animals under domestication, but whether under nature I will not venture to say, one sex may lose characters proper to it, and may thus come to resemble to a certain extent the opposite sex; for instance, the males of some breeds of the fowl have lost their masculine plumes and hackles. On the other hand the differences between the sexes may be increased under domestication, as with merino sheep, in which the ewes have lost their horns. Again, characters proper to one sex may suddenly appear in the other sex; as with those sub-breeds of the fowl in which the hens whilst young acquire spurs; or, as in certain Polish sub-breeds, in which the females, as there is reason to believe, originally acquired a crest, and subsequently transferred it to the males. All these cases are intelligible on the hypothesis of pangenesis; for they depend on the gemmules of certain units of the body, although present in both sexes, becoming through the influence of domestication dormant in the one sex; or if naturally dormant, becoming developed.

There is one difficult question which it will be convenient to defer to a future chapter; namely, whether a character at first developed in both sexes, can be rendered through selection limited in its development to one sex alone. If, for instance, a breeder observed that some of his pigeons (in which species characters are usually transferred in an equal degree to both sexes) varied into pale blue; could he by long-continued selection make a breed, in which the males alone should be of this tint, whilst the females remained unchanged? I will here only say, that this, though perhaps not impossible, would be extremely difficult; for the natural result of breeding from the pale-blue males would be to change his whole stock, including both sexes, into this tint. If, however, variations of the desired tint appeared, which were from the first limited in their development to the male sex, there would not be the least difficulty in making a breed characterised by the two sexes being of a different colour, as indeed has been effected with a Belgian breed, in which the males alone are streaked with black. In a similar manner, if any variation appeared in a female pigeon, which was from the first sexually limited in its development, it would be easy to make a breed with the females alone thus characterised; but if the variation was not thus originally limited, the process would be extremely difficult, perhaps impossible.

On the Relation between the period of Development of a Character and its transmission to one sex or to both sexes.—Why certain characters should be inherited by both sexes, and other characters by one sex alone, namely by that sex in which the character first appeared, is in most cases quite unknown. We cannot even conjecture why with certain sub-breeds of the pigeon, black striæ, though transmitted through the female, should be developed in the male alone, whilst every other character is equally transferred to both sexes. Why, again, with cats, the tortoise-shell colour should, with rare exceptions, be developed in the female alone. The very same characters, such as deficient or supernumerary digits, colour-blindness, &c., may with mankind be inherited by the males alone of one family, and in another family by the females alone, though in both cases transmitted through the opposite as well as the same sex.[23] Although we are thus ignorant, two rules often hold good, namely

[23] References are given in my 'Variation of Animals under Domestication,' vol. ii. p. 72.

that variations which first appear in either sex at a late period of life, tend to be developed in the same sex alone; whilst variations which first appear early in life in either sex tend to be developed in both sexes. I am, however, far from supposing that this is the sole determining cause. As I have not elsewhere discussed this subject, and as it has an important bearing on sexual selection, I must here enter into lengthy and somewhat intricate details.

It is in itself probable that any character appearing at an early age would tend to be inherited equally by both sexes, for the sexes do not differ much in constitution, before the power of reproduction is gained. On the other hand, after this power has been gained and the sexes have come to differ in constitution, the gemmules (if I may again use the language of pangenesis) which are cast off from each varying part in the one sex would be much more likely to possess the proper affinities for uniting with the tissues of the same sex, and thus becoming developed, than with those of the opposite sex.

I was first led to infer that a relation of this kind exists, from the fact that whenever and in whatever manner the adult male has come to differ from the adult female, he differs in the same manner from the young of both sexes. The generality of this fact is quite remarkable: it holds good with almost all mammals, birds, amphibians, and fishes; also with many crustaceans, spiders and some few insects, namely certain orthoptera and libellulæ. In all these cases the variations, through the accumulation of which the male acquired his proper masculine characters, must have occurred at a somewhat late period of life; otherwise the young males would have been similarly characterised; and conformably with our rule, they are transmitted to and developed in the adult males alone. When, on the other hand, the adult male closely resembles the young of both sexes (these, with rare exceptions, being alike), he generally resembles the adult female; and in most of these cases the variations through which the young and old acquired their present characters, probably occurred in conformity with our rule during youth. But there is here room for doubt, as characters are sometimes transferred to the offspring at an earlier age than that at which they first appeared in the parents, so that the parents may have varied when adult, and have transferred their characters to their offspring whilst young. There are, moreover, many animals, in which the two sexes closely resemble each other, and yet both differ from their young; and here the characters of the adults must have been acquired late in life; nevertheless, these characters in apparent contradiction to our rule, are transferred to both sexes. We must not, however, overlook the possibility or even probability of successive variations of the same nature sometimes occurring, under exposure to similar conditions, simultaneously in both sexes at a rather late period of life; and in this case the variations would be transferred to the offspring of both sexes at a corresponding late age; and there would be no real contradiction to our rule of the variations which occur late in life being transferred exclusively to the sex in which they first appeared. This latter rule seems to hold true more generally than the second rule, namely, that variations which occur in either sex early in life tend to be transferred to both sexes. As it was obviously impossible even to estimate in how large a number of cases throughout the animal kingdom these two propositions hold good, it occurred to me to investigate some striking or crucial instances, and to rely on the result.

An excellent case for investigation is afforded by the Deer Family. In all the species, excepting one, the horns are developed in the male alone, though certainly transmitted through the female, and capable of occasional abnormal development in her. In the reindeer, on the other hand, the female is provided with horns; so that in this species, the horns ought, according to our rule, to appear early in life, long before the two sexes had arrived at maturity and had come to differ much in constitution. In all the other species of deer the horns ought to appear later in life, leading to their development in that sex alone, in which they first appeared in the progenitor of the whole Family. Now in seven species, belonging to distinct sections of the family and inhabiting different regions, in which the stags alone bear horns, I find that the horns first appear at periods varying from nine months after birth in the roebuck to ten or twelve or even more months in the stags of the six other larger species.[24] But with the reindeer the case is widely different, for as I hear from Prof. Nilsson, who kindly made special enquiries for me in Lapland, the horns appear in the young animals within four or five weeks after birth, and at the same time in both sexes. So that here we have a structure, developed at a most unusually early age in one species of the family, and common to both sexes in this one species.

In several kinds of antelopes the males alone are

---

[24] I am much obliged to Mr. Cupples for having made enquiries for me in regard to the Roebuck and Red Deer of Scotland from Mr. Robertson, the experienced head-forester to the Marquis of Breadalbane. In regard to Fallow-deer, I am obliged to Mr. Eyton and others for information. For the *Cervus alces* of N. America, see 'Land and Water,' 1868, p. 221 and 254; and for the *C. Virginianus* and *strongyloceros* of the same continent, see J. D. Caton, in 'Ottawa Acad. of Nat. Sc.' 1868, p. 13. For *Cervus Eldi* of Pegu, see Lieut. Beavan, 'Proc. Zoolog. Soc.' 1867, p. 762.

provided with horns, whilst in the greater number both sexes have horns. With respect to the period of development, Mr. Blyth informs me that there lived at one time in the Zoological Gardens a young koodoo (*Ant. strepsiceros*), in which species the males alone are horned, and the young of a closely-allied species, viz. the eland (*Ant. oreas*), in which both sexes are horned. Now in strict conformity with our rule, in the young male koodoo, although arrived at the age of ten months, the horns were remarkably small considering the size ultimately attained by them: whilst in the young male eland, although only three months old, the horns were already very much larger than in the koodoo. It is also worth notice that in the prong-horned antelope,[25] in which species the horns, though present in both sexes, are almost rudimentary in the female, they do not appear until about five or six months after birth. With sheep, goats and cattle, in which the horns are well developed in both sexes, though not quite equal in size, they can be felt, or even seen, at birth or soon afterwards.[26] Our rule, however, fails in regard to some breeds of sheep, for instance merinos, in which the rams alone are horned; for I cannot find on enquiry,[27] that

---

[25] *Antilocapra Americana.* Owen, 'Anatomy of Vertebrates,' vol. iii. p. 627.

[26] I have been assured that the horns of the sheep in North Wales can always be felt, and are sometimes even an inch in length, at birth. With cattle Youatt says ('Cattle,' 1834, p. 277) that the prominence of the frontal bone penetrates the cutis at birth, and that the horny matter is soon formed over it.

[27] I am greatly indebted to Prof. Victor Carus for having made inquiries for me, from the highest authorities, with respect to the merino sheep of Saxony. On the Guinea coast of Africa there is a breed of sheep in which, as with merinos, the rams alone bear horns; and Mr. Winwood Reade informs me that in the one case observed, a young ram born on Feb. 10th first showed horns on March 6th, so that in this instance the development of the horns occurred at a later

the horns are developed later in life in this breed than in ordinary sheep in which both sexes are horned. But with domesticated sheep the presence or absence of horns is not a firmly fixed character; a certain proportion of the merino ewes bearing small horns, and some of the rams being hornless; whilst with ordinary sheep hornless ewes are occasionally produced.

In most of the species of the splendid family of the Pheasants, the males differ conspicuously from the females, and they acquire their ornaments at a rather late period of life. The eared pheasant (*Crossoptilon auritum*), however, offers a remarkable exception, for both sexes possess the fine caudal plumes, the large ear-tufts and the crimson velvet about the head; and I find on enquiry in the Zoological Gardens that all these characters, in accordance with our rule, appear very early in life. The adult male can, however, be distinguished from the adult female by one character, namely by the presence of spurs; and conformably with our rule, these do not begin to be developed, as I am assured by Mr. Bartlett, before the age of six months, and even at this age, can hardly be distinguished in the two sexes.[28] The male and female Peacock differ con-

---

period of life, conformably with our rule, than in the Welch sheep, in which both sexes are horned.

[28] In the common peacock (*Pavo cristatus*) the male alone possesses spurs, whilst both sexes of the Java peacock (*P. muticus*) offer the unusual case of being furnished with spurs. Hence I fully expected that in the latter species they would have been developed earlier in life than in the common peacock; but M. Hegt of Amsterdam informs me, that with young birds of the previous year, belonging to both species, compared on April 23rd, 1869, there was no difference in the development of the spurs. The spurs, however, were as yet represented merely by slight knobs or elevations. I presume that I should have been informed if any difference in the rate of development had subsequently been observed.

spicuously from each other in almost every part of their plumage, except in the elegant head-crest, which is common to both sexes; and this is developed very early in life, long before the other ornaments which are confined to the male. The wild-duck offers an analogous case, for the beautiful green speculum on the wings is common to both sexes, though duller and somewhat smaller in the female, and it is developed early in life, whilst the curled tail-feathers and other ornaments peculiar to the male are developed later.[29] Between such extreme cases of close sexual resemblance and wide dissimilarity, as those of the Crossoptilon and peacock, many intermediate ones could be given, in which the characters follow in their order of development our two rules.

As most insects emerge from their pupal state in a mature condition, it is doubtful whether the period of development determines the transference of their characters to one or both sexes. But we do not know that the coloured scales, for instance, in two species of butterflies, in one of which the sexes differ in colour, whilst in the other they are alike, are developed at the same relative age in the cocoon. Nor do we know whether all the scales are simultaneously developed on the wings

---

[29] In some other species of the Duck Family the speculum in the two sexes differs in a greater degree; but I have not been able to discover whether its full development occurs later in life in the males of such species, than in the male of the common duck, as ought to be the case according to our rule. With the allied *Mergus cucullatus* we have, however, a case of this kind: the two sexes differ conspicuously in general plumage, and to a considerable degree in the speculum, which is pure white in the male and greyish-white in the female. Now the young males at first resemble, in all respects, the female, and have a greyish-white speculum, but this becomes pure white at an earlier age than that at which the adult male acquires his other more strongly-marked sexual differences in plumage: see Audubon, 'Ornithological Biography,' vol. iii. 1835, p. 249-250.

of the same species of butterfly, in which certain coloured marks are confined to one sex, whilst other marks are common to both sexes. A difference of this kind in the period of development is not so improbable as it may at first appear; for with the Orthoptera, which assume their adult state, not by a single metamorphosis, but by a succession of moults, the young males of some species at first resemble the females, and acquire their distinctive masculine characters only during a later moult. Strictly analogous cases occur during the successive moults of certain male crustaceans.

We have as yet only considered the transference of characters, relatively to their period of development, with species in a natural state; we will now turn to domesticated animals; first touching on monstrosities and diseases. The presence of supernumerary digits, and the absence of certain phalanges, must be determined at an early embryonic period—the tendency to profuse bleeding is at least congenital, as is probably colour-blindness—yet these peculiarities, and other similar ones, are often limited in their transmission to one sex; so that the rule that characters which are developed at an early period tend to be transmitted to both sexes, here wholly fails. But this rule, as before remarked, does not appear to be nearly so generally true as the converse proposition, namely, that characters which appear late in life in one sex are transmitted exclusively to the same sex. From the fact of the above abnormal peculiarities becoming attached to one sex, long before the sexual functions are active, we may infer that there must be a difference of some kind between the sexes at an extremely early age. With respect to sexually-limited diseases, we know too little of the period at which they originate, to draw any fair conclusion. Gout, however, seems to fall under our rule; for it is generally caused by intemperance after early youth, and is transmitted from the father to his sons in a much more marked manner than to his daughters.

In the various domestic breeds of sheep, goats, and cattle, the males differ from their respective females in the shape or development of their horns, forehead, mane, dewlap, tail, and hump on the shoulders; and these peculiarities, in accordance with our rule, are not fully developed until rather late in life. With dogs, the sexes do not differ, except that in certain breeds, especially in the Scotch deer-hound, the male is much larger and heavier than the female; and as we shall see in a future chapter, the male goes on increasing in size to an unusually late period of life, which will account, according to our rule, for his increased size being transmitted to his male offspring alone. On the other hand, the tortoise-shell colour of the hair, which is confined to female cats, is quite distinct at birth, and this case violates our rule. There is a breed of pigeons in which the males alone are streaked with black, and the streaks can be detected even in the nestlings; but they become more conspicuous at each successive moult, so that this case partly opposes and partly supports the rule. With the English Carrier and Pouter pigeon the full development of the wattle and the crop occurs rather late in life, and these characters, conformably with our rule, are transmitted in full perfection to the males alone. The following cases perhaps come within the class previously alluded to, in which the two sexes have varied in the same manner at a rather late period of life, and have consequently transferred their new characters to both sexes at a corresponding late period; and if so, such cases are not opposed to our rule. Thus there are sub-breeds of the pigeon, described by Neumeister,[30] both sexes of which change colour after moulting twice or thrice, as does likewise the Almond Tumbler; nevertheless these changes, though occurring rather late in life, are common to both sexes. One variety of the Canary-bird, namely the London Prize, offers a nearly analogous case.

With the breeds of the Fowl the inheritance of various characters by one sex or by both sexes, seems generally determined by the period at which such characters are developed. Thus in all the many breeds in which the adult male differs greatly in colour from the female and from the adult male parent-species, he differs from the young male, so that the newly acquired characters must have appeared at a rather late period of life. On the other hand with most of the breeds in which the two sexes resemble each other, the young are coloured in nearly the same manner as their parents, and this renders it probable that their colours first appeared early in life. We have instances of this fact in all black and white breeds, in which the young and old of both sexes are alike; nor can it be maintained that there is something peculiar in a black or white plumage, leading to its transference to both sexes; for the males alone of many natural species are either black or white, the females being very differently coloured. With the so-called Cuckoo sub-breeds of the fowl, in which the feathers are transversely pencilled with dark stripes, both sexes and the chickens are coloured in nearly the same manner. The laced plumage of the Sebright bantam is the same in both sexes, and in the chickens the feathers are tipped with black, which makes a near approach to lacing. Spangled Hamburghs, however, offer a partial exception,

30 'Das Ganze der Taubenzucht,' 1837, s. 21, 24. For the case of the streaked pigeons, see Dr. Chapuis, 'Le Pigeon Voyageur Belge,' 1865, p. 87.

for the two sexes, though not quite alike, resemble each other more closely than do the sexes of the aboriginal parent-species, yet they acquire their characteristic plumage late in life, for the chickens are distinctly pencilled. Turning to other characters besides colour: the males alone of the wild parent-species and of most domestic breeds possess a fairly well developed comb, but in the young of the Spanish fowl it is largely developed at a very early age, and apparently in consequence of this it is of unusual size in the adult females. In the Game breeds pugnacity is developed at a wonderfully early age, of which curious proofs could be given; and this character is transmitted to both sexes, so that the hens, from their extreme pugnacity, are now generally exhibited in separate pens. With the Polish breeds the bony protuberance of the skull which supports the crest is partially developed even before the chickens are hatched, and the crest itself soon begins to grow, though at first feebly;[31] and in this breed a great bony protuberance and an immense crest characterise the adults of both sexes.

Finally, from what we have now seen of the relation which exists in many natural species and domesticated races, between the period of the development of their characters and the manner of their transmission—for example the striking fact of the early growth of the horns in the reindeer, in which both sexes have horns, in comparison with their much later growth in the other species in which the male alone bears horns—we may conclude that one cause, though not the sole

[31] For full particulars and references on all these points respecting the several breeds of the Fowl, see 'Variation of Animals and Plants under Domestication,' vol. i. p. 250, 256. In regard to the higher animals, the sexual differences which have arisen under domestication are described in the same work under the head of each species.

cause, of characters being exclusively inherited by one sex, is their development at a late age. And secondly, that one, though apparently a less efficient, cause of characters being inherited by both sexes is their development at an early age, whilst the sexes differ but little in constitution. It appears, however, that some difference must exist between the sexes even during an early embryonic period, for characters developed at this age not rarely become attached to one sex.

Summary and concluding remarks.—From the foregoing discussion on the various laws of inheritance, we learn that characters often or even generally tend to become developed in the same sex, at the same age, and periodically at the same season of the year, in which they first appeared in the parents. But these laws, from unknown causes, are very liable to change. Hence the successive steps in the modification of a species might readily be transmitted in different ways; some of the steps being transmitted to one sex, and some to both; some to the offspring at one age, and some at all ages. Not only are the laws of inheritance extremely complex, but so are the causes which induce and govern variability. The variations thus caused are preserved and accumulated by sexual selection, which is in itself an extremely complex affair, depending, as it does, on ardour in love, courage, and the rivalry of the males, and on the powers of perception, taste, and will of the female. Sexual selection will also be dominated by natural selection for the general welfare of the species. Hence the manner in which the individuals of either sex or of both sexes are affected through sexual selection cannot fail to be complex in the highest degree.

When variations occur late in life in one sex, and are

transmitted to the same sex at the same age, the other sex and the young are necessarily left unmodified. When they occur late in life, but are transmitted to both sexes at the same age, the young alone are left unmodified. Variations, however, may occur at any period of life in one sex or in both, and be transmitted to both sexes at all ages, and then all the individuals of the species will be similarly modified. In the following chapters it will be seen that all these cases frequently occur under nature.

Sexual selection can never act on any animal whilst young, before the age for reproduction has arrived. From the great eagerness of the male it has generally acted on this sex and not on the females. The males have thus become provided with weapons for fighting with their rivals, or with organs for discovering and securely holding the female, or for exciting and charming her. When the sexes differ in these respects, it is also, as we have seen, an extremely general law that the adult male differs more or less from the young male; and we may conclude from this fact that the successive variations, by which the adult male became modified, cannot have occurred much before the age for reproduction. How then are we to account for this general and remarkable coincidence between the period of variability and that of sexual selection,—principles which are quite independent of each other? I think we can see the cause: it is not that the males have never varied at an early age, but that such variations have commonly been lost, whilst those occurring at a later age have been preserved.

All animals produce more offspring than can survive to maturity; and we have every reason to believe that death falls heavily on the weak and inexperienced young. If then a certain proportion of the offspring were to vary at birth or soon afterwards, in some manner which at this age was of no service to them, the chance of the preservation of such variations would be small. We have good evidence under domestication how soon variations of all kinds are lost, if not selected. But variations which occurred at or near maturity, and which were of immediate service to either sex, would probably be preserved; as would similar variations occurring at an earlier period in any individuals which happened to survive. As this principle has an important bearing on sexual selection, it may be advisable to give an imaginary illustration. We will take a pair of animals, neither very fertile nor the reverse, and assume that after arriving at maturity they live on an average for five years, producing each year five young. They would thus produce 25 offspring; and it would not, I think, be an unfair estimate to assume that 18 or 20 out of the 25 would perish before maturity, whilst still young and inexperienced; the remaining seven or five sufficing to keep up the stock of mature individuals. If so, we can see that variations which occurred during youth, for instance in brightness, and which were not of the least service to the young, would run a good chance of being utterly lost. Whilst similar variations, which occurring at or near maturity in the comparatively few individuals surviving to this age, and which immediately gave an advantage to certain males, by rendering them more attractive to the females, would be likely to be preserved. No doubt some of the variations in brightness which occurred at an earlier age would by chance be preserved, and eventually give to the male the same advantage as those which appeared later; and this will account for the young males commonly partaking to a certain extent (as may be observed with many birds) of the bright colours of their adult male parents. If only a few of the successive variations in brightness were to occur at a late age, the adult male would be only a little brighter than the young male; and such cases are common.

In this illustration I have assumed that the young varied in a manner which was of no service to them; but many characters proper to the adult male would be actually injurious to the young,—as bright colours from making them conspicuous, or horns of large size from expending much vital force. Such variations in the young would promptly be eliminated through natural selection. With the adult and experienced males, on the other hand, the advantage thus derived in their rivalry with other males would often more than counterbalance exposure to some degree of danger. Thus we can understand how it is that variations which must originally have appeared rather late in life have alone or in chief part been preserved for the development of secondary sexual characters; and the remarkable coincidence between the periods of variability and of sexual selection is intelligible.

As variations which give to the male an advantage in fighting with other males, or in finding, securing, or charming the female, would be of no use to the female, they will not have been preserved in this sex either during youth or maturity. Consequently such variations would be extremely liable to be lost; and the female, as far as these characters are concerned, would be left unmodified, excepting in so far as she may have received them by transference from the male. No doubt if the female varied and transferred serviceable characters to her male offspring, these would be favoured through sexual selection; and then both sexes would thus far be modified in the same manner. But I shall hereafter have to recur to these more intricate contingencies.

In the following chapters, I shall treat of the secondary sexual characters in animals of all classes, and shall endeavour in each case to apply the principles explained in the present chapter. The lowest classes will detain us for a very short time, but the higher animals, especially birds, must be treated at considerable length. It should be borne in mind that for reasons already assigned, I intend to give only a few illustrative instances of the innumerable structures by the aid of which the male finds the female, or, when found, holds her. On the other hand, all structures and instincts by which the male conquers other males, and by which he allures or excites the female, will be fully discussed, as these are in many ways the most interesting.

Supplement on the proportional numbers of the two sexes in animals belonging to various classes.

As no one, as far as I can discover, has paid attention to the relative numbers of the two sexes throughout the animal kingdom, I will here give such materials as I have been able to collect, although they are extremely imperfect. They consist in only a few instances of actual enumeration, and the numbers are not very large. As the proportions are known with certainty on a large scale in the case of man alone, I will first give them, as a standard of comparison.

Man.—In England during ten years (from 1857 to 1866) 707,120 children on an annual average have been born alive, in the proportion of 104·5 males to 100 females. But in 1857 the male births throughout England were as 105·2, and in 1865 as 104·0 to 100. Looking to separate districts, in Buckinghamshire (where on an average 5000 children are annually born) the *mean* proportion of male to female births, during the whole period of the above ten years, was as 102·8 to 100; whilst in N. Wales (where the average annual births are 12,873) it was as high as 106·2 to 100. Taking a still smaller district, viz., Rutlandshire (where the annual births average only 739), in 1864 the male births were as 114·6, and in 1862 as 97·0 to 100; but even in this small district the average of the 7385 births during the whole ten years was as 104·5 to 100; that is in the same ratio as throughout England.[32] The proportions are sometimes slightly disturbed by unknown causes; thus Prof. Faye states "that in "some districts of Norway there has been during a "decennial period a steady deficiency of boys, whilst "in others the opposite condition has existed." In France during forty-four years the male to the female births have been as 106·2 to 100; but during this period it has occurred five times in one department, and six times in another, that the female births have exceeded the males. In Russia the average proportion is as high as 108·9 to 100.[33] It is a singular fact that with Jews the proportion of male births is decidedly larger than with Christians: thus in Prussia the proportion is as 113, in Breslau as 114, and in Livonia as 120 to 100; the Christian births in these countries being the same as usual, for instance, in Livonia as 104 to 100.[34] It is a still more singular fact that in different nations, under different conditions and climates, in Naples, Prussia, Westphalia, France and England, the

[32] 'Twenty-ninth Annual Report of the Registrar-General for 1866.' In this report (p. xii) a special decennial table is given.

[33] For Norway and Russia, see abstract of Prof. Faye's researches, in 'British and Foreign Medico-Chirurg. Review,' April, 1867, p. 343, 345. For France, the 'Annuaire pour l'An 1867,' p. 213.

[34] In regard to the Jews, see M. Thury, 'La Loi de Production des Sexes,' 1863, p. 25.

excess of male over female births is less when they are illegitimate than when legitimate.[35]

In various parts of Europe, according to Prof. Faye and other authors, "a still greater preponderance of "males would be met with, if death struck both sexes "in equal proportion in the womb and during birth. "But the fact is, that for every 100 still-born females, "we have in several countries from 134·6 to 144·9 "still-born males." Moreover during the first four or five years of life more male children die than females; "for example in England, during the first year, 126 "boys die for every 100 girls,—a proportion which in "France is still more unfavourable."[36] As a consequence of this excess in the death-rate of male children, and of the exposure of men when adult to various dangers, and of their tendency to emigrate, the females in all old-settled countries, where statistical records have been kept,[37] are found to preponderate considerably over the males.

It has often been supposed that the relative ages of the parents determine the sex of the offspring; and Prof. Leuckart[38] has advanced what he considers

---

35 Babbage, 'Edinburgh Journal of Science,' 1829, vol. i. p. 88; also p. 90, on still-born children. On illegitimate children in England, see 'Report of Registrar-General for 1866,' p. xv.

36 'British and Foreign Medico-Chirurg. Review,' April, 1867, p. 343. Dr. Stark also remarks ('Tenth Annual Report of Births, Deaths, &c., in Scotland,' 1867, p. xxviii) that "These examples may suffice "to shew that, at almost every stage of life, the males in Scotland "have a greater liability to death and a higher death-rate than the "females. The fact, however, of this peculiarity being most strongly "developed at that infantile period of life when the dress, food, and "general treatment of both sexes are alike, seems to prove that the "higher male death-rate is an impressed, natural, and constitutional "peculiarity due to sex alone."

37 With the savage Guaranys of Paraguay, according to the accurate Azara ('Voyages dans l'Amérique mérid.' tom. ii. 1809, p. 60, 179), the women in proportion to the men are as 14 to 13.

38 Leuckart (in Wagner, 'Handwörterbuch der Phys.' B. iv. 1853, s. 774.

sufficient evidence, with respect to man and certain domesticated animals, to shew that this is one important factor in the result. So again the period of impregnation has been thought to be the efficient cause; but recent observations discountenance this belief. Again, with mankind polygamy has been supposed to lead to the birth of a greater proportion of female infants; but Dr. J. Campbell[39] carefully attended to this subject in the harems of Siam, and he concludes that the proportion of male to female births is the same as from monogamous unions. Hardly any animal has been rendered so highly polygamous as our English race-horses, and we shall immediately see that their male and female offspring are almost exactly equal in number.

Horses.—Mr. Tegetmeier has been so kind as to tabulate for me from the 'Racing Calendar' the births of race-horses during a period of twenty-one years, viz. from 1846 to 1867; 1849 being omitted, as no returns were that year published. The total births have been 25,560,[40] consisting of 12,763 males and 12,797 females, or in the proportion of 99·7 males to 100 females. As these numbers are tolerably large, and as they are drawn from all parts of England, during several years, we may with much confidence conclude that with the domestic horse, or at least with the race-horse, the two sexes are produced in almost equal numbers. The fluctuations in the proportions during successive years are closely like those which occur with mankind, when a small and thinly-populated area is considered: thus in 1856 the male horses were as 107·1, and in 1867 as only 92·6 to 100 females. In the tabulated returns the proportions vary in cycles, for the males exceeded the females during six successive years; and the females exceeded the males during two

---

[39] Anthropological Review, April, 1870, p. cviii.

[40] During the last eleven years a record has been kept of the number of mares which have proved barren or prematurely slipped their foals; and it deserves notice, as shewing how infertile these highly-nurtured and rather closely-interbred animals have become, that not far from one-third of the mares failed to produce living foals. Thus during 1866, 809 male colts and 816 female colts were born, and 743 mares failed to produce offspring. During 1867, 836 males and 902 females were born, and 794 mares failed.

periods each of four years: this, however, may be accidental; at least I can detect nothing of the kind with man in the decennial table in the Registrar's Report for 1866. I may add that certain mares, and this holds good with certain cows and with women, tend to produce more of one sex than of the other; Mr. Wright of Yeldersley House, informs me that one of his Arab mares, though put seven times to different horses, produced seven fillies.

Dogs.—During a period of twelve years, from 1857 to 1868, the births of a large number of greyhounds, throughout England, have been sent to the 'Field' newspaper; and I am again indebted to Mr. Tegetmeier for carefully tabulating the results. The recorded births have been 6878, consisting of 3605 males and 3273 females, that is, in the proportion of 110·1 males to 100 females. The greatest fluctuations occurred in 1864, when the proportion was as 95·3 males, and in 1867, as 116·3 males to 100 females. The above average proportion of 110·1 to 100 is probably nearly correct in the case of the greyhound, but whether it would hold with other domesticated breeds is in some degree doubtful. Mr. Cupples has enquired from several great breeders of dogs, and finds that all without exception believe that females are produced in excess; he suggests that this belief may have arisen from females being less valued and the consequent disappointment producing a stronger impression on the mind.

Sheep.—The sexes of sheep are not ascertained by agriculturists until several months after birth, at the period when the males are castrated; so that the following returns do not give the proportions at birth. Moreover, I find that several great breeders in Scotland, who annually raise some thousand sheep, are firmly convinced that a larger proportion of males than of females die during the first one or two years; therefore the proportion of males would be somewhat greater at birth than at the age of castration. This is a remarkable coincidence with what occurs, as we have seen, with mankind, and both cases probably depend on some common cause. I have received returns from four gentlemen in England who have bred lowland sheep, chiefly Leicesters, during the last ten or sixteen years; they amount altogether to 8965 births, consisting of 4407 males and 4558 females; that is in the proportion of 96·7 males to 100 females. With respect to Cheviot and black-faced sheep bred in Scotland, I have received returns from six breeders, two of them on a large scale, chiefly for the years 1867–1869, but some of the returns extending back to 1862. The total number recorded amounts to 50,685, consisting of 25,071 males and 25,614 females, or in the proportion of 97·9 males to 100 females. If we take the English and Scotch returns together, the total number amounts to 59,650, consisting of 29,478 males and 30,172 females, or as 97·7 to 100. So that with sheep at the age of castration the females are certainly in excess of the males; but whether this would hold good at birth is doubtful, owing to the greater liability in the males to early death.[41]

Of *Cattle* I have received returns from nine gentlemen of 982 births, too few to be trusted; these consisted of 477 bull-calves and 505 cow-calves; *i.e.* in the proportion of 94·4 males to 100 females. The Rev. W. D. Fox informs me that in 1867 out of 34 calves born on a farm in Derbyshire only one was a bull. Mr. Harrison Weir writes to me that he has enquired from several breeders of *Pigs*, and most of them estimate the male to the female births as about 7 to 6. This same gentleman has bred *Rabbits* for many years, and has noticed that a far greater number of bucks are produced than does.

Of mammalia in a state of nature I have been able to learn very little. In regard to the common rat, I have received conflicting statements. Mr. R. Elliot of Laighwood, informs me that a rat-catcher assured him that he had always found the males in great excess, even with the young in the nest. In consequence of this, Mr. Elliot himself subsequently examined some hundred old ones, and found the statement true. Mr. F. Buckland has bred a large number of white rats, and he also believes that the males greatly exceed the females. In regard to Moles, it is said that "the males are much more numerous than the females;"[42] and as the catching of these animals is a special occupation, the statement may perhaps be trusted. Sir A. Smith, in describing an antelope of S. Africa[43] (*Kobus ellipsiprymnus*), remarks, that in the herds of this and other species, the males are few in number compared with the females: the natives believe that they are born in this proportion; others believe that the younger males are expelled from the herds, and Sir A. Smith says, that though he has himself never seen herds consisting of young males alone, others affirm that this does occur. It appears probable that the young males when expelled from the herd, would be likely to fall a prey to the many beasts of prey of the country.

---

[41] I am much indebted to Mr. Cupples for having procured for me the above returns from Scotland, as well as some of the following returns on cattle. Mr. R. Elliot, of Laighwood, first called my attention to the premature deaths of the males,—a statement subsequently confirmed by Mr. Aitchison and others. To this latter gentleman, and to Mr. Payan, I owe my thanks for the larger returns on sheep.

[42] Bell, 'History of British Quadrupeds,' p. 100.

[43] 'Illustrations of the Zoology of S. Africa,' 1849, pl. 29.

BIRDS.

With respect to the *Fowl*, I have received only one account, namely, that out of 1001 chickens of a highly-bred stock of Cochins, reared during eight years by Mr. Stretch, 487 proved males and 514 females: *i. e.* as 94·7 to 100. In regard to domestic pigeons there is good evidence that the males are produced in excess, or that their lives are longer; for these birds invariably pair, and single males, as Mr. Tegetmeier informs me, can always be purchased cheaper than females. Usually the two birds reared from the two eggs laid in the same nest consist of a male and female; but Mr. Harrison Weir, who has been so large a breeder, says that he has often bred two cocks from the same nest, and seldom two hens; moreover the hen is generally the weaker of the two, and more liable to perish.

With respect to birds in a state of nature, Mr. Gould and others[44] are convinced that the males are generally the more numerous; and as the young males of many species resemble the females, the latter would naturally appear to be the most numerous. Large numbers of pheasants are reared by Mr. Baker of Leadenhall from eggs laid by wild birds, and he informs Mr. Jenner Weir that four or five males to one female are generally produced. An experienced observer remarks[45] that in Scandinavia the broods of the capercailzie and black-cock contain more males than females; and that with the Dal-ripa (a kind of ptarmigan) more males than females attend the *leks* or places of courtship; but this latter circumstance is accounted for by some observers by a greater number of hen birds being killed by vermin. From various facts given by White of Selborne,[46] it seems clear that the males of the partridge must be in considerable excess in the south of England; and I have been assured that this is the case in Scotland. Mr. Weir on enquiring from the dealers who receive at certain seasons large numbers of ruffs (*Machetes pugnax*) was told that the males are much the most numerous. This same naturalist has also enquired for me from the bird-catchers, who annually catch an astonishing number of various small species alive for the London market, and he was unhesitatingly answered by an old and trustworthy man, that with the chaffinch the males are in large excess; he thought as high as 2 males to

---

[44] Brehm ('Illust. Thierleben,' B. iv. s. 990) comes to the same conclusion.

[45] On the authority of L. Lloyd, 'Game Birds of Sweden,' 1867, p. 12, 132.

[46] 'Nat. Hist. of Selbourne,' letter xxix. edit. of 1825, vol. i. p. 139.

1 female, or at least as high as 5 to 3.[47] The males of the blackbird, he likewise maintained, were by far the most numerous, whether caught by traps or by netting at night. These statements may apparently be trusted, because the same man said that the sexes are about equal with the lark, the twite (*Linaria montana*), and goldfinch. On the other hand he is certain that with the common linnet, the females preponderate greatly, but unequally during different years; during some years he has found the females to the males as four to one. It should, however, be borne in mind, that the chief season for catching birds does not begin till September, so that with some species partial migrations may have begun, and the flocks at this period often consist of hens alone. Mr. Salvin paid particular attention to the sexes of the humming-birds in Central America, and he is convinced that with most of the species the males are in excess; thus one year he procured 204 specimens belonging to ten species, and these consisted of 166 males and of 38 females. With two other species the females were in excess: but the proportions apparently vary either during different seasons or in different localities; for on one occasion the males of *Campylopterus hemileucurus* were to the females as five to two, and on another occasion[48] in exactly the reversed ratio. As bearing on this latter point, I may add, that Mr. Powys found in Corfu and Epirus the sexes of the chaffinch keeping apart, and "the females "by far the most numerous;" whilst in Palestine Mr. Tristram found "the male flocks appearing greatly to exceed the female in "number."[49] So again with the *Quiscalus major*, Mr. G. Taylor[50] says, that in Florida there were "very few females in proportion to "the males," whilst in Honduras the proportion was the other way, the species there having the character of a polygamist.

FISH.

With Fish the proportional numbers of the sexes can be ascertained only by catching them in the adult or nearly adult state; and there

---

[47] Mr. Jenner Weir received similar information, on making enquiries during the following year. To shew the number of chaffinches caught, I may mention that in 1869 there was a match between two experts; and one man caught in a day 62, and another 40, male chaffinches. The greatest number ever caught by one man in a single day was 70.

[48] 'Ibis,' vol. ii. p. 260, as quoted in Gould's 'Trochilidæ,' 1861, p. 52. For the foregoing proportions, I am indebted to Mr. Salvin for a table of his results.

[49] 'Ibis,' 1860, p. 137; and 1867, p. 369.

[50] 'Ibis,' 1862, p. 137.

are many difficulties in arriving at any just conclusion.[51] Infertile females might readily be mistaken for males, as Dr. Günther has remarked to me in regard to trout. With some species the males are believed to die soon after fertilising the ova. With many species the males are of much smaller size than the females, so that a large number of males would escape from the same net by which the females were caught. M. Carbonnier,[52] who has especially attended to the natural history of the pike (*Esox lucius*) states that many males, owing to their small size, are devoured by the larger females; and he believes that the males of almost all fish are exposed from the same cause to greater danger than the females. Nevertheless in the few cases in which the proportional numbers have been actually observed, the males appear to be largely in excess. Thus Mr. R. Buist, the superintendent of the Stormontfield experiments, says that in 1865, out of 70 salmon first landed for the purpose of obtaining the ova, upwards of 60 were males. In 1867 he again "calls attention to the vast disproportion of the "males to the females. We had at the outset at least ten males "to one female." Afterwards sufficient females for obtaining ova were procured. He adds, "from the great proportion of the "males, they are constantly fighting and tearing each other on the "spawning-beds."[53] This disproportion, no doubt, can be accounted for in part, but whether wholly is very doubtful, by the males ascending the rivers before the females. Mr. F. Buckland remarks in regard to trout, that "it is a curious fact that the males prepon-"derate very largely in number over the females. It *invariably* "happens that when the first rush of fish is made to the net, there "will be at least seven or eight males to one female found captive. "I cannot quite account for this; either the males are more numer-"ous than the females, or the latter seek safety by concealment "rather than flight." He then adds, that by carefully searching the banks, sufficient females for obtaining ova can be found.[54] Mr. H. Lee informs me that out of 212 trout, taken for this purpose in Lord Portsmouth's park, 150 were males and 62 females.

With the Cyprinidæ the males likewise seem to be in excess; but several members of this Family, viz., the carp, tench, bream and minnow, appear regularly to follow the practice, rare in the

---

[51] Leuckart quotes Bloch (Wagner, 'Handwörterbuch der Phys.' B. iv. 1853, s. 775), that with fish there are twice as many males as females.

[52] Quoted in the 'Farmer,' March 18, 1869, p. 369.

[53] 'The Stormontfield Piscicultural Experiments,' 1866, p. 23. The 'Field' newspaper, June 29th, 1867.

[54] 'Land and Water,' 1868, p. 41.

animal kingdom, of polyandry; for the female whilst spawning is always attended by two males, one on each side, and in the case of the bream by three or four males. This fact is so well known, that it is always recommended to stock a pond with two male tenches to one female, or at least with three males to two females. With the minnow, an excellent observer states, that on the spawning-beds the males are ten times as numerous as the females; when a female comes amongst the males, "she is immediately pressed closely "by a male on each side; and when they have been in that situation for a time, are superseded by other two males."[55]

INSECTS.

In this class, the Lepidoptera alone afford the means of judging of the proportional numbers of the sexes; for they have been collected with special care by many good observers, and have been largely bred from the egg or caterpillar state. I had hoped that some breeders of silk-moths might have kept an exact record, but after writing to France and Italy, and consulting various treatises, I cannot find that this has ever been done. The general opinion appears to be that the sexes are nearly equal, but in Italy as I hear from Professor Canestrini, many breeders are convinced that the females are produced in excess. The same naturalist, however, informs me, that in the two yearly broods of the Ailanthus silk-moth (*Bombyx cynthia*), the males greatly preponderate in the first, whilst in the second the two sexes are nearly equal, or the females rather in excess.

In regard to Butterflies in a state of nature, several observers have been much struck by the apparently enormous preponderance of the males.[56] Thus Mr. Bates,[57] in speaking of the species, no less than about a hundred in number, which inhabit the Upper Amazons, says that the males are much more numerous than the females, even in the proportion of a hundred to one. In North America, Edwards, who had great experience, estimates in the genus Papilio the males to the females as four to one; and Mr.

---

[55] Yarrell, 'Hist. British Fishes,' vol. i. 1836, p. 307; on the *Cyprinus carpio*, p. 331; on the *Tinca vulgaris*, p. 331; on the *Abramis brama*, p. 336. See, for the minnow (*Leuciscus phoxinus*), 'Loudon's Mag. of Nat. Hist.' vol. v. 1832, p. 682.

[56] Leuckart quotes Meinecke (Wagner, 'Handwörterbuch der Phys.' B. iv. 1853, s. 775) that with Butterflies the males are three or four times as numerous as the females.

[57] 'The Naturalist on the Amazons,' vol. ii. 1863, p. 228, 347.

Walsh, who informed me of this statement, says that with *P. turnus* this is certainly the case. In South Africa, Mr. R. Trimen found the males in excess in 19 species;[58] and in one of these, which swarms in open places, he estimated the number of males as fifty to one female. With another species, in which the males are numerous in certain localities, he collected during seven years only five females. In the island of Bourbon, M. Maillard states that the males of one species of Papilio are twenty times as numerous as the females.[59] Mr. Trimen informs me that as far as he has himself seen, or heard from others, it is rare for the females of any butterfly to exceed in number the males; but this is perhaps the case with three South African species. Mr. Wallace[60] states that the females of *Ornithoptera crœsus*, in the Malay archipelago, are more common and more easily caught than the males; but this is a rare butterfly. I may here add, that in Hyperythra, a genus of moths, Guenée says, that from four to five females are sent in collections from India for one male.

When this subject of the proportional numbers of the sexes of insects was brought before the Entomological Society,[61] it was generally admitted that the males of most Lepidoptera, in the adult or imago state, are caught in greater numbers than the females; but this fact was attributed by various observers to the more retiring habits of the females, and to the males emerging earlier from the cocoon. This latter circumstance is well known to occur with most Lepidoptera, as well as with other insects. So that, as M. Personnat remarks, the males of the domesticated *Bombyx Yamamai*, are lost at the beginning of the season, and the females at the end, from the want of mates.[62] I cannot however persuade myself that these causes suffice to explain the great excess of males in the cases, above given, of butterflies which are extremely common in their native countries. Mr. Stainton, who has paid such close attention during many years to the smaller moths, informs me that when he collected them in the imago state, he thought that the males were ten times as numerous as the females, but that since he has reared them on a large scale from the caterpillar state, he is convinced that the females are the most

---

58 Four of these cases are given by Mr. Trimen in his 'Rhopalocera Africæ Australis.'

59 Quoted by Trimen, 'Transact. Ent. Soc.' vol. v. part iv. 1866, p. 330.

60 'Transact. Linn. Soc.' vol. xxv. p. 37.

61 'Proc. Entomolog. Soc.' Feb. 17th, 1868.

62 Quoted by Dr. Wallace in 'Proc. Ent. Soc.' 3rd series, vol. v. 1867, p. 487.

numerous. Several entomologists concur in this view. Mr. Doubleday, however, and some others, take an opposite view, and are convinced that they have reared from the egg and caterpillar states a larger proportion of males than of females.

Besides the more active habits of the males, their earlier emergence from the cocoon, and their frequenting in some cases more open stations, other causes may be assigned for an apparent or real difference in the proportional numbers of the sexes of Lepidoptera, when captured in the imago state, and when reared from the egg or caterpillar state. It is believed by many breeders in Italy, as I hear from Professor Canestrini, that the female caterpillar of the silk-moth suffers more from the recent disease than the male; and Dr. Staudinger informs me that in rearing Lepidoptera more females die in the cocoon than males. With many species the female caterpillar is larger than the male, and a collector would naturally choose the finest specimens, and thus unintentionally collect a larger number of females. Three collectors have told me that this was their practice; but Dr. Wallace is sure that most collectors take all the specimens which they can find of the rarer kinds, which alone are worth the trouble of rearing. Birds when surrounded by caterpillars would probably devour the largest; and Professor Canestrini informs me that in Italy some breeders believe, though on insufficient evidence, that in the first brood of the Ailanthus silk-moth, the wasps destroy a larger number of the female than of the male caterpillars. Dr. Wallace further remarks that female caterpillars, from being larger than the males, require more time for their development and consume more food and moisture; and thus they would be exposed during a longer time to danger from ichneumons, birds, &c., and in times of scarcity would perish in greater numbers. Hence it appears quite possible that, in a state of nature, fewer female Lepidoptera may reach maturity than males; and for our special object we are concerned with the numbers at maturity, when the sexes are ready to propagate their kind.

The manner in which the males of certain moths congregate in extraordinary numbers round a single female, apparently indicates a great excess of males, though this fact may perhaps be accounted for by the earlier emergence of the males from their cocoons. Mr. Stainton informs me that from twelve to twenty males may often be seen congregated round a female *Elachista rufocinerea*. It is well known that if a virgin *Lasiocampa quercus* or *Saturnia carpini* be exposed in a cage, vast numbers of males collect round her, and if confined in a room will even come down the chimney to her.

Mr. Doubleday believes that he has seen from fifty to a hundred males of both these species attracted in the course of a single day by a female under confinement. Mr. Trimen exposed in the Isle of Wight a box in which a female of the Lasiocampa had been confined on the previous day, and five males soon endeavoured to gain admittance. M. Verreaux, in Australia, having placed the female of a small Bombyx in a box in his pocket, was followed by a crowd of males, so that about 200 entered the house with him.[63]

Mr. Doubleday has called my attention to Dr. Staudinger's[64] list of Lepidoptera, which gives the prices of the males and females of 300 species or well-marked varieties of (Rhopalocera) butterflies. The prices for both sexes of the very common species are of course the same; but with 114 of the rarer species they differ; the males being in all cases, excepting one, the cheapest. On an average of the prices of the 113 species, the price of the male to that of the female is as 100 to 149; and this apparently indicates that inversely the males exceed the females in number in the same proportion. About 2000 species or varieties of moths (Heterocera) are catalogued, those with wingless females being here excluded on account of the difference in habits of the two sexes: of these 2000 species, 141 differ in price according to sex, the males of 130 being cheaper, and the males of only 11 being dearer than the females. The average price of the males of the 130 species, to that of the females, is as 100 to 143. With respect to the butterflies in this priced list, Mr. Doubleday thinks (and no man in England has had more experience), that there is nothing in the habits of the species which can account for the difference in the prices of the two sexes, and that it can be accounted for only by an excess in the numbers of the males. But I am bound to add that Dr. Staudinger himself, as he informs me, is of a different opinion. He thinks that the less active habits of the females and the earlier emergence of the males will account for his collectors securing a larger number of males than of females, and consequently for the lower prices of the former. With respect to specimens reared from the caterpillar-state, Dr. Staudinger believes, as previously stated, that a greater number of females than of males die under confinement in the cocoons. He adds that with certain species one sex seems to preponderate over the other during certain years.

Of direct observations on the sexes of Lepidoptera, reared either

---

63 Blanchard, 'Metamorphoses, Mœurs des Insectes,' 1868, p. 225-226.

64 'Lepidopteren-Doubblettren Liste,' Berlin, No. x. 1866.

from eggs or caterpillars, I have received only the few following cases:—

	Males.	Females.
The Rev. J. Hellins[65] of Exeter reared, during 1868, imagos of 73 species, which consisted of	153	137
Mr. Albert Jones of Eltham reared, during 1868, imagos of 9 species, which consisted of	159	126
During 1869 he reared imagos from 4 species, consisting of	114	112
Mr. Buckler of Emsworth, Hants, during 1869, reared imagos from 74 species, consisting of	180	169
Dr. Wallace of Colchester reared from one brood of Bombyx cynthia	52	48
Dr. Wallace raised, from cocoons of Bombyx Pernyi sent from China, during 1869	224	123
Dr. Wallace raised, during 1868 and 1869, from two lots of cocoons of Bombyx yama-mai	52	46
Total	934	761

So that in these eight lots of cocoons and eggs, males were produced in excess. Taken together the proportion of males is as 122·7 to 100 females. But the numbers are hardly large enough to be trustworthy.

On the whole, from the above various sources of evidence, all pointing to the same direction, I infer that with most species of Lepidoptera, the males in the imago state generally exceed the females in number, whatever the proportions may be at their first emergence from the egg.

With reference to the other Orders of insects, I have been able to collect very little reliable information. With the stag-beetle (*Lucanus cervus*) "the males appear to be much more numerous "than the females;" but when, as Cornelius remarked during 1867, an unusual number of these beetles appeared in one part of Germany, the females appeared to exceed the males as six so one. With one of the Elateridæ, the males are said to be much more numerous than the females, and "two or three are often found "united with one female;[66] so that here polyandry seems to prevail.

[65] This naturalist has been so kind as to send me some results from former years, in which the females seemed to preponderate; but so many of the figures were estimates, that I found it impossible to tabulate them.

[66] Günther's 'Record of Zoological Literature,' 1867, p. 260. On the excess of female Lucanus, ibid. p. 250. On the males of Lucanus in England, Westwood, 'Modern Class. of Insects,' vol. i. p. 187. On the Siagonium, ibid. p. 172.

With Siagonium (Staphylinidæ), in which the males are furnished with horns, "the females are far more numerous than the opposite "sex." Mr. Janson stated at the Entomological Society that the females of the bark-feeding *Tomicus villosus* are so common as to be a plague, whilst the males are so rare as to be hardly known. In other Orders, from unknown causes, but apparently in some instances owing to parthenogenesis, the males of certain species have never been discovered or are excessively rare, as with several of the Cynipidæ.[67] In all the gall-making Cynipidæ known to Mr. Walsh, the females are four or five times as numerous as the males; and so it is, as he informs me, with the gall-making Cecidomyiiæ (Diptera). With some common species of Saw-flies (Tenthredinæ) Mr. F. Smith has reared hundreds of specimens from larvæ of all sizes, but has never reared a single male: on the other hand Curtis says,[68] that with certain species (Athalia), bred by him, the males to the females were as six to one; whilst exactly the reverse occurred with the mature insects of the same species caught in the fields. With the Neuroptera, Mr. Walsh states that in many, but by no means in all, the species of the Odonatous groups (Ephemerina), there is a great overplus of males: in the genus Hetærina, also, the males are generally at least four times as numerous as the females. In certain species in the genus Gomphus the males are equally numerous, whilst in two other species, the females are twice or thrice as numerous as the males. In some European species of Psocus thousands of females may be collected without a single male, whilst with other species of the same genus both sexes are common.[69] In England, Mr. MacLachlan has captured hundreds of the female *Apatania muliebris*, but has never seen the male; and of *Boreus hyemalis* only four or five males have been here seen.[70] With most of these species (excepting, as I have heard, with the Tenthredinæ) there is no reason to suppose that the females are subject to parthenogenesis; and thus we see how ignorant we are on the causes of the apparent discrepancy in the proportional numbers of the two sexes.

In the other Classes of the Articulata I have been able to collect still less information. With Spiders, Mr. Blackwall, who has carefully attended to this class during many years, writes to me that the males from their more erratic habits are more commonly seen,

---

67 Walsh, in 'The American Entomologist,' vol. i. 1869, p. 103. F. Smith, 'Record of Zoological Literature,' 1867, p. 328.

68 'Farm Insects,' p. 45-46.

69 'Observations on N. American Neuroptera,' by H. Hagen and B. D. Walsh, 'Proc. Ent. Soc. Philadelphia,' Oct. 1863, p. 168, 223, 239.

70 'Proc. Ent. Soc. London,' Feb. 17, 1868.

and therefore appear to be the more numerous. This is actually the case with a few species; but he mentions several species in six genera, in which the females appear to be much more numerous than the males.[71] The small size of the males in comparison with the females, which is sometimes carried to an extreme degree, and their widely different appearance, may account in some instances for their rarity in collections.[72]

Some of the lower Crustaceans are able to propagate their kind asexually, and this will account for the extreme rarity of the males. With some other forms (as with Tanais and Cypris) there is reason to believe, as Fritz Müller informs me, that the male is much shorter-lived than the female, which, supposing the two sexes to be at first equal in number, would explain the scarcity of the males. On the other hand this same naturalist has invariably taken, on the shores of Brazil, far more males than females of the Diastylidæ and of Cypridina; thus with a species in the latter genus, 63 specimens caught the same day, included 57 males; but he suggests that this preponderance may be due to some unknown difference in the habits of the two sexes. With one of the higher Brazilian crabs, namely a Gelasimus, Fritz Müller found the males to be more numerous than the females. The reverse seems to be the case, according to the large experience of Mr. C. Spence Bate, with six common British crabs, the names of which he has given me.

On the Power of Natural Selection to regulate the proportional Numbers of the Sexes, and General Fertility.— In some peculiar cases, an excess in the number of one sex over the other might be a great advantage to a species, as with the sterile females of social insects, or with those animals in which more than one male is requisite to fertilise the female, as with certain cirripedes and perhaps certain fishes. An inequality between the sexes in these cases might have been acquired through natural selection, but from their rarity they need not here be further considered. In all ordinary

71 Another great authority in this class, Prof. Thorell of Upsala ('On European Spiders,' 1869-70, part i. p. 205) speaks as if female spiders were generally commoner than the males.

72 See, on this subject, Mr. Pickard-Cambridge, as quoted in 'Quarterly Journal of Science,' 1868, p. 429.

cases an inequality would be no advantage or disadvantage to certain individuals more than to others; and therefore it could hardly have resulted from natural selection. We must attribute the inequality to the direct action of those unknown conditions, which with mankind lead to the males being born in a somewhat larger excess in certain countries than in others, or which cause the proportion between the sexes to differ slightly in legitimate and illegitimate births.

Let us now take the case of a species producing from the unknown causes just alluded to, an excess of one sex—we will say of males—these being superfluous and useless, or nearly useless. Could the sexes be equalised through natural selection? We may feel sure, from all characters being variable, that certain pairs would produce a somewhat less excess of males over females than other pairs. The former, supposing the actual number of the offspring to remain constant, would necessarily produce more females, and would therefore be more productive. On the doctrine of chances a greater number of the offspring of the more productive pairs would survive; and these would inherit a tendency to procreate fewer males and more females. Thus a tendency towards the equalisation of the sexes would be brought about. But our supposed species would by this process be rendered, as just remarked, more productive; and this would in many cases be far from an advantage; for whenever the limit to the numbers which exist, depends, not on destruction by enemies, but on the amount of food, increased fertility will lead to severer competition and to most of the survivors being badly fed. In this case, if the sexes were equalised by an increase in the number of the females, a simultaneous decrease in the total number of the offspring would be beneficial, or even necessary, for the existence of the species; and this, I believe, could be effected through natural selection in the manner hereafter to be described. The same train of reasoning is applicable in the above, as well as in the following case, if we assume that females instead of males are produced in excess, for such females from not uniting with males would be superfluous and useless. So it would be with polygamous species, if we assume the excess of females to be inordinately great.

An excess of either sex, we will again say of the males, could, however, apparently be eliminated through natural selection in another and indirect manner, namely by an actual diminution of the males, without any increase of the females, and consequently without any increase in the productiveness of the species. From the variability of all characters, we may feel assured that some pairs, inhabiting any locality, would produce a rather smaller excess of superfluous males, but an equal number of productive females. When the offspring from the more and the less male-productive parents were all mingled together, none would have any direct advantage over the others; but those that produced few superfluous males would have one great indirect advantage, namely that their ova or embryos would probably be larger and finer, or their young better nurtured in the womb and afterwards. We see this principle illustrated with plants; as those which bear a vast number of seed produce small ones; whilst those which bear comparatively few seeds, often produce large ones well-stocked with nutriment for the use of the seedlings.[73] Hence the offspring of the parents which

[73] I have often been struck with the fact, that in several species of Primula the seeds in the capsules which contained only a few were very much larger than the numerous seeds in the more productive capsules.

had wasted least force in producing superfluous males would be the most likely to survive, and would inherit the same tendency not to produce superfluous males, whilst retaining their full fertility in the production of females. So it would be with the converse case of the female sex. Any slight excess, however, of either sex could hardly be checked in so indirect a manner. Nor indeed has a considerable inequality between the sexes been always prevented, as we have seen in some of the cases given in the previous discussion. In these cases the unknown causes which determine the sex of the embryo, and which under certain conditions lead to the production of one sex in excess over the other, have not been mastered by the survival of those varieties which were subjected to the least waste of organised matter and force by the production of superfluous individuals of either sex. Nevertheless we may conclude that natural selection will always tend, though sometimes inefficiently, to equalise the relative numbers of the two sexes.

Having said this much on the equalisation of the sexes, it may be well to add a few remarks on the regulation through natural selection of the ordinary fertility of species. Mr. Herbert Spencer has shewn in an able discussion [74] that with all organisms a ratio exists between what he calls individuation and genesis; whence it follows that beings which consume much matter or force in their growth, complicated structure or activity, or which produce ova and embryos of large size, or which expend much energy in nurturing their young, cannot be so productive as beings of an opposite nature. Mr. Spencer further shews that minor differences in fertility will be regulated through natural selection. Thus

[74] 'Principles of Biology,' vol. ii. 1867, chaps. ii.-xi.

the fertility of each species will tend to increase, from the more fertile pairs producing a larger number of offspring, and these from their mere number will have the best chance of surviving, and will transmit their tendency to greater fertility. The only check to a continued augmentation of fertility in each organism seems to be either the expenditure of more power and the greater risks run by the parents that produce a more numerous progeny, or the contingency of very numerous eggs and young being produced of smaller size, or less vigorous, or subsequently not so well nurtured. To strike a balance in any case between the disadvantages which follow from the production of a numerous progeny, and the advantages (such as the escape of at least some individuals from various dangers) is quite beyond our power of judgment.

When an organism has once been rendered extremely fertile, how its fertility can be reduced through natural selection is not so clear as how this capacity was first acquired. Yet it is obvious that if individuals of a species, from a decrease of their natural enemies, were habitually reared in larger numbers than could be supported, all the members would suffer. Nevertheless the offspring from the less fertile parents would have no direct advantage over the offspring from the more fertile parents, when all were mingled together in the same district. All the individuals would mutually tend to starve each other. The offspring indeed of the less fertile parents would lie under one great disadvantage, for from the simple fact of being produced in smaller numbers, they would be the most liable to extermination. Indirectly, however, they would partake of one great advantage; for under the supposed condition of severe competition, when all were pressed for food, it is extremely probable that those individuals which from some variation in their constitution produced fewer eggs or young, would produce them of greater size or vigour; and the adults reared from such eggs or young would manifestly have the best chance of surviving, and would inherit a tendency towards lessened fertility. The parents, moreover, which had to nourish or provide for fewer offspring would themselves be exposed to a less severe strain in the struggle for existence, and would have a better chance of surviving. By thes steps, and by no others as far as I can see, natural selection under the above conditions of severe competition for food, would lead to the formation of a new race less fertile, but better adapted for survival, than the parent-race.

CHAPTER IX.

SECONDARY SEXUAL CHARACTERS IN THE LOWER CLASSES OF THE ANIMAL KINGDOM.

These characters absent in the lowest classes — Brilliant colours — Mollusca — Annelids — Crustacea, secondary sexual characters strongly developed; dimorphism; colour; characters not acquired before maturity — Spiders, sexual colours of; stridulation by the males—Myriapoda.

IN the lowest classes the two sexes are not rarely united in the same individual, and therefore secondary sexual characters cannot be developed. In many cases in which the two sexes are separate, both are permanently attached to some support, and the one cannot search or struggle for the other. Moreover it is almost certain that these animals have too imperfect senses and much too low mental powers to feel mutual rivalry, or to appreciate each other's beauty or other attractions.

Hence in these classes, such as the Protozoa, Cœlenterata, Echinodermata, Scolecida, true secondary sexual characters do not occur; and this fact agrees with the belief that such characters in the higher classes have been acquired through sexual selection, which depends on the will, desires, and choice of either sex. Nevertheless some few apparent exceptions occur; thus, as I hear from Dr. Baird, the males of certain Entozoa, or internal parasitic worms, differ slightly in colour from the females; but we have no reason to suppose that such differences have been augmented through sexual selection.

Many of the lower animals, whether hermaphrodites or with the sexes separate, are ornamented with the most brilliant tints, or are shaded and striped in an elegant manner. This is the case with many corals and sea-anemonies (Actineæ), with some jelly-fish (Medusæ, Porpita, &c.), with some Planariæ, Ascidians, numerous Star-fishes, Echini, &c.; but we may conclude from the reasons already indicated, namely the union of the two sexes in some of these animals, the permanently affixed condition of others, and the low mental powers of all, that such colours do not serve as a sexual attraction, and have not been acquired through sexual selection. With the higher animals the case is very different; for with them when one sex is much more brilliantly or conspicuously coloured than the other, and there is no difference in the habits of the two sexes which will account for this difference, we have reason to believe in the influence of sexual selection; and this belief is strongly confirmed when the more ornamented individuals, which are almost always the males, display their attractions before the other sex. We may also extend this conclusion to both sexes, when coloured alike, if their colours are plainly analogous to those of one sex alone in certain other species of the same group.

How, then, are we to account for the beautiful or even gorgeous colours of many animals in the lowest classes? It appears very doubtful whether such colours usually serve as a protection; but we are extremely liable to err in regard to characters of all kinds in relation to protection, as will be admitted by every one who has read Mr. Wallace's excellent essay on this subject. It would not, for instance, at first occur to any one that the perfect transparency of the Medusæ, or jelly-fishes, was of the highest service to them as a protection; but when we are reminded by Häckel that not only the medusæ but many floating mollusca, crustaceans, and even small oceanic fishes partake of this same glass-like structure, we can hardly doubt that they thus escape the notice of pelagic birds and other enemies.

Notwithstanding our ignorance how far colour in many cases serves as a protection, the most probable view in regard to the splendid tints of many of the lowest animals seems to be that their colours are the direct result either of the chemical nature or the minute structure of their tissues, independently of any benefit thus derived. Hardly any colour is finer than that of arterial blood; but there is no reason to suppose that the colour of the blood is in itself any advantage; and though it adds to the beauty of the maiden's cheek, no one will pretend that it has been acquired for this purpose. So again with many animals, especially the lower ones, the bile is richly coloured; thus the extreme beauty of the Eolidæ (naked sea-slugs) is chiefly due, as I am informed by Mr. Hancock, to the biliary glands seen through the translucent integuments; this beauty being probably of no service to these animals. The tints of the decaying leaves in an American forest are described by every one as gorgeous; yet no one supposes that these tints are of the least advantage to the trees. Bearing in mind how many substances closely analogous to natural organic compounds have been recently formed by chemists, and which exhibit the most splendid colours, it would have been a strange fact if substances similarly coloured had not often originated, independently of any useful end being thus gained, in the complex laboratory of living organisms.

The sub-kingdom of the Mollusca.—Throughout this great division (taken in its largest acceptation) of the animal kingdom, secondary sexual characters, such as we are here considering, never, as far as I can discover, occur. Nor could they be expected in the three lowest classes, namely in the Ascidians, Polyzoa, and Brachiopods (constituting the Molluscoida of Huxley), for most of these animals are permanently affixed to a support or have their sexes united in the same individual. In the Lamellibranchiata, or bivalve shells, hermaphroditism is not rare. In the next higher class of the Gasteropoda, or marine univalve shells, the sexes are either united or separate. But in this latter case the males never possess special organs for finding, securing, or charming the females, or for fighting with other males. The sole external difference between the sexes consists, as I am informed by Mr. Gwyn Jeffreys, in the shell sometimes differing a little in form; for instance, the shell of the male periwinkle (*Littorina littorea*) is narrower and has a more elongated spire than that of the female. But differences of this nature, it may be presumed, are directly connected with the act of reproduction or with the development of the ova.

The Gasteropoda, though capable of locomotion and furnished with imperfect eyes, do not appear to be endowed with sufficient mental powers for the members of the same sex to struggle together in rivalry, and thus to acquire secondary sexual characters. Nevertheless with the pulmoniferous gasteropods, or land-shells, the pairing is preceded by courtship; for these animals, though hermaphrodites, are compelled by their structure to pair together. Agassiz remarks,[1] "Qui-"conque a eu l'occasion d'observer les amours des lima-

[1] 'De l'Espèce et de la Class.' &c., 1869, p. 106.

"çons, ne saurait mettre en doute la séduction. déployée "dans les mouvements et les allures qui préparent et "accomplissent le double embrassement de ces her-"maphrodites." These animals appear also susceptible of some degree of permanent attachment: an accurate observer, Mr. Lonsdale, informs me that he placed a pair of land-shells (*Helix pomatia*), one of which was weakly, into a small and ill-provided garden. After a short time the strong and healthy individual disappeared, and was traced by its track of slime over a wall into an adjoining well-stocked garden. Mr. Lonsdale concluded that it had deserted its sickly mate; but after an absence of twenty-four hours it returned, and apparently communicated the result of its successful exploration, for both then started along the same track and disappeared over the wall.

Even in the highest class of the Mollusca, namely the Cephalopoda or cuttle-fishes, in which the sexes are separate, secondary sexual characters of the kind which we are here considering, do not, as far as I can discover, occur. This is a surprising circumstance, as these animals possess highly-developed sense-organs and have considerable mental powers, as will be admitted by every one who has watched their artful endeavours to escape from an enemy.[2] Certain Cephalopoda, however, are characterised by one extraordinary sexual character, namely, that the male element collects within one of the arms or tentacles, which is then cast off, and, clinging by its sucking-discs to the female, lives for a time an independent life. So completely does the cast-off arm resemble a separate animal, that it was described by Cuvier as a parasitic worm under the name

---

[2] See, for instance, the account which I have given in my 'Journal of Researches,' 1845, p. 7.

of Hectocotyle. But this marvellous structure may be classed as a primary rather than as a secondary sexual character.

Although with the Mollusca sexual selection does not seem to have come into play; yet many univalve and bivalve shells, such as volutes, cones, scallops, &c., are beautifully coloured and shaped. The colours do not appear in most cases to be of any use as a protection; they are probably the direct result, as in the lowest classes, of the nature of the tissues; the patterns and the sculpture of the shell depending on its manner of growth. The amount of light seems to a certain extent to be influential; for although, as repeatedly stated by Mr. Gwyn Jeffreys, the shells of some species living at a profound depth are brightly coloured, yet we generally see the lower surfaces and the parts covered by the mantle less highly coloured than the upper and exposed surfaces.[3] In some cases, as with shells living amongst corals or brightly-tinted sea-weeds, the bright colours may serve as a protection. But many of the nudibranch mollusca, or sea-slugs, are as beautifully coloured as any shells, as may be seen in Messrs. Alder and Hancock's magnificent work; and from information kindly given me by Mr. Hancock, it is extremely doubtful whether these colours usually serve as a protection. With some species this may be the case, as with one which lives on the green leaves of algæ, and is itself bright-green. But many brightly-coloured, white or otherwise conspicuous species, do not seek concealment; whilst again some equally conspicuous species, as well as other dull-coloured kinds, live under stones and in

[3] I have given ('Geolog. Observations on Volcanic Islands,' 1844, p. 53) a curious instance of the influence of light on the colours of a frondescent incrustation, deposited by the surf on the coast-rocks of Ascension, and formed by the solution of triturated sea-shells.

dark recesses. So that with these nudibranch molluscs, colour apparently does not stand in any close relation to the nature of the places which they inhabit.

These naked sea-slugs are hermaphrodites, yet they pair together, as do land-snails, many of which have extremely pretty shells. It is conceivable that two hermaphrodites, attracted by each others' greater beauty, might unite and leave offspring which would inherit their parents' greater beauty. But with such lowly-organised creatures this is extremely improbable. Nor is it at all obvious how the offspring from the more beautiful pairs of hermaphrodites would have any advantage, so as to increase in numbers, over the offspring of the less beautiful, unless indeed vigour and beauty generally coincided. We have not here a number of males becoming mature before the females, and the more beautiful ones selected by the more vigorous females. If, indeed, brilliant colours were beneficial to an hermaphrodite animal in relation to its general habits of life, the more brightly-tinted individuals would succeed best and would increase in number; but this would be a case of natural and not of sexual selection.

Sub-kingdom of the Vermes or Annulosa: Class, *Annelida* (*or Sea-worms*).—In this class, although the sexes (when separate) sometimes differ from each other in characters of such importance that they have been placed under distinct genera or even families, yet the differences do not seem of the kind which can be safely attributed to sexual selection. These animals, like those in the preceding classes, apparently stand too low in the scale, for the individuals of either sex to exert any choice in selecting a partner, or for the individuals of the same sex to struggle together in rivalry.

Sub-kingdom of the Arthropoda: Class, *Crustacea.*—In this great class we first meet with undoubted secondary sexual characters, often developed in a remarkable manner. Unfortunately the habits of crustaceans are very imperfectly known, and we cannot explain the uses of many structures peculiar to one sex. With the lower parasitic species the males are of small size, and they alone are furnished with perfect swimming-legs, antennæ and sense-organs; the females being destitute of these organs, with their bodies often consisting of a mere distorted mass. But these extraordinary differences between the two sexes are no doubt related to their widely different habits of life, and consequently do not concern us. In various crustaceans, belonging to distinct families, the anterior antennæ are furnished with peculiar thread-like bodies, which are believed to act as smelling-organs, and these are much more numerous in the males than in the females. As the males, without any unusual development of their olfactory organs, would almost certainly be able sooner or later to find the females, the increased number of the smelling-threads has probably been acquired through sexual selection, by the better provided males having been the most successful in finding partners and in leaving offspring. Fritz Müller has described a remarkable dimorphic species of Tanais, in which the male is represented by two distinct forms, never graduating into each other. In the one form the male is furnished with more numerous smelling-threads, and in the other form with more powerful and more elongated chelæ or pincers which serve to hold the female. Fritz Müller suggests that these differences between the two male forms of the same species must have originated in certain individuals having varied in the number of the smelling-threads, whilst other individuals varied in the shape and size of their chelæ; so that of the former, those which were best able to find the female, and of the latter, those which were best able to hold her when found, have left the greater number of progeny to inherit their respective advantages.[4]

In some of the lower crustaceans, the right-hand anterior antenna of the male differs greatly in structure from the left-hand one, the latter resembling in its simple tapering joints the antennæ of the female. In the male the modified antenna is either swollen in the middle or angularly bent, or converted (fig. 3) into an elegant, and sometimes wonderfully complex, prehensile organ.[5] It serves, as I hear from Sir J. Lubbock, to hold the female, and for this same purpose one of the two posterior legs (*b*) on the same side of the body is converted into a forceps. In another family the inferior or posterior antennæ are "curiously zigzagged" in the males alone.

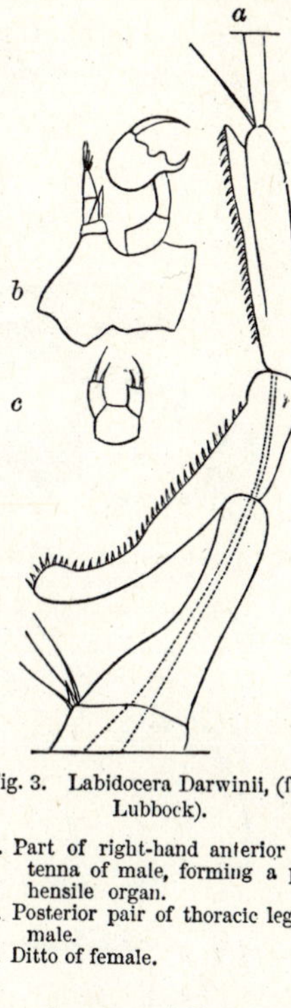

Fig. 3. Labidocera Darwinii, (from Lubbock).

a. Part of right-hand anterior antenna of male, forming a prehensile organ.
b. Posterior pair of thoracic legs of male.
c. Ditto of female.

---

[4] 'Facts and Arguments for Darwin,' English translat. 1869, p. 20. See the previous discussion on the olfactory threads. Sars has described a somewhat analogous case (as quoted in 'Nature,' 1870, p. 455) in a Norwegian crustacean, the *Pontoporeia affinis*.

[5] See Sir J. Lubbock in 'Annals. and Mag. of Nat. Hist.' vol. xi. 1853, pl. i. and x.; and vol. xii. (1853) pl. vii. See also Lubbock in 'Transact. Ent. Soc.' vol. iv. new series, 1856-1858, p. 8. With respect to the zigzagged antennæ mentioned below, see Fritz Müller, 'Facts and Arguments for Darwin' 1869, p. 40, foot-note.

In the higher crustaceans the anterior legs form a pair of chelæ or pincers, and these are generally larger in the male than in the female. In many species the chelæ on the opposite sides of the body are of unequal size, the right-hand one being, as I am in-

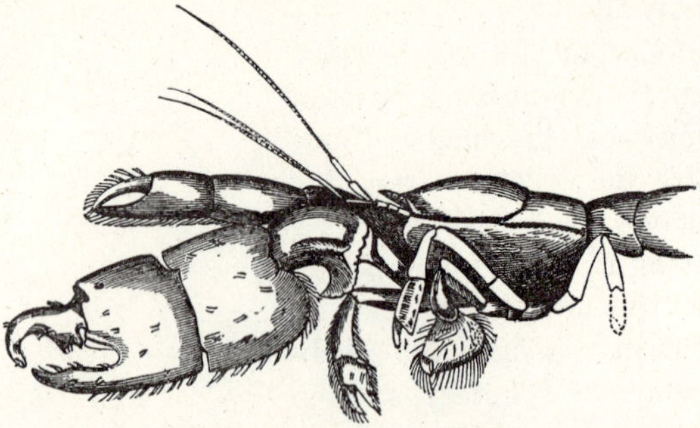

Fig. 4. Anterior part of body of Callianassa (from Milne-Edwards), showing the unequal and differently-constructed right and left-hand chelæ of the male.

N.B.—The artist by mistake has reversed the drawing, and made the left-hand chela the largest.]

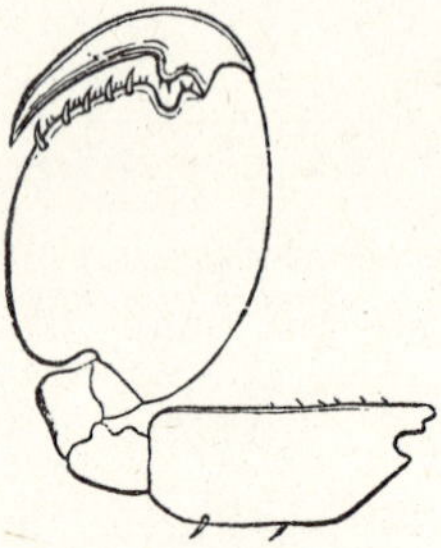

Fig. 5.

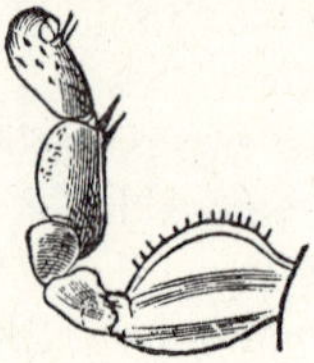

Fig. 6.

Fig. 5. Second leg of male Orchestia Tucuratinga (from Fritz Müller).
Fig. 6. Ditto of female.

formed by Mr. C. Spence Bate, generally, though not invariably, the largest. This inequality is often much greater in the male than in the female. The two chelæ also often differ in structure (figs. 4 and 5), the smaller one resembling those of the female. What advantage is gained by their inequality in size on the opposite sides of the body, and by the inequality being much greater in the male than in the female; and why, when they are of equal size, both are often much larger in the male than in the female, is not known. The chelæ are sometimes of such length and size that they cannot possibly be used, as I hear from Mr. Spence Bate, for carrying food to the mouth. In the males of certain fresh-water prawns (Palæmon) the right leg is actually longer than the whole body.[6] It is probable that the great size of one leg with its chelæ may aid the male in fighting with his rivals; but this use will not account for their inequality in the female on the opposite sides of the body. In Gelasimus, according to a statement quoted by Milne-Edwards,[7] the male and female live in the same burrow, which is worth notice, as shewing that they pair, and the male closes the mouth of the burrow with one of its chelæ, which is enormously developed; so that here it indirectly serves as a means of defence. Their main use, however, probably is to seize and to secure the female, and this in some instances, as with Gammarus, is known to be the case. The sexes, however, of the common shore-crab (*Carcinus mænas*), as Mr. Spence Bate informs me, unite directly after the female has moulted her hard shell, and when she is so soft that she would be injured if seized by the strong pincers of the male; but as she is caught and carried about by the male previously to the act of moulting, she could then be seized with impunity.

Fritz Müller states that certain species of Melita are

---

[6] See a paper by Mr. C. Spence Bate, with figures, in 'Proc. Zoolog. Soc.' 1868, p. 363; and on the nomenclature of the genus, ibid. p. 585. I am greatly indebted to Mr. Spence Bate for nearly all the above statements with respect to the chelæ of the higher crustaceans.

[7] 'Hist. Nat. des Crust.' tom. ii. 1837, p. 50.

distinguished from all other amphipods by the females having "the coxal lamellæ of the penultimate pair of "feet produced into hook-like processes, of which the "males lay hold with the hands of the first pair." The development of these hook-like processes probably resulted from those females which were the most securely held during the act of reproduction, having left the largest number of offspring. Another Brazilian amphipod (*Orchestia Darwinii*, fig. 7) is described by Fritz Müller, as presenting a case of dimorphism, like that of Tanais; for there are two male forms, which differ in the structure of their chelæ.[8] As chelæ of either shape would certainly have sufficed to hold the female, for both are now used for this purpose, the two male forms probably originated, by some having varied in one manner and some in another; both forms having derived certain special, but nearly equal advantages, from their differently shaped organs.

It is not known that male crustaceans fight together for the possession of the females, but this is probable; for with most animals when the male is larger than the female, he seems to have acquired his greater size by having conquered during many generations other males. Now Mr. Spence Bate informs me that in most of the crustacean orders, especially in the highest or the Brachyura, the male is larger than the female; the parasitic genera, however, in which the sexes follow different habits of life, and most of the Entomostraca must be excepted. The chelæ of many crustaceans are weapons well adapted for fighting. Thus a Devil-crab (*Portunus puber*) was seen by a son of Mr. Bate fighting with a *Carcinus mænas*, and the latter was soon thrown on its back, and had every limb torn from its body.

[8] Fritz Müller, 'Facts and Arguments for Darwin,' 1869, p. 25-28.

When several males of a Brazilian Gelasimus, a species furnished with immense pincers, were placed together by Fritz Müller in a glass vessel, they mutilated and killed each other. Mr. Bate put a large male *Carcinus*

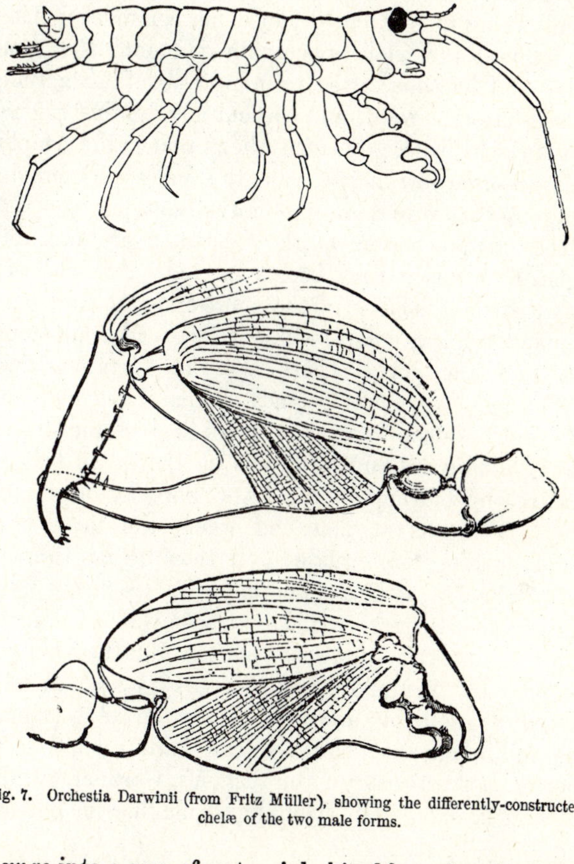

Fig. 7. Orchestia Darwinii (from Fritz Müller), showing the differently-constructed chelæ of the two male forms.

mænas into a pan of water, inhabited by a female paired with a smaller male; the latter was soon dispossessed, but, as Mr. Bate adds, "if they fought, the victory "was a bloodless one, for I saw no wounds." This same naturalist separated a male sand-skipper (so common on our sea-shores), *Gammarus marinus*, from its female, both of which were imprisoned in the same vessel with many individuals of the same species. The female being thus divorced joined her comrades. After an interval the male was again put into the same vessel and he then, after swimming about for a time, dashed into the crowd, and without any fighting at once took away his wife. This fact shews that in the Amphipoda, an order low in the scale, the males and females recognise each other, and are mutually attached.

The mental powers of the Crustacea are probably higher than might have been expected. Any one who has tried to catch one of the shore-crabs, so numerous on many tropical coasts, will have perceived how wary and alert they are. There is a large crab (*Birgos latro*), found on coral islands, which makes at the bottom of a deep burrow a thick bed of the picked fibres of the cocoa-nut. It feeds on the fallen fruit of this tree by tearing off the husk, fibre by fibre; and it always begins at that end where the three eye-like depressions are situated. It then breaks through one of these eyes by hammering with its heavy front pincers, and turning round, extracts the albuminous core with its narrow posterior pincers. But these actions are probably instinctive, so that they would be performed as well by a young as by an old animal. The following case, however, can hardly be so considered: a trustworthy naturalist, Mr. Gardner,[9] whilst watching a shore-crab (Gelasimus) making its burrow,

[9] 'Travels in the Interior of Brazil,' 1846, p. 111. I have given, in my 'Journal of Researches,' p. 463, an account of the habits of the Birgos.

threw some shells towards the hole. One rolled in, and three other shells remained within a few inches of the mouth. In about five minutes the crab brought out the shell which had fallen in, and carried it away to the distance of a foot; it then saw the three other shells lying near, and evidently thinking that they might likewise roll in, carried them to the spot where it had laid the first. It would, I think, be difficult to distinguish this act from one performed by man by the aid of reason.

With respect to colour which so often differs in the two sexes of animals belonging to the higher classes, Mr. Spence Bate does not know of any well-marked instances with our British crustaceans. In some cases, however, the male and female differ slightly in tint, but Mr. Bate thinks not more than may be accounted for by their different habits of life, such as by the male wandering more about and being thus more exposed to the light. In a curious Bornean crab, which inhabits sponges, Mr. Bate could always distinguish the sexes by the male not having the epidermis so much rubbed off. Dr. Power tried to distinguish by colour the sexes of the species which inhabit the Mauritius, but always failed, except with one species of Squilla, probably the *S. stylifera*, the male of which is described as being "of a beautiful blueish-green," with some of the appendages cherry-red, whilst the female is clouded with brown and grey, "with the red about her much "less vivid than in the male."[10] In this case, we may suspect the agency of sexual selection. With Saphirina (an oceanic genus of Entomostraca, and therefore low in the scale) the males are furnished with

[10] Mr. Ch. Fraser, in 'Proc. Zoolog. Soc.' 1869, p. 3. I am indebted to Mr. Bate for the statement from Dr. Power.

minute shields or cell-like bodies, which exhibit beautiful changing colours; these being absent in the females, and in the case of one species in both sexes.[11] It would, however, be extremely rash to conclude that these curious organs serve merely to attract the females. In the female of a Brazilian species of Gelasimus, the whole body, as I am informed by Fritz Müller, is of a nearly uniform greyish-brown. In the male the posterior part of the cephalo-thorax is pure white, with the anterior part of a rich green, shading into dark brown; and it is remarkable that these colours are liable to change in the course of a few minutes—the white becoming dirty grey or even black, the green "losing much of its brilliancy." The males apparently are much more numerous than the females. It deserves especial notice that they do not acquire their bright colours until they become mature. They differ also from the females in the larger size of their chelæ. In some species of the genus, probably in all, the sexes pair and inhabit the same burrow. They are also, as we have seen, highly intelligent animals. From these various considerations it seems highly probable that the male in this species has become gaily ornamented in order to attract or excite the female.

It has just been stated that the male Gelasimus does not acquire his conspicuous colours until mature and nearly ready to breed. This seems the general rule in the whole class with the many remarkable differences in structure between the two sexes. We shall hereafter find the same law prevailing throughout the great sub-kingdom of the Vertebrata, and in all cases it is eminently distinctive of characters which have been

[11] Claus, 'Die freilebenden Copepoden,' 1863, s. 35.

acquired through sexual selection. Fritz Müller[12] gives some striking instances of this law; thus the male sand-hopper (Orchestia) does not acquire his large claspers, which are very differently constructed from those of the female, until nearly full-grown; whilst young his claspers resemble those of the female. Thus, again, the male Brachyscelus possesses, like all other amphipods, a pair of posterior antennæ; the female, and this is a most extraordinary circumstance, is destitute of them, and so is the male as long as he remains immature.

Class, *Arachnida* (Spiders).—The males are often darker, but sometimes lighter than the females, as may be seen in Mr. Blackwall's magnificent work.[13] In some species the sexes differ conspicuously from each other in colour; thus the female of *Sparassus smaragdulus* is dullish-green; whilst the adult male has the abdomen of a fine yellow, with three longitudinal stripes of rich red. In some species of Thomisus the two sexes closely resemble each other; in others they differ much; thus in *T. citreus* the legs and body of the female are pale-yellow or green, whilst the front legs of the male are reddish-brown: in *T. floricolens*, the legs of the female are pale-green, those of the male being ringed in a conspicuous manner with various tints. Numerous analogous cases could be given in the genera Epeira, Nephila, Philodromus, Theridion, Linyphia, &c. It is often difficult to say which of the two sexes departs most from the ordinary coloration of the genus to which the species belong; but Mr. Blackwall

12 'Facts and Arguments,' &c., p. 79.

13 'A History of the Spiders of Great Britain,' 1861-64. For the following facts, see p. 102, 77, 88.

thinks that, as a general rule, it is the male. Both sexes whilst young, as I am informed by the same author, usually resemble each other; and both often undergo great changes in colour during their successive moults before arriving at maturity. In other cases the male alone appears to change colour. Thus the male of the above-mentioned brightly-coloured Sparassus at first resembles the female and acquires his peculiar tints only when nearly adult. Spiders are possessed of acute senses, and exhibit much intelligence. The females often shew, as is well known, the strongest affection for their eggs, which they carry about enveloped in a silken web. On the whole it appears probable that well-marked differences in colour between the sexes have generally resulted from sexual selection, either on the male or female side. But doubts may be entertained on this head from the extreme variability in colour of some species, for instance of *Theridion lineatum*, the sexes of which differ when adult; this great variability indicates that their colours have not been subjected to any form of selection.

Mr. Blackwall does not remember to have seen the males of any species fighting together for the possession of the female. Nor, judging from analogy, is this probable; for the males are generally much smaller than the females, sometimes to an extraordinary degree.[14] Had the males been in the habit of fighting together, they would, it is probable, have gradually

[14] Aug. Vinson ('Aranéides des Iles de la Réunion,' pl. vi. figs. 1 and 2) gives a good instance of the small size of the male in *Epeira nigra*. In this species, as I may add, the male is testaceous and the female black with legs banded with red. Other even more striking cases of inequality in size between the sexes have been recorded ('Quarterly Journal of Science,' 1868, July, p. 429); but I have not seen the original accounts.

acquired greater size and strength. Mr. Blackwall has sometimes seen two or more males on the same web with a single female; but their courtship is too tedious and prolonged an affair to be easily observed. The male is extremely cautious in making his advances, as the female carries her coyness to a dangerous pitch. De Geer saw a male that "in the midst of his preparatory "caresses was seized by the object of his attractions, "enveloped by her in a web and then devoured, a "sight which, as he adds, filled him with horror and "indignation."[15]

Westring has made the interesting discovery that the males of several species of Theridion[16] have the power of making a stridulating sound (like that made by many beetles and other insects, but feebler), whilst the females are quite mute. The apparatus consists of a serrated ridge at the base of the abdomen, against which the hard hinder part of the thorax is rubbed; and of this structure not a trace could be detected in the females. From the analogy of the Orthoptera and Homoptera, to be described in the next chapter, we may feel almost sure that the stridulation serves, as Westring remarks, either to call or to excite the female; and this is the first case in the ascending scale of the animal kingdom, known to me, of sounds emitted for this purpose.

Class, *Myriapoda.*—In neither of the two orders in this class, including the millipedes and centipedes,

---

[15] Kirby and Spence, 'Introduction to Entomology,' vol. i. 1818, p. 280.

[16] Theridion (Asagena, Sund.) serratipes, 4-punctatum et guttatum; see Westring, in Kroyer, 'Naturhist. Tidskrift,' vol. iv. 1842-1843, p. 349; and vol. ii. 1846-1849, p. 342. See, also, for other species, 'Araneæ Svecicæ,' p. 184.

can I find any well-marked instances of sexual differences such as more particularly concern us. In *Glomeris limbata*, however, and perhaps in some few other species, the males differ slightly in colour from the females; but this Glomeris is a highly variable species. In the males of the Diplopoda, the legs belonging to one of the anterior segments of the body, or to the posterior segment, are modified into prehensile hooks which serve to secure the female. In some species of Iulus the tarsi of the male are furnished with membranous suckers for the same purpose. It is a much more unusual circumstance, as we shall see when we treat of Insects, that it is the female in Lithobius which is furnished with prehensile appendages at the extremity of the body for holding the male.[17]

---

[17] Walckenaer et P. Gervais, 'Hist. Nat. des Insectes: Aptères,' tom. iv. 1847, p. 17, 19, 68.

CHAPTER X.

SECONDARY SEXUAL CHARACTERS OF INSECTS.

Diversified structures possessed by the males for seizing the females — Differences between the sexes, of which the meaning is not understood — Difference in size between the sexes — Thysanura — Diptera — Hemiptera — Homoptera, musical powers possessed by the males alone — Orthoptera, musical instruments of the males, much diversified in structure; pugnacity; colours — Neuroptera, sexual differences in colour—Hymenoptera, pugnacity and colours — Coleoptera, colours; furnished with great horns, apparently as an ornament; battles; stridulating organs generally common to both sexes.

IN the immense class of insects the sexes sometimes differ in their organs for locomotion, and often in their sense-organs, as in the pectinated and beautifully plumose antennæ of the males of many species. In one of the Ephemeræ, namely Chloëon, the male has great pillared eyes, of which the female is entirely destitute.[1] The ocelli are absent in the females of certain other insects, as in the Mutillidæ, which are likewise destitute of wings. But we are chiefly concerned with structures by which one male is enabled to conquer another, either in battle or courtship, through his strength, pugnacity, ornaments, or music. The innumerable contrivances, therefore, by which the male is able to seize the female, may be briefly passed over. Besides the complex structures at the apex of the abdomen, which ought perhaps to be ranked as primary

[1] Sir J. Lubbock, 'Transact. Linnean Soc.' vol. xxv. 1866, p. 484. With respect to the Mutillidæ see Westwood, 'Modern Class. of Insects,' vol. ii. p. 213.

organs,[2] "it is astonishing," as Mr. B. D. Walsh[3] has remarked, "how many different organs are worked in "by nature, for the seemingly insignificant object of "enabling the male to grasp the female firmly." The mandibles or jaws are sometimes used for this purpose; thus the male *Corydalis cornutus* (a neuropterous insect in some degree allied to the Dragon-flies, &c.) has immense curved jaws, many times longer than those of the female; and they are smooth instead of being toothed, by which means he is enabled to seize her without injury.[4] One of the stag-beetles of North America (*Lucanus elaphus*) uses his jaws, which are much larger than those of the female, for the same purpose, but probably likewise for fighting. In one of the sand-wasps (*Ammophila*) the jaws in the two sexes are closely alike, but are used for widely different purposes; the males, as Professor Westwood observes, "are exceed- "ingly ardent, seizing their partners round the neck "with their sickle-shaped jaws;"[5] whilst the females use

---

2 These organs in the male often differ in closely-allied species, and afford excellent specific characters. But their importance, under a functional point of view, as Mr. R. MacLachlan has remarked to me, has probably been overrated. It has been suggested, that slight differences in these organs would suffice to prevent the intercrossing of well-marked varieties or incipient species, and would thus aid in their development. That this can hardly be the case, we may infer from the many recorded cases (see for instance, Bronn, 'Geschichte der Natur,' B. ii. 1843, s. 164; and Westwood, 'Transact. Ent. Soc.' vol. iii. 1842, p. 195) of distinct species having been observed in union. Mr. MacLachlan informs me (vide 'Stett. Ent. Zeitung,' 1867, s. 155) that when several species of Phryganidæ, which present strongly-pronounced differences of this kind, were confined together by Dr. Aug. Meyer, *they coupled*, and one pair produced fertile ova.

3 'The Practical Entomologist,' Philadelphia, vol. ii. May, 1867, p. 88.

4 Mr. Walsh, ibid. p. 107.

5 'Modern Classification of Insects,' vol. ii. 1840, p. 206, 205. Mr. Walsh, who called my attention to this double use of the jaws, says that he has repeatedly observed this fact.

these organs for burrowing in sand-banks and making their nests.

The tarsi of the front-legs are dilated in many male beetles, or are furnished with broad cushions of hairs; and in many genera of water-beetles they are armed with a round flat sucker, so that the male may adhere to the slippery body of the female. It is a much more unusual circumstance that the females of some water-beetles (Dytiscus) have their elytra deeply grooved, and in *Acilius sulcatus* thickly set with hairs, as an aid to the male. The females of some other water-beetles (Hydroporus) have their elytra punctured for the same object.[6] In the male of *Crabro cribrarius* (fig. 8.), it is the tibia which is dilated into a broad horny plate, with minute membraneous dots, giving to it a singular appearance like that of a riddle.[7] In the male of Penthe (a genus of beetles) a few of the middle joints of the antennæ are dilated and furnished on the inferior surface

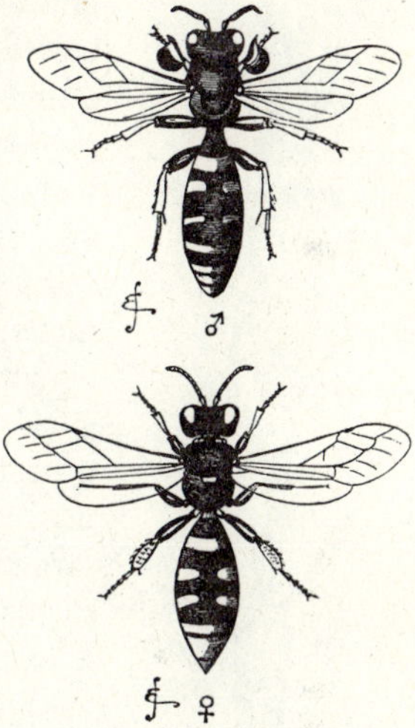

Fig. 8. Crabro cribrarius. Upper figure, male; lower figure, female.

[6] We have here a curious and inexplicable case of dimorphism, for some of the females of four European species of Dytiscus, and of certain species of Hydroporus, have their elytra smooth; and no intermediate gradations between sulcated or punctured and quite smooth elytra have been observed. See Dr. H. Schaum, as quoted in the 'Zoologist,' vol. v.-vi. 1847-48, p. 1896. Also Kirby and Spence, 'Introduction to Entomology,' vol. iii. 1826, p. 305.

[7] Westwood, 'Modern Class.' vol. ii. p. 193. The following statement about Penthe, and others in inverted commas, are taken from Mr. Walsh, 'Practical Entomologist,' Philadelphia, vol. ii. p. 88.

with cushions of hair, exactly like those on the tarsi of the Carabidæ, "and obviously for the same end." In male dragon-flys, "the appendages at the tip of the tail "are modified in an almost infinite variety of curious "patterns to enable them to embrace "the neck of the female." Lastly in the males of many insects, the legs are furnished with peculiar spines, knobs or spurs; or the whole leg is bowed or thickened, but this is by no means invariably a sexual character; or one pair, or all three pairs are elongated, sometimes to an extravagant length.[8]

Fig. 9. Taphroderes distortus (much enlarged). Upper figure, male; lower figure, female.

In all the orders, the sexes of many species present differences, of which the meaning is not understood. One curious case is that of a beetle (fig. 9), the male of which has the left mandible much enlarged; so that the mouth is greatly distorted. In another Carabidous beetle, the Eurygnathus,[9] we have the unique case, as far as known to Mr. Wollaston, of the head of the female being much broader and larger, though in a variable degree, than that of the male. Any number of such cases could be given. They abound in the Lepidoptera: one of the most extraordinary is that certain male butterflies have their fore-legs more or

---

[8] Kirby and Spence, 'Introduct.' &c., vol. iii. p. 332-336.
[9] 'Insecta Maderensia,' 1854, p. 20.

less atrophied, with the tibiæ and tarsi reduced to mere rudimentary knobs. The wings, also, in the two sexes often differ in neuration,[10] and sometimes considerably in outline, as in the *Aricoris epitus*, which was shown to me in the British Museum by Mr. A. Butler. The males of certain South American butterflies have tufts of hair on the margins of the wings, and horny excrescences on the discs of the posterior pair.[11] In several British butterflies, the males alone, as shewn by Mr. Wonfor, are in parts clothed with peculiar scales.

The purpose of the luminosity in the female glow-worm is likewise not understood; for it is very doubtful whether the primary use of the light is to guide the male to the female. It is no serious objection to this latter belief that the males emit a feeble light; for secondary sexual characters proper to one sex are often developed in a slight degree in the other sex. It is a more valid objection that the larvæ shine, and in some species brilliantly: Fritz Müller informs me that the most luminous insect which he ever beheld in Brazil, was the larva of some beetle. Both sexes of certain luminous species of Elater emit light. Kirby and Spence suspect that the phosphorescence serves to frighten and drive away enemies.

Difference in Size between the Sexes.—With insects of all kinds the males are commonly smaller than the females;[12] and this difference can often be detected even in the larval state. So considerable is the difference

---

[10] E. Doubleday, 'Annals and Mag. of Nat. Hist.' vol. i. 1848, p. 379. I may add that the wings in certain Hymenoptera (see Shuckard, 'Fossorial Hymenop.' 1837, p. 39-43) differ in neuration according to sex.

[11] H. W. Bates, in 'Journal of Proc. Linn. Soc.' vol. vi. 1862, p. 74. Mr. Wonfor's observations are quoted in 'Popular Science Review,' 1868, p. 343.

[12] Kirby and Spence, 'Introduction to Entomology,' vol. iii. p. 299.

between the male and female cocoons of the silk-moth (*Bombyx mori*), that in France they are separated by a particular mode of weighing.[13] In the lower classes of the animal kingdom, the greater size of the females seems generally to depend on their developing an enormous number of ova; and this may to a certain extent hold good with insects. But Dr. Wallace has suggested a much more probable explanation. He finds, after carefully attending to the development of the caterpillars of *Bombyx cynthia* and *yamamai*, and especially of some dwarfed caterpillars reared from a second brood on unnatural food, "that in proportion as the individual moth is finer, so is the time required for its metamorphosis longer; and for this reason the female, which is the larger and heavier insect, from having to carry her numerous eggs, will be preceded by the male, which is smaller and has less to mature."[14] Now as most insects are short-lived, and as they are exposed to many dangers, it would manifestly be advantageous to the female to be impregnated as soon as possible. This end would be gained by the males being first matured in large numbers ready for the advent of the females; and this again would naturally follow, as Mr. A. R. Wallace has remarked,[15] through natural selection; for the smaller males would be first matured, and thus would procreate a large number of offspring which would inherit the reduced size of their male parents, whilst the larger males from being matured later would leave fewer offspring.

There are, however, exceptions to the rule of male insects being smaller than the females; and some of

---

[13] Robinet, 'Vers à Soie,' 1848, p. 207.
[14] 'Transact. Ent. Soc.' 3rd series, vol. v. p. 486.
[15] 'Journal of Proc. Ent. Soc.' Feb. 4th, 1867, p. lxxi.

these exceptions are intelligible. Size and strength would be an advantage to the males, which fight for the possession of the female; and in these cases the males, as with the stag-beetle (Lucanus), are larger than the females. There are, however, other beetles which are not known to fight together, of which the males exceed the females in size; and the meaning of this fact is not known; but in some of these cases, as with the huge Dynastes and Megasoma, we can at least see that there would be no necessity for the males to be smaller than the females, in order to be matured before them, for these beetles are not short-lived, and there would be ample time for the pairing of the sexes. So, again, male dragon-flies (Libellulidæ) are sometimes sensibly larger, and never smaller, than the females;[16] and they do not, as Mr. MacLachlan believes, generally pair with the females, until a week or fortnight has elapsed, and until they have assumed their proper masculine colours. But the most curious case, shewing on what complex and easily-overlooked relations, so trifling a character as a difference in size between the sexes may depend, is that of the aculeate Hymenoptera; for Mr. F. Smith informs me that throughout nearly the whole of this large group the males, in accordance with the general rule, are smaller than the females and emerge about a week before them; but amongst the Bees, the males of *Apis mellifica, Anthidium manicatum* and *Anthophora acervorum*, and amongst the Fossores, the males of the *Methoca ichneumonides*, are larger than the females. The explanation of this anomaly is that a marriage-flight is absolutely necessary

---

[16] For this and other statements on the size of the sexes, see Kirby and Spence, ibid. vol. iii. p. 300; on the duration of life in insects, see p. 344.

with these species, and the males require great strength and size in order to carry the females through the air. Increased size has here been acquired in opposition to the usual relation between size and the period of development, for the males, though larger, emerge before the smaller females.

We will now review the several Orders, selecting such facts as more particularly concern us. The Lepidoptera (Butterflies and Moths) will be retained for a separate chapter.

Order, *Thysanura.*—The members of this Order are lowly organised for their class. They are wingless, dull-coloured, minute insects, with ugly, almost misshapen heads and bodies. The sexes do not differ; but they offer one interesting fact, by shewing that the males pay sedulous court to their females even low down in the animal scale. Sir J. Lubbock[17] in describing the *Smynthurus luteus*, says: "it is very amusing to see these "little creatures coquetting together. The male, which "is much smaller than the female, runs round her, and "they butt one another, standing face to face, and "moving backward and forward like two playful lambs. "Then the female pretends to run away and the male "runs after her with a queer appearance of anger, gets "in front and stands facing her again; then she turns "coyly round, but he, quicker and more active, scuttles "round too, and seems to whip her with his antennæ; "then for a bit they stand face to face, play with their "antennæ, and seem to be all in all to one another."

Order, *Diptera* (Flies).—The sexes differ little in colour. The greatest difference, known to Mr. F. Walker,

---

17 'Transact. Linnean Soc.' vol. xxvi. 1868, p. 296.

is in the genus Bibio, in which the males are blackish or quite black, and the females obscure brownish-orange. The genus Elaphomyia, discovered by Mr. Wallace [18] in New Guinea, is highly remarkable, as the males are furnished with horns, of which the females are quite destitute. The horns spring from beneath the eyes, and curiously resemble those of stags, being either branched or palmated. They equal in length the whole of the body in one of the species. They might be thought to serve for fighting, but as in one species they are of a beautiful pink colour, edged with black, with a pale central stripe, and as these insects have altogether a very elegant appearance, it is perhaps more probable that the horns serve as ornaments. That the males of some Diptera fight together is certain; for Prof. Westwood [19] has several times seen this with some species of Tipula or Harry-long-legs. Many observers believe that when gnats (Culicidæ) dance in the air in a body, alternately rising and falling, the males are courting the females. The mental faculties of the Diptera are probably fairly well developed, for their nervous system is more highly developed than in most other Orders of insects.[20]

Order, *Hemiptera* (Field-Bugs).—Mr. J. W. Douglas, who has particularly attended to the British species, has kindly given me an account of their sexual differences. The males of some species are furnished with wings, whilst the females are wingless; the sexes differ in the form of the body and elytra; in the second joints of their antennæ and in their tarsi; but as the signification

---

[18] 'The Malay Archipelago,' vol. ii. 1869, p. 313.

[19] 'Modern Classification of Insects,' vol. ii. 1840, p. 526.

[20] See Mr. B. T. Lowne's very interesting work, 'On the Anatomy of the Blow-Fly, Musca vomitoria,' 1870, p. 14.

of these differences is quite unknown, they may be here passed over. The females are generally larger and more robust than the males. With British, and, as far as Mr. Douglas knows, with exotic species, the sexes do not commonly differ much in colour; but in about six British species the male is considerably darker than the female, and in about four other species the female is darker than the male. Both sexes of some species are beautifully marked with vermilion and black. It is doubtful whether these colours serve as a protection. If in any species the males had differed from the females in an analogous manner, we might have been justified in attributing such conspicuous colours to sexual selection with transference to both sexes.

Some species of Reduvidæ make a stridulating noise; and, in the case of *Pirates stridulus*, this is said[21] to be effected by the movement of the neck within the pro-thoracic cavity. According to Westring, *Reduvius personatus* also stridulates. But I have not been able to learn any particulars about these insects; nor have I any reason to suppose that they differ sexually in this respect.

Order, *Homoptera.*—Every one who has wandered in a tropical forest must have been astonished at the din made by the male Cicadæ. The females are mute; as the Grecian poet Xenarchus says, "Happy the "Cicadas live, since they all have voiceless wives." The noise thus made could be plainly heard on board the "Beagle," when anchored at a quarter of a mile from the shore of Brazil; and Captain Hancock says it can be heard at the distance of a mile. The Greeks formerly kept, and the Chinese now keep, these insects

---

[21] Westwood, 'Modern Class. of Insects,' vol. ii. p. 473.

in cages for the sake of their song, so that it must be pleasing to the ears of some men.[22] The Cicadidæ usually sing during the day; whilst the Fulgoridæ appear to be night-songsters. The sound, according to Landois,[23] who has recently studied the subject, is produced by the vibration of the lips of the spiracles, which are set into motion by a current of air emitted from the tracheæ. It is increased by a wonderfully complex resounding apparatus, consisting of two cavities covered by scales. Hence the sound may truly be called a voice. In the female the musical apparatus is present, but very much less developed than in the male, and is never used for producing sound.

With respect to the object of the music, Dr. Hartman in speaking of the *Cicada septemdecim* of the United States, says,[24] "the drums are now (June 6th and 7th, "1851) heard in all directions. This I believe to be the "marital summons from the males. Standing in thick "chestnut sprouts about as high as my head, where "hundreds were around me, I observed the females "coming around the drumming males." He adds, "this "season (Aug. 1868) a dwarf pear-tree in my garden "produced about fifty larvæ of *Cic. pruinosa;* and I "several times noticed the females to alight near a "male while he was uttering his clanging notes." Fritz Müller writes to me from S. Brazil that he has often listened to a musical contest between two or three males of a Cicada, having a particularly loud voice, and seated at a considerable distance from each other. As

---

22 These particulars are taken from Westwood's 'Modern Class. of Insects,' vol. ii. 1840, p. 422. See, also, on the Fulgoridæ, Kirby and Spence, 'Introduct.' vol. ii. p. 401.

23 'Zeitschrift für wissenschaft. Zoolog.' B. xvii. 1867, s. 152-158.

24 I am indebted to Mr. Walsh for having sent me this extract from a 'Journal of the Doings of Cicada septemdecim,' by Dr. Hartman.

soon as the first had finished his song, a second immediately began; and after he had concluded, another began, and so on. As there is so much rivalry between the males, it is probable that the females not only discover them by the sounds emitted, but that, like female birds, they are excited or allured by the male with the most attractive voice.

I have not found any well-marked cases of ornamental differences between the sexes of the Homoptera. Mr. Douglas informs me that there are three British species, in which the male is black or marked with black bands, whilst the females are pale-coloured or obscure.

Order, *Orthoptera.*—The males in the three saltatorial families belonging to this Order are remarkable for their musical powers, namely the Achetidæ or crickets, the Locustidæ for which there is no exact equivalent name in English, and the Acridiidæ or grasshoppers. The stridulation produced by some of the Locustidæ is so loud that it can be heard during the night at the distance of a mile;[25] and that made by certain species is not unmusical even to the human ear, so that the Indians on the Amazons keep them in wicker cages. All observers agree that the sounds serve either to call or excite the mute females. But it has been noticed[26] that the male migratory locust of Russia (one of the Acridiidæ) whilst coupled with the female, stridulates from anger or jealousy when approached by another male. The house-cricket when surprised at night uses its voice to warn its fellows.[27] In North America the Katy-did (*Platyphyllum concavum,*

[25] L. Guilding, 'Transact. Linn. Soc.' vol. xv. p. 154.
[26] Köppen, as quoted in the 'Zoological Record,' for 1867, p. 460.
[27] Gilbert White, 'Nat. Hist. of Selborne,' vol. ii. 1825, p. 262.

one of the Locustidæ) is described[28] as mounting on the upper branches of a tree, and in the evening beginning "his noisy babble, while rival notes issue from the neighbouring trees, and the groves resound with the call of "*Katy-did-she-did*, the live-long night." Mr. Bates, in speaking of the European field-cricket (one of the Achetidæ), says, "the male has been observed to place itself "in the evening at the entrance of its burrow, and "stridulate until a female approaches, when the louder "notes are succeeded by a more subdued tone, whilst "the successful musician caresses with his antennæ "the mate he has won."[29] Dr. Scudder was able to excite one of these insects to answer him, by rubbing on a file with a quill.[30] In both sexes a remarkable auditory apparatus has been discovered by Von Siebold, situated in the front legs.[31]

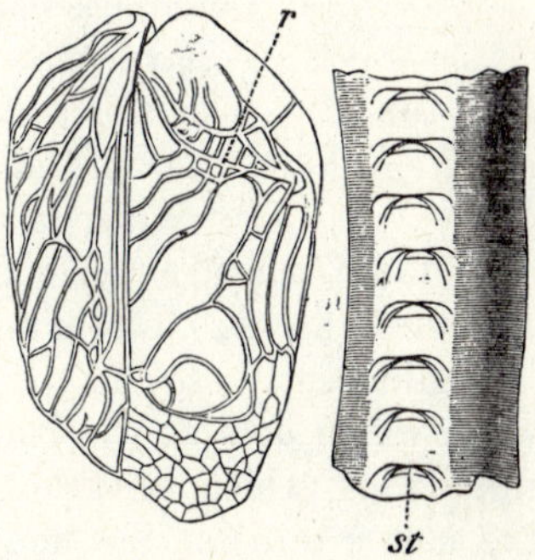

Fig. 10. Gryllus campestris (from Landois).
Right-hand figure, under side of part of the wing-nervure, much magnified, showing the teeth, *st*.
Left-hand figure, upper surface of wing-cover, with the projecting, smooth nervure, *r*., across which the teeth (*st*) are scraped.

In the three Families the sounds are differently produced. In the males of the Achetidæ both wing-covers have the same structure; and this in the field-cricket (*Gryllus campestris*, fig. 10) consists, as de-

[28] Harris, 'Insects of New England,' 1842, p. 128.

[29] 'The Naturalist on the Amazons,' vol. i. 1863, p. 252. Mr. Bates gives a very interesting discussion on the gradations in the musical apparatus of the three families. See also Westwood, 'Modern Class.' vol. ii. p. 445 and 453.

[30] 'Proc. Boston Soc. of Nat. Hist.' vol. xi. April, 1868.

[31] 'Nouveau Manuel d'Anat. Comp.' (French translat.), tom. i. 1850 p. 567.

scribed by Landois,[32] of from 131 to 138 sharp, transverse ridges or teeth (*st*) on the under side of one of the nervures of the wing-cover. This toothed nervure is rapidly scraped across a projecting, smooth, hard nervure (*r*) on the upper surface of the opposite wing. First one wing is rubbed over the other, and then the movement is reversed. Both wings are raised a little at the same time, so as to increase the resonance. In some species the wing-covers of the males are furnished at the base with a talc-like plate.[33] I have here given a drawing (fig. 11) of the teeth on the under side of the nervure of another species of Gryllus, viz. *G. domesticus*.

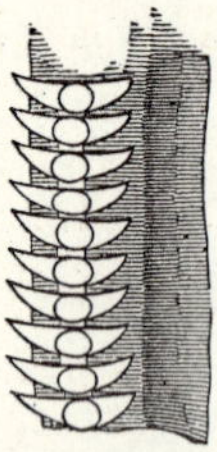

Fig. 11. Teeth of Nervure of Gryllus domesticus (from Landois).

In the Locustidæ the opposite wing-covers differ in structure (fig. 12), and cannot, as in the last family, be indifferently used in a reversed manner. The left wing, which acts as the bow of the fiddle, lies over the right wing which serves as the fiddle itself. One of the nervures (*a*) on the under surface of the former is finely serrated, and is scraped across the prominent nervures on the upper surface of the opposite or right wing. In our British *Phasgonura viridissima* it appeared to me that the serrated nervure is rubbed against the rounded hind corner of the opposite wing, the edge of which is thickened, coloured brown, and very sharp. In the right wing, but not in the left, there is a little plate, as transparent as talc, surrounded by nervures, and called the speculum. In *Ephippiger vitium*, a member of this same family, we have a curious

32 'Zeitschrift für wissenschaft. Zoolog.' B. xvii. 1867, s. 117.

33 Westwood, 'Modern Class. of Insects,' vol. i. p. 440.

subordinate modification; for the wing-covers are greatly reduced in size, but "the posterior part of the pro-thorax "is elevated into a kind of dome over the wing-covers, "and which has probably the effect of increasing the "sound."[34]

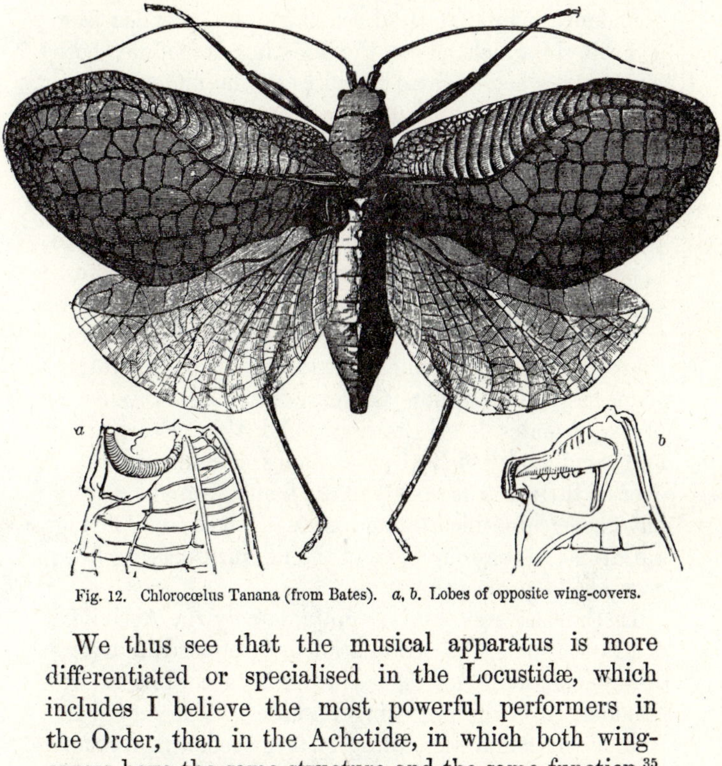

Fig. 12. Chlorocœlus Tanana (from Bates). *a, b.* Lobes of opposite wing-covers.

We thus see that the musical apparatus is more differentiated or specialised in the Locustidæ, which includes I believe the most powerful performers in the Order, than in the Achetidæ, in which both wing-covers have the same structure and the same function.[35] Landois, however, detected in one of the Locustidæ, namely in Decticus, a short and narrow row of small

---

[34] Westwood, 'Modern Class. of Insects,' vol. i. p. 453.

[35] Landois, ibid. s. 121, 122.

teeth, mere rudiments, on the inferior surface of the right wing-cover, which underlies the other and is never used as the bow. I observed the same rudimentary structure on the under side of the right wing-cover in *Phasgonura viridissima.* Hence we may with confidence infer that the Locustidæ are descended from a form, in which, as in the existing Achetidæ, both wing-covers had serrated nervures on the under surface, and could be indifferently used as the bow; but that in the Locustidæ the two wing-covers gradually became differentiated and perfected, on the principle of the division of labour, the one to act exclusively as the bow and the other as the fiddle. By what steps the more simple apparatus in the Achetidæ originated, we do not know, but it is probable that the basal portions of the wing-covers overlapped each other formerly as at present, and that the friction of the nervures produced a grating sound, as I find is now the case with the wing-covers of the females.[36] A grating sound thus occasionally and accidentally made by the males, if it served them ever so little as a love-call to the females, might readily have been intensified through sexual selection by fitting variations in the roughness of the nervures having been continually preserved.

In the last and third Family, namely the Acridiidæ or grasshoppers, the stridulation is produced in a very different manner, and is not so shrill, according to Dr. Scudder, as in the preceding Families. The inner surface of the femur (fig. 13, *r*) is furnished with a longitudinal row of minute, elegant, lancet-shaped, elastic teeth, from 85 to 93 in number;[37] and these are scraped

36 Mr. Walsh also informs me that he has noticed that the female of the *Platyphyllum concavum*, "when captured makes a feeble grating "noise by shuffling her wing-covers together."

37 Landois, ibid. s. 113.

across the sharp, projecting nervures on the wing-covers, which are thus made to vibrate and resound. Harris[38] says that when one of the males begins to play, he first "bends the shank "of the hind-leg beneath "the thigh, where it is "lodged in a furrow de-"signed to receive it, "and then draws the leg "briskly up and down. "He does not play both "fiddles together, but al-"ternately first upon one "and then on the other." In many species, the base of the abdomen is hollowed out into a great cavity which is believed to act as a resounding board. In Pneumora (fig. 14), a S. African genus belonging to this same family, we meet with a new and remarkable modification: in the males a small notched ridge projects obliquely from each side of the abdomen, against which the hind femora are rubbed.[39] As the male is furnished with wings, the female being wingless, it is remarkable that the thighs are not rubbed in the usual manner against the wing-covers; but this may perhaps be accounted for by the unusually small size of the hind-legs. I have not been able to examine the inner surface of the thighs, which, judging from analogy, would be finely serrated. The species of Pneumora have been more profoundly modified for the sake of stridulation than any other orthopterous insect; for

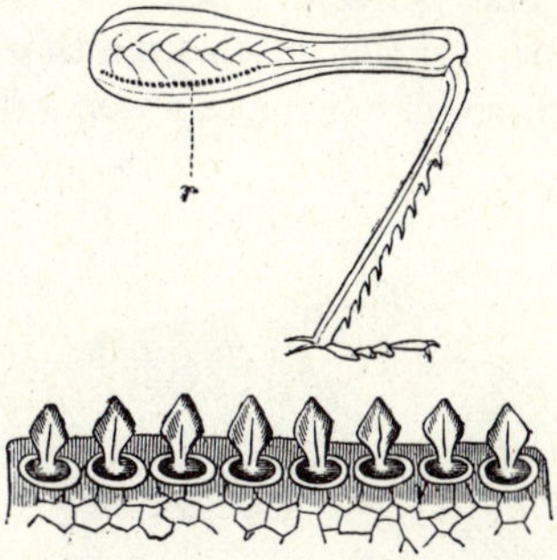

Fig. 13. Hind-leg of Stenobothrus pratorum: *r*, the stridulating ridge; lower figure, the teeth, forming the ridge, much magnified (from Landois).

[38] 'Insects of New England,' 1842, p. 133.
[39] Westwood, 'Modern Classification,' vol. i. p. 462.

in the male the whole body has been converted into a musical instrument, being distended with air, like a great pellucid bladder, so as to increase the resonance. Mr. Trimen informs me that at the Cape of Good Hope these insects make a wonderful noise during the night.

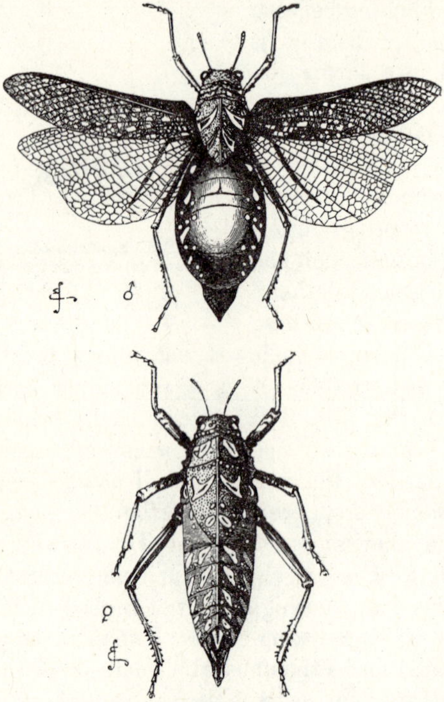

Fig. 14. Pneumora (from specimens in the British Museum). Upper figure, male; lower figure, female.

There is one exception to the rule that the females in these three Families are destitute of an efficient musical apparatus; for both sexes of Ephippiger (Locustidæ) are said[40] to be thus provided. This case may

[40] Westwood, ibid. vol. i. p. 453.

be compared with that of the reindeer, in which species alone both sexes possess horns. Although the female orthoptera are thus almost invariably mute, yet Landois[41] found rudiments of the stridulating organs on the femora of the female Acridiidæ, and similar rudiments on the under surface of the wing-covers of the female Achetidæ; but he failed to find any rudiments in the females of Decticus, one of the Locustidæ. In the Homoptera the mute females of Cicada, have the proper musical apparatus in an undeveloped state; and we shall hereafter meet in other divisions of the animal kingdom with innumerable instances of structures proper to the male being present in a rudimentary condition in the female. Such cases appear at first sight to indicate that both sexes were primordially constructed in the same manner, but that certain organs were subsequently lost by the females. It is, however, a more probable view, as previously explained, that the organs in question were acquired by the males and partially transferred to the females.

Landois has observed another interesting fact, namely that in the females of the Acridiidæ, the stridulating teeth on the femora remain throughout life in the same condition in which they first appear in both sexes during the larval state. In the males, on the other hand, they become fully developed and acquire their perfect structure at the last moult, when the insect is mature and ready to breed.

From the facts now given, we see that the means by which the males produce their sounds are extremely diversified in the Orthoptera, and are altogether different from those employed by the Homoptera. But throughout the animal kingdom we incessantly find the

---

41 Landois, ibid. s. 115, 116, 120, 122.

same object gained by the most diversified means; this being due to the whole organisation undergoing in the course of ages multifarious changes; and as part after part varies, different variations are taken advantage of for the same general purpose. The diversification of the means for producing sound in the three families of the Orthoptera and in the Homoptera, impresses the mind with the high importance of these structures to the males, for the sake of calling or alluring the females. We need feel no surprise at the amount of modification which the Orthoptera have undergone in this respect, as we now know, from Dr. Scudder's remarkable discovery,[42] that there has been more than ample time. This naturalist has lately found a fossil insect in the Devonian formation of New Brunswick, which is furnished with "the well-known tympanum or stridulating apparatus "of the male Locustidæ." This insect, though in most respects related to the Neuroptera, appears to connect, as is so often the case with very ancient forms, the two Orders of the Neuroptera and Orthoptera which are now generally ranked as quite distinct.

I have but little more to say on the Orthoptera. Some of the species are very pugnacious: when two male field-crickets (*Gryllus campestris*) are confined together, they fight till one kills the other; and the species of Mantis are described as manœuvring with their sword-like front-limbs, like hussars with their sabres. The Chinese keep these insects in little bamboo cages and match them like game-cocks.[43] With respect to colour, some exotic locusts are beautifully ornamented; the posterior wings being marked with red,

---

[42] 'Transact. Ent. Soc.' 3rd series, vol. ii. ('Journal of Proceedings, p. 117.)

[43] Westwood, 'Modern Class. of Insects,' vol. i. p. 427; for crickets, p. 445.

blue, and black; but as throughout the Order the two sexes rarely differ much in colour, it is doubtful whether they owe these bright tints to sexual selection. Conspicuous colours may be of use to these insects as a protection, on the principle to be explained in the next chapter, by giving notice to their enemies that they are unpalatable. Thus it has been observed[44] that an Indian brightly-coloured locust was invariably rejected when offered to birds and lizards. Some cases, however, of sexual differences in colour in this Order are known. The male of an American cricket[45] is described as being as white as ivory, whilst the female varies from almost white to greenish-yellow or dusky. Mr. Walsh informs me that the adult male of *Spectrum femoratum* (one of the Phasmidæ) "is of a shining "brownish-yellow colour; the adult female being of "a dull, opaque, cinereous-brown; the young of both "sexes being green." Lastly, I may mention that the male of one curious kind of cricket[46] is furnished with "a long membranous appendage, which falls over the "face like a veil;" but whether this serves as an ornament is not known.

Order, *Neuroptera.*—Little need here be said, except in regard to colour. In the Ephemeridæ the sexes often differ slightly in their obscure tints;[47] but it is not probable that the males are thus rendered attractive to the females. The Libellulidæ or dragon-flies are ornamented with splendid green, blue, yellow, and

44 Mr. Ch. Horne, in 'Proc. Ent. Soc.' May 3, 1869, p. xii.

45 The *Oecanthus nivalis.* Harris, 'Insects of New England,' 1842, p. 124.

46 Platyblemnus: Westwood, 'Modern. Class.' vol. i. p. 447.

47 B. D. Walsh, the Pseudo-neuroptera of Illinois, in 'Proc. Ent. Soc. of Philadelphia,' 1862, p. 361.

vermilion metallic tints; and the sexes often differ. Thus, the males of some of the Agrionidæ, as Prof. Westwood remarks,[48] "are of a rich blue with black "wings, whilst the females are fine green with colourless "wings." But in *Agrion Ramburii* these colours are exactly reversed in the two sexes.[49] In the extensive N. American genus of Hetærina, the males alone have a beautiful carmine spot at the base of each wing. In *Anax junius* the basal part of the abdomen in the male is a vivid ultra-marine blue, and in the female grass-green. In the allied genus Gomphus, on the other hand, and in some other genera, the sexes differ but little in colour. Throughout the animal kingdom, similar cases of the sexes of closely-allied forms either differing greatly, or very little, or not at all, are of frequent occurrence. Although with many Libellulidæ there is so wide a difference in colour between the sexes, it is often difficult to say which is the most brilliant; and the ordinary coloration of the two sexes is exactly reversed, as we have just seen, in one species of Agrion. It is not probable that their colours in any case have been gained as a protection. As Mr. MacLachlan, who has closely attended to this family, writes to me, dragon-flies—the tyrants of the insect-world—are the least liable of any insect to be attacked by birds or other enemies. He believes that their bright colours serve as a sexual attraction. It deserves notice, as bearing on this subject, that certain dragon-flies appear to be attracted by particular colours: Mr. Patterson observed[50] that the species of Agrionidæ, of which the males are blue, settled in numbers on the blue float of a fishing

48 'Modern Class.' vol. ii. p. 37.

49 Walsh, ibid. p. 381. I am indebted to this naturalist for the following facts on Hetærina, Anax, and Gomphus.

50 'Transact. Ent. Soc.' vol. i. 1836, p. lxxxi.

line; whilst two other species were attracted by shining white colours.

It is an interesting fact, first observed by Schelver, that the males, in several genera belonging to two sub-families, when they first emerge from the pupal state are coloured exactly like the females; but that their bodies in a short time assume a conspicuous milky-blue tint, owing to the exudation of a kind of oil, soluble in ether and alcohol. Mr. MacLachlan believes that in the male of *Libellula depressa* this change of colour does not occur until nearly a fortnight after the metamorphosis, when the sexes are ready to pair.

Certain species of Neurothemis present, according to Brauer [51] a curious case of dimorphism, some of the females having their wings netted in the usual manner; whilst other females have them "very richly netted as in "the males of the same species." Brauer "explains "the phenomenon on Darwinian principles by the "supposition that the close netting of the veins is a "secondary sexual character in the males." This latter character is generally developed in the males alone, but being, like every other masculine character, latent in the female, is occasionally developed in them. We have here an illustration of the manner in which the two sexes of many animals have probably come to resemble each other, namely by variations first appearing in the males, being preserved in them, and then transmitted to and developed in the females; but in this particular genus a complete transference is occasionally and abruptly effected. Mr. MacLachlan informs me of another case of dimorphism occurring in several species of Agrion in which a certain number of individuals are found of an orange colour, and these are

---

[51] See abstract in the 'Zoological Record' for 1867, p. 450.

invariably females. This is probably a case of reversion, for in the true Libellulæ, when the sexes differ in colour, the females are always orange or yellow, so that supposing Agrion to be descended from some primordial form having the characteristic sexual colours of the typical Libellulæ, it would not be surprising that a tendency to vary in this manner should occur in the females alone.

Although many dragon-flies are such large, powerful, and fierce insects, the males have not been observed by Mr. MacLachlan to fight together, except, as he believes, in the case of some of the smaller species of Agrion. In another very distinct group in this Order, namely in the Termites or white ants, both sexes at the time of swarming may be seen running about, "the "male after the female, sometimes two chasing one "female, and contending with great eagerness who shall "win the prize."[52]

Order, *Hymenoptera.*—That inimitable observer, M. Fabre,[53] in describing the habits of Cerceris, a wasp-like insect, remarks that "fights frequently ensue "between the males for the possession of some particular female, who sits an apparently unconcerned "beholder of the struggle for supremacy, and when the "victory is decided, quietly flies away in company "with the conqueror." Westwood[54] says that the males of one of the saw-flies (Tenthredinæ) "have been "found fighting together, with their mandibles locked." As M. Fabre speaks of the males of Cerceris striving to obtain a particular female, it may be well to bear in

52 Kirby and Spence, 'Introduct. to Entomology,' vol. ii. 1818, p. 35.

53 See an interesting article, "The Writings of Fabre," in 'Nat. Hist. Review,' April, 1862, p. 122.

54 'Journal of Proc. of Entomolog. Soc.' Sept. 7th, 1863, p. 169.

mind that insects belonging to this Order have the power of recognising each other after long intervals of time, and are deeply attached. For instance, Pierre Huber, whose accuracy no one doubts, separated some ants, and when after an interval of four months they met others which had formerly belonged to the same community, they mutually recognised and caressed each other with their antennæ. Had they been strangers they would have fought together. Again, when two communities engage in a battle, the ants on the same side in the general confusion sometimes attack each other, but they soon perceive their mistake, and the one ant soothes the other.[55]

In this Order slight differences in colour, according to sex, are common, but conspicuous differences are rare except in the family of Bees; yet both sexes of certain groups are so brilliantly coloured—for instance in Chrysis, in which vermilion and metallic greens prevail—that we are tempted to attribute the result to sexual selection. In the Ichneumonidæ, according to Mr. Walsh,[56] the males are almost universally lighter coloured than the females. On the other hand, in the Tenthredinidæ the males are generally darker than the females. In the Siricidæ the sexes frequently differ; thus the male of *Sirex juvencus* is banded with orange, whilst the female is dark purple; but it is difficult to say which sex is the most ornamented. In *Tremex columbæ* the female is much brighter coloured than the male. With ants, as I am informed by Mr. F. Smith, the males of several species are black, the females being testaceous. In the family of Bees, especially in

---

55 P. Huber, 'Recherches sur les Mœurs des Fourmis,' 1810, p. 150, 165.

56 'Proc. Entomolog. Soc. of Philadelphia,' 1866, p. 238-239.

the solitary species, as I hear from the same distinguished entomologist, the sexes often differ in colour. The males are generally the brightest, and in Bombus as well as in Apathus, much more variable in colour than the females. In *Anthophora retusa* the male is of a rich fulvous-brown, whilst the female is quite black: so are the females of several species of Xylocopa, the males being bright yellow. In an Australian bee (*Lestis bombylans*), the female is of an extremely brilliant steel-blue, sometimes tinted with vivid green; the male being of a bright brassy colour clothed with rich fulvous pubescence. As in this group the females are provided with excellent defensive weapons in their stings, it is not probable that they have come to differ in colour from the males for the sake of protection.

Mutilla Europæa emits a stridulating noise; and according to Goureau [57] both sexes have this power. He attributes the sound to the friction of the third and preceding abdominal segments; and I find that these surfaces are marked with very fine concentric ridges, but so is the projecting thoracic collar, on which the head articulates; and this collar, when scratched with the point of a needle, emits the proper sound. It is rather surprising that both sexes should have the power of stridulating, as the male is winged and the female wingless. It is notorious that Bees express certain emotions, as of anger, by the tone of their humming, as do some dipterous insects; but I have not referred to these sounds, as they are not known to be in any way connected with the act of courtship.

Order, *Coleoptera* (Beetles). — Many beetles are coloured so as to resemble the surfaces which they

---

57 Quoted by Westwood, 'Modern Class. of Insects,' vol. ii. p. 214.

habitually frequent. Other species are ornamented with gorgeous metallic tints,—for instance, many Carabidæ, which live on the ground and have the power of defending themselves by an intensely acrid secretion, —the splendid diamond-beetles which are protected by an extremely hard covering,—many species of Chrysomela, such as *C. cerealis,* a large species beautifully striped with various colours, and in Britain confined to the bare summit of Snowdon,—and a host of other species. These splendid colours, which are often arranged in stripes, spots, crosses and other elegant patterns, can hardly be beneficial, as a protection, except in the case of some flower-feeding species; and we cannot believe that they are purposeless. Hence the suspicion arises, that they serve as a sexual attraction; but we have no evidence on this head, for the sexes rarely differ in colour. Blind beetles, which cannot of course behold each other's beauty, never exhibit, as I hear from Mr. Waterhouse, jun., bright colours, though they often have polished coats: but the explanation of their obscurity may be that blind insects inhabit caves and other obscure stations.

Some Longicorns, however, especially certain Prionidæ, offer an exception to the common rule that the sexes of beetles do not differ in colour. Most of these insects are large and splendidly coloured. The males in the genus Pyrodes,[58] as I saw in Mr. Bates' collection, are

[58] Pyrodes pulcherrimus, in which the sexes differ conspicuously, has been described by Mr. Bates in 'Transact. Ent. Soc.' 1869, p. 50. I will specify the few other cases in which I have heard of a difference in colour between the sexes of beetles. Kirby and Spence ('Introduct. to Entomology,' vol. iii. p. 301) mention a Cantharis, Meloe, Rhagium, and the *Leptura testacea*; the male of the latter being testaceous, with a black thorax, and the female of a dull red all over. These two latter beetles belong to the Order of Longicorns. Messrs. R. Trimen and Waterhouse, junr., inform me of two Lamellicorns, viz., a Peri-

generally redder but rather duller than the females, the latter being coloured of a more or less splendid golden green. On the other hand, in one species the male is golden-green, the female being richly tinted with red and purple. In the genus Esmeralda the sexes differ so greatly in colour that they have been ranked as distinct species: in one species both are of a beautiful shining green, but the male has a red thorax. On the whole, as far as I could judge, the females of those Prionidæ, in which the sexes differ, are coloured more richly than the males; and this does not accord with the common rule in regard to colour when acquired through sexual selection.

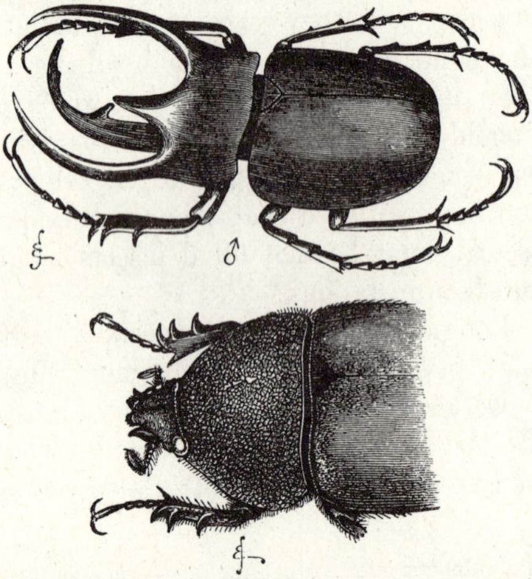

Fig. 15. Chalcosoma atlas. Upper figure, male (reduced); lower figure, female (nat. size).

trichia and Trichius, the male of the latter being more obscurely coloured than the female. In *Tillus elongatus* the male is black, and the female always, as it is believed, of a dark blue colour with a red thorax. The male, also, of *Orsodacna atra*, as I hear from Mr. Walsh, is black, the female (the so-called *O. ruficollis*) having a rufous thorax.

Fig. 16. Copris isidis. (Left-hand figures, males.)

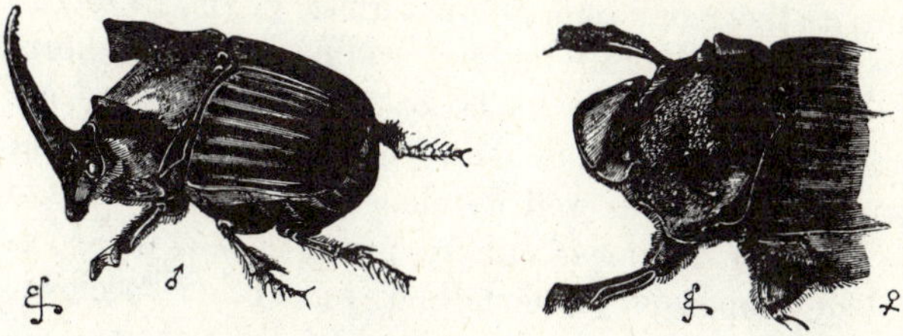

Fig. 17. Phanæus faunus.

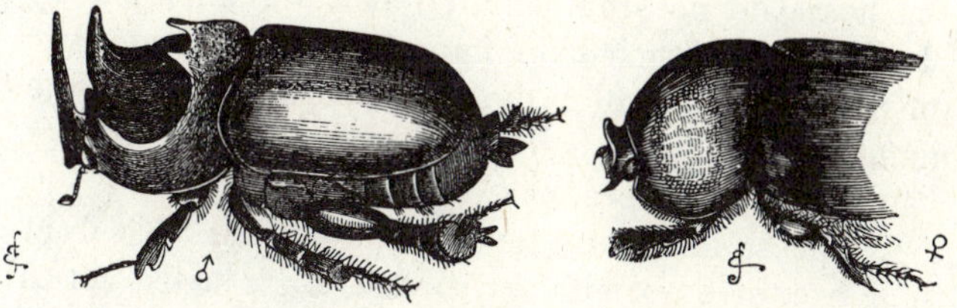

Fig. 18. Dipelicus cantori.

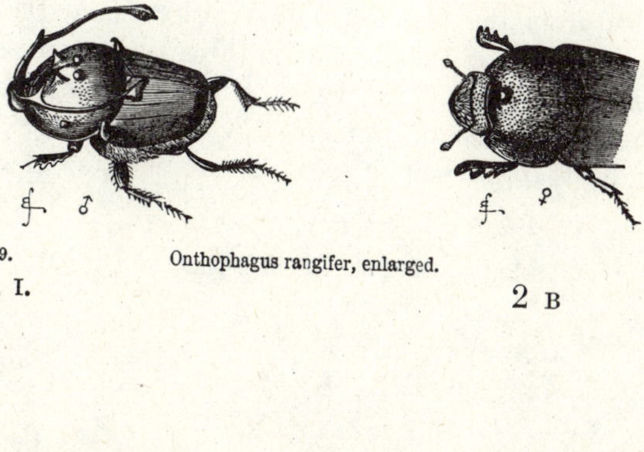

Fig. 19. Onthophagus rangifer, enlarged.

A most remarkable distinction between the sexes of many beetles is presented by the great horns which rise from the head, thorax, or clypeus of the males; and in some few cases from the under surface of the body. These horns, in the great family of the Lamellicorns, resemble those of various quadrupeds, such as stags, rhinoceroses, &c., and are wonderful both from their size and diversified shapes. Instead of describing them, I have given figures of the males and females of some of the more remarkable forms. (Figs. 15 to 19.) The females generally exhibit rudiments of the horns in the form of small knobs or ridges; but some are destitute of even a rudiment. On the other hand, the horns are nearly as well developed in the female as in the male of *Phanæus lancifer*; and only a little less well developed in the females of some other species of the same genus and of Copris. In the several subdivisions of the family, the differences in structure of the horns do not run parallel, as I am informed by Mr. Bates, with their more important and characteristic differences; thus within the same natural section of the genus Onthophagus, there are species which have either a single cephalic horn, or two distinct horns.

In almost all cases, the horns are remarkable from their excessive variability; so that a graduated series can be formed, from the most highly developed males to others so degenerate that they can barely be distinguished from the females. Mr Walsh[59] found that in *Phanæus carnifex* the horns were thrice as long in some males as in others. Mr. Bates, after examining above a hundred males of *Onthophagus rangifer* (fig. 19), thought that he had at last discovered a species in

59 'Proc. Entomolog. Soc. of Philadelphia,' 1864, p. 228.

which the horns did not vary; but further research proved the contrary.

The extraordinary size of the horns, and their widely different structure in closely-allied forms, indicate that they have been formed for some important purpose; but their excessive variability in the males of the same species leads to the inference that this purpose cannot be of a definite nature. The horns do not show marks of friction, as if used for any ordinary work. Some authors suppose[60] that as the males wander much more than the females, they require horns as a defence against their enemies; but in many cases the horns do not seem well adapted for defence, as they are not sharp. The most obvious conjecture is that they are used by the males for fighting together; but they have never been observed to fight; nor could Mr. Bates, after a careful examination of numerous species, find any sufficient evidence in their mutilated or broken condition of their having been thus used. If the males had been habitual fighters, their size would probably have been increased through sexual selection, so as to have exceeded that of the female; but Mr. Bates, after comparing the two sexes in above a hundred species of the Copridæ, does not find in well-developed individuals any marked difference in this respect. There is, moreover, one beetle, belonging to the same great division of the Lamellicorns, namely Lethrus, the males of which are known to fight, but they are not provided with horns, though their mandibles are much larger than those of the female.

The conclusion, which best agrees with the fact of the horns having been so immensely yet not fixedly developed,—as shewn by their extreme variability in

---

60 Kirby and Spence, 'Introduct. Entomolog.' vol. iii. p. 300.

the same species and by their extreme diversity in closely-allied species—is that they have been acquired as ornaments. This view will at first appear extremely improbable; but we shall hereafter find with many animals, standing much higher in the scale, namely fishes, amphibians, reptiles and birds, that various kinds of crests, knobs, horns and combs have been developed apparently for this sole purpose.

The males of *Onitis furcifer* (fig. 20) are furnished with singular projections on their anterior femora, and with a great fork or pair of horns on the lower surface of the thorax. This situation seems extremely ill adapted for the display of these projections, and they may be of some real service; but no use can at present be assigned to them. It is a highly remarkable fact, that although the males do not exhibit even a trace of horns on the upper surface of the body, yet in the females a rudiment of a single horn on the head (fig. 21, *a*), and of a crest (*b*) on the thorax, are plainly visible. That the slight thoracic crest in the

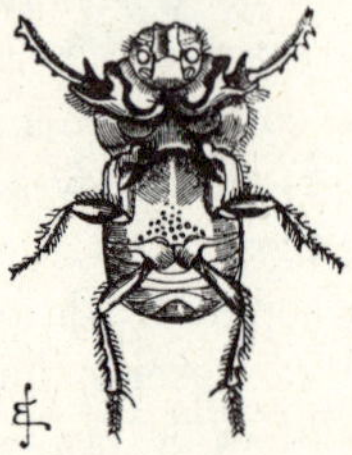

Fig. 20. Onitis furcifer, male, viewed from beneath.

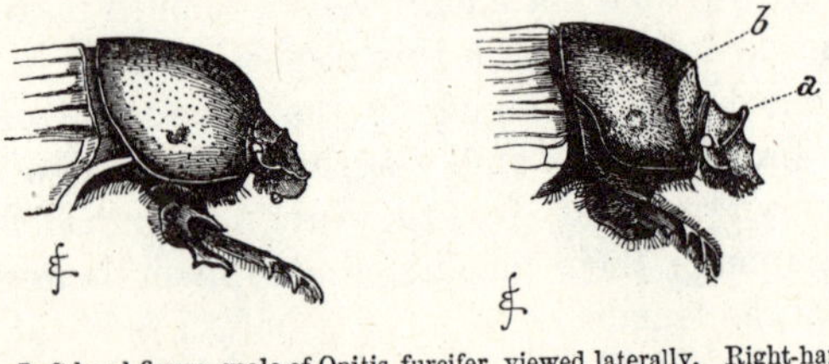

Fig. 21. Left-hand figure, male of Onitis furcifer, viewed laterally. Right-hand figure, female. *a*. Rudiment of cephalic horn. *b*. Trace of thoracic horn or crest.

female is a rudiment of a projection proper to the male, though entirely absent in the male of this particular species, is clear: for the female of *Bubas bison* (a form which comes next to Onitis) has a similar slight crest on the thorax, and the male has in the same situation a great projection. So again there can be no doubt that the little point (*a*) on the head of the female *Onitis furcifer*, as well of the females of two or three allied species, is a rudimentary representative of the cephalic horn, which is common to the males of so many lamellicorn beetles, as in Phanæus, fig. 17. The males indeed of some unnamed beetles in the British Museum, which are believed actually to belong to the genus Onitis, are furnished with a similar horn. The remarkable nature of this case will be best perceived by an illustration: the Ruminant quadrupeds run parallel with the lamellicorn beetles, in some females possessing horns as large as those of the male, in others having them much smaller, or existing as mere rudiments (though this is as rare with ruminants as it is common with Lamellicorns), or in having none at all. Now if a new species of deer or sheep were discovered with the female bearing distinct rudiments of horns, whilst the head of the male was absolutely smooth, we should have a case like that of *Onitis furcifer*.

In this case the old belief of rudiments having been created to complete the scheme of nature is so far from holding good, that all ordinary rules are completely broken through. The view which seems the most probable is that some early progenitor of Onitis acquired, like other Lamellicorns, horns on the head and thorax, and then transferred them, in a rudimentary condition, as with so many existing species, to the female, by whom they have ever since been retained. The subsequent loss of the horns by the male may have resulted through the principle of compensation from the development of the projections on the lower surface, whilst the female has not been thus affected, as she is not furnished with these projections, and consequently has retained the rudiments of the horns on the upper surface. Although this view is supported by the case of Bledius immediately to be given, yet the projections on the lower surface differ greatly in structure and development in the males of the several species of Onitis, and are even rudimentary in some; nevertheless the upper surface in all these species is quite destitute of horns. As secondary sexual characters are so eminently variable, it is possible that the projections on the lower surface may have been first acquired by some progenitor of Onitis and produced their effect through compensation, and then have been in certain cases almost completely lost.

All the cases hitherto given refer to the Lamellicorns, but the males of some few other beetles, belonging to two widely distinct groups, namely, the Curculionidæ and Staphylinidæ, are furnished with horns,—in the former on the lower surface of the body,[61] in the latter on the upper surface of the head and thorax. In the Staphylinidæ the horns of the males in the same species are extraordinarily variable, just as we have seen with the Lamellicorns. In Siagonium

Fig. 22. Bledius taurus, magnified. Left-hand figure, male; right-hand figure, female.

we have a case of dimorphism, for the males can be divided into two sets, differing greatly in the size of their bodies, and in the development of their horns, without any intermediate gradations. In a species of Bledius (fig. 22), also belonging to the Staphylinidæ, male specimens can be found in the same locality, as

[61] Kirby and Spence, ibid. vol. iii. p. 329.

Professor Westwood states, "in which the central horn "of the thorax is very large, but the horns of the head "quite rudimental; and others, in which the thoracic "horn is much shorter, whilst the protuberances on "the head are long."[62] Here, then, we apparently have an instance of compensation of growth, which throws light on the curious case just given of the loss of the upper horns by the males of *Onitis furcifer*.

Law of Battle.—Some male beetles, which seem ill fitted for fighting, nevertheless engage in conflicts for the possession of the females. Mr. Wallace[63] saw two males of *Leptorhynchus angustatus*, a linear beetle with a much elongated rostrum, "fighting for a female, who "stood close by busy at her boring. They pushed at "each other with their rostra, and clawed and thumped, "apparently in the greatest rage." The smaller male, however, "soon ran away, acknowledging himself van-"quished." In some few cases the males are well adapted for fighting, by possessing great toothed mandibles, much larger than those of the females. This is the case with the common stag-beetle (*Lucanus cervus*), the males of which emerge from the pupal state about a week before the other sex, so that several may often be seen pursuing the same female. At this period they engage in fierce conflicts. When Mr. A. H. Davis[64] enclosed two males with one female in a box, the larger male severely pinched the smaller one, until he resigned his pretensions. A friend informs me

---

[62] 'Modern Classification of Insects,' vol. i. p. 172. On the same page there is an account of Siagonium. In the British Museum I noticed one male specimen of Siagonium in an intermediate condition, so that the dimorphism is not strict.

[63] 'The Malay Archipelago,' vol. ii. 1869, p. 276.

[64] 'Entomological Magazine,' vol. i. 1833, p. 82. See also on the conflicts of this species, Kirby and Spence, ibid. vol. iii. p. 314; and Westwood, ibid. vol. i. p. 187.

that when a boy he often put the males together to see them fight, and he noticed that they were much bolder and fiercer than the females, as is well known to be the case with the higher animals. The males would seize hold of his finger, if held in front, but not so the females. With many of the Lucanidæ, as well as with the above-mentioned Leptorhynchus, the males are larger and more powerful insects than the females. The two sexes of *Lethrus cephalotes* (one of the Lamellicorns) inhabit the same burrow; and the male has larger mandibles than the female. If, during the breeding-season, a strange male attempts to enter the burrow, he is attacked; the female does not remain passive, but closes the mouth of the burrow, and encourages her mate by continually pushing him on from behind. The action does not cease until the aggressor is killed or runs away.[65] The two sexes of another lamellicorn beetle, the *Ateuchus cicatricosus* live in pairs, and seem much attached to each other; the male excites the female to roll the balls of dung in which the ova are deposited; and if she is removed, he becomes much agitated. If the male is removed, the female ceases all work, and as M. Brulerie[66] believes, would remain on the spot until she died.

The great mandibles of the male Lucanidæ are extremely variable both in size and structure, and in this respect resemble the horns on the head and thorax of many male Lamellicorns and Staphylinidæ. A perfect series can be formed from the best-provided to the worst-provided or degenerate males. Although the mandibles of the common stag-beetle, and probably of

---

65 Quoted from Fischer, in 'Dict. Class. d'Hist. Nat.' tom. x. p. 324.

66 'Ann. Soc. Entomolog. France,' 1866, as quoted in 'Journal of Travel,' by A. Murray, 1868, p. 135.

many other species, are used as efficient weapons for fighting, it is doubtful whether their great size can thus be accounted for. We have seen that with the *Lucanus elaphus* of N. America they are used for seizing the female. As they are so conspicuous and so elegantly branched, the suspicion has sometimes crossed my mind that they may be serviceable to the males as an ornament, in the same manner as the horns on the head and thorax of the various above described species. The male *Chiasognathus grantii* of S. Chile—a splendid beetle belonging to the same family—has enormously-developed mandibles (fig. 23); he is bold and pugnacious; when threatened on any side he faces round, opening his great jaws, and at the same time stridulating loudly; but the mandibles were not strong enough to pinch my finger so as to cause actual pain.

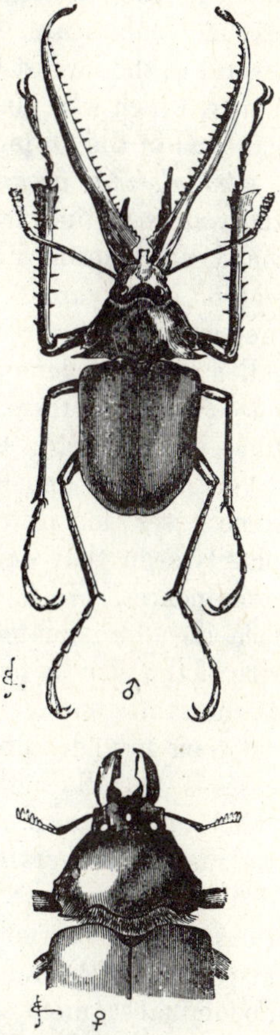

Fig. 23. Chiasognathus grantii, reduced. Upper figure, male; lower figure, female.

Sexual selection, which implies the possession of considerable perceptive powers and of strong passions, seems to have been more effective with the Lamellicorns than with any other family of the Coleoptera or beetles. With some species the males are provided with weapons for fighting; some live in pairs and show mutual affection;

many have the power of stridulating when excited; many are furnished with the most extraordinary horns, apparently for the sake of ornament; some which are diurnal in their habits are gorgeously coloured; and, lastly, several of the largest beetles in the world belong to this family, which was placed by Linnæus and Fabricius at the head of the Order of the Coleoptera.[67]

Stridulating organs.—Beetles belonging to many and widely distinct families possess these organs. The sound can sometimes be heard at the distance of several feet or even yards,[68] but is not comparable with that produced by the Orthoptera. The part which may be called the rasp generally consists of a narrow slightly-raised surface, crossed by very fine, parallel ribs, sometimes so fine as to cause iridescent colours, and having a very elegant appearance under the microscope. In some cases, for instance, with Typhæus, it could be plainly seen that extremely minute, bristly, scale-like prominences, which cover the whole surrounding surface in approximately parallel lines, give rise to the ribs of the rasp by becoming confluent and straight, and at the same time more prominent and smooth. A hard ridge on any adjoining part of the body, which in some cases is specially modified for the purpose, serves as the scraper for the rasp. The scraper is rapidly moved across the rasp, or conversely the rasp across the scraper.

These organs are situated in widely different positions. In the carrion-beetles (Necrophorus) two parallel rasps (*r*, fig. 24) stand on the dorsal surface of the fifth abdominal segment, each rasp being crossed, as described by Landois,[69] by from 126 to 140 fine ribs. These

---

67 Westwood, 'Modern Class.' vol. i. p. 184.

68 Wollaston, On certain musical Curculionidæ, 'Annals and Mag. of Nat. Hist.' vol. vi. 1860, p. 14.

69 'Zeitschrift für wiss. Zoolog.' B. xvii. 1867, s. 127.

ribs are scraped by the posterior margins of the elytra, a small portion of which projects beyond the general outline. In many Crioceridæ, and in *Clythra 4-punctata*

Fig. 24. Necrophorus (from Landois). *r*. The two rasps. Left-hand figure, part of the rasp highly magnified.

(one of the Chrysomelidæ), and in some Tenebrionidæ, &c.,[70] the rasp is seated on the dorsal apex of the abdomen, on the pygidium or pro-pygidium, and is scraped as above by the elytra. In Heterocerus, which belongs to another family, the rasps are placed on the sides of the first abdominal segment, and are scraped by ridges on the femora.[71] In certain Curculionidæ and Carabidæ,[72] the parts are completely reversed in position,

---

[70] I am greatly indebted to Mr. G. R. Crotch for having sent me numerous prepared specimens of various beetles belonging to these three families and others, as well as for valuable information of all kinds. He believes that the power of stridulation in the Clythra has not been previously observed. I am also much indebted to Mr. E. W. Janson, for information and specimens. I may add that my son, Mr. F. Darwin, finds that *Dermestes murinus* stridulates, but he searched in vain for the apparatus. Scolytus has lately been described by Mr. Algen as a stridulator, in the 'Edinburgh Monthly Magazine,' 1869, Nov., p. 130.

[71] Schiödte, translated in 'Annals and Mag. of Nat. Hist.' vol. xx. 1867, p. 37.

[72] Westring has described (Kroyer, 'Naturhist. Tidskrift,' B. ii. 1848-49, p. 334) the stridulating organs in these two, as well as in other families. In the Carabidæ I have examined *Elaphrus uliginosus* and *Blethisa multipunctata*, sent to me by Mr. Crotch. In Blethisa the transverse ridges on the furrowed border of the abdominal segment do not come into play, as far as I could judge, in scraping the rasps on the elytra.

for the rasps are seated on the inferior surface of the elytra, near their apices, or along their outer margins, and the edges of the abdominal segments serve as the scrapers. In *Pelobius hermanni* (one of Dytiscidæ or water-beetles) a strong ridge runs parallel and near to the sutural margin of the elytra, and is crossed by ribs, coarse in the middle part, but becoming gradually finer at both ends, especially at the upper end; when this insect is held under water or in the air, a stridulating noise is produced by scraping the extreme horny margin of the abdomen against the rasp. In a great number of long-horned beetles (Longicornia) the organs are altogether differently situated, the rasp being on the mesothorax, which is rubbed against the pro-thorax; Landois counted 238 very fine ribs on the rasp of *Cerambyx heros*.

Many Lamellicorns have the power of stridulating, and the organs differ greatly in position. Some species stridulate very loudly, so that when Mr. F. Smith caught a *Trox sabulosus*, a gamekeeper who stood by thought that he had caught a mouse; but I failed to discover the proper organs in this beetle. In Geotrupes and Typhæus a narrow ridge runs obliquely across (*r*, fig. 25) the coxa of each hind-leg, having in *G. stercorarius* 84 ribs, which are scraped by a specially-projecting part of one of the abdominal segments. In the nearly allied *Copris lunaris*, an excessively narrow fine rasp runs along the sutural margin of the elytra, with another short rasp near the basal outer margin; but in some other Coprini

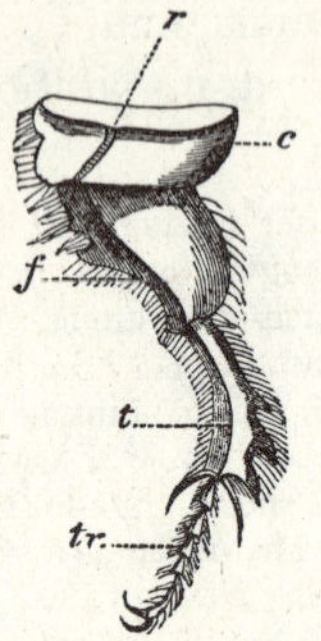

8

Fig. 25. Hind-leg of Geotrupes stercorarius (from Landois).

r. Rasp. *c*. Coxa. *f*. Femur. *t*. Tibia. *tr*. Tarsi.

the rasp is seated, according to Leconte,[73] on the dorsal surface of the abdomen. In Oryctes it is seated on the pro-pygidium, and in some other Dynastini, according to the same entomologist, on the under surface of the elytra. Lastly, Westring states that in *Omaloplia brunnea* the rasp is placed on the pro-sternum, and the scraper on the meta-sternum, the parts thus occupying the under surface of the body, instead of the upper surface as in the Longicorns.

We thus see that the stridulating organs in the different coleopterous families are wonderfully diversified in position, but not much in structure. Within the same family some species are provided with these organs, and some are quite destitute of them. This diversity is intelligible, if we suppose that originally various species made a shuffling or hissing noise by the rubbing together of the hard and rough parts of their bodies which were in contact; and that from the noise thus produced being in some way useful, the rough surfaces were gradually developed into regular stridulating organs. Some beetles as they move, now produce, either intentionally or unintentionally, a shuffling noise, without possessing any proper organs for the purpose. Mr. Wallace informs me that the *Euchirus longimanus* (a Lamellicorn, with the anterior legs wonderfully elongated in the male) "makes, whilst moving, "a low hissing sound by the protrusion and contraction "of the abdomen; and when seized it produces a "grating sound by rubbing its hind-legs against the "edges of the elytra." The hissing sound is clearly due to a narrow rasp running along the sutural margin of each elytron; and I could likewise make the grating

[73] I am indebted to Mr. Walsh, of Illinois, for having sent me extracts from Leconte's 'Introduction to Entomology,' p. 101, 143.

sound by rubbing the shagreened surface of the femur against the granulated margin of the corresponding elytron; but I could not here detect any proper rasp; nor is it likely that I could have overlooked it in so large an insect. After examining Cychrus and reading what Westring has written in his two papers about this beetle, it seems very doubtful whether it possesses any true rasp, though it has the power of emitting a sound.

From the analogy of the Orthoptera and Homoptera, I expected to find that the stridulating organs in the Coleoptera differed according to sex; but Landois, who has carefully examined several species, observed no such difference; nor did Westring; nor did Mr. G. R. Crotch in preparing the numerous specimens which he had the kindness to send me for examination. Any slight sexual difference, however, would be difficult to detect, on account of the great variability of these organs. Thus in the first pair of the *Necrophorus humator* and of the *Pelobius* which I examined, the rasp was considerably larger in the male than in the female; but not so with succeeding specimens. In *Geotrupes stercorarius* the rasp appeared to me thicker, opaquer, and more prominent in three males than in the same number of females; consequently my son, Mr. F. Darwin, in order to discover whether the sexes differed in their power of stridulating, collected 57 living specimens, which he separated into two lots, according as they made, when held in the same manner, a greater or lesser noise. He then examined their sexes, but found that the males were very nearly in the same proportion to the females in both lots. Mr. F. Smith has kept alive numerous specimens of *Mononychus pseudacori* (Curculionidæ), and is satisfied that both sexes stridulate, and apparently in an equal degree.

Nevertheless the power of stridulating is certainly a

sexual character in some few Coleoptera. Mr. Crotch has discovered that the males alone of two species of Heliopathes (Tenebrionidæ) possess stridulating organs. I examined five males of *H. gibbus,* and in all these there was a well-developed rasp, partially divided into two, on the dorsal surface of the terminal abdominal segment; whilst in the same number of females there was not even a rudiment of the rasp, the membrane of this segment being transparent and much thinner than in the male. In *H. cribratostriatus* the male has a similar rasp, excepting that it is not partially divided into two portions, and the female is completely destitute of this organ; but in addition the male has on the apical margins of the elytra, on each side of the suture, three or four short longitudinal ridges, which are crossed by extremely fine ribs, parallel to and resembling those on the abdominal rasp; whether these ridges serve as an independent rasp, or as a scraper for the abdominal rasp, I could not decide: the female exhibits no trace of this latter structure.

Again, in three species of the Lamellicorn genus Oryctes, we have a nearly parallel case. In the females of *O. gryphus* and *nasicornis* the ribs on the rasp of the pro-pygidium are less continuous and less distinct than in the males; but the chief difference is that the whole upper surface of this segment, when held in the proper light, is seen to be clothed with hairs, which are absent or are represented by excessively fine down in the males. It should be noticed that in all Coleoptera the effective part of the rasp is destitute of hairs. In *O. senegalensis* the difference between the sexes is more strongly marked, and this is best seen when the proper segment is cleaned and viewed as a transparent object. In the female the whole surface is covered with little separate crests, bearing spines; whilst in the male these crests become, in proceeding towards the apex, more and more confluent, regular, and naked; so that three-fourths of the segment is covered with extremely fine parallel ribs, which are quite absent in the female. In the females, however, of all three species of Oryctes, when the abdomen of a softened specimen is pushed backwards and forwards, a slight grating or stridulating sound can be produced.

In the case of the Heliopathes and Oryctes there can hardly be a doubt that the males stridulate in order to call or to excite the females; but with most beetles the stridulation apparently serves both sexes as a mutual call. This view is not rendered improbable from beetles stridulating under various emotions; we know that birds use their voices for many purposes besides singing to their mates. The great Chiasognathus stridulates in anger or defiance; many species do the same from distress or fear, when held so that they cannot escape; Messrs. Wollaston and Crotch were able, by striking the hollow stems of trees in the Canary Islands, to discover the presence of beetles belonging to the genus Acalles by their stridulation. Lastly the male Ateuchus stridulates to encourage the female in her work, and from distress when she is removed.[74] Some naturalists believe that beetles make this noise to frighten away their enemies; but I cannot think that the quadrupeds and birds which are able to devour the larger beetles with their extremely hard coats, would be frightened by so slight a grating sound. The belief that the stridulation serves as a sexual call is supported by the fact that death-ticks (*Anobium tesselatum*) are well known to answer each other's ticking, or, as I have

---

74 M. P. de la Brulerie, as quoted in 'Journal of Travel,' A. Murray, vol. i. 1868, p. 135.

myself observed, a tapping noise artificially made; and Mr. Doubelday informs me that he has twice or thrice observed a female ticking,[75] and in the course of an hour or two has found her united with a male, and on one occasion surrounded by several males. Finally, it seems probable that the two sexes of many kinds of beetles were at first enabled to find each other by the slight shuffling noise produced by the rubbing together of the adjoining parts of their hard bodies; and that as the males or females which made the greatest noise succeeded best in finding partners, the rugosities on various parts of their bodies were gradually developed by means of sexual selection into true stridulating organs.

[75] Mr. Doubleday informs me that "the noise is produced by the "insect raising itself on its legs as high as it can, and then striking its "thorax five or six times, in rapid succession, against the substance "upon which it is sitting." For references on this subject see Landois, 'Zeitschrift für wissen. Zoolog.' B. xvii. s. 131. Olivier says (as quoted by Kirby and Spence, 'Introduct.' vol. ii. p. 395) that the female of *Pimelia striata* produces a rather loud sound by striking her abdomen against any hard substance, "and that the male, obedient to this call, "soon attends her and they pair."

CHAPTER XI.

Insects, *continued.*—Order Lepidoptera.

Courtship of butterflies — Battles — Ticking noise — Colours common to both sexes, or more brilliant in the males — Examples — Not due to the direct action of the conditions of life — Colours adapted for protection — Colours of moths — Display — Perceptive powers of the Lepidoptera — Variability — Causes of the difference in colour between the males and females — Mimickry, female butterflies more brilliantly coloured than the males — Bright colours of caterpillars — Summary and concluding remarks on the secondary sexual characters of insects — Birds and insects compared.

In this great Order the most interesting point for us is the difference in colour between the sexes of the same species, and between the distinct species of the same genus. Nearly the whole of the following chapter will be devoted to this subject; but I will first make a few remarks on one or two other points. Several males may often be seen pursuing and crowding round the same female. Their courtship appears to be a prolonged affair, for I have frequently watched one or more males pirouetting round a female until I became tired, without seeing the end of the courtship. Although butterflies are such weak and fragile creatures, they are pugnacious, and an Emperor butterfly [1] has been captured with the tips of its wings broken from a conflict with another male. Mr. Collingwood in speaking of the frequent battles

[1] Apatura Iris: 'The Entomologist's Weekly Intelligencer,' 1859, p. 139. For the Bornean Butterflies see C. Collingwood, 'Rambles of a Naturalist,' 1868, p. 183.

between the butterflies of Borneo says, "They whirl "round each other with the greatest rapidity, and appear "to be incited by the greatest ferocity." One case is known of a butterfly, namely the *Ageronia feronia*, which makes a noise like that produced by a toothed wheel passing under a spring catch, and which could be heard at the distance of several yards. At Rio de Janeiro this sound was noticed by me, only when two were chasing each other in an irregular course, so that it is probably made during the courtship of the sexes; but I neglected to attend to this point.[2]

Every one has admired the extreme beauty of many butterflies and of some moths; and we are led to ask, how has this beauty been acquired? Have their colours and diversified patterns simply resulted from the direct action of the physical conditions to which these insects have been exposed, without any benefit being thus derived? Or have successive variations been accumulated and determined either as a protection or for some unknown purpose, or that one sex might be rendered attractive to the other? And, again, what is the meaning of the colours being widely different in the males and females of certain species, and alike in the two sexes of other species? Before attempting to answer these questions a body of facts must be given.

With most of our English butterflies, both those which are beautiful, such as the admiral, peacock, and painted lady (Vanessæ), and those which are plain-coloured, such as the meadow-browns (Hipparchiæ), the sexes are alike. This is also the case with the magnificent Heliconidæ and Danaidæ of the tropics. But in certain

---

[2] See my 'Journal of Researches,' 1845, p. 33. Mr. Doubleday has detected ('Proc. Ent. Soc.' March 3rd, 1845, p. 123) a peculiar membranous sac at the base of the front wings, which is probably connected with the production of the sound.

other tropical groups, and with some of our English butterflies, as the purple emperor, orange-tip, &c. (*Apatura Iris* and *Anthocharis cardamines*), the sexes differ either greatly or slightly in colour. No language suffices to describe the splendour of the males of some tropical species. Even within the same genus we often find species presenting an extraordinary difference between the sexes, whilst others have their sexes closely alike. Thus in the South American genus Epicalia, Mr. Bates, to whom I am much indebted for most of the following facts and for looking over this whole discussion, informs me that he knows twelve species, the two sexes of which haunt the same stations (and this is not always the case with butterflies), and therefore cannot have been differently affected by external conditions.[3] In nine of these species the males rank amongst the most brilliant of all butterflies, and differ so greatly from the comparatively plain females that they were formerly placed in distinct genera. The females of these nine species resemble each other in their general type of coloration, and likewise resemble both sexes in several allied genera, found in various parts of the world. Hence in accordance with the descent-theory we may infer that these nine species, and probably all the others of the genus, are descended from an ancestral form which was coloured in nearly the same manner. In the tenth species the female still retains the same general colouring, but the male resembles her, so that he is coloured in a much less gaudy and contrasted manner than the males of the previous species. In the eleventh and twelfth species, the females depart from the type of colouring which

[3] See also Mr. Bates' paper in 'Proc. Ent. Soc. of Philadelphia,' 1865, p. 206. Also Mr. Wallace on the same subject, in regard to Diadema, in 'Transact. Entomolog. Soc. of London,' 1869, p. 278.

is usual with their sex in this genus, for they are gaily decorated in nearly the same manner as the males, but in a somewhat less degree. Hence in these two species the bright colours of the males seem to have been transferred to the females; whilst the male of the tenth species has either retained or recovered the plain colours of the female as well as of the parent-form of the genus; the two sexes being thus rendered in both cases, though in an opposite manner, nearly alike. In the allied genus Eubagis, both sexes of some of the species are plain-coloured and nearly alike; whilst with the greater number the males are decorated with beautiful metallic tints, in a diversified manner, and differ much from their females. The females throughout the genus retain the same general style of colouring, so that they commonly resemble each other much more closely than they resemble their own proper males.

In the genus Papilio, all the species of the Æneas group are remarkable for their conspicuous and strongly contrasted colours, and they illustrate the frequent tendency to gradation in the amount of difference between the sexes. In a few species, for instance in *P. ascanius*, the males and females are alike; in others the males are a little or very much more superbly coloured than the females. The genus Junonia allied to our Vanessæ offers a nearly parallel case, for although the sexes of most of the species resemble each other and are destitute of rich colours, yet in certain species, as in *J. œnone*, the male is rather more brightly coloured than the female, and in a few (for instance *J. andremiaja*) the male is so different from the female that he might be mistaken for an entirely distinct species.

Another striking case was pointed out to me in the British museum by Mr. A. Butler, namely one of the Tropical American Theclæ, in which both sexes are nearly alike and wonderfully splendid; in another, the male is coloured in a similarly gorgeous manner, whilst the whole upper surface of the female is of a dull uniform brown. Our common little English blue butterflies of the genus Lycæna, illustrate the various differences in colour between the sexes, almost as well, though not in so striking a manner, as the above exotic genera. In *Lycæna agestis* both sexes have wings of a brown colour, bordered with small ocellated orange spots, and are consequently alike. In *L. œgon* the wings of the male are of a fine blue, bordered with black; whilst the wings of the female are brown, with a similar border, and closely resemble those of *L. agestis.* Lastly, in *L. arion* both sexes are of a blue colour and nearly alike, though in the female the edges of the wings are rather duskier, with the black spots plainer; and in a bright blue Indian species both sexes are still more closely alike.

I have given the foregoing cases in some detail in order to shew, in the first place, that when the sexes of butterflies differ, the male as a general rule is the most beautiful, and departs most from the usual type of colouring of the group to which the species belongs. Hence in most groups the females of the several species resemble each other much more closely than do the males. In some exceptional cases, however, to which I shall hereafter allude, the females are coloured more splendidly than the males. In the second place these cases have been given to bring clearly before the mind that within the same genus, the two sexes frequently present every gradation from no difference in colour to so great a difference that it was long before the two were placed by entomologists in the same genus. In the third place, we have seen that when the sexes nearly resemble each other, this apparently may be due either to the male having transferred his colours to the female, or to the male having retained, or perhaps recovered, the primordial colours of the genus to which the species belongs. It also deserves notice that in those groups in which the sexes present any difference of colour, the females usually resemble the males to a certain extent, so that when the males are beautiful to an extraordinary degree, the females almost invariably exhibit some degree of beauty. From the numerous cases of gradation in the amount of difference between the sexes, and from the prevalence of the same general type of coloration throughout the whole of the same group, we may conclude that the causes, whatever they may be, which have determined the brilliant colouring of the males alone of some species, and of both sexes in a more or less equal degree of other species, have generally been the same.

As so many gorgeous butterflies inhabit the tropics, it has often been supposed that they owe their colours to the great heat and moisture of these zones; but Mr. Bates[4] has shewn by the comparison of various closely-allied groups of insects from the temperate and tropical regions, that this view cannot be maintained; and the evidence becomes conclusive when brilliantly-coloured males and plain-coloured females of the same species inhabit the same district, feed on the same food, and follow exactly the same habits of life. Even when the sexes resemble each other, we can hardly believe that their brilliant and beautifully-arranged colours are the purposeless result of the nature of the tissues, and the action of the surrounding conditions.

With animals of all kinds, whenever colour has been modified for some special purpose, this has been, as far

---

[4] 'The Naturalist on the Amazons,' vol. i. 1863, p. 19.

as we can judge, either for protection or as an attraction between the sexes. With many species of butterflies the upper surfaces of the wings are obscurely coloured, and this in all probability leads to their escaping observation and danger. But butterflies when at rest would be particularly liable to be attacked by their enemies; and almost all the kinds when resting raise their wings vertically over their backs, so that the lower sides alone are exposed to view. Hence it is this side which in many cases is obviously coloured so as to imitate the surfaces on which these insects commonly rest. Dr. Rössler, I believe, first noticed the similarity of the closed wings of certain Vanessæ and other butterflies to the bark of trees. Many analogous and striking facts could be given. The most interesting one is that recorded by Mr. Wallace[5] of a common Indian and Sumatran butterfly (Kallima), which disappears like magic when it settles in a bush; for it hides its head and antennæ between its closed wings, and these in form, colour, and veining cannot be distinguished from a withered leaf together with the footstalk. In some other cases the lower surfaces of the wings are brilliantly coloured, and yet are protective; thus in *Thecla rubi* the wings when closed are of an emerald green and resemble the young leaves of the bramble, on which this butterfly in the spring may often be seen seated.

Although the obscure tints of the upper or under surface of many butterflies no doubt serve to conceal them, yet we cannot possibly extend this view to the brilliant and conspicuous colours of many kinds, such as our admiral and peacock Vanessæ, our white

[5] See the interesting article in the 'Westminster Review,' July, 1867, p. 10. A woodcut of the Kallima is given by Mr. Wallace in 'Hardwicke's 'Science Gossip,' Sept. 1867, p. 196.

cabbage-butterflies (Pieris), or the great swallow-tail Papilio which haunts the open fens—for these butterflies are thus rendered visible to every living creature. With these species both sexes are alike; but in the common brimstone butterfly (*Gonepteryx rhamni*), the male is of an intense yellow, whilst the female is much paler; and in the orange-tip (*Anthocharis cardamines*) the males alone have the bright orange tips to their wings. In these cases the males and females are equally conspicuous, and it is not credible that their difference in colour stands in any relation to ordinary protection. Nevertheless it is possible that the conspicuous colours of many species may be in an indirect manner beneficial, as will hereafter be explained, by leading their enemies at once to recognise them as unpalatable. Even in this case it does not certainly follow that their bright colours and beautiful patterns were acquired for this special purpose. In some other remarkable cases, beauty has been gained for the sake of protection, through the imitation of other beautiful species, which inhabit the same district and enjoy an immunity from attack by being in some way offensive to their enemies.

The female of our orange-tip buttterfly, above referred to, and of an American species (*Anth. genutia*) probably shew us, as Mr. Walsh has remarked to me, the primordial colours of the parent-species of the genus; for both sexes of four or five widely-distributed species are coloured in nearly the same manner. We may infer here, as in several previous cases, that it is the males of *Anth. cardamines* and *genutia* which have departed from the usual type of colouring of their genus. In the *Anth. sara* from California, the orange-tips have become partially developed in the female; for her wings are tipped with reddish-orange, but paler than in the male, and slightly different in some other respects. In an allied Indian form, the *Iphias glaucippe*, the orange-tips are fully developed in both sexes. In this Iphias the under surface of the wings marvellously resembles, as pointed out to me by Mr. A. Butler, a pale-coloured leaf; and in our English orange-tip, the under surface resembles the flower-head of the wild parsley, on which it may be seen going to rest at night.[6] The same reasoning power which compels us to believe that the lower surfaces have here been coloured for the sake of protection, leads us to deny that the wings have been tipped, especially when this character is confined to the males, with bright orange for the same purpose.

Turning now to Moths: most of these rest motionless with their wings depressed during the whole or greater part of the day; and the upper surfaces of their wings are often shaded and coloured in an admirable manner, as Mr. Wallace has remarked, for escaping detection. With most of the Bombycidæ and Noctuidæ,[7] when at rest, the front-wings overlap and conceal the hind-wings; so that the latter might be brightly coloured without much risk; and they are thus coloured in many species of both families. During the act of flight, moths would often be able to escape from their enemies; nevertheless, as the hind-wings are then fully exposed to view, their bright colours must generally have been acquired at the cost of some little risk. But the following fact shews us how cautious we ought to be in drawing conclusions on this head. The common yellow under-wings

[6] See the interesting observations by Mr. T. W. Wood, 'The Student,' Sept. 1868, p. 81.

[7] Mr. Wallace in 'Hardwicke's Science Gossip,' Sept. 1867, p. 193.

(Triphaena) often fly about during the day or early evening, and are then conspicuous from the colour of their hind-wings. It would naturally be thought that this would be a source of danger; but Mr. J. Jenner Weir believes that it actually serves them as a means of escape, for birds strike at these brightly coloured and fragile surfaces, instead of at the body. For instance, Mr. Weir turned into his aviary a vigorous specimen of *Triphaena pronuba*, which was instantly pursued by a robin; but the bird's attention being caught by the coloured wings, the moth was not captured until after about fifty attempts, and small portions of the wings were repeatedly broken off. He tried the same experiment, in the open air, with a *T. fimbria* and swallow; but the large size of this moth probably interfered with its capture.[8] We are thus reminded of a statement made by Mr. Wallace,[9] namely, that in the Brazilian forests and Malayan islands, many common and highly-decorated butterflies are weak flyers, though furnished with a broad expanse of wings; and they "are "often captured with pierced and broken wings, as if "they had been seized by birds, from which they had "escaped: if the wings had been much smaller in pro- "portion to the body, it seems probable that the insect "would more frequently have been struck or pierced in "a vital part, and thus the increased expanse of the "wings may have been indirectly beneficial."

Display.—The bright colours of butterflies and of some moths are specially arranged for display, whether or not they serve in addition as a protection. Bright

---

[8] See also, on this subject, Mr. Weir's paper in 'Transact. Ent. Soc.' 1869, p. 23.

[9] 'Westminster Review,' July, 1867, p. 16.

colours would not be visible during the night; and there can be no doubt that moths, taken as a body, are much less gaily decorated than butterflies, all of which are diurnal in their habits. But the moths in certain families, such as the Zygænidæ, various Sphingidæ, Uraniidæ, some Arctiidæ and Saturniidæ, fly about during the day or early evening, and many of these are extremely beautiful, being far more brightly coloured than the strictly nocturnal kinds. A few exceptional cases, however, of brightly-coloured nocturnal species have been recorded.[10]

There is evidence of another kind in regard to display. Butterflies, as before remarked, elevate their wings when at rest, and whilst basking in the sunshine often alternately raise and depress them, thus exposing to full view both surfaces; and although the lower surface is often coloured in an obscure manner as a protection, yet in many species it is as highly coloured as the upper surface, and sometimes in a very different manner. In some tropical species the lower surface is even more brilliantly coloured than the upper.[11] In one English fritillary, the *Argynnis aglaia*, the lower surface alone is ornamented with shining silver discs. Nevertheless, as a general rule, the upper surface, which is probably the most fully exposed, is coloured more brightly and in a more diversified manner than the lower. Hence the lower surface generally affords

---

[10] For instance, Lithosia; but Prof. Westwood ('Modern Class. of Insects,' vol. ii. p. 390) seems surprised at this case. On the relative colours of diurnal and nocturnal Lepidoptera, see ibid. p. 333 and 392; also Harris, 'Treatise on the Insects of New England,' 1842, p. 315.

[11] Such differences between the upper and lower surfaces of the wings of several species of Papilio, may be seen in the beautiful plates to Mr. Wallace's Memoir on the Papilionidæ of the Malayan Region, in 'Transact. Linn. Soc.' vol. xxv. part i. 1865.

to entomologists the most useful character for detecting the affinities of the various species.

Now if we turn to the enormous group of moths, which do not habitually expose to full view the under surface of their wings, this side is very rarely, as I hear from Mr. Stainton, coloured more brightly than the upper side, or even with equal brightness. Some exceptions to the rule, either real or apparent, must be noticed, as that of Hypopira, specified by Mr. Wormald.[12] Mr. R. Trimen informs me that in Guenée's great work, three moths are figured, in which the under surface is much the most brilliant. For instance, in the Australian Gastrophora the upper surface of the fore-wing is pale greyish-ochreous, while the lower surface is magnificently ornamented by an ocellus of cobalt-blue, placed in the midst of a black mark, surrounded by orange-yellow, and this by bluish-white. But the habits of these three moths are unknown; so that no explanation can be given of their unusual style of colouring. Mr. Trimen also informs me that the lower surface of the wings in certain other Geometræ[13] and quadrifid Noctuæ are either more variegated or more brightly-coloured than the upper surface; but some of these species have the habit of "holding their wings quite erect over their "backs, retaining them in this position for a considerable "time," and thus exposing to view the under surface. Other species when settled on the ground or herbage have the habit of now and then suddenly and slightly lifting up their wings. Hence the lower surface of the wings being more brightly-coloured than the upper sur-

---

12 'Proc. Ent. Soc.' March 2nd, 1868.

13 See also an account of the S. American genus Erateina (one of the Geometræ) in 'Transact. Ent. Soc.' new series, vol. v. pl. xv. and xvi.

face in certain moths is not so anomalous a circumstance as it at first appears. The Saturniidæ include some of the most beautiful of all moths, their wings being decorated, as in our British Emperor moth, with fine ocelli; and Mr. T. W. Wood [14] observes that they resemble butterflies in some of their movements; "for "instance, in the gentle waving up and down of the "wings, as if for display, which is more characteristic "of diurnal than of nocturnal Lepidoptera."

It is a singular fact that no British moths, nor as far as I can discover hardly any foreign species, which are brilliantly coloured, differ much in colour according to sex; though this is the case with many brilliant butterflies. The male, however, of one American moth, the *Saturnia Io*, is described as having its fore-wings deep yellow, curiously marked with purplish-red spots; whilst the wings of the female are purple-brown, marked with grey lines.[15] The British moths which differ sexually in colour are all brown, or various tints of dull yellow, or nearly white. In several species the males are much darker than the females,[16] and these belong to groups which generally fly about during the afternoon. On the other hand, in many genera, as Mr. Stainton informs me,

---

[14] 'Proc. Ent. Soc. of London,' July 6, 1868, p. xxvii.

[15] Harris, 'Treatise,' &c., edited by Flint, 1862, p. 395.

[16] For instance, I observe in my son's cabinet that the males are darker than the females in the *Lasiocampa quercus*, *Odonestis potatoria*, *Hypogymna dispar*, *Dasychira pudibunda*, and *Cycnia mendica*. In this latter species the difference in colour between the two sexes is strongly marked; and Mr. Wallace informs me that we here have, as he believes, an instance of protective mimickry confined to one sex, as will hereafter be more fully explained. The white female of the Cycnia resembles the very common *Spilosoma menthrasti*, both sexes of which are white; and Mr. Stainton observed that this latter moth was rejected with utter disgust by a whole brood of young turkeys, which were fond of eating other moths; so that if the Cycnia was commonly mistaken by British birds for the Spilosoma, it would escape being devoured, and its white deceptive colour would thus be highly beneficial.

the males have the hind-wings whiter than those of the female—of which fact *Agrotis exclamationis* offers a good instance. The males are thus rendered more conspicuous than the females, whilst flying about in the dusk. In the Ghost Moth (*Hepialus humuli*) the difference is more strongly marked; the males being white and the females yellow with darker markings. It is difficult to conjecture what the meaning can be of these differences between the sexes in the shades of darkness or lightness; but we can hardly suppose that they are the result of mere variability with sexually-limited inheritance, independently of any benefit thus derived.

From the foregoing statements it is impossible to admit that the brilliant colours of butterflies and of some few moths, have commonly been acquired for the sake of protection. We have seen that their colours and elegant patterns are arranged and exhibited as if for display. Hence I am led to suppose that the females generally prefer, or are most excited by the more brilliant males; for on any other supposition the males would be ornamented, as far as we can see, for no purpose. We know that ants and certain lamellicorn beetles are capable of feeling an attachment for each other, and that ants recognise their fellows after an interval of several months. Hence there is no abstract improbability in the Lepidoptera, which probably stand nearly or quite as high in the scale as these insects, having sufficient mental capacity to admire bright colours. They certainly discover flowers by colour, and, as I have elsewhere shewn, the plants which are fertilised exclusively by the wind never have a conspicuously-coloured corolla. The Humming-bird Sphinx may often be seen to swoop down from a distance on a bunch of flowers in the midst of green foliage;

and I have been assured by a friend, that these moths repeatedly visited flowers painted on the walls of a room in the South of France. The common white butterfly, as I hear from Mr. Doubleday, often flies down to a bit of paper on the ground, no doubt mistaking it for one of its own species. Mr. Collingwood[17] in speaking of the difficulty of collecting certain butterflies in the Malay Archipelago, states that "a dead specimen pinned upon "a conspicuous twig will often arrest an insect of the "same species in its headlong flight, and bring it down "within easy reach of the net, especially if it be of the "opposite sex."

The courtship of butterflies is a prolonged affair. The males sometimes fight together in rivalry; and many may be seen pursuing or crowding round the same female. If, then, the females do not prefer one male to another, the pairing must be left to mere chance, and this does not appear to me a probable event. If, on the other hand, the females habitually, or even occasionally, prefer the more beautiful males, the colours of the latter will have been rendered brighter by degrees, and will have been transmitted to both sexes or to one sex, according to which law of inheritance prevailed. The process of sexual selection will have been much facilitated, if the conclusions arrived at from various kinds of evidence in the supplement to the ninth chapter can be trusted; namely that the males of many Lepidoptera, at least in the imago state, greatly exceed in number the females.

Some facts, however, are opposed to the belief that female butterflies prefer the more beautiful males; thus, as I have been assured by several observers, fresh females may frequently be seen paired with battered, faded or

---

[17] 'Rambles of a Naturalist in the Chinese Seas,' 1868, p. 182.

dingy males; but this is a circumstance which could hardly fail often to follow from the males emerging from their cocoons earlier than the females. With moths of the family of the Bombycidæ, the sexes pair immediately after assuming the imago state; for they cannot feed, owing to the rudimentary condition of their mouths. The females, as several entomologists have remarked to me, lie in an almost torpid state, and appear not to evince the least choice in regard to their partners. This is the case with the common silk-moth (*B. mori*), as I have been told by some continental and English breeders. Dr. Wallace, who has had such immense experience in breeding *Bombyx cynthia*, is convinced that the females evince no choice or preference. He has kept above 300 of these moths living together, and has often found the most vigorous females mated with stunted males. The reverse apparently seldom occurs; for, as he believes, the more vigorous males pass over the weakly females, being attracted by those endowed with most vitality. Although we have been indirectly induced to believe that the females of many species prefer the more beautiful males, I have no reason to suspect, either with moths or butterflies, that the males are attracted by the beauty of the females. If the more beautiful females had been continually preferred, it is almost certain, from the colours of butterflies being so frequently transmitted to one sex alone, that the females would often have been rendered more beautiful than their male partners. But this does not occur except in a few instances; and these can be explained, as we shall presently see, on the principle of mimickry and protection.

As sexual selection primarily depends on variability, a few words must be added on this subject. In respect to colour there is no difficulty, as any number of highly variable Lepidoptera could be named. One good instance will suffice. Mr. Bates shewed me a whole series of specimens of *Papilio sesostris* and *childrenæ;* in the latter the males varied much in the extent of the beautifully enamelled green patch on the fore-wings, and in the size of the white mark, as well as of the splendid crimson stripe on the hind-wings; so that there was a great contrast between the most and least gaudy males. The male of *Papilio sesostris,* though a beautiful insect, is much less so than *P. childrenæ.* It likewise varies a little in the size of the green patch on the fore-wings, and in the occasional appearance of a small crimson stripe on the hind-wings, borrowed, as it would seem, from its own female; for the females of this and of many other species in the Æneas group possess this crimson stripe. Hence between the brightest specimens of *P. sesostris* and the least bright of *P. childrenæ,* there was but a small interval; and it was evident that as far as mere variability is concerned, there would be no difficulty in permanently increasing by means of selection the beauty of either species. The variability is here almost confined to the male sex; but Mr. Wallace and Mr. Bates have shewn[18] that the females of some other species are extremely variable, the males being nearly constant. As I have before mentioned the Ghost Moth (*Hepialus humuli*) as one of the best instances in Britain of a difference in colour between the sexes of moths, it may be worth adding[19] that in the Shetland

---

18 Wallace on the Papilionidæ of the Malayan Region, in 'Transact. Linn. Soc.' vol. xxv. 1865, p. 8, 36. A striking case of a rare variety, strictly intermediate between two other well-marked female varieties, is given by Mr. Wallace. See also Mr. Bates, in 'Proc. Entomolog. Soc.' Nov. 19th, 1866, p. xl.

19 Mr. R. MacLachlan, 'Transact. Ent. Soc.' vol. ii. part 6th, 3rd series, 1866, p. 459.

Islands, males are frequently found which closely resemble the females. In a future chapter I shall have occasion to shew that the beautiful eye-like spots or ocelli, so common on the wings of many Lepidoptera, are eminently variable.

On the whole, although many serious objections may be urged, it seems probable that most of the species of Lepidoptera which are brilliantly coloured, owe their colours to sexual selection, excepting in certain cases, presently to be mentioned, in which conspicuous colours are beneficial as a protection. From the ardour of the male throughout the animal kingdom, he is generally willing to accept any female; and it is the female which usually exerts a choice. Hence if sexual selection has here acted, the male, when the sexes differ, ought to be the most brilliantly coloured; and this undoubtedly is the ordinary rule. When the sexes are brilliantly coloured and resemble each other, the characters acquired by the males appear to have been transmitted to both sexes. But will this explanation of the similarity and dissimilarity in colour between the sexes suffice?

The males and females of the same species of butterfly are known[20] in several cases to inhabit different stations, the former commonly basking in the sunshine, the latter haunting gloomy forests. It is therefore possible that different conditions of life may have acted directly on the two sexes; but this is not probable,[21] as in the adult state they are exposed during a very short period to different conditions; and the larvæ of both are exposed to the same conditions. Mr. Wallace believes

[20] H. W. Bates, 'The Naturalist on the Amazons,' vol. ii. 1863, p. 228. A. R. Wallace, in 'Transact. Linn. Soc.' vol. xxv. 1865, p. 10.

[21] On this whole subject see 'The Variation of Animals and Plants under Domestication,' vol. ii. 1868, chap. xxiii.

that the less brilliant colours of the female have been specially gained in all or almost all cases for the sake of protection. On the contrary it seems to me more probable that the males alone, in the large majority of cases, have acquired their bright colours through sexual selection, the females having been but little modified. Consequently the females of distinct but allied species ought to resemble each other much more closely than do the males of the same species; and this is the general rule. The females thus approximately show us the primordial colouring of the parent-species of the group to which they belong. They have, however, almost always been modified to a certain extent by some of the successive steps of variation, through the accumulation of which the males were rendered beautiful, having been transferred to them. The males and females of allied though distinct species will also generally have been exposed during their prolonged larval state to different conditions, and may have been thus indirectly affected; though with the males any slight change of colour thus caused will often have been completely masked by the brilliant tints gained through sexual selection. When we treat of Birds, I shall have to discuss the whole question whether the differences in colour between the males and females have been in part specially gained by the latter as a protection; so that I will here only give unavoidable details.

In all cases when the more common form of equal inheritance by both sexes has prevailed, the selection of bright-coloured males would tend to make the females bright-coloured; and the selection of dull-coloured females would tend to make the males dull. If both processes were carried on simultaneously, they would tend to neutralise each other. As far as I can see, it would be extremely difficult to change through selection the one form of inheritance into the other. But by the selection of successive variations, which were from the first sexually limited in their transmission, there would not be the slightest difficulty in giving bright colours to the males alone, and at the same time or subsequently, dull colours to the females alone. In this latter manner female butterflies and moths may, as I fully admit, have been rendered inconspicuous for the sake of protection, and widely different from their males.

Mr. Wallace[22] has argued with much force in favour of his view that when the sexes differ, the female has been specially modified for the sake of protection; and that this has been effected by one form of inheritance, namely, the transmission of characters to both sexes, having been changed through the agency of natural selection into the other form, namely, transmission to one sex. I was at first strongly inclined to accept this view; but the more I have studied the various classes throughout the animal kingdom, the less probable it has appeared. Mr. Wallace urges that both sexes of the *Heliconidæ, Danaidæ, Acroeidæ* are equally brilliant because both are protected from the attacks of birds and other enemies, by their offensive odour; but that in other groups, which do not possess this immunity, the females have been rendered inconspicuous, from having more need of protection than the males. This supposed difference in the "need of protection by the "two sexes" is rather deceptive, and requires some discussion. It is obvious that brightly-coloured individuals, whether males or females, would equally attract, and obscurely-coloured individuals equally escape, the

22 A. R. Wallace, in 'The Journal of Travel,' vol. i. 1868, p. 88. 'Westminster Review,' July, 1867, p. 37. See also Messrs. Wallace and Bates in 'Proc. Ent. Soc.' Nov. 19th, 1866, p. xxxix.

attention of their enemies. But we are concerned with the effects of the destruction or preservation of certain individuals of either sex, on the character of the race. With insects, after the male has fertilised the female, and after the latter has laid her eggs, the greater or less immunity from danger of either sex could not possibly have any effect on the offspring. Before the sexes have performed their proper functions, if they existed in equal numbers and if they strictly paired (all other circumstances being the same), the preservation of the males and females would be equally important for the existence of the species and for the character of the offspring. But with most animals, as is known to be the case with the domestic silk-moth, the male can fertilise two or three females; so that the destruction of the males would not be so injurious to the species as that of the females. On the other hand, Dr. Wallace believes that with moths the progeny from a second or third fertilisation is apt to be weakly, and therefore would not have so good chance of surviving. When the males exist in much greater numbers than the females, no doubt many males might be destroyed with impunity to the species; but I cannot see that the results of ordinary selection for the sake of protection would be influenced by the sexes existing in unequal numbers; for the same proportion of the more conspicuous individuals, whether males or females, would probably be destroyed. If indeed the males presented a greater range of variation in colour, the result would be different; but we need not here follow out such complex details. On the whole I cannot perceive that an inequality in the numbers of the two sexes would influence in any marked manner the effects of ordinary selection on the character of the offspring.

Female Lepidoptera require, as Mr. Wallace insists,

some days to deposit their fertilised ova and to search for a proper place; during this period (whilst the life of the male was of no importance) the brighter-coloured females would be exposed to danger and would be liable to be destroyed. The duller-coloured females on the other hand would survive, and thus would influence, it might be thought, in a marked manner the character of the species,—either of both sexes or of one sex, according to which form of inheritance prevailed. But it must not be forgotten that the males emerge from the cocoon-state some days before the females, and during this period, whilst the unborn females were safe, the brighter-coloured males would be exposed to danger; so that ultimately both sexes would probably be exposed during a nearly equal length of time to danger, and the elimination of conspicuous colours would not be much more effective in the one than the other sex.

It is a more important consideration that female Lepidoptera, as Mr. Wallace remarks, and as is known to every collector, are generally slower flyers than the males. Consequently the latter, if exposed to greater danger from being conspicuously coloured, might be able to escape from their enemies, whilst the similarly-coloured females would be destroyed; and thus the females would have the most influence in modifying the colour of their progeny.

There is one other consideration: bright colours, as far as sexual selection is concerned, are commonly of no service to the females; so that if the latter varied in brightness, and the variations were sexually limited in their transmission, it would depend on mere chance whether the females had their bright colours increased; and this would tend throughout the Order to diminish the number of species with brightly-coloured females in comparison with the species having brightly-coloured males. On the other hand, as bright colours are supposed to be highly serviceable to the males in their love-struggles, the brighter males (as we shall see in the chapter on Birds) although exposed to rather greater danger, would on an average procreate a greater number of offspring than the duller males. In this case, if the variations were limited in their transmission to the male sex, the males alone would be rendered more brilliantly coloured; but if the variations were not thus limited, the preservation and augmentation of such variations would depend on whether more evil was caused to the species by the females being rendered conspicuous, than good to the males by certain individuals being successful over their rivals.

As there can hardly be a doubt that both sexes of many butterflies and moths have been rendered dull-coloured for the sake of protection, so it may have been with the females alone of some species in which successive variations towards dullness first appeared in the female sex and were from the first limited in their transmission to the same sex. If not thus limited, both sexes would become dull-coloured. We shall immediately see, when we treat of mimickry, that the females alone of certain butterflies have been rendered extremely beautiful for the sake of protection, without any of the successive protective variations having been transferred to the male, to whom they could not possibly have been in the least degree injurious, and therefore could not have been eliminated through natural selection. Whether in each particular species, in which the sexes differ in colour, it is the female which has been specially modified for the sake of protection; or whether it is the male which has been specially modified for the sake of sexual attraction, the female having retained her primordial colouring only slightly changed through the agencies before alluded to; or whether again both sexes have been modified, the female for protection and the male for sexual attraction, can only be definitely decided when we know the life-history of each species.

Without distinct evidence, I am unwilling to admit that a double process of selection has long been going on with a multitude of species,—the males having been rendered more brilliant by beating their rivals; and the females more dull-coloured by having escaped from their enemies. We may take as an instance the common brimstone butterfly (Gonepteryx), which appears early in the spring before any other kind. The male of this species is of a far more intense yellow than the female, though she is almost equally conspicuous; and in this case it does not seem probable that she specially acquired her pale tints as a protection, though it is probable that the male acquired his bright colours as a sexual attraction. The female of *Anthocharis cardamines* does not possess the beautiful orange tips to her wings with which the male is ornamented; consequently she closely resembles the white butterflies (Pieris) so common in our gardens; but we have no evidence that this resemblance is beneficial. On the contrary, as she resembles both sexes of several species of the same genus inhabiting various quarters of the world, it is more probable that she has simply retained to a large extent her primordial colours.

Various facts support the conclusion that with the greater number of brilliantly-coloured Lepidoptera, it is the male which has been modified; the two sexes having come to differ from each other, or to resemble each other, according to which form of inheritance has prevailed. Inheritance is governed by so many unknown laws or conditions, that they seem to us to be most capricious in their action;[23] and we can so far understand how it is that with closely-allied species the sexes of some differ to an astonishing degree, whilst the sexes of others are identical in colour. As the successive steps in the process of variation are necessarily all transmitted through the female, a greater or less number of such steps might readily become developed in her; and thus we can understand the frequent gradations from an extreme difference to no difference at all between the sexes of the species within the same group. These cases of gradation are much too common to favour the supposition that we here see females actually undergoing the process of transition and losing their brightness for the sake of protection; for we have every reason to conclude that at any one time the greater number of species are in a fixed condition. With respect to the differences between the females of the species in the same genus or family, we can perceive that they depend, at least in part, on the females partaking of the colours of their respective males. This is well illustrated in those groups in which the males are ornamented to an extraordinary degree; for the females in these groups generally partake to a certain extent of the splendour of their male partners. Lastly, we continually find, as already remarked, that the females of almost all the species in the same genus, or even family, resemble each other much more closely in colour than do the males; and this indicates that the males have undergone a greater amount of modification than the females.

---

[23] 'The Variation of Animals and Plants under Domestication,' vol. ii. chap. xii. p. 17.

Mimickry.—This principle was first made clear in an admirable paper by Mr. Bates,[24] who thus threw a flood of light on many obscure problems. It had previously been observed that certain butterflies in S. America belonging to quite distinct families, resembled the Heliconidæ so closely in every stripe and shade of colour that they could not be distinguished except by an experienced entomologist. As the Heliconidæ are coloured in their usual manner, whilst the others depart from the usual colouring of the groups to which they belong, it is clear that the latter are the imitators, and the Heliconidæ the imitated. Mr. Bates further observed that the imitating species are comparatively rare, whilst the imitated swarm in large numbers; the two sets living mingled together. From the fact of the Heliconidæ being conspicuous and beautiful insects, yet so numerous in individuals and species, he concluded that they must be protected from the attacks of birds by some secretion or odour; and this hypothesis has now been confirmed by a considerable body of curious evidence.[25] From these considerations Mr. Bates inferred that the butterflies which imitate the protected species had acquired their present marvellously deceptive appearance, through variation and natural selection, in order to be mistaken for the protected kinds and thus to escape being devoured. No explanation is here attempted of the brilliant colours of the imitated, but only of the imitating butterflies. We must account for the colours of the former in the same general manner, as in the cases previously discussed in this chapter. Since the publication of Mr. Bates' paper, similar and equally striking facts have been observed

[24] 'Transact. Linn. Soc.' vol. xxiii. 1862, p. 495.

[25] 'Proc. Ent. Soc.' Dec. 3rd, 1866, p. xlv.

by Mr. Wallace[26] in the Malayan region, and by Mr. Trimen in South Africa.

As some writers[27] have felt much difficulty in understanding how the first steps in the process of mimickry could have been effected through natural selection, it may be well to remark that the process probably has never commenced with forms widely dissimilar in colour. But with two species moderately like each other, the closest resemblance if beneficial to either form could readily be thus gained; and if the imitated form was subsequently and gradually modified through sexual selection or any other means, the imitating form would be led along the same track, and thus be modified to almost any extent, so that it might ultimately assume an appearance or colouring wholly unlike that of the other members of the group to which it belonged. As extremely slight variations in colour would not in many cases suffice to render a species so like another protected species as to lead to its preservation, it should be remembered that many species of Lepidoptera are liable to considerable and abrupt variations in colour. A few instances have been given in this chapter; but under this point of view Mr. Bates' original paper on mimickry, as well as Mr. Wallace's papers, should be consulted.

In the foregoing cases both sexes of the imitating species resemble the imitated; but occasionally the

---

26 'Transact. Linn. Soc.' vol. xxv. 1865, p. 1; also 'Transact. Ent. Soc.' vol. iv. (3rd series), 1867, p. 301.

27 See an ingenious article entitled, "Difficulties of the Theory of Natural Selection," in the 'Month,' 1869. The writer strangely supposes that I attribute the variations in colour of the Lepidoptera, by which certain species belonging to distinct families have come to resemble others, to reversion to a common progenitor; but there is no more reason to attribute these variations to reversion than in the case of any ordinary variation.

female alone mocks a brilliantly-coloured and protected species inhabiting the same district. Consequently the female differs in colour from her own male, and, which is a rare and anomalous circumstance, is the more brightly-coloured of the two. In all the few species of Pieridæ, in which the female is more conspicuously coloured than the male, she imitates, as I am informed by Mr. Wallace, some protected species inhabiting the same region. The female of *Diadema anomala* is rich purple-brown with almost the whole surface glossed with satiny blue, and she closely imitates the *Euplœa midamus*, "one of the commonest butterflies of the East;" whilst the male is bronzy or olive-brown, with only a slight blue gloss on the outer parts of the wings.[28] Both sexes of this Diadema and of *D. bolina* follow the same habits of life, so that the differences in colour between the sexes cannot be accounted for by exposure to different conditions;[29] even if this explanation were admissible in other instances.[30]

The above cases of female butterflies which are more brightly-coloured than the males, shew us, firstly, that variations have arisen in a state of nature in the female sex, and have been transmitted exclusively, or almost exclusively, to the same sex; and, secondly, that this form of inheritance has not been determined through natural selection. For if we assume that the females, before they became brightly coloured in imitation of some protected kind, were exposed during each season for a longer period to danger than the males; or if we assume that

---

28 Wallace, "Notes on Eastern Butterflies," 'Transact. Ent. Soc.' 1869, p. 287.

29 Wallace, in 'Westminster Review,' July, 1867, p. 37; and in 'Journal of Travel and Nat. Hist.' vol. i. 1868, p. 88.

30 See remarks by Messrs. Bates and Wallace, in 'Proc. Ent. Soc.' Nov. 19, 1866, p. xxxix.

they could not escape so swiftly from their enemies, we can understand how they alone might originally have acquired through natural selection and sexually-limited inheritance their present protective colours. But except on the principle of these variations having been transmitted exclusively to the female offspring, we cannot understand why the males should have remained dull-coloured; for it would surely not have been in any way injurious to each individual male to have partaken by inheritance of the protective colours of the female, and thus to have had a better chance of escaping destruction. In a group in which brilliant colours are so common as with butterflies, it cannot be supposed that the males have been kept dull-coloured through sexual selection by the females rejecting the individuals which were rendered as beautiful as themselves. We may, therefore, conclude that in these cases inheritance by one sex is not due to the modification through natural selection of a tendency to equal inheritance by both sexes.

It may be well here to give an analogous case in another Order, of characters acquired only by the female, though not in the least injurious, as far as we can judge, to the male. Amongst the Phasmidæ, or spectre-insects, Mr. Wallace states that "it is often the females alone "that so strikingly resemble leaves, while the males show "only a rude approximation." Now, whatever may be the habits of these insects, it is highly improbable that it could be disadvantageous to the males to escape detection by resembling leaves.[31] Hence we may conclude

[31] See Mr. Wallace in 'Westminster Review,' July, 1867, p. 11 and 37. The male of no butterfly, as Mr. Wallace informs me, is known to differ in colour, as a protection, from the female; and he asks me how I can explain this fact on the principle that one sex alone has varied and has transmitted its variations exclusively to the same sex, without

that the females alone in this latter as in the previous cases originally varied in certain characters; these characters having been preserved and augmented through ordinary selection for the sake of protection and from the first transmitted to the female offspring alone.

Bright Colours of Caterpillars.—Whilst reflecting on the beauty of many butterflies, it occurred to me that some caterpillars were splendidly coloured, and as sexual selection could not possibly have here acted, it appeared rash to attribute the beauty of the mature insect to this agency, unless the bright colours of their larvæ could be in some manner explained. In the first place it may be observed that the colours of caterpillars do not stand in any close correlation with those of the mature insect. Secondly, their bright colours do not

---

the aid of selection to check the variations being inherited by the other sex. No doubt if it could be shewn that the females of very many species had been rendered beautiful through protective mimickry, but that this has never occurred with the males, it would be a serious difficulty. But the number of cases as yet known hardly suffices for a fair judgment. We can see that the males, from having the power of flying more swiftly, and thus escaping danger, would not be so likely as the females to have had their colours modified for the sake of protection; but this would not in the least have interfered with their receiving protective colours through inheritance from the females. In the second place, it is probable that sexual selection would actually tend to prevent a beautiful male from becoming obscure, for the less brilliant individuals would be less attractive to the females. Supposing that the beauty of the male of any species had been mainly acquired through sexual selection, yet if this beauty likewise served as a protection, the acquisition would have been aided by natural selection. But it would be quite beyond our power to distinguish between the two processes of sexual and ordinary selection. Hence it is not likely that we should be able to adduce cases of the males having been rendered brilliant exclusively through protective mimickry, though this is comparatively easy with the females, which have rarely or never been rendered beautiful, as far as we can judge, for the sake of sexual attraction, although they have often received beauty through inheritance from their male parents.

serve in any ordinary manner as a protection. As an instance of this, Mr. Bates informs me that the most conspicuous caterpillar which he ever beheld (that of a Sphinx) lived on the large green leaves of a tree on the open llanos of South America; it was about four inches in length, transversely banded with black and yellow, and with its head, legs, and tail of a bright red. Hence it caught the eye of any man who passed by at the distance of many yards, and no doubt of every passing bird.

I then applied to Mr. Wallace, who has an innate genius for solving difficulties. After some consideration he replied: "Most caterpillars require protection, as "may be inferred from some kinds being furnished "with spines or irritating hairs, and from many being "coloured green like the leaves on which they feed, "or curiously like the twigs of the trees on which they "live." I may add as another instance of protection, that there is a caterpillar of a moth, as I am informed by Mr. J. Mansel Weale, which lives on the mimosas in South Africa, and fabricates for itself a case, quite undistinguishable from the surrounding thorns. From such considerations Mr. Wallace thought it probable that conspicuously-coloured caterpillars were protected by having a nauseous taste; but as their skin is extremely tender, and as their intestines readily protrude from a wound, a slight peck from the beak of a bird would be as fatal to them as if they had been devoured. Hence, as Mr. Wallace remarks, "distastefulness alone "would be insufficient to protect a caterpillar unless "some outward sign indicated to its would-be destroyer "that its prey was a disgusting morsel." Under these circumstances it would be highly advantageous to a caterpillar to be instantaneously and certainly recognised as unpalatable by all birds and other animals.

Thus the most gaudy colours would be serviceable, and might have been gained by variation and the survival of the most easily-recognised individuals.

This hypothesis appears at first sight very bold; but when it was brought before the Entomological Society [32] it was supported by various statements; and Mr. J. Jenner Weir, who keeps a large number of birds in an aviary, has made, as he informs me, numerous trials, and finds no exception to the rule, that all caterpillars of nocturnal and retiring habits with smooth skins, all of a green colour, and all which imitate twigs, are greedily devoured by his birds. The hairy and spinose kinds are invariably rejected, as were four conspicuously-coloured species. When the birds rejected a caterpillar, they plainly shewed, by shaking their heads and cleansing their beaks, that they were disgusted by the taste.[33] Three conspicuous kinds of caterpillars and moths were also given by Mr. A. Butler to some lizards and frogs, and were rejected; though other kinds were eagerly eaten. Thus the probable truth of Mr. Wallace's view is confirmed, namely, that certain caterpillars have been made conspicuous for their own good, so as to be easily recognised by their enemies, on nearly the same principle that certain poisons are coloured by druggists for the good of man. This view will, it is probable, be hereafter extended to many animals, which are coloured in a conspicuous manner.

Summary and Concluding Remarks on Insects.— Looking back to the several Orders, we have seen that the sexes often differ in various characters, the meaning

[32] 'Proc. Entomolog. Soc.' Dec. 3rd, 1866, p. xlv., and March 4th, 1867, p. lxxx.

[33] See Mr. J. Jenner Weir's paper on insects and insectivorous birds, in 'Transact. Ent. Soc.' 1869, p. 21; also Mr. Butler's paper, ibid. p. 27.

of which is not understood. The sexes, also, often differ in their organs of sense or locomotion, so that the males may quickly discover or reach the females, and still oftener in the males possessing diversified contrivances for retaining the females when found. But we are not here much concerned with sexual differences of these kinds.

In almost all the Orders, the males of some species, even of weak and delicate kinds, are known to be highly pugnacious; and some few are furnished with special weapons for fighting with their rivals. But the law of battle does not prevail nearly so widely with insects as with the higher animals. Hence probably it is that the males have not often been rendered larger and stronger than the females. On the contrary they are usually smaller, in order that they may be developed within a shorter time, so as to be ready in large numbers for the emergence of the females.

In two families of the Homoptera the males alone possess, in an efficient state, organs which may be called vocal; and in three families of the Orthoptera the males alone possess stridulating organs. In both cases these organs are incessantly used during the breeding-season, not only for calling the females, but for charming or exciting them in rivalry with other males. No one who admits the agency of natural selection, will dispute that these musical instruments have been acquired through sexual selection. In four other Orders the members of one sex, or more commonly of both sexes, are provided with organs for producing various sounds, which apparently serve merely as call-notes. Even when both sexes are thus provided, the individuals which were able to make the loudest or most continuous noise would gain partners before those which were less noisy, so that their organs have probably been gained through sexual selection. It is instructive to reflect on the wonderful diversity of the means for producing sound, possessed by the males alone or by both sexes in no less than six Orders, and which were possessed by at least one insect at an extremely remote geological epoch. We thus learn how effectual sexual selection has been in leading to modifications of structure, which sometimes, as with the Homoptera, are of an important nature.

From the reasons assigned in the last chapter, it is probable that the great horns of the males of many lamellicorn, and some other beetles, have been acquired as ornaments. So perhaps it may be with certain other peculiarities confined to the male sex. From the small size of insects, we are apt to undervalue their appearance. If we could imagine a male Chalcosoma (fig. 15) with its polished, bronzed coat of mail, and vast complex horns, magnified to the size of a horse or even of a dog, it would be one of the most imposing animals in the world.

The colouring of insects is a complex and obscure subject. When the male differs slightly from the female, and neither are brilliantly coloured, it is probable that the two sexes have varied in a slightly different manner, with the variations transmitted to the same sex, without any benefit having been thus derived or evil suffered. When the male is brilliantly coloured and differs conspicuously from the female, as with some dragon-flies and many butterflies, it is probable that he alone has been modified, and that he owes his colours to sexual selection; whilst the female has retained a primordial or very ancient type of colouring, slightly modified by the agencies before explained, and has therefore not been rendered obscure, at least in most cases, for the sake of protection. But the female alone has sometimes been coloured brilliantly so as to imitate other protected species inhabiting the same district. When the sexes resemble each other and both are obscurely coloured, there is no doubt that they have been in a multitude of cases coloured for the sake of protection. So it is in some instances when both are brightly coloured, causing them to resemble surrounding objects such as flowers, or other protected species, or indirectly by giving notice to their enemies that they are of an unpalatable nature. In many other cases in which the sexes resemble each other and are brilliantly coloured, especially when the colours are arranged for display, we may conclude that they have been gained by the male sex as an attraction, and have been transferred to both sexes. We are more especially led to this conclusion whenever the same type of coloration prevails throughout a group, and we find that the males of some species differ widely in colour from the females, whilst both sexes of other species are quite alike, with intermediate gradations connecting these extreme states.

In the same manner as bright colours have often been partially transferred from the males to the females, so it has been with the extraordinary horns of many lamellicorn and some other beetles. So, again, the vocal or instrumental organs proper to the males of the Homoptera and Orthoptera have generally been transferred in a rudimentary, or even in a nearly perfect condition to the females; yet not sufficiently perfect to be used for producing sound. It is also an interesting fact, as bearing on sexual selection, that the stridulating organs of certain male Orthoptera are not fully developed until the last moult; and that the colours of certain male dragon-flies are not fully developed until some little time after their emergence from the pupal state, and when they are ready to breed.

Sexual selection implies that the more attractive individuals are preferred by the opposite sex; and as with insects, when the sexes differ, it is the male which, with rare exceptions, is the most ornamented and departs most from the type to which the species belongs;—and as it is the male which searches eagerly for the female, we must suppose that the females habitually or occasionally prefer the more beautiful males, and that these have thus acquired their beauty. That in most or all the orders the females have the power of rejecting any particular male, we may safely infer from the many singular contrivances possessed by the males, such as great jaws, adhesive cushions, spines, elongated legs, &c., for seizing the female; for these contrivances shew that there is some difficulty in the act. In the case of unions between distinct species, of which many instances have been recorded, the female must have been a consenting party. Judging from what we know of the perceptive powers and affections of various insects, there is no antecedent improbability in sexual selection having come largely into action; but we have as yet no direct evidence on this head, and some facts are opposed to the belief. Nevertheless, when we see many males pursuing the same female, we can hardly believe that the pairing is left to blind chance—that the female exerts no choice, and is not influenced by the gorgeous colours or other ornaments, with which the male alone is decorated.

If we admit that the females of the Homoptera and Orthoptera appreciate the musical tones emitted by their male partners, and that the various instruments for this purpose have been perfected through sexual selection, there is little improbability in the females of other insects appreciating beauty in form or colour, and consequently in such characters having been thus gained by the males. But from the circumstance of colour being so variable, and from its having been so often modified for the sake of protection, it is extremely difficult to decide in how large a proportion of cases sexual selection has come into play. This is more especially difficult in those Orders, such as the Orthoptera, Hymenoptera, and Coleoptera, in which the two sexes rarely differ much in colour ; for we are thus cut off from our best evidence of some relation between the reproduction of the species and colour. With the Coleoptera, however, as before remarked, it is in the great lamellicorn group, placed by some authors at the head of the Order, and in which we sometimes see a mutual attachment between the sexes, that we find the males of some species possessing weapons for sexual strife, others furnished with wonderful horns, many with stridulating organs, and others ornamented with splendid metallic tints. Hence it seems probable that all these characters have been gained through the same means, namely sexual selection.

When we treat of Birds, we shall see that they present in their secondary sexual characters the closest analogy with insects. Thus, many male birds are highly pugnacious, and some are furnished with special weapons for fighting with their rivals. They possess organs which are used during the breeding-season for producing vocal and instrumental music. They are frequently ornamented with combs, horns, wattles and plumes of the most diversified kinds, and are decorated with beautiful colours, all evidently for the sake of display. We shall find that, as with insects, both sexes, in certain groups, are equally beautiful, and are equally provided with ornaments which are usually confined to the male sex. In other groups both sexes are equally plain-coloured and unornamented. Lastly, in some few anomalous cases, the females are more beautiful than the males. We shall often find, in the same group of birds, every gradation from no difference between the sexes, to an extreme difference. In the latter case we shall see that the females, like female insects, often possess more or less plain traces of the characters which properly belong to the males. The analogy, indeed, in all these respects between birds and insects, is curiously close. Whatever explanation applies to the one class probably applies to the other; and this explanation, as we shall hereafter attempt to shew, is almost certainly sexual selection.

END OF VOL. I.

LONDON:
PRINTED BY W. CLOWES AND SONS, STAMFORD STREET,
AND CHARING CROSS.

THE

DESCENT OF MAN,

AND

SELECTION IN RELATION TO SEX.

By CHARLES DARWIN, M.A., F.R.S., &c.

IN TWO VOLUMES.—Vol. II.

WITH ILLUSTRATIONS.

LONDON:
JOHN MURRAY, ALBEMARLE STREET.
1871.

ERRATA.

VOL. I.

Page	line		For		read
27	13	..	kaolo	..	koala.
31	6	..	prostratica	..	prostatica.
59, *note* 42	2	..	speech	..	species.
74, *note* 7	–	..	Browne	..	Brown.
118, *note* 28	–	..	Vol. I.	..	Vol. II.
128, *note* 45	4	..	*Before* vol. xiv. *insert* 'Proc. Royal Soc.		
208	2	..	prostratica	..	prostatica.
322	5	..	Actineæ	..	Actiniæ.
324	30	..	land-shells	..	land-snails.
330	16	..	figs. 4 and 5	..	figs. 4, 5, and 6.
334	17	..	Birgos	..	Birgus.
339	8	..	attractions	..	attentions.
341	3	..	dragon-flys	..	dragon-flies.
378	17	..	Typhæus	..	Typhœus.
384	31	..	tesselatum	..	tessellatum.
397	9	..	Hypopira	..	Hypopyra.
405	21	..	Acrœidæ	..	Acræidæ.

VOL. II.

32	30	..	chamelion	..	chameleon.
115	4	..	mail	..	male.
178	23	..	Chloehaga	..	Chloephaga.
227, *note* 53	–	..	Ramphaston	..	Ramphastos.
240, *note* 2	–	..	Mr. H. Brown	..	Mr. R. Brown.
— *note* 3	2	..	elephus	..	elaphas.
242	14	..	walruses	..	narwhals.
339	27	..	Durfur	..	Darfur.

CONTENTS.

PART II.

SEXUAL SELECTION—*continued.*

CHAPTER XII.

SECONDARY SEXUAL CHARACTERS OF FISHES, AMPHIBIANS, AND REPTILES.

CHAPTER XIII.

SECONDARY SEXUAL CHARACTERS OF BIRDS.

CHAPTER XIV.

BIRDS—*continued.*

CHAPTER XV.

BIRDS—*continued.*

CHAPTER XVI.

BIRDS—*concluded.*

CHAPTER XVII.

SECONDARY SEXUAL CHARACTERS OF MAMMALS.

CHAPTER XVIII.

SECONDARY SEXUAL CHARACTERS OF MAMMALS—*continued.*

CHAPTER XIX.

SECONDARY SEXUAL CHARACTERS OF MAN.

CHAPTER XX.

Secondary Sexual Characters of Man—*continued.*

CHAPTER XXI.

General Summary and Conclusion.

POSTSCRIPT

Vol. I. pp. 297–299.—I have fallen into a serious and unfortunate error, in relation to the sexual differences of animals, in attempting to explain what seemed to me a singular coincidence in the late period of life at which the necessary variations have arisen in many cases, and the late period at which sexual selection acts. The explanation given is wholly erroneous, as I have discovered by working out an illustration in figures. Moreover, the supposed coincidence of period is far from general, and is not remarkable; for, as I have elsewhere attempted to show, variations arising early in life have often been accumulated through sexual selection, being then commonly transmitted to both sexes. On the other hand, variations arising late in life cannot fail to coincide approximately in period with that of the process of sexual selection. Allusions to these erroneous views reappear in Vol. II. pp. 161 and 237.

SEXUAL SELECTION.

CHAPTER XII.

Secondary Sexual Characters of Fishes, Amphibians, and Reptiles.

Fishes: Courtship and battles of the males — Larger size of the females — Males, bright colours and ornamental appendages; other strange characters — Colours and appendages acquired by the males during the breeding-season alone — Fishes with both sexes brilliantly coloured — Protective colours — The less conspicuous colours of the female cannot be accounted for on the principle of protection — Male fishes building nests, and taking charge of the ova and young. Amphibians: Differences in structure and colour between the sexes — Vocal organs. Reptiles: Chelonians — Crocodiles — Snakes, colours in some cases protective — Lizards, battles of — Ornamental appendages — Strange differences in structure between the sexes — Colours — Sexual differences almost as great as with birds.

We have now arrived at the great sub-kingdom of the Vertebrata, and will commence with the lowest class, namely Fishes. The males of Plagiostomous fishes (sharks, rays) and of Chimæroid fishes are provided with claspers which serve to retain the female, like the various structures possessed by so many of the lower animals. Besides the claspers, the males of many rays have clusters of strong sharp spines on their heads, and several rows along "the upper outer surface of their pectoral fins." These are present in the males of some species, which have the other parts of their bodies smooth. They are only temporarily developed during the breeding-season; and Dr. Günther suspects that they are brought into action as prehensile organs by the doubling inwards and downwards of the two sides of the body. It is a remarkable fact that the females and not the males of some species, as of *Raia clavata*, have their backs studded with large hook-formed spines.[1]

Owing to the element which fishes inhabit, little is known about their courtship, and not much about their battles. The male stickleback (*Gasterosteus leiurus*) has been described as "mad with delight" when the female comes out of her hiding-place and surveys the nest which he has made for her. "He darts round "her in every direction, then to his accumulated ma-"terials for the nest, then back again in an instant; "and as she does not advance he endeavours to push "her with his snout, and then tries to pull her by the "tail and side-spine to the nest."[2] The males are said to be polygamists;[3] they are extraordinarily bold and pugnacious, whilst "the females are quite pacific." Their battles are at times desperate; "for these puny "combatants fasten tight on each other for several "seconds, tumbling over and over again, until their "strength appears completely exhausted." With the rough-tailed stickleback (*G. trachurus*) the males whilst fighting swim round and round each other, biting and endeavouring to pierce each other with their raised lateral spines. The same writer adds,[4] "the bite of these little

---

1 Yarrell's 'Hist. of British Fishes,' vol. ii. 1836, p. 417, 425, 436. Dr. Günther informs me that the spines in *R. clavata* are peculiar to the female.

2 See Mr. R. Warington's interesting articles in 'Annals and Mag. of Nat. Hist.' Oct. 1852 and Nov. 1855.

3 Noel Humphreys, 'River Gardens,' 1857.

4 Loudon's 'Mag. of Natural History,' vol. iii. 1830, p. 331.

"furies is very severe. They also use their lateral spines "with such fatal effect, that I have seen one during a "battle absolutely rip his opponent quite open, so that "he sank to the bottom and died." When a fish is conquered, "his gallant bearing forsakes him; his gay "colours fade away; and he hides his disgrace among "his peaceable companions, but is for some time the "constant object of his conqueror's persecution."

The male salmon is as pugnacious as the little stickleback; and so is the male trout, as I hear from Dr. Günther. Mr. Shaw saw a violent contest between two male salmons which lasted the whole day; and Mr. R. Buist, Superintendent of Fisheries, informs me that he has often watched from the bridge at Perth the males driving away their rivals whilst the females were spawning. The males "are constantly fighting and tearing "each other on the spawning-beds, and many so injure "each other as to cause the death of numbers, many "being seen swimming near the banks of the river in "a state of exhaustion, and apparently in a dying state."[5] The keeper of the Stormontfield breeding-ponds visited, as Mr. Buist informs me, in June, 1868, the northern Tyne, and found about 300 dead salmon, all of which with one exception were males; and he was convinced that they had lost their lives by fighting.

The most curious point about the male salmon is that during the breeding-season, besides a slight change in colour, "the lower jaw elongates, and a cartilaginous "projection turns upwards from the point, which, when "the jaws are closed, occupies a deep cavity between

---

[5] 'The Field,' June 29th, 1867. For Mr. Shaw's statement, see 'Edinburgh Review,' 1843. Another experienced observer (Scrope's 'Days of Salmon Fishing,' p. 60) remarks that the male would, if he could, keep, like the stag, all other males away.

"the intermaxillary bones of the upper jaw."[6] (Figs. 26 and 27.) In our salmon this change of structure lasts only during the breeding-season; but in the *Salmo lycaodon* of N.W. America the change, as Mr. J. K.

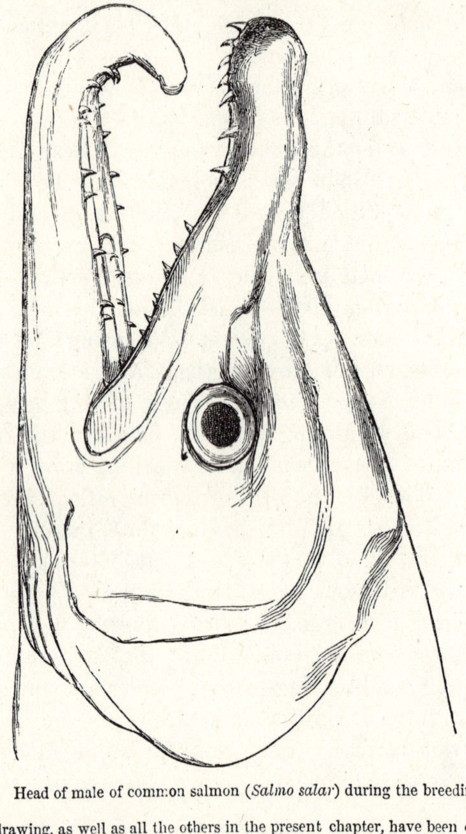

Fig. 26. Head of male of common salmon (*Salmo salar*) during the breeding-season.

[This drawing, as well as all the others in the present chapter, have been executed by the well-known artist, Mr. G. Ford, under the kind superintendence of Dr. Günther, from specimens in the British Museum.]

---

6 Yarrell, 'History of British Fishes,' vol. ii. 1836, p. 10.

Lord[7] believes, is permanent and best marked in the older males which have previously ascended the rivers. In these old males the jaws become developed into immense hook-like projections, and the teeth grow into

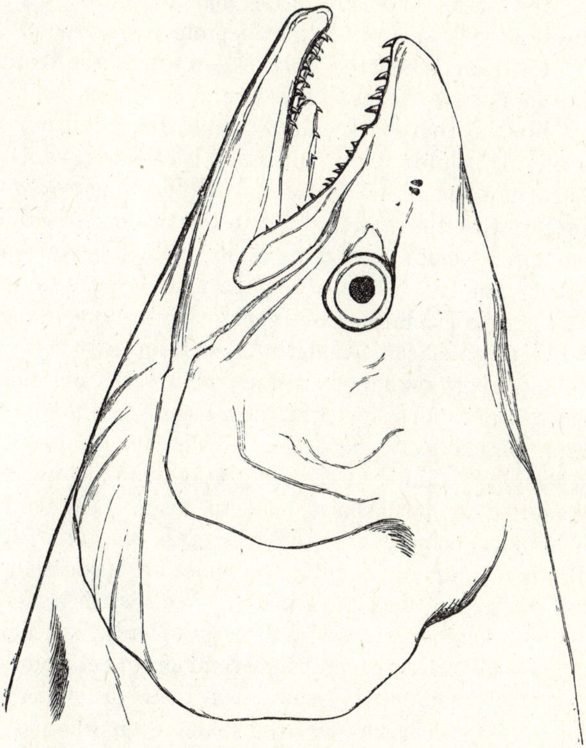

Fig. 27. Head of female salmon.

regular fangs, often more than half an inch in length. With the European salmon, according to Mr. Lloyd,[8]

[7] 'The Naturalist in Vancouver's Island,' vol. i. 1866, p. 54.
[8] 'Scandinavian Adventures,' vol. i. 1854, p. 100, 104.

the temporary hook-like structure serves to strengthen and protect the jaws, when one male charges another with wonderful violence; but the greatly developed teeth of the male American salmon may be compared with the tusks of many male mammals, and they indicate an offensive rather than a protective purpose.

The salmon is not the only fish in which the teeth differ in the two sexes. This is the case with many rays. In the thornback (*Raia clavata*) the adult male has sharp, pointed teeth, directed backwards, whilst those of the female are broad and flat, forming a pavement; so that these teeth differ in the two sexes of the same species more than is usual in distinct genera of the same family. The teeth of the male become sharp only when he is adult: whilst young they are broad and flat like those of the female. As so frequently occurs with secondary sexual characters, both sexes of some species of rays, for instance *R. batis*, possess, when adult, sharp, pointed teeth; and here a character, proper to and primarily gained by the male, appears to have been transmitted to the offspring of both sexes. The teeth are likewise pointed in both sexes of *R. maculata*, but only when completely adult; the males acquiring them at an earlier age than the females. We shall hereafter meet with analogous cases with certain birds, in which the male acquires the plumage common to both adult sexes, at a somewhat earlier age than the female. With other species of rays the males even when old never possess sharp teeth, and consequently both sexes when adult are provided with broad, flat teeth like those of the young, and of the mature females of the above-mentioned species.[9] As the rays are bold,

---

9 See Yarrell's account of the Rays in his 'Hist. of British Fishes,' vol. ii. 1836, p. 416, with an excellent figure, and p. 422, 432.

strong and voracious fishes, we may suspect that the males require their sharp teeth for fighting with their rivals; but as they possess many parts modified and adapted for the prehension of the female, it is possible that their teeth may be used for this purpose.

In regard to size, M. Carbonnier[10] maintains that with almost all fishes the female is larger than the male; and Dr. Günther does not know of a single instance in which the male is actually larger than the female. With some Cyprinodonts the male is not even half as large as the female. As with many kinds of fishes the males habitually fight together; it is surprising that they have not generally become through the effects of sexual selection larger and stronger than the females. The males suffer from their small size, for according to M. Carbonnier they are liable to be devoured by the females of their own species when carnivorous, and no doubt by other species. Increased size must be in some manner of more importance to the females, than strength and size are to the males for fighting with other males; and this perhaps is to allow of the production of a vast number of ova.

In many species the male alone is ornamented with bright colours; or these are much brighter in the male than the female. The male, also, is sometimes provided with appendages which appear to be of no more use to him for the ordinary purposes of life than are the tail-feathers to the peacock. I am indebted for most of the following facts to the great kindness of Dr. Günther. There is reason to suspect that many tropical fishes differ sexually in colour and structure; and there are some striking cases with our British fishes. The male *Callionymus lyra* has been called the *gemmeous dragonet*

---

[10] As quoted in 'The Farmer,' 1868, p. 369.

"from its brilliant gem-like colours." When freshly taken from the sea the body is yellow of various shades, striped and spotted with vivid blue on the head; the dorsal fins are pale brown with dark longitudinal bands; the ventral, caudal and anal fins being bluish-black. The female, or sordid dragonet, was considered by Linnæus and by many subsequent naturalists as a distinct

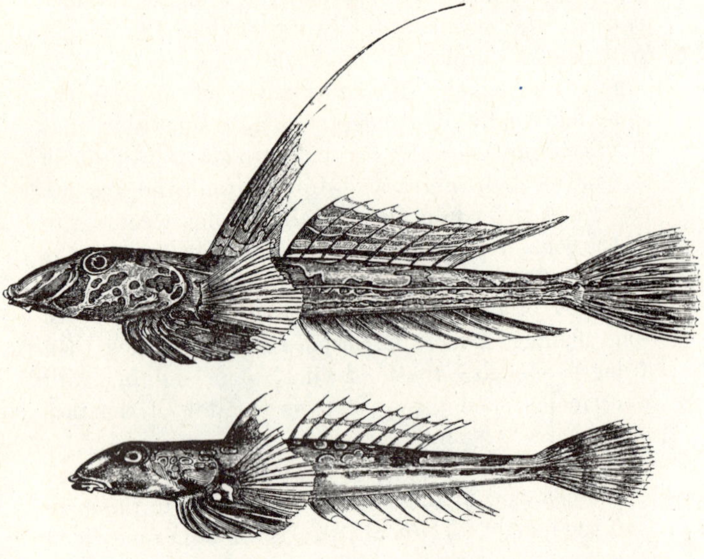

Fig. 28. Callionymus lyra. Upper figure, male; lower figure, female.

species; it is of a dingy reddish-brown, with the dorsal fin brown and the other fins white. The sexes differ also in the proportional size of the head and mouth, and in the position of the eyes;[11] but the most striking difference is the extraordinary elongation in the male (fig. 28)

[11] I have drawn up this description from Yarrell's 'British Fishe,' vol. i. 1836, p. 261 and 266.

of the dorsal fin. The young males resemble in structure and colour the adult females. Throughout the genus Callionymus,[12] the male is generally much more brightly spotted than the female, and in several species, not only the dorsal, but the anal fin of the male is much elongated.

The male of the *Cottus scorpius*, or sea-scorpion, is more slender and smaller than the female. There is also a great difference in colour between them. It is difficult, as Mr. Lloyd[13] remarks, "for any one, who has not seen "this fish during the spawning-season, when its hues are "brightest, to conceive the admixture of brilliant colours "with which it, in other respects so ill-favoured, is at "that time adorned." Both sexes of the *Labrus mixtus*, although very different in colour, are beautiful; the male being orange with bright-blue stripes, and the female bright-red with some black spots on the back.

In the very distinct family of the Cyprinodontidæ—inhabitants of the fresh waters of foreign lands—the sexes sometimes differ much in various characters. In the male of the *Mollienesia petenensis*,[14] the dorsal fin is greatly developed and is marked with a row of large, round, ocellated, bright-coloured spots; whilst the same fin in the female is smaller, of a different shape, and marked only with irregularly-curved brown spots. In the male the basal margin of the anal fin is also a little produced and dark-coloured. In the male of an allied form, the *Xiphophorus Hellerii* (fig. 29), the inferior margin of the anal fin is developed into a long filament,

---

[12] 'Catalogue of Acanth. Fishes in the British Museum,' by Dr. Günther, 1861, p. 138-151.

[13] 'Game Birds of Sweden,' &c., 1867, p. 466.

[14] With respect to this and the following species I am indebted to Dr. Günther for information: see also his paper on the Fishes of Central America, in 'Transact. Zoolog. Soc.' vol. vi. 1868, p. 485.

which is striped, as I hear from Dr. Günther, with bright colours. This filament does not contain any muscles, and apparently cannot be of any direct use to the fish. As in the case of the Callionymus, the males whilst young resemble in colour and structure the adult females. Sexual differences such as these may be strictly compared with those which are so frequent with gallinaceous birds.[15]

Fig. 29. Xiphophorus Hellerii. Upper figure, male; lower figure, female.

In a siluroid fish, inhabiting the fresh waters of South America, namely the *Plecostomus barbatus*[16] (fig. 30), the male has its mouth and interoperculum fringed with a beard of stiff hairs, of which the female shews hardly a trace. These hairs are of the nature of scales. In another species of the same genus, soft flexible tentacles project from the front part of the head of the

[15] Dr. Günther makes this remark; 'Catalogue of Fishes in the British Museum,' vol. iii. 1861, p. 141.

[16] See Dr. Günther on this genus, in 'Proc. Zoolog. Soc.' 1868, p. 232.

Fig. 30. Plecostomus barbatus. Upper figure, head of male; lower figure, female.

male, which are absent in the female. These tentacles are prolongations of the true skin, and therefore are not homologous with the stiff hairs of the former species; but it can hardly be doubted that both serve the same purpose. What this purpose may be it is difficult to conjecture; ornament does not here seem probable, but we can hardly suppose that stiff hairs and flexible filaments can be useful in any ordinary way to the males alone. The *Monacanthus scopas,* which was shewn to me in the British Museum by Dr. Günther, presents a nearly analogous case. The male has a cluster of stiff, straight spines, like those of a comb, on the sides of the tail; and these in a specimen six inches long were nearly an inch and a half in length; the female has on the same place a cluster of bristles, which may be compared with those of a tooth-brush. In another species, the *M. peronii,* the male has a brush like that possessed by the female of the last species, whilst the sides of the tail in the female are smooth. In some other species the same part of the tail can be perceived to be a little roughened in the male and perfectly smooth in the female; and lastly in others, both sexes have smooth sides. In that strange monster, the *Chimæra monstrosa,* the male has a hook-shaped bone on the top of the head, directed forwards, with its rounded end covered with sharp spines; in the female "this crown is altogether absent," but what its use may be is utterly unknown.[17]

The structures as yet referred to are permanent in the male after he has arrived at maturity; but with some Blennies and in another allied genus[18] a crest is developed on the head of the male only during the breed-

---

17 F. Buckland, in 'Land and Water,' July, 1868, p. 377, with a figure.

18 Dr. Günther, 'Catalogue of Fishes,' vol. iii. p. 221 and 240.

ing-season, and their bodies at the same time become more brightly-coloured. There can be little doubt that this crest serves as a temporary sexual ornament, for the female does not exhibit a trace of it. In other species of the same genus both sexes possess a crest, and in at least one species neither sex is thus provided. In this case and in that of the Monacanthus, we have good instances to how great an extent the sexual characters of closely-allied forms may differ. In many of the Chromidæ, for instance in Geophagus and especially in Cichla, the males, as I hear from Professor Agassiz,[19] have a conspicuous protuberance on the forehead, which is wholly wanting in the females and in the young males. Professor Agassiz adds, "I have often observed these fishes "at the time of spawning when the protuberance is "largest, and at other seasons when it is totally wanting "and the two sexes shew no difference whatever in the "outline of the profile of the head. I never could "ascertain that it subserves any special function, and "the Indians on the Amazon know nothing about its "use." These protuberances in their periodical appearance resemble the fleshy caruncles on the heads of certain birds; but whether they serve as ornaments must remain at present doubtful.

The males of those fishes, which differ permanently in colour from the females, often become more brilliant, as I hear from Professor Agassiz and Dr. Günther, during the breeding-season. This is likewise the case with a multitude of fishes, the sexes of which at all other seasons of the year are identical in colour. The tench, roach, and perch may be given as instances. The male salmon at this season is "marked on the cheeks with

---

[19] See also 'A Journey in Brazil,' by Prof. and Mrs. Agassiz, 1868, p. 220.

"orange-coloured stripes, which give it the appearance "of a Labrus, and the body partakes of a golden-orange "tinge. The females are dark in colour, and are com- "monly called black-fish."[20] An analogous and even greater change takes place with the *Salmo eriox* or bull-trout; the males of the char (*S. umbla*) are likewise at this season rather brighter in colour than the females.[21] The colours of the pike (*Esox reticulatus*) of the United States, especially of the male, become, during the breeding-season, exceedingly intense, brilliant, and iridescent.[22] Another striking instance out of many is afforded by the male stickleback (*Gasterosteus leiurus*), which is described by Mr. Warington,[23] as being then "beautiful beyond description." The back and eyes of the female are simply brown, and the belly white. The eyes of the male, on the other hand, are "of the most "splendid green, having a metallic lustre like the "green feathers of some humming-birds. The throat "and belly are of a bright crimson, the back of an "ashy-green, and the whole fish appears as though it "were somewhat translucent and glowed with an in- "ternal incandescence." After the breeding-season these colours all change, the throat and belly become of a paler red, the back more green, and the glowing tints subside.

That with fishes there exists some close relation between their colours and their sexual functions we can clearly see;—firstly, from the adult males of certain species being differently coloured from the females, and often much more brilliantly;—secondly, from these same

---

20 Yarrell, 'British Fishes,' vol. ii. 1836, p. 10, 12, 35.

21 W. Thompson, in 'Annals and Mag. of Nat. History,' vol. vi. 1841, p. 440.

22 'The American Agriculturist,' 1868, p. 100.

23 'Annals and Mag. of Nat. Hist.' Oct. 1852.

males, whilst immature, resembling the mature females; —and, lastly, from the males, even of those species which at all other times of the year are identical in colour with the females, often acquiring brilliant tints during the spawning-season. We know that the males are ardent in their courtship and sometimes fight desperately together. If we may assume that the females have the power of exerting a choice and of selecting the more highly-ornamented males, all the above facts become intelligible through the principle of sexual selection. On the other hand, if the females habitually deposited and left their ova to be fertilised by the first male which chanced to approach, this fact would be fatal to the efficiency of sexual selection; for there could be no choice of a partner. But, as far as is known, the female never willingly spawns except in the close presence of a male, and the male never fertilises the ova except in the close presence of a female. It is obviously difficult to obtain direct evidence with respect to female fishes selecting their partners. An excellent observer,[24] who carefully watched the spawning of minnows (*Cyprinus phoxinus*), remarks that owing to the males, which were ten times as numerous as the females, crowding closely round them, he could "speak only doubtfully on their operations. When a female came among a number of "males they immediately pursued her; if she was not "ready for shedding her spawn, she made a precipitate "retreat; but if she was ready, she came boldly in "among them, and was immediately pressed closely by "a male on each side; and when they had been in that "situation a short time, were superseded by other two, "who wedged themselves in between them and the

[24] Loudon's 'Mag. of Nat. Hist.' vol. v. 1832, p. 681.

"female, who appeared to treat all her lovers with "the same kindness." Notwithstanding this last statement, I cannot, from the several previous considerations, give up the belief that the males which are the most attractive to the females, from their brighter colours or other ornaments, are commonly preferred by them; and that the males have thus been rendered more beautiful in the course of ages.

We have next to inquire whether this view can be extended, through the law of the equal transmission of characters to both sexes, to those groups in which the males and females are brilliant in the same or nearly the same degree and manner. In such a genus as Labrus, which includes some of the most splendid fishes in the world, for instance, the Peacock Labrus (*L. pavo*), described,[25] with pardonable exaggeration, as formed of polished scales of gold encrusting lapis-lazuli, rubies, sapphires, emeralds and amethysts, we may, with much probability, accept this belief; for we have seen that the sexes in at least one species differ greatly in colour. With some fishes, as with many of the lowest animals, splendid colours may be the direct result of the nature of their tissues and of the surrounding conditions, without any aid from selection. The gold-fish (*Cyprinus auratus*), judging from the analogy of the golden variety of the common carp, is, perhaps, a case in point, as it may owe its splendid colours to a single abrupt variation, due to the conditions to which this fish has been subjected under confinement. It is, however, more probable that these colours have been intensified through artificial selection, as this species has been carefully bred in China from a remote

25 Bory de Saint Vincent, in 'Dict. Class. d'Hist. Nat.' tom. ix. 1826, p. 151.

period.[26] Under natural conditions it does not seem probable that beings so highly organised as fishes, and which live under such complex relations, should become brilliantly coloured without suffering some evil or receiving some benefit from so great a change, and consequently without the intervention of natural selection.

What, then, must we conclude in regard to the many fishes, both sexes of which are splendidly coloured? Mr. Wallace[27] believes that the species which frequent reefs, where corals and other brightly-coloured organisms abound, are brightly coloured in order to escape detection by their enemies; but according to my recollection they were thus rendered highly conspicuous. In the fresh-waters of the Tropics there are no brilliantly-coloured corals or other organisms for the fishes to resemble; yet many species in the Amazons are beautifully coloured, and many of the carnivorous Cyprinidæ in India are ornamented with "bright longitudinal lines of various tints."[28] Mr. M'Clelland, in describing these fishes goes so far as to suppose that "the peculiar brilliancy of their colours" serves as "a better mark for king-fishers, terns, and other birds which are destined to keep the number of these fishes in check;" but at the present day few naturalists will

---

26 Owing to some remarks on this subject, made in my work 'On the Variation of Animals under Domestication,' Mr. W. F. Mayers ('Chinese Notes and Queries,' Aug. 1868, p. 123) has searched the ancient Chinese encyclopedias. He finds that gold-fish were first reared in confinement during the Sung Dynasty, which commenced A.D. 960. In the year 1129 these fishes abounded. In another place it is said that since the year 1548 there has been "produced at Hang-chow a variety called the fire-fish, from its intensely red colour. It is universally admired, and there is not a household where it is not cultivated, *in rivalry as to its colour*, and as a source of profit."

27 'Westminster Review,' July, 1867, p. 7.

28 "Indian Cyprinidæ," by Mr. J. M'Clelland, 'Asiatic Researches,' vol. xix. part ii. 1839, p. 230.

admit that any animal has been made conspicuous as an aid to its own destruction. It is possible that certain fishes may have been rendered conspicuous in order to warn birds and beasts of prey (as explained when treating of caterpillars) that they were unpalatable; but it is not, I believe, known that any fish, at least any fresh-water fish, is rejected from being distasteful to fish-devouring animals. On the whole, the most probable view in regard to the fishes, of which both sexes are brilliantly coloured, is that their colours have been acquired by the males as a sexual ornament, and have been transferred in an equal or nearly equal degree to the other sex.

We have now to consider whether, when the male differs in a marked manner from the female in colour or in other ornaments, he alone has been modified, with the variations inherited only by his male offspring; or whether the female has been specially modified and rendered inconspicuous for the sake of protection, such modifications being inherited only by the females. It is impossible to doubt that colour has been acquired by many fishes as a protection: no one can behold the speckled upper surface of a flounder, and overlook its resemblance to the sandy bed of the sea on which it lives. One of the most striking instances ever recorded of an animal gaining protection by its colour (as far as can be judged in preserved specimens) and by its form, is that given by Dr. Günther[29] of a pipe-fish, which, with its reddish streaming filaments, is hardly distinguishable from the sea-weed to which it clings with its prehensile tail. But the question now under consideration is whether the females alone have been modified for this object. Fishes offer valuable

[29] 'Proc. Zoolog. Soc.' 1865, p. 327, pl. xiv. and xv.

evidence on this head. We can see that one sex will not be modified through natural selection for the sake of protection more than the other, supposing both to vary, unless one sex is exposed for a longer period to danger, or has less power of escaping from such danger than the other sex; and it does not appear that with fishes the sexes differ in these respects. As far as there is any difference, the males, from being generally of smaller size, and from wandering more about, are exposed to greater danger than the females; and yet, when the sexes differ, the males are almost always the most conspicuously coloured. The ova are fertilised immediately after being deposited, and when this process lasts for several days, as in the case of the salmon,[30] the female, during the whole time, is attended by the male. After the ova are fertilised they are, in most cases, left unprotected by both parents, so that the males and females, as far as oviposition is concerned, are equally exposed to danger, and both are equally important for the production of fertile ova; consequently the more or less brightly-coloured individuals of either sex would be equally liable to be destroyed or preserved, and both would have an equal influence on the colours of their offspring or the race.

Certain fishes, belonging to several families, make nests; and some of these fishes take care of their young when hatched. Both sexes of the brightly-coloured *Crenilabrus massa* and *melops* work together in building their nests with sea-weed, shells, &c.[31] But the males of certain fishes do all the work, and afterwards take exclusive charge of the young. This is the case

[30] Yarrell, 'British Fishes,' vol. ii. p. 11.

[31] According to the observations of M. Gerbe; see Günther's 'Record of Zoolog. Literature,' 1865, p. 194.

with the dull-coloured gobies,[32] in which the sexes are not known to differ in colour, and likewise with the sticklebacks (Gasterosteus), in which the males become brilliantly coloured during the spawning-season. The male of the smooth-tailed stickleback (*G. leiurus*) performs during a long time the duties of a nurse with exemplary care and vigilance, and is continually employed in gently leading back the young to the nest when they stray too far. He courageously drives away all enemies, including the females of his own species. It would indeed be no small relief to the male if the female, after depositing her eggs, were immediately devoured by some enemy, for he is forced incessantly to drive her from the nest.[33]

The males of certain other fishes inhabiting South America and Ceylon, and belonging to two distinct orders, have the extraordinary habit of hatching the eggs laid by the females within their mouths or branchial cavities.[34] With the Amazonian species which follow this habit, the males, as I am informed by the kindness of Professor Agassiz, "not only are generally brighter "than the females, but the difference is greater at "the spawning-season than at any other time." The species of Geophagus act in the same manner; and in this genus, a conspicuous protuberance becomes developed on the forehead of the males during the breeding-season. With the various species of Chromids, as Professor Agassiz likewise informs me, sexual differences

---

32 Cuvier, 'Règne Animal,' vol. ii. 1829, p. 242.

33 See Mr. Warington's most interesting description of the habits of the *Gasterosteus leiurus*, in 'Annals and Mag. of Nat. Hist.' November, 1855.

34 Prof. Wyman, in 'Proc. Boston Soc. of Nat. Hist.' Sept. 15, 1857. Also W. Turner, in 'Journal of Anatomy and Phys.' Nov. 1, 1866, p. 78. Dr. Günther has likewise described other cases.

in colour may be observed, "whether they lay their "eggs in the water among aquatic plants, or deposit "them in holes, leaving them to come out without "further care, or build shallow nests in the river-mud, "over which they sit, as our Promotis does. It ought "also to be observed that these sitters are among the "brightest species in their respective families; for "instance, Hygrogonus is bright green, with large "black ocelli, encircled with the most brilliant red." Whether with all the species of Chromids it is the male alone which sits on the eggs is not known. It is, however, manifest that the fact of the eggs being protected or unprotected, has had little or no influence on the differences in colour between the sexes. It is further manifest, in all the cases in which the males take exclusive charge of the nests and young, that the destruction of the brighter-coloured males would be far more influential on the character of the race, than the destruction of the brighter-coloured females; for the death of the male during the period of incubation or nursing would entail the death of the young, so that these could not inherit his peculiarities; yet, in many of these very cases the males are more conspicuously coloured than the females.

In most of the Lophobranchii (Pipe-fish, Hippocampi, &c.) the males have either marsupial sacks or hemispherical depressions on the abdomen, in which the ova laid by the female are hatched. The males also shew great attachment to their young.[35] The sexes do not commonly differ much in colour; but Dr. Günther believes that the male Hippocampi are rather brighter than the females. The genus Solenostoma,

---

[35] Yarrell, 'Hist. of British Fishes,' vol. ii. 1836, p. 329, 338.

however, offers a very curious exceptional case,[36] for the female is much more vividly coloured and spotted than the male, and she alone has a marsupial sack and hatches the eggs; so that the female of Solenostoma differs from all the other Lophobranchii in this latter respect, and from almost all other fishes, in being more brightly coloured than the male. It is improbable that this remarkable double inversion of character in the female should be an accidental coincidence. As the males of several fishes which take exclusive charge of the eggs and young are more brightly coloured than the females, and as here the female Solenostoma takes the same charge and is brighter than the male, it might be argued that the conspicuous colours of the sex which is the most important of the two for the welfare of the offspring must serve, in some manner, as a protection. But from the multitude of fishes, the males of which are either permanently or periodically brighter than the females, but whose life is not at all more important than that of the female for the welfare of the species, this view can hardly be maintained. When we treat of birds we shall meet with analogous cases in which there has been a complete inversion of the usual attributes of the two sexes, and we shall then give what appears to be the probable explanation, namely, that the males have selected the more attractive females, instead of the latter having selected, in accordance with the usual rule throughout the animal kingdom, the more attractive males.

On the whole we may conclude, that with most fishes, in which the sexes differ in colour or in other orna-

---

36 Dr. Günther, since publishing an account of this species in 'The Fishes of Zanzibar,' by Col. Playfair, 1866, p. 137, has re-examined the specimens, and has given me the above information.

mental characters, the males originally varied, with their variations transmitted to the same sex, and accumulated through sexual selection by attracting or exciting the females. In many cases, however, such characters have been transferred, either partially or completely, to the females. In other cases, again, both sexes have been coloured alike for the sake of protection; but in no instance does it appear that the female alone has had her colours or other characters specially modified for this purpose.

The last point which need be noticed is that in many parts of the world fishes are known to make peculiar noises, which are described in some cases as being musical. Very little has been ascertained with respect to the means by which such sounds are produced, and even less about their purpose. The drumming of the Umbrinas in the European seas is said to be audible from a depth of twenty fathoms. The fishermen of Rochelle assert "that the males alone make the noise "during the spawning-time; and that it is possible by "imitating it, to take them without bait."[37] If this statement is trustworthy, we have an instance in this, the lowest class of the Vertebrata, of what we shall find prevailing throughout the other vertebrate classes, and which prevails, as we have already seen, with insects and spiders; namely, that vocal and instrumental sounds so commonly serve as a love-call or as a love-charm, that the power of producing them was probably first developed in connection with the propagation of the species.

---

[37] The Rev. C. Kingsley, in 'Nature,' May, 1870, p. 40.

AMPHIBIANS.

Urodela.—First for the tailed amphibians. The sexes of salamanders or newts often differ much both in colour and structure. In some species prehensile claws are developed on the fore-legs of the males during the breeding-season; and at this season in the male *Triton palmipes* the hind-feet are provided with a swimming web, which is almost completely absorbed during the winter; so that their feet then

Fig. 31. Triton cristatus (half natural size, from Bell's 'British Reptiles'). Upper figure, male during the breeding-season; lower figure, female.

resemble those of the female.[38] This structure no doubt aids the male in his eager search and pursuit of the female. With our common newts (*Triton punctatus* and *cristatus*) a deep, much-indented crest is developed along the back and tail of the male during the breeding-season, being absorbed during the winter. It is not furnished, as Mr. St. George Mivart informs me,

---

[38] Bell, 'History of British Reptiles,' 2nd edit. 1849, p. 156-159.

with muscles, and therefore cannot be used for locomotion. As during the season of courtship it becomes edged with bright colours, it serves, there can hardly be a doubt, as a masculine ornament. In many species the body presents strongly contrasted, though lurid tints; and these become more vivid during the breeding-season. The male, for instance, of our common little newt (*Triton punctatus*) is "brownish-grey "above, passing into yellow beneath, which in the "spring becomes a rich bright orange, marked every-"where with round dark spots." The edge of the crest also is then tipped with bright red or violet. The female is usually of a yellowish-brown colour with scattered brown dots; and the lower surface is often quite plain.[39] The young are obscurely tinted. The ova are fertilised during the act of deposition and are not subsequently tended by either parent. We may therefore conclude that the males acquired their strongly-marked colours and ornamental appendages through sexual selection; these being transmitted either to the male offspring alone or to both sexes.

Anura or *Batrachia.*—With many frogs and toads the colours evidently serve as a protection, such as the bright green tints of tree-frogs and the obscure mottled shades of many terrestrial species. The most conspicuously coloured toad which I ever saw, namely the *Phryniscus nigricans*,[40] had the whole upper surface of the body as black as ink, with the soles of the feet and parts of the abdomen spotted with the brightest vermilion. It crawled about the bare sandy or open grassy plains of La Plata under a scorching sun, and

---

39 Bell, ibid. p. 146, 151.

40 'Zoology of the Voyage of the "Beagle,"' 1843. "Reptiles," by Mr. Bell, p. 49.

could not fail to catch the eye of every passing creature. These colours may be beneficial by making this toad known to all birds of prey as a nauseous mouthful; for it is familiar to every one that these animals emit a poisonous secretion, which causes the mouth of a dog to froth, as if attacked by hydrophobia. I was the more struck with the conspicuous colours of this toad, as close by I found a lizard (*Proctotretus multimaculatus*) which, when frightened, flattened its body, closed its eyes, and then from its mottled tints could hardly be distinguishable from the surrounding sand.

With respect to sexual differences of colour, Dr. Günther knows of no striking instance with frogs or toads; yet he can often distinguish the male from the female, by the tints of the former being a little more intense. Nor does Dr. Günther know of any striking difference in external structure between the sexes, excepting the prominences which become developed during the breeding-season on the front-legs of the male, by which he is enabled to hold the female. The *Megalophrys montana*[41] (fig. 32) offers the best case of a certain amount of structural difference between the sexes; for in the male the tip of the nose and the eyelids are produced into triangular flaps of skin, and there is a little black tubercle on the back—characters which are absent, or only feebly developed, in the females. It is surprising that frogs and toads should not have acquired more strongly-marked sexual differences; for though cold-blooded, their passions are strong. Dr. Günther informs me that he has several times found an unfortunate female toad dead and smothered from having been so closely embraced by three or four males.

---

41 'The Reptiles of India,' by Dr. A. Günther, Ray Soc. 1864, p. 413.

These animals, however, offer one interesting sexual difference, namely in the musical powers possessed by the males; but to speak of music, when applied to the discordant and overwhelming sounds emitted by male bull-frogs and some other species, seems, according to our taste, a singularly inappropriate expression. Nevertheless certain frogs sing in a decidedly pleasing manner. Near Rio de Janeiro I used often to sit in the evening to listen to a number of little Hylæ, which,

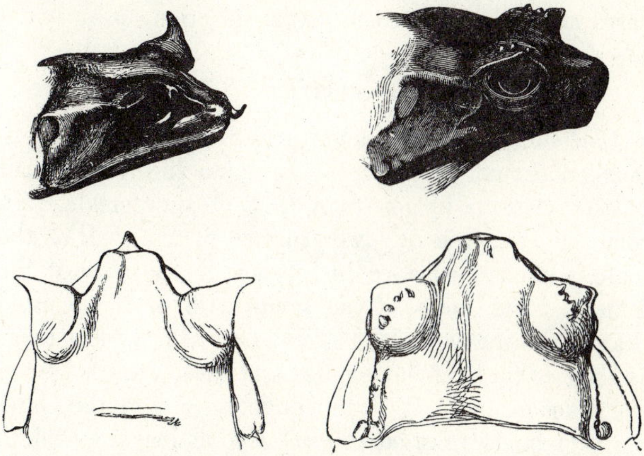

Fig. 32. Megalophrys montana. The two left-hand figures, the male; the two right-hand figures, the female.

perched on blades of grass close to the water, sent forth sweet chirping notes in harmony. The various sounds are emitted chiefly by the males during the breeding-season, as in the case of the croaking of our common frog.[42] In accordance with this fact the vocal organs of the males are more highly developed than those of the females. In some genera the males alone

[42] Bell, 'History of British Reptiles,' 1849, p. 93.

are provided with sacs which open into the larynx.[43] For instance, in the edible frog (*Rana esculenta*) "the "sacs are peculiar to the males, and become, when filled "with air in the act of croaking, large globular bladders, standing out one on each side of the head, near "the corners of the mouth." The croak of the male is thus rendered exceedingly powerful; whilst that of the female is only a slight groaning noise.[44] The vocal organs differ considerably in structure in the several genera of the family; and their development in all cases may be attributed to sexual selection.

REPTILES.

Chelonia.—Tortoises and turtles do not offer well-marked sexual differences. In some species, the tail of the male is longer than that of the female. In some, the plastron or lower surface of the shell of the male is slightly concave in relation to the back of the female. The male of the mud-turtle of the United States (*Chrysemys picta*) has claws on its front-feet twice as long as those of the female; and these are used when the sexes unite.[45] With the huge tortoise of the Galapagos Islands (*Testudo nigra*) the males are said to grow to a larger size than the females: during the pairing-season, and at no other time, the male utters a hoarse, bellowing noise, which can be heard at the distance of more than a hundred yards; the female, on the other hand, never uses her voice.[46]

Crocodilia.—The sexes apparently do not differ in

---

43 J. Bishop, in 'Todd's Cyclop. of Anat. and Phys.' vol. iv. p. 1503.

44 Bell, ibid. p. 112-114.

45 Mr. C. J. Maynard, 'The American Naturalist,' Dec. 1869, p. 555.

46 See my 'Journal of Researches during the Voyage of the "Beagle,"' 1845, p. 384.

colour; nor do I know that the males fight together, though this is probable, for some kinds make a prodigious display before the females. Bartram [47] describes the male alligator as striving to win the female by splashing and roaring in the midst of a lagoon, "swollen "to an extent ready to burst, with his head and tail "lifted up, he spins or twirls round on the surface of "the water, like an Indian chief rehearsing his feats "of war." During the season of love, a musky odour is emitted by the submaxillary glands of the crocodile, and pervades their haunts.[48]

Ophidia.—I have little to say about Snakes. Dr. Günther informs me that the males are always smaller than the females, and generally have longer and slenderer tails; but he knows of no other difference in external structure. In regard to colour, Dr. Günther can almost always distinguish the male from the female by his more strongly-pronounced tints; thus the black zigzag band on the back of the male English viper is more distinctly defined than in the female. The difference is much plainer in the Rattle-snakes of N. America, the male of which, as the keeper in the Zoological Gardens shewed me, can instantly be distinguished from the female by having more lurid yellow about its whole body. In S. Africa the *Bucephalus capensis* presents an analogous difference, for the female "is never so fully "variegated with yellow on the sides, as the male."[49] The male of the Indian *Dipsas cynodon,* on the other hand, is blackish-brown, with the belly partly black, whilst the female is reddish or yellowish-olive with the belly either uniform yellowish or marbled with black.

---

47 'Travels through Carolina,' &c., 1791, p. 128.
48 Owen, 'Anatomy of Vertebrates,' vol. i. 1866, p. 615.
49 Sir Andrew Smith, 'Zoolog. of S. Africa: Reptilia,' 1849, pl. x.

In the *Tragops dispar* of the same country, the male is bright green, and the female bronze-coloured.[50] No doubt the colours of some snakes serve as a protection, as the green tints of tree-snakes and the various mottled shades of the species which live in sandy places; but it is doubtful whether the colours of many kinds, for instance of the common English snake or viper, serve to conceal them; and this is still more doubtful with the many foreign species which are coloured with extreme elegance.

During the breeding-season their anal scent-glands are in active function;[51] and so it is with the same glands in lizards, and as we have seen with the submaxillary glands of crocodiles. As the males of most animals search for the females, these odoriferous glands probably serve to excite or charm the female, rather than to guide her to the spot where the male may be found.[52] Male snakes, though appearing so sluggish, are amorous; for many have been observed crowding round the same female, and even round the dead body of a female. They are not known to fight together from rivalry. Their intellectual powers are higher than might have been anticipated. An excellent observer in Ceylon, Mr. E. Layard,[53] saw a Cobra thrust its head through a narrow hole and swallow a toad. "With

---

50 Dr. A. Günther, 'Reptiles of British India,' Ray Soc. 1864, p. 304, 308.

51 Owen, 'Anatomy of Vertebrates,' vol. i. 1866, p. 615.

52 The celebrated botanist Schleiden incidently remarks ('Ueber den Darwinismus: Unsere Zeit,' 1869, s. 269), that Rattle-snakes use their rattles as a sexual call, by which the two sexes find each other. I do not know whether this suggestion rests on any direct observations. These snakes pair in the Zoological Gardens, but the keepers have never observed that they use their rattles at this season more than at any other.

53 "Rambles in Ceylon," 'Annals and Mag. of Nat. Hist.' 2nd series, vol. ix. 1852, p. 333.

"this incumbrance he could not withdraw himself; "finding this, he reluctantly disgorged the precious "morsel, which began to move off; this was too much "for snake philosophy to bear, and the toad was again "seized, and again was the snake, after violent efforts "to escape, compelled to part with its prey. This time, "however, a lesson had been learnt, and the toad was "seized by one leg, withdrawn, and then swallowed in "triumph."

It does not, however, follow because snakes have some reasoning power and strong passions, that they should likewise be endowed with sufficient taste to admire brilliant colours in their partners, so as to lead to the adornment of the species through sexual selection. Nevertheless it is difficult to account in any other manner for the extreme beauty of certain species; for instance, of the coral-snakes of S. America, which are of a rich red with black and yellow transverse bands. I well remember how much surprise I felt at the beauty of the first coral-snake which I saw gliding across a path in Brazil. Snakes coloured in this peculiar manner, as Mr. Wallace states on the authority of Dr. Günther,[54] are found nowhere else in the world except in S. America, and here no less than four genera occur. One of these, Elaps, is venomous; a second and widely-distinct genus is doubtfully venomous, and the two others are quite harmless. The species belonging to these distinct genera inhabit the same districts, and are so like each other, that no one "but a naturalist would distin-"guish the harmless from the poisonous kinds." Hence, as Mr. Wallace believes, the innocuous kinds have probably acquired their colours as a protection, on the principle of imitation; for they would naturally be

54 'Westminster Review,' July 1st, 1867, p. 32.

thought dangerous by their enemies. The cause, however, of the bright colours of the venomous Elaps remains to be explained, and this may perhaps be sexual selection.

Lacertilia.—The males of some, probably of many kinds of lizards fight together from rivalry. Thus the arboreal *Anolis cristatellus* of S. America is extremely pugnacious: "During the spring and early part of the "summer, two adult males rarely meet without a con-"test. On first seeing one another, they nod their heads "up and down three or four times, at the same time "expanding the frill or pouch beneath the throat; their "eyes glisten with rage, and after waving their tails "from side to side for a few seconds, as if to gather "energy, they dart at each other furiously, rolling over "and over, and holding firmly with their teeth. The "conflict generally ends in one of the combatants losing "his tail, which is often devoured by the victor." The male of this species is considerably larger than the female;[55] and this, as far as Dr. Günther has been able to ascertain, is the general rule with lizards of all kinds.

The sexes often differ greatly in various externa characters. The male of the above-mentioned Anolis is furnished with a crest, which runs along the back and tail, and can be erected at pleasure; but of this crest the female does not exhibit a trace. In the Indian *Cophotis ceylanica*, the female possesses a dorsal crest, though much less developed than in the male; and so it is, as Dr. Günther informs me, with the females of many Iguanas, Chamelions and other lizards. In some species, however, the crest is equally developed in both sexes, as in the *Iguana tuberculata*. In the genus

55 Mr. N. L. Austen kept these animals alive for a considerable time see 'Land and Water,' July, 1867, p. 9.

Sitana, the males alone are furnished with a large throat-pouch (fig. 33), which can be folded up like a fan, and is coloured blue, black, and red; but these splendid colours are exhibited only during the pairing-season. The female does not possess even a rudiment of this appendage. In the *Anolis cristatellus*, according to Mr. Austen, the throat-pouch, which is bright red marbled with yellow, is present, though in a rudimental condition, in the female. Again, in certain other lizards, both sexes are equally well provided with throat-pouches. Here, as in so many previous cases, we see with species belonging to the same group, the same character confined to the males, or more largely developed in the males than in the females, or equally developed in both sexes. The little lizards of the genus Draco, which glide through the air on their rib-supported parachutes, and which in the beauty of their colours baffle description, are furnished with skinny appendages to the throat, "like the wattles of gallinaceous birds." These become erected when the animal is excited. They occur in both sexes, but are best developed in the male when arrived at maturity, at which age the middle appendage is sometimes twice as long as the head. Most of the species likewise have a low crest running along the neck; and this is much more developed in the full-grown males, than in the females or young males.[56]

Fig. 33. Sitana minor. Male, with the gular pouch expanded (from Günther's 'Reptiles of India').

[56] All these statements and quotations, in regard to Cophotis, Sitana and Draco, as well as the following facts in regard to Ceratophora, are

There are other and much more remarkable differences between the sexes of certain lizards. The male of *Ceratophora aspera* bears on the extremity of his snout an appendage half as long as the head. It is cylindrical, covered with scales, flexible, and apparently capable of erection: in the female it is quite rudimental. In a second species of the same genus a terminal scale forms a minute horn on the summit of the flexible appendage; and in a third species (*C. Stoddartii,* fig. 34) the whole appendage is converted into a horn, which is usually of a white colour, but assumes a purplish tint when the animal is excited. In the adult male of this latter species the horn is half an inch in length, but is of quite minute size in the female and in the young. These appendages, as Dr. Günther has remarked to me, may be compared with the combs of gallinaceous birds, and apparently serve as ornaments.

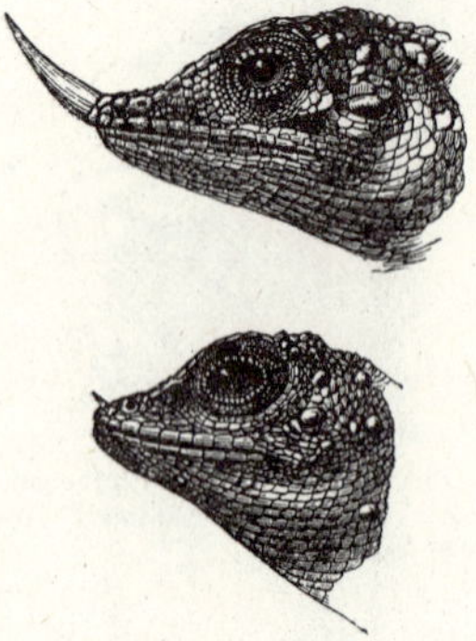

Fig. 34. Ceratophora Stoddartii. Upper figure, male; lower figure, female.

In the genus Chamæleon we come to the climax of difference between the sexes. The upper part of the skull of the male *C. bifurcus* (fig. 35), an inhabitant of Madagascar, is produced into two great, solid, bony projections, covered with scales like the rest of the head; and of this wonderful modification of structure the female exhibits only a rudiment. Again, in *Chamæleon Owenii* (fig. 36), from the West Coast of Africa, the male bears

taken from Dr. Günther's magnificent work on the 'Reptiles of British India,' Ray Soc. 1864, p. 122, 130, 135.

on his snout and forehead three curious horns, of which the female has not a trace. These horns consist of an excrescence of bone covered with a smooth sheath, forming part of the general integuments of the body, so that they are identical in structure with those of a

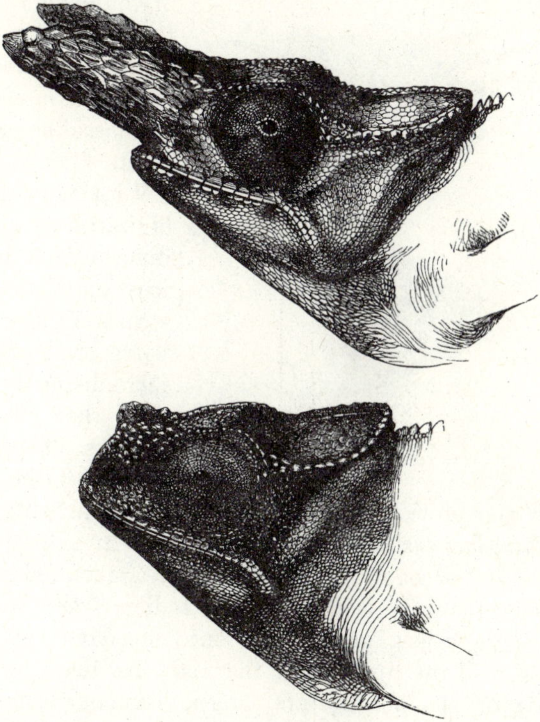

Fig. 35. Chamæleon bifurcus. Upper figure, male; lower figure, female.

bull, goat, or other sheath-horned ruminant. Although the three horns differ so much in appearance from the two great prolongations of the skull in *C. bifurcus,* we can hardly doubt that they serve the same general purpose in the economy of these two animals. The first conjecture which will occur to every one is that they are used by the males for fighting together; but Dr. Günther, to whom I am indebted for the foregoing details, does not believe that such peacable creatures would ever become pugnacious. Hence we are driven to infer that these almost monstrous deviations of structure serve as masculine ornaments.

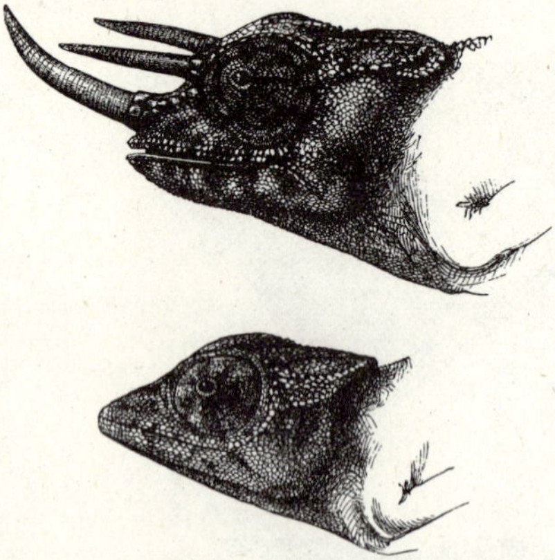

Fig. 36. Chamæleon Owenii. Upper figure, male; lower figure, female.

With many kinds of lizards, the sexes differ slightly in colour, the tints and stripes of the males being brighter and more distinctly defined than in the females. This, for instance, is the case with the previously-mentioned Cophotis and with the *Acanthodactylus capensis* of S. Africa. In a Cordylus of the latter country, the male is either much redder or greener than the female. In the Indian *Calotes nigrilabris* there is a greater difference in colour between the sexes; the lips also of the male are black, whilst those of the female are green. In our common little viviparous lizard (*Zootoca vivipara*) "the under "side of the body and base of the tail in the male are "bright orange, spotted with black; in the female "these parts are pale greyish-green without spots."[57] We have seen that the males alone of Sitana possess a

---

[57] Bell, 'History of British Reptiles,' 2nd edit. 1849, p. 40.

throat-pouch; and this is splendidly tinted with blue, black, and red. In the *Proctotretus tenuis* of Chile the male alone is marked with spots of blue, green, and coppery-red.[58] I collected in S. America fourteen species of this genus, and though I neglected to record the sexes, I observed that certain individuals alone were marked with emerald-like green spots, whilst others had orange-coloured gorges; and these in both cases no doubt were the males.

In the foregoing species, the males are more brightly coloured than the females, but with many lizards both sexes are coloured in the same elegant or even magnificent manner; and there is no reason to suppose that such conspicuous colours are protective. With some lizards, however, the green tints no doubt serve for concealment; and an instance has already been incidently given of one species of Proctotretus which closely resembles the sand on which it lives. On the whole we may conclude with tolerable safety that the beautiful colours of many lizards, as well as various appendages and other strange modifications of structure, have been gained by the males through sexual selection for the sake of ornament, and have been transmitted either to their male offspring alone or to both sexes. Sexual selection, indeed, seems to have played almost as important a part with reptiles as with birds. But the less conspicuous colours of the females in comparison with those of the males cannot be accounted for, as Mr. Wallace believes to be the case with birds, by the exposure of the females to danger during incubation.

---

58 For Proctotretus see 'Zoology of the Voyage of the "Beagle:" Reptiles,' by Mr. Bell, p. 8. For the Lizards of S. Africa, see 'Zoology of S. Africa: Reptiles,' by Sir Andrew Smith, pl. 25 and 39. For the Indian Calotes, see 'Reptiles of British India,' by Dr. Günther, p. 143.

CHAPTER XIII.

SECONDARY SEXUAL CHARACTERS OF BIRDS.

Sexual differences — Law of battle — Special weapons — Vocal organs — Instrumental music — Love-antics and dances — Decorations, permanent and seasonal — Double and single annual moults — Display of ornaments by the males.

SECONDARY sexual characters are more diversified and conspicuous in birds, though not perhaps entailing more important changes of structure, than in any other class of animals. I shall, therefore, treat the subject at considerable length. Male birds sometimes, though rarely, possess special weapons for fighting with each other. They charm the females by vocal or instrumental music of the most varied kinds. They are ornamented by all sorts of combs, wattles, protuberances, horns, air-distended sacs, topknots, naked shafts, plumes and lengthened feathers gracefully springing from all parts of the body. The beak and naked skin about the head, and the feathers are often gorgeously coloured. The males sometimes pay their court by dancing, or by fantastic antics performed either on the ground or in the air. In one instance, at least, the male emits a musky odour which we may suppose serves to charm or excite the female; for that excellent observer, Mr. Ramsay,[1] says of the Australian musk-duck (*Biziura lobata*) that "the "smell which the male emits during the summer "months is confined to that sex, and in some indi-"viduals is retained throughout the year; I have never

[1] 'Ibis,' vol. iii. (new series) 1867, p. 414.

"even in the breeding-season, shot a female which had "any smell of musk." So powerful is this odour during the pairing-season, that it can be detected long before the bird can be seen.[2] On the whole, birds appear to be the most æsthetic of all animals, excepting of course man, and they have nearly the same taste for the beautiful as we have. This is shewn by our enjoyment of the singing of birds, and by our women, both civilised and savage, decking their heads with borrowed plumes, and using gems which are hardly more brilliantly coloured than the naked skin and wattles of certain birds.

Before treating of the characters with which we are here more particularly concerned, I may just allude to certain differences between the sexes which apparently depend on differences in their habits of life; for such cases, though common in the lower, are rare in the higher classes. Two humming-birds belonging to the genus Eustephanus, which inhabit the island of Juan Fernandez, were long thought to be specifically distinct, but are now known, as Mr. Gould informs me, to be the sexes of the same species, and they differ slightly in the form of the beak. In another genus of humming-birds (*Grypus*), the beak of the male is serrated along the margin and hooked at the extremity, thus differing much from that of the female. In the curious Neomorpha of New Zealand, there is a still wider difference in the form of the beak; and Mr. Gould has been informed that the male with his "straight and stout beak" tears off the bark of trees, in order that the female may feed on the uncovered larvæ with her weaker and more curved beak. Something of the same kind may be observed with our goldfinch (*Carduelis elegans*), for I

---

[2] Gould, 'Handbook to the Birds of Australia,' 1865, vol. ii. p. 383.

am assured by Mr. J. Jenner Weir that the bird-catchers can distinguish the males by their slightly longer beaks. The flocks of males, as an old and trustworthy bird-catcher asserted, are commonly found feeding on the seeds of the teazle (Dipsacus) which they can reach with their elongated beaks, whilst the females more commonly feed on the seeds of the betony or Scrophularia. With a slight difference of this nature as a foundation, we can see how the beaks of the two sexes might be made to differ greatly through natural selection. In all these cases, however, especially in that of the quarrelsome humming-birds, it is possible that the differences in the beaks may have been first acquired by the males in relation to their battles, and afterwards led to slightly changed habits of life.

Law of Battle.—Almost all male birds are extremely pugnacious, using their beaks, wings, and legs for fighting together. We see this every spring with our robins and sparrows. The smallest of all birds, namely the humming-bird, is one of the most quarrelsome. Mr. Gosse[3] describes a battle, in which a pair of humming-birds seized hold of each other's beaks, and whirled round and round, till they almost fell to the ground; and M. Montes de Oca, in speaking of another genus, says that two males rarely meet without a fierce aerial encounter: when kept in cages "their fighting has mostly ended "in the splitting of the tongue of one of the two, which "then surely dies from being unable to feed."[4] With Waders, the males of the common water-hen (*Gallinula chloropus*) "when pairing, fight violently for the females: "they stand nearly upright in the water and strike "with their feet." Two were seen to be thus engaged

[3] Quoted by Mr. Gould, 'Introduction to the Trochilidæ,' 1861, p. 29.
[4] Gould, ibid. p. 52.

for half an hour, until one got hold of the head of the other which would have been killed, had not the observer interfered; the female all the time looking on as a quiet spectator.[5] The males of an allied bird (*Gallicrex cristatus*), as Mr. Blyth informs me, are one third larger than the females, and are so pugnacious during the breeding-season, that they are kept by the natives of Eastern Bengal for the sake of fighting. Various other birds are kept in India for the same purpose, for instance the Bulbuls (*Pycnonotus hæmorrhous*) which "fight with great spirit."[6]

The polygamous Ruff (*Machetes pugnax*, fig. 37) is notorious for his extreme pugnacity; and in the spring, the males, which are considerably larger than the females, congregate day after day at a particular spot, where the females propose to lay their eggs. The fowlers discover these spots by the turf being trampled somewhat bare. Here they fight very much like game-cocks, seizing each other with their beaks and striking with their wings. The great ruff of feathers round the neck is then erected, and according to Col. Montagu "sweeps the ground as a shield to defend the more "tender parts;" and this is the only instance known to me in the case of birds, of any structure serving as a shield. The ruff of feathers, however, from its varied and rich colours probably serves in chief part as an ornament. Like most pugnacious birds, they seem always ready to fight, and when closely confined often kill each other; but Montagu observed that their pugnacity becomes greater during the spring, when the long feathers on their necks are fully developed; and at this period the least movement by any one bird

[5] W. Thompson, 'Nat. Hist. of Ireland: Birds,' vol. ii. 1850, p. 327

[6] Jerdon, 'Birds of India,' 1863, vol. ii. p. 96.

provokes a general battle.[7] Of the pugnacity of web-footed birds, two instances will suffice: in Guiana "bloody "fights occur during the breeding-season between the

Fig 37. The Ruff or Machetes pugnax (from Brehm's 'Thierleben').

[7] Macgillivray, 'Hist. Brit. Birds,' vol. iv. 1852, p. 177-181.

"males of the wild musk-duck (*Cairina moschata*); "and where these fights have occurred the river "is covered for some distance with feathers."[8] Birds which seem ill-adapted for fighting engage in fierce conflicts; thus with the pelican the stronger males drive away the weaker ones, snapping with their huge beaks and giving heavy blows with their wings. Male snipes fight together, "tugging and pushing each "other with their bills in the most curious manner "imaginable." Some few species are believed never to fight; this is the case, according to Audubon, with one of the woodpeckers of the United States (*Picus auratus*), although "the hens are followed by even half a dozen "of their gay suitors."[9]

The males of many birds are larger than the females, and this no doubt is an advantage to them in their battles with their rivals, and has been gained through sexual selection. The difference in size between the two sexes is carried to an extreme point in several Australian species; thus the male musk-duck (Biziura) and the male *Cincloramphus cruralis* (allied to our pipits) are by measurement actually twice as large as their respective females.[10] With many other birds the females are larger than the males; and as formerly remarked, the explanation often given, namely that the females have most of the work in feeding their young, will not suffice. In some few cases, as we shall hereafter see, the females apparently have acquired their greater size and strength for the sake of conquering other females and obtaining possession of the males.

---

8 Sir R. Schomburgk, in 'Journal of R. Geograph. Soc.' vol. xiii. 1843, p. 31.

9 'Ornithological Biography,' vol. i. p. 191. For pelicans and snipes, see vol. iii. p. 381, 477.

10 Gould, 'Handbook of Birds of Australia,' vol. i. p. 395; vol. ii. p. 383.

The males of many gallinaceous birds, especially of the polygamous kinds, are furnished with special weapons for fighting with their rivals, namely spurs, which can be used with fearful effect. It has been recorded by a trustworthy writer[11] that in Derbyshire a kite struck at a game-hen accompanied by her chickens, when the cock rushed to the rescue and drove his spur right through the eye and skull of the aggressor. The spur was with difficulty drawn from the skull, and as the kite though dead retained his grasp, the two birds were firmly locked together; but the cock when disentangled was very little injured. The invincible courage of the game-cock is notorious: a gentleman who long ago witnessed the following brutal scene, told me that a bird had both its legs broken by some accident in the cock-pit, and the owner laid a wager that if the legs could be spliced so that the bird could stand upright, he would continue fighting. This was effected on the spot, and the bird fought with undaunted courage until he received his death-stroke. In Ceylon a closely-allied and wild species, the *Gallus Stanleyi*, is known to fight desperately "in "defence of his seraglio," so that one of the combatants is frequently found dead.[12] An Indian partridge (*Ortygornis gularis*), the male of which is furnished with strong and sharp spurs, is so quarrelsome, "that the "scars of former fights disfigure the breast of almost "every bird you kill."[13]

The males of almost all gallinaceous birds, even those which are not furnished with spurs, engage during the breeding-season in fierce conflicts. The Capercailzie and

---

[11] Mr. Hewitt in the 'Poultry Book by Tegetmeier,' 1866, p. 137.

[12] Layard, 'Annals and Mag. of Nat. Hist.' vol. xiv. 1854, p. 63.

[13] Jerdon, 'Birds of India,' vol. iii. p. 574.

Black-cock (*Tetrao urogallus* and *T. tetrix*), which are both polygamists, have regular appointed places, where during many weeks they congregate in numbers to fight together and to display their charms before the females. M. W. Kowalevsky informs me that in Russia he has seen the snow all bloody on the arenas where the Capercailzie have fought; and the Black-cocks "make the feathers fly in every direction," when several "engage in a battle royal." The elder Brehm gives a curious account of the Balz, as the love-dance and love-song of the Black-cock is called in Germany. The bird utters almost continuously the most strange noises: "he holds his tail up and spreads it out like a "fan, he lifts up his head and neck with all the feathers "erect, and stretches his wings from the body. Then "he takes a few jumps in different directions, some-"times in a circle, and presses the under part of his "beak so hard against the ground that the chin-feathers "are rubbed off. During these movements he beats "his wings and turns round and round. The more "ardent he grows the more lively he becomes, until at "last the bird appears like a frantic creature." At such times the black-cocks are so absorbed that they become almost blind and deaf, but less so than the capercailzie: hence bird after bird may be shot on the same spot, or even caught by the hand. After performing these antics the males begin to fight: and the same black-cock, in order to prove his strength over several antagonists, will visit in the course of one morning several Balz-places, which remain the same during successive years.[14]

[14] Brehm, 'Illust. Thierleben,' 1867, B. iv. s. 351. Some of the foregoing statements are taken from L. Lloyd, 'The Game Birds of Sweden,' &c., 1867, p. 79.

The peacock with his long train appears more like a dandy than a warrior, but he sometimes engages in fierce contests: the Rev. W. Darwin Fox informs me that two peacocks became so excited whilst fighting at some little distance from Chester that they flew over the whole city, still fighting, until they alighted on the top of St. John's tower.

The spur, in those gallinaceous birds which are thus provided, is generally single; but Polyplectron (see fig. 51, p. 90) has two or more on each leg; and one of the Blood-pheasants (*Ithaginis cruentus*) has been seen with five spurs. The spurs are generally confined to the male, being represented by mere knobs or rudiments in the female; but the females of the Java peacock (*Pavo muticus*) and, as I am informed by Mr. Blyth, of the small fire-backed pheasant (*Euplocamus erythropthalmus*) possess spurs. In Galloperdix it is usual for the males to have two spurs, and for the females to have only one on each leg.[15] Hence spurs may safely be considered as a masculine character, though occasionally transferred in a greater or less degree to the females. Like most other secondary sexual characters, the spurs are highly variable both in number and development in the same species.

Various birds have spurs on their wings. But the Egyptian goose (*Chenalopex ægyptiacus*) has only "bare "obtuse knobs," and these probably shew us the first steps by which true spurs have been developed in other allied birds. In the spur-winged goose, *Plectropterus gambensis*, the males have much larger spurs than the females; and they use them, as I am informed by Mr. Bartlett, in fighting together, so that, in this case, the

---

[15] Jerdon, 'Birds of India:' on Ithaginis, vol. iii. p. 523; on Galloperdix, p. 541.

wing-spurs serve as sexual weapons; but according to Livingstone, they are chiefly used in the defence of the young. The Palamedea (fig. 38) is armed with a pair of

Fig. 38. Palamedea cornuta (from Brehm), shewing the double-wing-spurs, and the filament on the head.

spurs on each wing; and these are such formidable weapons that a single blow has driven a dog howling away. But it does not appear that the spurs in this case, or in that of some of the spur-winged rails, are larger in the male than in the female.[16] In certain plovers, however, the wing-spurs must be considered as a sexual character. Thus in the male of our common peewit (*Vanellus cristatus*) the tubercle on the shoulder of the wing becomes more prominent during the breeding-season, and the males are known to fight together. In some species of Lobivanellus a similar tubercle becomes developed during the breeding-season "into a short horny spur." In the Australian *L. lobatus* both sexes have spurs, but these are much larger in the males than in the females. In an allied bird, the *Hoplopterus armatus*, the spurs do not increase in size during the breeding-season; but these birds have been seen in Egypt to fight together, in the same manner as our peewits, by turning suddenly in the air and striking sideways at each other, sometimes with a fatal result. Thus also they drive away other enemies.[17]

The season of love is that of battle; but the males of some birds, as of the game-fowl and ruff, and even the young males of the wild turkey and grouse,[18] are ready to fight whenever they meet. The presence of the female is the *teterrima belli causa*. The Bengali

---

16 For the Egyptian goose, see Macgillivray, 'British Birds,' vol. iv. p. 639. For Plectropterus, 'Livingstone's Travels,' p. 254. For Palamedea, Brehm's 'Thierleben,' B. iv. s. 740. See also on this bird Azara, 'Voyages dans l'Amérique mérid.' tom. iv. 1809, p. 179, 253.

17 See, on our peewit, Mr. R. Carr in 'Land and Water,' Aug. 8th, 1868, p. 46. In regard to Lobivanellus, see Jerdon's 'Birds of India,' vol. iii. p. 647, and Gould's 'Handbook of Birds of Australia,' vol. ii. p. 220. For the Holopterus, see Mr. Allen in the 'Ibis,' vol. v. 1863, p. 156.

18 Audubon, 'Ornith. Biography,' vol. ii. p. 492; vol. i. p. 4-13.

baboos make the pretty little males of the amadavat (*Estrelda amandava*) fight together by placing three small cages in a row, with a female in the middle; after a little time the two males are turned loose, and immediately a desperate battle ensues.[19] When many males congregate at the same appointed spot and fight together, as in the case of grouse and various other birds, they are generally attended by the females,[20] which afterwards pair with the victorious combatants. But in some cases the pairing precedes instead of succeeding the combat: thus, according to Audubon,[21] several males of the Virginian goat-sucker (*Caprimulgus Virginianus*) "court, in a highly entertaining "manner, the female, and no sooner has she made her "choice, than her approved gives chase to all intruders, "and drives them beyond his dominions." Generally the males try with all their power to drive away or kill their rivals before they pair. It does not, however, appear that the females invariably prefer the victorious males. I have indeed been assured by M. W. Kowalevsky that the female capercailzie sometimes steals away with a young male who has not dared to enter the arena with the older cocks; in the same manner as occasionally happens with the does of the red-deer in Scotland. When two males contend in presence of a single female, the victor, no doubt, commonly gains his

19 Mr. Blyth, 'Land and Water,' 1867, p. 212.

20 Richardson, on Tetrao umbellus, 'Fauna Bor. Amer.: Birds,' 1831, p. 343. L. Lloyd, 'Game Birds of Sweden,' 1867, p. 22, 79, on the capercailzie and black-cock. Brehm, however, asserts ('Thierleben,' &c., B. iv. s. 352) that in Germany the grey-hens do not generally attend the Balzen of the black-cocks, but this is an exception to the common rule; possibly the hens may lie hidden in the surrounding bushes, as is known to be the case with the grey-hens in Scandinavia, and with other species in N. America.

21 'Ornithological Biography,' vol. ii. p. 275.

desire; but some of these battles are caused by wandering males trying to distract the peace of an already mated pair.[22]

Even with the most pugnacious species it is probable that the pairing does not depend exclusively on the mere strength and courage of the male: for such males are generally decorated with various ornaments, which often become more brilliant during the breeding-season, and which are sedulously displayed before the females. The males also endeavour to charm or excite their mates by love-notes, songs, and antics; and the courtship is, in many instances, a prolonged affair. Hence it is not probable that the females are indifferent to the charms of the opposite sex, or that they are invariably compelled to yield to the victorious males. It is more probable that the females are excited, either before or after the conflict, by certain males, and thus unconsciously prefer them. In the case of *Tetrao umbellus*, a good observer[23] goes so far as to believe that the battles of the males "are all a sham, performed "to show themselves to the greatest advantage before "the admiring females who assemble around; for I "have never been able to find a maimed hero, and "seldom more than a broken feather." I shall have to recur to this subject, but I may here add that with the *Tetrao cupido* of the United States, about a score of males assemble at a particular spot, and strutting about make the whole air resound with their extraordinary noises. At the first answer from a female the males begin to fight furiously, and the weaker give way; but then, according to Audubon, both the victors and vanquished search for the female, so that the females must

[22] Brehm, 'Thierleben,' &c., B. iv. 1867, p. 990. Audubon, 'Ornith. Biography,' vol. ii. p. 492.

[23] 'Land and Water,' July 25th, 1868, p. 14.

either then exert a choice, or the battle must be renewed. So, again, with one of the Field-starlings of the United States (*Sturnella ludoviciana*) the males engage in fierce conflicts, "but at the sight of a female "they all fly after her, as if mad."[24]

Vocal and instrumental Music.—With birds the voice serves to express various emotions, such as distress, fear, anger, triumph, or mere happiness. It is apparently sometimes used to excite terror, as with the hissing noise made by some nestling-birds. Audubon[25] relates that a night-heron (*Ardea nycticorax*, Linn.) which he kept tame, used to hide itself when a cat approached, and then "suddenly start up uttering one of the most "frightful cries, apparently enjoying the cat's alarm "and flight." The common domestic cock clucks to the hen, and the hen to her chickens, when a dainty morsel is found. The hen, when she has laid an egg, "repeats the same note very often, and concludes with "the sixth above, which she holds for a longer time;"[26] and thus she expresses her joy. Some social birds apparently call to each other for aid; and as they flit from tree to tree, the flock is kept together by chirp answering chirp. During the nocturnal migrations of geese and other water-fowl, sonorous clangs from the van may be heard in the darkness overhead, answered by clangs in the rear. Certain cries serve as danger-signals, which, as the sportsman knows to his cost, are well understood by the same species and by others. The domestic cock crows, and the humming-bird chirps, in triumph over a defeated rival. The true song, how-

[24] Audubon's 'Ornitholog. Biography;' on Tetrao cupido, vol. ii. p. 492; on the Sturnus, vol. ii. p. 219.

[25] 'Ornithological Biograph.' vol. v. p. 601.

[26] The Hon. Daines Barrington, 'Philosoph. Transact.' 1773, p. 252.

ever, of most birds and various strange cries are chiefly uttered during the breeding-season, and serve as a charm, or merely as a call-note, to the other sex.

Naturalists are much divided with respect to the object of the singing of birds. Few more careful observers ever lived than Montagu, and he maintained that the "males "of song-birds and of many others do not in general "search for the female, but, on the contrary, their "business in the spring is to perch on some conspicuous "spot breathing out their full and amorous notes, which, "by instinct, the female knows, and repairs to the spot to "choose her mate."[27] Mr. Jenner Weir informs me that this is certainly the case with the nightingale. Bechstein, who kept birds during his whole life, asserts, "that "the female canary always chooses the best singer, and "that in a state of nature the female finch selects that "male out of a hundred whose notes please her most."[28] There can be no doubt that birds closely attend to each other's song. Mr. Weir has told me of the case of a bullfinch which had been taught to pipe a German waltz, and who was so good a performer that he cost ten guineas; when this bird was first introduced into a room where other birds were kept and he began to sing, all the others, consisting of about twenty linnets and canaries, ranged themselves on the nearest side of their cages, and listened with the greatest interest to the new performer. Many naturalists believe that the singing of birds is almost exclusively "the effect of ri- "valry and emulation," and not for the sake of charming their mates. This was the opinion of Daines Barrington and White of Selborne, who both especially attended to

---

27 'Ornithological Dictionary,' 1833, p. 475.

28 'Naturgeschichte der Stubenvögel,' 1840, s. 4. Mr. Harrison Weir likewise writes to me:—"I am informed that the best singing males "generally get a mate first when they are bred in the same room."

this subject.[29] Barrington, however, admits that "superiority in song gives to birds an amazing ascendancy "over others, as is well known to bird-catchers."

It is certain that there is an intense degree of rivalry between the males in their singing. Bird-fanciers match their birds to see which will sing longest; and I was told by Mr. Yarrell that a first-rate bird will sometimes sing till he drops down almost dead, or, according to Bechstein,[30] quite dead from rupturing a vessel in the lungs. Whatever the cause may be, male birds, as I hear from Mr. Weir, often die suddenly during the season of song. That the habit of singing is sometimes quite independent of love is clear, for a sterile hybrid canary-bird has been described[31] as singing whilst viewing itself in a mirror, and then dashing at its own image; it likewise attacked with fury a female canary when put into the same cage. The jealousy excited by the act of singing is constantly taken advantage of by bird-catchers; a male, in good song, is hidden and protected, whilst a stuffed bird, surrounded by limed twigs, is exposed to view. In this manner a man, as Mr. Weir informs me, has caught, in the course of a single day, fifty, and in one instance seventy, male chaffinches. The power and inclination to sing differ so greatly with birds that although the price of an ordinary male chaffinch is only sixpence, Mr. Weir saw one bird for which the bird-catcher asked three pounds; the test of a really good singer being that it will continue to sing whilst the cage is swung round the owner's head.

That birds should sing from emulation as well as for

---

29 'Philosophical Transactions,' 1773, p. 263. White's 'Natural History of Selborne,' vol. i. 1825, p. 246.

30 'Naturges. der Stubenvögel,' 1840, s. 252.

31 Mr. Bold, 'Zoologist,' 1843-44, p. 659.

the sake of charming the female, is not at all incompatible; and, indeed, might have been expected to go together, like decoration and pugnacity. Some authors, however, argue that the song of the male cannot serve to charm the female, because the females of some few species, such as the canary, robin, lark, and bullfinch, especially, as Bechstein remarks, when in a state of widowhood, pour forth fairly melodious strains. In some of these cases the habit of singing may be in part attributed to the females having been highly fed and confined,[32] for this disturbs all the usual functions connected with the reproduction of the species. Many instances have already been given of the partial transference of secondary masculine characters to the female, so that it is not at all surprising that the females of some species should possess the power of song. It has also been argued, that the song of the male cannot serve as a charm, because the males of certain species, for instance, of the robin, sing during the autumn.[33] But nothing is more common than for animals to take pleasure in practising whatever instinct they follow at other times for some real good. How often do we see birds which fly easily, gliding and sailing through the air obviously for pleasure. The cat plays with the captured mouse, and the cormorant with the captured fish. The weaver-bird (Ploceus), when confined in a cage, amuses itself by neatly weaving blades of grass between the wires of its cage. Birds which habitually fight during the breeding-season are generally ready to fight at all times; and the males of the capercailzie sometimes hold their *balzens* or *leks* at the usual place of

[32] D. Barrington, 'Phil. Transact.' 1773, p. 262. Bechstein, 'Stubenvögel,' 1840, s. 4.

[33] This is likewise the case with the water-ouzel, see Mr. Hepburn in the 'Zoologist,' 1845-1846, p. 1068.

assemblage during the autumn.[34] Hence it is not at all surprising that male birds should continue singing for their own amusement after the season for courtship is over.

Singing is to a certain extent, as shewn in a previous chapter, an art, and is much improved by practice. Birds can be taught various tunes, and even the unmelodious sparrow has learnt to sing like a linnet. They acquire the song of their foster-parents,[35] and sometimes that of their neighbours.[36] All the common songsters belong to the Order of Insessores, and their vocal organs are much more complex than those of most other birds; yet it is a singular fact that some of the Insessores, such as ravens, crows, and magpies, possess the proper apparatus,[37] though they never sing, and do not naturally modulate their voices to any great extent. Hunter asserts[38] that with the true songsters the muscles of the larynx are stronger in the males than in the females; but with this slight exception there is no difference in the vocal organs of the two sexes, although the males of most species sing so much better and more continuously than the females.

It is remarkable that only small birds properly sing. The Australian genus Menura, however, must be excepted; for the *Menura Alberti*, which is about the size of a half-grown turkey, not only mocks other birds, but "its own whistle is exceedingly beautiful and varied." The males congregate and form "*corroborying* places," where they sing, raising and spreading their tails like

34 L. Lloyd, 'Game Birds of Sweden,' 1867, p. 25.

35 Barrington, ibid. p. 264. Bechstein, ibid. s. 5.

36 Dureau de la Malle gives a curious instance ('Annales des Sc. Nat.' 3rd series, Zoolog. tom. x. p. 118) of some wild blackbirds in his garden in Paris which naturally learnt from a caged bird a republican air.

37 Bishop, in 'Todd's Cyclop. of Anat. and Phys.' vol. iv. p. 1496.

38 As stated by Barrington in 'Philosoph. Transact.' 1773, p. 262.

peacocks and drooping their wings.[39] It is also remarkable that the birds which sing are rarely decorated with brilliant colours or other ornaments. Of our British birds, excepting the bullfinch and goldfinch, the best songsters are plain-coloured. The king-fisher, bee-eater, roller, hoopee, woodpeckers, &c., utter harsh cries; and the brilliant birds of the tropics are hardly ever songsters.[40] Hence bright colours and the power of song seem to replace each other. We can perceive that if the plumage did not vary in brightness, or if bright colours were dangerous to the species, other means would have to be employed to charm the females; and the voice being rendered melodious would offer one such means.

In some birds the vocal organs differ greatly in the two sexes. In the *Tetrao cupido* (fig. 39) the male has two bare, orange-coloured sacks, one on each side of the neck; and these are largely inflated when the male, during the breeding-season, makes a curious hollow sound, audible at a great distance. Audubon proved that the sound was intimately connected with this apparatus, which reminds us of the air-sacks on each side of the mouth of certain male frogs, for he found that the sound was much diminished when one of the sacks of a tame bird was pricked, and when both were pricked it was altogether stopped. The female has "a some-" what similar, though smaller, naked space of skin on " the neck; but this is not capable of inflation."[41] The

39 Gould, 'Handbook to the Birds of Australia,' vol. i. 1865, p. 308-310. See also Mr. T. W. Wood in the 'Student,' April, 1870, p. 125.

40 See remarks to this effect in Gould's 'Introduction to the Trochilidæ,' 1861, p. 22.

41 'The Sportsman and Naturalist in Canada,' by Major W. Ross King, 1866, p. 144-146. Mr. T. W. Wood gives in the 'Student' (April, 1870, p. 116) an excellent account of the attitude and habits of this bird during its courtship. He states that the ear-tufts or neck-plumes are erected, so that they meet over the crown of the head.

male of another kind of grouse (*Tetrao urophasianus*), whilst courting the female, has his "bare yellow œso- "phagus inflated to a prodigious size, fully half as large "as the body;" and he then utters various grating,

Fig. 39. Tetrao cupido: male. (From Brehm.)

deep hollow tones. With his neck-feathers erect, his wings lowered and buzzing on the ground, and his long pointed tail spread out like a fan, he displays a variety of grotesque attitudes. The œsophagus of the female is not in any way remarkable.[42]

It seems now well made out that the great throat-pouch of the European male bustard (*Otis tarda*), and of at least four other species, does not serve, as was formerly supposed, to hold water, but is connected with the utterance during the breeding-season of a peculiar sound resembling "ock." The bird whilst uttering this sound throws himself into the most extraordinary attitudes. It is a singular fact that with the males of the same species the sack is not developed in all the individuals.[43] A crow-like bird inhabiting South America (*Cephalopterus ornatus*, fig. 40) is called the umbrella-bird, from its immense top-knot, formed of bare white quills surmounted by dark-blue plumes, which it can elevate into a great dome no less than five inches in diameter, covering the whole head. This bird has on its neck a long, thin, cylindrical, fleshy appendage, which is thickly clothed with scale-like blue feathers. It probably serves in part as an ornament, but likewise as a resounding apparatus, for Mr. Bates found that it is connected "with an unusual development of the trachea "and vocal organs." It is dilated when the bird utters its singularly deep, loud, and long-sustained fluty note.

---

42 Richardson, 'Fauna Bor. Americana: Birds,' 1831, p. 359. Audubon, ibid. vol. iv. p. 507.

43 The following papers have been lately written on this subject:—Prof. A. Newton, in the 'Ibis,' 1862, p. 107; Dr. Cullen, ibid. 1865, p. 145; Mr. Flower, in 'Proc. Zool. Soc.' 1865, p. 747; and Dr. Murie, in 'Proc. Zool. Soc.' 1868, p. 471. In this latter paper an excellent figure is given of the male Australian Bustard in full display with the sack distended.

The head-crest and neck-appendage are rudimentary in the female.[44]

The vocal organs of various web-footed and wading birds are extraordinarily complex, and differ to a certain extent in the two sexes. In some cases the trachea is

Fig. 40. The Umbrella-Bird or Cephalopterus ornatus (male, from Brehm).

convoluted, like a French horn, and is deeply embedded in the sternum. In the wild swan (*Cygnus ferus*) it is

[44] Bates, 'The Naturalist on the Amazons,' 1863, vol. ii. p. 284; Wallace, in 'Proc. Zool. Soc.' 1850, p. 206. A new species, with a still larger neck-appendage (*C. penduliger*), has lately been discovered, see 'Ibis,' vol. i. p. 457.

more deeply embedded in the adult male than in the female or young male. In the male Merganser the enlarged portion of the trachea is furnished with an additional pair of muscles.[45] But the meaning of these differences between the sexes of many Anatidæ is not at all understood; for the male is not always the more vociferous; thus with the common duck, the male hisses, whilst the female utters a loud quack.[46] In both sexes of one of the cranes (*Grus virgo*) the trachea penetrates the sternum, but presents "certain sexual modifications." In the male of the black stork there is also a well-marked sexual difference in the length and curvature of the bronchi.[47] So that highly important structures have in these cases been modified according to sex.

It is often difficult to conjecture whether the many strange cries and notes, uttered by male birds during the breeding-season, serve as a charm or merely as a call to the female. The soft cooing of the turtle-dove and of many pigeons, it may be presumed, pleases the female. When the female of the wild turkey utters her call in the morning, the male answers by a different note from the gobbling noise which he makes, when with erected feathers, rustling wings and distended wattles, he puffs and struts before her.[48] The *spel* of the black-cock certainly serves as a call to the female, for it has been known to bring four or five females

---

45 Bishop, in Todd's 'Cyclop. of Anat. and Phys.' vol. iv. p. 1499.

46 The spoonbill (Platalea) has its trachea convoluted into a figure of eight, and yet this bird (Jerdon, 'Birds of India,' vol. iii. p. 763) is mute; but Mr. Blyth informs me that the convolutions are not constantly present, so that perhaps they are now tending towards abortion.

47 'Elements of Comp. Anat.' by R. Wagner, Eng. translat. 1845, p. 111. With respect to the swan, as given above, Yarrell's 'Hist. of British Birds,' 2nd edit. 1845, vol. iii. p. 193.

48 C. L. Bonaparte, quoted in the 'Naturalist Library: Birds,' vol. xiv. p. 126.

from a distance to a male under confinement; but as the black-cock continues his *spel* for hours during successive days, and in the case of the capercailzie "with an agony of passion," we are led to suppose that the females which are already present are thus charmed.[49] The voice of the common rook is known to alter during the breeding-season, and is therefore in some way sexual.[50] But what shall we say about the harsh screams of, for instance, some kinds of macaws; have these birds as bad taste for musical sounds as they apparently have for colour, judging by the inharmonious contrast of their bright yellow and blue plumage? It is indeed possible that the loud voices of many male birds may be the result, without any advantage being thus gained, of the inherited effects of the continued use of their vocal organs, when they are excited by the strong passions of love, jealousy, and rage; but to this point we shall recur when we treat of quadrupeds.

We have as yet spoken only of the voice, but the males of various birds practise, during their courtship, what may be called instrumental music. Peacocks and Birds of Paradise rattle their quills together, and the vibratory movement apparently serves merely to make a noise, for it can hardly add to the beauty of their plumage. Turkey-cocks scrape their wings against the ground, and some kinds of grouse thus produce a buzzing sound. Another North American grouse, the *Tetrao umbellus*, when with his tail erect, his ruffs displayed, "he shows off his finery to the "females, who lie hid in the neighbourhood," drums rapidly with his "lowered wings on the trunk of a

---

[49] L. Lloyd, 'The Game Birds of Sweden,' &c., 1867, p. 22, 81.
[50] Jenner, 'Philosoph. Transactions,' 1824, p. 20.

"fallen tree," or, according to Audubon, against his own body; the sound thus produced is compared by some to distant thunder, and by others to the quick roll of a drum. The female never drums, "but flies directly to "the place where the male is thus engaged." In the Himalayas the male of the Kalij-pheasant "often makes "a singular drumming noise with his wings, not unlike "the sound produced by shaking a stiff piece of cloth." On the west coast of Africa the little black-weavers (Ploceus?) congregate in a small party on the bushes round a small open space, and sing and glide through the air with quivering wings, "which make a rapid "whirring sound like a child's rattle." One bird after another thus performs for hours together, but only during the courting-season. At this same season the males of certain night-jars (Caprimulgus) make a most strange noise with their wings. The various species of wood-peckers strike a sonorous branch with their beaks, with so rapid a vibratory movement that "the head appears "to be in two places at once." The sound thus produced is audible at a considerable distance, but cannot be described; and I feel sure that its cause would never be conjectured by any one who heard it for the first time. As this jarring sound is made chiefly during the breeding-season, it has been considered as a love-song; but it is perhaps more strictly a love-call. The female, when driven from her nest, has been observed thus to call her mate, who answered in the same manner and soon appeared. Lastly the male Hoopoe (*Upupa epops*) combines vocal and instrumental music; for during the breeding-season this bird, as Mr. Swinhoe saw, first draws in air and then taps the end of its beak perpendicularly down against a stone or the trunk of a tree, "when the breath being forced down the "tubular bill produces the correct sound." When the male utters its cry without striking his beak the sound is quite different.[51]

In the foregoing cases sounds are made by the aid of structures already present and otherwise necessary; but in the following cases certain feathers have been specially modified for the express purpose of producing the sounds. The drumming, or bleating, or neighing, or thundering noise, as expressed by different observers, which is made by the common snipe (*Scolopax gallinago*) must have surprised every one who has ever heard it. This bird, during the pairing-season, flies to "perhaps a "thousand feet in height," and after zig-zagging about for a time descends in a curved line, with outspread tail and quivering pinions, with surprising velocity to the

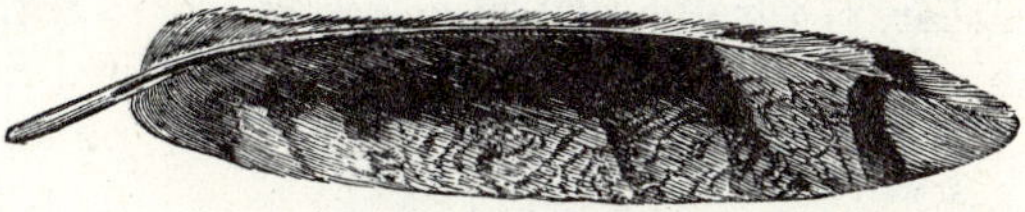

Fig. 41. Outer tail-feather of Scolopax gallinago (from Proc. Zool. Soc. 1858).

earth. The sound is emitted only during this rapid descent. No one was able to explain the cause, until M. Meves observed that on each side of the tail the outer feathers are peculiarly formed (fig. 41), having a stiff sabre-shaped shaft, with the oblique barbs of unusual length, the outer webs being strongly bound together.

[51] For the foregoing several facts see, on Birds of Paradise, Brehm, 'Thierleben,' Band iii. s. 325. On Grouse, Richardson, 'Fauna Bor. Americ.: Birds,' p. 343 and 359; Major W. Ross King, 'The Sportsman in Canada,' 1866, p. 156; Audubon, 'American Ornitholog. Biograph.' vol. i. p. 216. On the Kalij-pheasant, Jerdon, 'Birds of India,' vol. iii. p. 533. On the Weavers, 'Livingstone's Expedition to the Zambesi,' 1865, p. 425. On Woodpeckers, Macgillivray, 'Hist. of British Birds,' vol. iii. 1840, p. 84, 88, 89, and 95. On the Hoopoe, Mr. Swinhoe, in 'Proc. Zoolog. Soc.' June 23, 1863. On the Night-Jar, Audubon, ibid. vol. ii. p. 255. The English Night-Jar likewise makes in the spring a curious noise during its rapid flight.

He found that by blowing on these feathers, or by fastening them to a long thin stick and waving them rapidly through the air, he could exactly reproduce the drumming noise made by the living bird. Both sexes are furnished with these feathers, but they are generally larger in the male than in the female, and emit a deeper note. In some species, as in *S. frenata* (fig. 42), four feathers, and in *S. javensis* (fig. 43), no less than eight on each side of the tail are greatly modified. Different tones are emitted by the feathers of the different species when waved through the air; and the *Scolopax Wilsonii* of the United States makes a switching noise whilst descending rapidly to the earth.[52]

Fig. 42. Outer tail-feather of Scolopax frenata.

Fig. 43. Outer tail-feather of Scolopax javensis.

In the male of the *Chamæpetes unicolor* (a large gallinaceous bird of America) the first primary wing-feather is arched towards the tip and is much more attenuated than in the female. In an allied bird, the *Penelope nigra*, Mr. Salvin observed a male, which, whilst it flew downwards "with outstretched wings, gave forth "a kind of crashing, rushing noise," like the falling of a tree.[53] The male alone of one of the Indian bustards (*Sypheotides auritus*) has its primary wing-feathers greatly acuminated; and the male of an allied

52 See M. Meves' interesting paper in 'Proc. Zool. Soc.' 1858, p. 199. For the habits of the snipe, Macgillivray, 'Hist. British Birds,' vol. iv. p. 371. For the American snipe, Capt. Blakiston, 'Ibis,' vol. v. 1863, p. 131.

53 Mr. Salvin, in 'Proc. Zool. Soc.' 1867, p. 160. I am much indebted to this distinguished ornithologist for sketches of the feathers of the Chamæpetes, and for other information.

species is known to make a humming noise whilst courting the female.[54] In a widely different group of birds, namely the Humming-birds, the males alone of certain kinds have either the shafts of their primary wing-feathers broadly dilated, or the webs abruptly excised towards the extremity. The male, for instance, of *Selasphorus platycercus*, when adult, has the first primary wing-feather (fig. 44), excised in this manner. Whilst flying from flower to flower he makes "a shrill, almost whistling, noise;"[55] but it did not appear to Mr. Salvin that the noise was intentionally made.

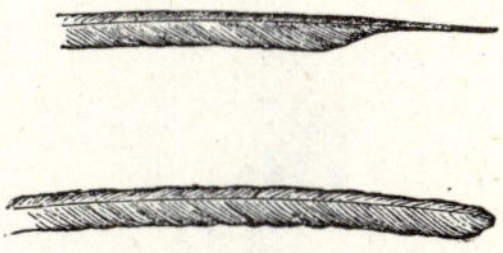

Fig. 44. Primary wing-feather of a Humming-bird, the *Selasphorus platycerus* (from a sketch by Mr. Salvin). Upper figure, that of male; lower figure, corresponding feather of female.

Lastly, in several species of a sub-genus of Pipra or Manakin, the males have their *secondary* wing-feathers modified, as described by Mr. Sclater, in a still more remarkable manner. In the brilliantly-coloured *P. deliciosa* the first three secondaries are thick-stemmed and curved towards the body; in the fourth and fifth (fig. 45, *a*) the change is greater; and in the sixth and seventh (*b*, *c*) the shaft "is thickened to an "extraordinary degree, forming a solid horny lump." The barbs also are greatly changed in shape, in comparison with the corresponding feathers (*d*, *e*, *f*) in the female. Even the bones of the wing which support these singular feathers in the male are said by Mr. Fraser to be much thickened. These little birds make

[54] Jerdon, 'Birds of India,' vol. iii. p. 618, 621.

[55] Gould, 'Introduction to the Trochilidæ,' 1861, p. 49. Salvin, 'Proc. Zoolog. Soc.' 1867, p. 160.

an extraordinary noise, the first "sharp note being not "unlike the crack of a whip."[56]

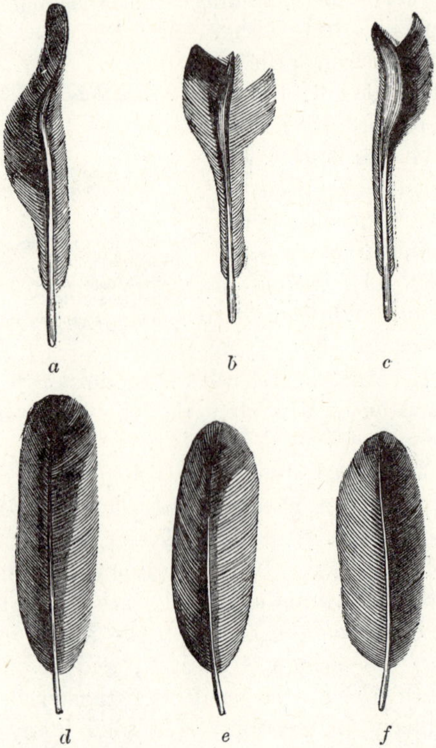

Fig. 45. Secondary wing-feathers of *Pipra deliciosa* (from Mr. Sclater, in Proc. Zool. Soc. 1860). The three upper feathers, *a, b, c*, from the male; the three lower corresponding feathers, *d, e, f*, from the female.

a. and *d.* Fifth secondary wing-feather of male and female, upper surface. *b* and *e.* Sixth secondary, upper surface. *c* and *f.* Seventh secondary, lower surface.

The diversity of the sounds, both vocal and instrumental, made by the males of many species during the

56 Sclater, in 'Proc. Zool. Soc.' 1860, p. 90, and in 'Ibis,' vol. iv. 1862, p. 175. Also Salvin, in 'Ibis,' 1860, p. 37.

breeding-season, and the diversity of the means for producing such sounds, are highly remarkable. We thus gain a high idea of their importance for sexual purposes, and are reminded of the same conclusion with respect to insects. It is not difficult to imagine the steps by which the notes of a bird, primarily used as a mere call or for some other purpose, might have been improved into a melodious love-song. This is somewhat more difficult in the case of the modified feathers, by which the drumming, whistling, or roaring noises are produced. But we have seen that some birds during their courtship flutter, shake, or rattle their unmodified feathers together; and if the females were led to select the best performers, the males which possessed the strongest or thickest, or most attenuated feathers, situated on any part of the body, would be the most successful; and thus by slow degrees the feathers might be modified to almost any extent. The females, of course, would not notice each slight successive alteration in shape, but only the sounds thus produced. It is a curious fact that in the same class of animals, sounds so different as the drumming of the snipe's tail, the tapping of the woodpecker's beak, the harsh trumpet-like cry of certain water-fowl, the cooing of the turtle-dove, and the song of the nightingale, should all be pleasing to the females of the several species. But we must not judge the tastes of distinct species by a uniform standard; nor must we judge by the standard of man's taste. Even with man, we should remember what discordant noises, the beating of tom-toms and the shrill notes of reeds, please the ears of savages. Sir S. Baker remarks,[57] that "as the stomach of the "Arab prefers the raw meat and reeking liver taken

[57] 'The Nile Tributaries of Abyssinia,' 1867, p. 203.

"hot from the animal, so does his ear prefer his equally "coarse and discordant music to all other."

Love-Antics and Dances.—The curious love-gestures of various birds, especially of the Gallinaceæ, have already been incidentally noticed; so that little need here be added. In Northern America, large numbers of a grouse, the *Tetrao phasianellus*, meet every morning during the breeding-season on a selected level spot, and here they run round and round in a circle of about fifteen or twenty feet in diameter, so that the ground is worn quite bare, like a fairy-ring. In these Partridge-dances, as they are called by the hunters, the birds assume the strangest attitudes, and run round, some to the left and some to the right. Audubon describes the males of a heron (*Ardea herodias*) as walking about on their long legs with great dignity before the females, bidding defiance to their rivals. With one of the disgusting carrion-vultures (*Cathartes jota*) the same naturalist states that "the gesticulations "and parade of the males at the beginning of the "love-season are extremely ludicrous." Certain birds perform their love-antics on the wing, as we have seen with the black African weaver, instead of on the ground. During the spring our little white-throat (*Sylvia cinerea*) often rises a few feet or yards in the air above some bush, and "flutters with a fitful and "fantastic motion, singing all the while, and then drops "to its perch." The great English bustard throws himself into indescribably odd attitudes whilst courting the female, as has been figured by Wolf. An allied Indian bustard (*Otis bengalensis*) at such times "rises "perpendicularly into the air with a hurried flapping "of his wings, raising his crest and puffing out the "feathers of his neck and breast, and then drops to the "ground;" he repeats this manœuvre several times successively, at the same time humming in a peculiar tone. Such females as happen to be near "obey this "saltatory summons," and when they approach he trails his wings and spreads his tail like a turkey-cock.[58]

But the most curious case is afforded by three allied genera of Australian birds, the famous Bower-birds, —no doubt the co-descendants of some ancient species which first acquired the strange instinct of constructing bowers for performing their love-antics. The bowers (fig. 46), which, as we shall hereafter see, are highly decorated with feathers, shells, bones and leaves, are built on the ground for the sole purpose of courtship, for their nests are formed in trees. Both sexes assist in the erection of the bowers, but the male is the principal workman. So strong is this instinct that it is practised under confinement, and Mr. Strange has described[59] the habits of some Satin Bower-birds, which he kept in his aviary in New South Wales. "At "times the male will chase the female all over the "aviary, then go to the bower, pick up a gay feather "or a large leaf, utter a curious kind of note, set all "his feathers erect, run round the bower and become "so excited that his eyes appear ready to start from "his head; he continues opening first one wing, and "then the other, uttering a low, whistling note, and,

[58] For Tetrao phasianellus, see Richardson, 'Fauna Bor. America,' p. 361, and for further particulars Capt. Blakiston, 'Ibis,' 1863, p. 125. For the Cathartes and Ardea, Audubon, 'Ornith. Biography,' vol. ii. p. 51, and vol iii. p. 89. On the White-throat, Macgillivray, 'Hist. British Birds,' vol. ii. p. 354. On the Indian Bustard, Jerdon, 'Birds of India,' vol. iii. p. 618.

[59] Gould, 'Handbook to the Birds of Australia,' vol. i. p. 444, 449, 455. The bower of the Satin Bower-bird may always be seen in the Zoological Society's Gardens, Regent's Park.

"like the domestic cock, seems to be picking up some-"thing from the ground, until at last the female goes "gently towards him." Captain Stokes has described the habits and "play-houses" of another species, the

Fig. 46. Bower-bird, Chlamydera maculata, with bower (from Brehm).

Great Bower-bird, which was seen "amusing itself by "flying backwards and forwards, taking a shell alter- "nately from each side, and carrying it through the "archway in its mouth." These curious structures, formed solely as halls of assemblages, where both sexes amuse themselves and pay their court, must cost the birds much labour. The bower, for instance, of the fawn-breasted species, is nearly four feet in length, eighteen inches in height, and is raised on a thick platform of sticks.

Decoration.—I will first discuss the cases in which the males are ornamented either exclusively or in a much higher degree than the females; and in a succeeding chapter those in which both sexes are equally ornamented, and finally the rare cases in which the female is somewhat more brightly-coloured than the male. As with the artificial ornaments used by savage and civilised men, so with the natural ornaments of birds, the head is the chief seat of decoration.[60] The ornaments, as mentioned at the commencement of this chapter, are wonderfully diversified. The plumes on the front or back of the head consist of variously-shaped feathers, sometimes capable of erection or expansion, by which their beautiful colours are fully displayed. Elegant ear-tufts (see fig. 39 ante) are occasionally present. The head is sometimes covered with velvety down like that of the pheasant; or is naked and vividly coloured; or supports fleshy appendages, filaments, and solid protuberances. The throat, also, is sometimes ornamented with a beard, or with wattles or caruncles. Such appendages are generally brightly coloured, and no doubt serve as

60 See remarks to this effect, on the "Feeling of Beauty among Animals," by Mr. J. Shaw, in the 'Athenæum,' Nov. 24th, 1866, p. 681.

ornaments, though not always ornamental in our eyes; for whilst the male is in the act of courting the female, they often swell and assume more vivid tints, as in the case of the male turkey. At such times the fleshy appendages about the head of the male Tragopan pheasant (*Ceriornis temminckii*) swell into a large lappet on the throat and into two horns, one on each side of the splendid top-knot; and these are then coloured of the most intense blue which I have ever beheld. The African hornbill (*Bucorax abyssinicus*) inflates the scarlet bladder-like wattle on its neck, and with its wings drooping and tail expanded "makes quite a grand "appearance."[61] Even the iris of the eye is sometimes more brightly coloured in the male than in the female; and this is frequently the case with the beak, for instance, in our common black-bird. In *Buceros corrugatus*, the whole beak and immense casque are coloured more conspicuously in the male than in the female; and "the oblique grooves upon the sides of "the lower mandible are peculiar to the male sex."[62]

The males are often ornamented with elongated feathers or plumes springing from almost every part of the body. The feathers on the throat and breast are sometimes developed into beautiful ruffs and collars. The tail-feathers are frequently increased in length; as we see in the tail-coverts of the peacock, and in the tail of the Argus pheasant. The body of this latter bird is not larger than that of a fowl; yet the length from the end of the beak to the extremity of the tail is no less than five feet three inches.[63] The wing-feathers are not elongated nearly so often as the tail-feathers; for their

---

61 Mr. Monteiro, 'Ibis,' vol. iv. 1862, p. 339.

62 'Land and Water,' 1868, p. 217.

63 Jardine's 'Naturalist Library: Birds,' vol. xiv. p. 166.

elongation would impede the act of flight. Yet the beautifully ocellated secondary wing-feathers of the male Argus pheasant are nearly three feet in length; and in a small African night-jar (*Cosmetornis vexillarius*) one of the primary wing-feathers, during the breeding-season, attains a length of twenty-six inches, whilst the bird itself is only ten inches in length. In another closely-allied genus of night-jars, the shafts of the elongated wing-feathers are naked, except at the extremity, where there is a disc.[64] Again, in another genus of nightjars, the tail-feathers are even still more prodigiously developed; so that we see the same kind of ornament gained by the males of closely-allied birds, through the development of widely different feathers.

It is a curious fact that the feathers of birds belonging to distinct groups have been modified in almost exactly the same peculiar manner. Thus the wing-feathers in one of the above-mentioned night-jars are bare along the shaft and terminate in a disc; or are, as they are sometimes called, spoon or racket-shaped. Feathers of this kind occur in the tail of a motmot (*Eumomota superciliaris*), of a king-fisher, finch, humming-bird, parrot, several Indian drongos (*Dicrurus* and *Edolius*, in one of which the disc stands vertically), and in the tail of certain Birds of Paradise. In these latter birds, similar feathers, beautifully ocellated, ornament the head, as is likewise the case with some gallinaceous birds. In an Indian bustard (*Sypheotides auritus*) the feathers forming the ear-tufts, which are about four inches in length, also terminate in discs.[65]

---

[64] Sclater, in the 'Ibis,' vol. vi. 1864, p. 114. Livingstone, 'Expedition to the Zambesi,' 1865, p. 66.

[65] Jerdon, 'Birds of India,' vol. iii. p. 620.

The barbs of the feathers in various widely-distinct birds are filamentous or plumose, as with some Herons, Ibises, Birds of Paradise and Gallinaceæ. In other cases the barbs disappear, leaving the shafts bare; and these in the tail of the *Paradisea apoda* attain a length of thirty-four inches.[66] Smaller feathers when thus denuded appear like bristles, as on the breast of the turkey-cock. As any fleeting fashion in dress comes to be admired by man, so with birds a change of almost any kind in the structure or colouring of the feathers in the male appears to have been admired by the female. The fact of the feathers in widely distinct groups, having been modified in an analogous manner, no doubt depends primarily on all the feathers having nearly the same structure and manner of development, and consequently tending to vary in the same manner. We often see a tendency to analogous variability in the plumage of our domestic breeds belonging to distinct species. Thus top-knots have appeared in several species. In an extinct variety of the turkey, the top-knot consisted of bare quills surmounted with plumes of down, so that they resembled, to a certain extent, the racket-shaped feathers above described. In certain breeds of the pigeon and fowl the feathers are plumose, with some tendency in the shafts to be naked. In the Sebastopol goose the scapular feathers are greatly elongated, curled, or even spirally twisted, with the margins plumose.[67]

In regard to colour hardly anything need here be said; for every one knows how splendid are the tints

---

66 Wallace, in 'Annals and Mag. of Nat. Hist.' vol. xx. 1857, p. 416; and in his 'Malay Archipelago,' vol. ii. 1869, p. 390.

67 See my work on 'The Variation of Animals and Plants under Domestication,' vol. i. p. 289, 293.

of birds, and how harmoniously they are combined. The colours are often metallic and iridescent. Circular spots are sometimes surrounded by one or more differently shaded zones, and are thus converted into ocelli.

Fig. 47. Paradisea rubra, male (from Brehm).

Nor need much be said on the wonderful differences between the sexes, or of the extreme beauty of the males of many birds. The common peacock offers a striking instance. Female Birds of Paradise are

Fig. 48. Lophornis ornatus, male and female (from Brehm).

obscurely coloured and destitute of all ornaments, whilst the males are probably the most highly decorated of all birds, and in so many ways, that they must be seen to be appreciated. The elongated and golden-

Fig. 49. Spathura underwoodi, male and female (from Brehm).

orange plumes which spring from beneath the wings of the *Paradisea apoda* (see fig. 47 of *P. rubra*, a much less beautiful species), when vertically erected and made to vibrate, are described as forming a sort of halo, in the centre of which the head "looks like a little "emerald sun with its rays formed by the two plumes."[68] In another most beautiful species the head is bald, "and of a rich cobalt blue, crossed by several lines of "black velvety feathers."[69]

Male humming-birds (figs. 48 and 49) almost vie with Birds of Paradise in their beauty, as every one will admit who has seen Mr. Gould's splendid volumes or his rich collection. It is very remarkable in how many different ways these birds are ornamented. Almost every part of the plumage has been taken advantage of and modified; and the modifications have been carried, as Mr. Gould shewed me, to a wonderful extreme in some species belonging to nearly every sub-group. Such cases are curiously like those which we see in our fancy breeds, reared by man for the sake of ornament: certain individuals originally varied in one character, and other individuals belonging to the same species in other characters; and these have been seized on by man and augmented to an extreme point—as the tail of the fantail-pigeon, the hood of the jacobin, the beak and wattle of the carrier, and so forth. The sole difference between these cases is that in the one the result is due to man's selection, whilst in the other, as with Humming-birds, Birds of Paradise, &c., it is due to sexual selection,—that is to the selection by the females of the more beautiful males.

---

68 Quoted from M. de Lafresnaye, in 'Annals and Mag. of Nat. Hist.' vol. xiii. 1854, p. 157: see also Mr. Wallace's much fuller account in vol. xx. 1857, p. 412, and in his Malay Archipelago.

69 Wallace, 'The Malay Archipelago,' vol. ii. 1869, p. 405.

I will mention only one other bird, remarkable from the extreme contrast in colour between the sexes, namely the famous Bell-bird (*Chasmorhynchus niveus*) of S. America, the note of which can be distinguished at the distance of nearly three miles, and astonishes every one who first hears it. The male is pure white, whilst the female is dusky-green; and the former colour with terrestrial species of moderate size and inoffensive habits is very rare. The male, also, as described by Waterton, has a spiral tube, nearly three inches in length, which rises from the base of the beak. It is jet-black, dotted over with minute downy feathers. This tube can be inflated with air, through a communication with the palate; and when not inflated hangs down on one side. The genus consists of four species, the males of which are very distinct, whilst the females, as described by Mr. Sclater in a most interesting paper, closely resemble each other, thus offering an excellent instance of the common rule that within the same group the males differ much more from each other than do the females. In a second species (*C. nudicollis*) the male is likewise snow-white, with the exception of a large space of naked skin on the throat and round the eyes, which during the breeding-season is of a fine green colour. In a third species (*C. tricarunculatus*) the head and neck alone of the male are white, the rest of the body being chesnut-brown, and the male of this species is provided with three filamentous projections half as long as the body—one rising from the base of the beak and the two others from the corners of the mouth.[70]

The coloured plumage and certain other ornaments of

[70] Mr. Sclater, 'Intellectual Observer,' Jan. 1867. 'Waterton's Wanderings,' p. 118. See also Mr. Salvin's interesting paper, with a plate, in the 'Ibis,' 1865, p. 90.

the males when adult are either retained for life or are periodically renewed during the summer and breeding-season. At this season the beak and naked skin about the head frequently change colour, as with some herons, ibises, gulls, one of the bell-birds just noticed, &c. In the white ibis, the cheeks, the inflatable skin of the throat, and the basal portion of the beak, then become crimson.[71] In one of the rails, *Gallicrex cristatus* a large red caruncle is developed during this same period on the head of the male. So it is with a thin horny crest on the beak of one of the pelicans, *P. erythrorhynchus*; for after the breeding-season, these horny crests are shed, like horns from the heads of stags, and the shore of an island in a lake in Nevada was found covered with these curious exuviæ.[72]

Changes of colour in the plumage according to the season depend firstly on a double annual moult, secondly on an actual change of colour in the feathers themselves, and thirdly on their dull-coloured margins being periodically shed, or on these three processes more or less combined. The shedding of the deciduary margins may be compared with the shedding by very young birds of their down; for the down in most cases arises from the summits of the first true feathers.[73]

With respect to the birds which annually undergo a double moult, there are, firstly, some kinds, for instance snipes, swallow-plovers (Glareolæ), and curlews, in which the two sexes resemble each other and do not change colour at any season. I do not know whether the winter-plumage is thicker and warmer than the

---

71 'Land and Water,' 1867, p. 394.

72 Mr. D. G. Elliot, in 'Proc. Zool. Soc.' 1869, p. 589.

73 'Nitzsch's Pterylography,' edited by P. L. Sclater. Ray Soc. 1867, p. 14.

summer-plumage, which seems, when there is no change of colour, the most probable cause of a double moult. Secondly, there are birds, for instance certain species of Totanus and other grallatores, the sexes of which resemble each other, but have a slightly different summer and winter plumage. The difference, however, in colour in these cases is so slight that it can hardly be an advantage to them; and it may, perhaps, be attributed to the direct action of the different conditions to which the birds are exposed during the two seasons. Thirdly, there are many other birds the sexes of which are alike, but which are widely different in their summer and winter plumage. Fourthly, there are birds, the sexes of which differ from each other in colour; but the females, though moulting twice, retain the same colours throughout the year, whilst the males undergo a change, sometimes, as with certain bustards, a great change of colour. Fifthly and lastly, there are birds the sexes of which differ from each other in both their summer and winter plumage, but the male undergoes a greater amount of change at each recurrent season than the female—of which the Ruff (*Machetes pugnax*) offers a good instance.

With respect to the cause or purpose of the differences in colour between the summer and winter plumage, this may in some instances, as with the ptarmigan,[74] serve during both seasons as a protection. When the difference between the two plumages is slight it may perhaps be attributed, as already remarked, to the

[74] The brown mottled summer plumage of the ptarmigan is of as much importance to it, as a protection, as the white winter plumage; for in Scandinavia, during the spring, when the snow has disappeared, this bird is known to suffer greatly from birds of prey, before it has acquired its summer dress: see Wilhelm von Wright, in Lloyd, 'Game Birds of Sweden,' 1867, p. 125.

direct action of the conditions of life. But with many birds there can hardly be a doubt that the summer plumage is ornamental, even when both sexes are alike. We may conclude that this is the case with many herons, egrets, &c., for they acquire their beautiful plumes only during the breeding-season. Moreover, such plumes, top-knots, &c., though possessed by both sexes, are occasionally a little more highly developed in the male than in the female; and they resemble the plumes and ornaments possessed by the males alone of other birds. It is also known that confinement, by affecting the reproductive system of male birds, frequently checks the development of their secondary sexual characters, but has no immediate influence on any other characters; and I am informed by Mr. Bartlett that eight or nine specimens of the Knot (*Tringa canutus*) retained their unadorned winter plumage in the Zoological Gardens throughout the year, from which fact we may infer that the summer plumage though common to both sexes partakes of the nature of the exclusively masculine plumage of many other birds.[75]

From the foregoing facts, more especially from neither sex of certain birds changing colour during either annual moult, or changing so slightly that the change can hardly be of any service to them, and from the females of other species moulting twice yet retaining the same colours throughout the year, we may conclude that the habit of moulting twice in the year has

---

[75] In regard to the previous statements on moulting, see, on snipes, &c., Macgillivray, 'Hist. Brit. Birds,' vol. iv. p. 371; on Glareolæ, curlews, and bustards, Jerdon, 'Birds of India,' vol. iii. p. 615, 630, 683; on Totanus, ibid, p. 700; on the plumes of herons, ibid, p. 738, and Macgillivray, vol. iv. p. 435 and 444, and Mr. Stafford Allen, in the 'Ibis,' vol. v. 1863, p. 33.

not been acquired in order that the male should assume during the breeding-season an ornamental character; but that the double moult, having been originally acquired for some distinct purpose, has subsequently been taken advantage of in certain cases for gaining a nuptial plumage.

It appears at first sight a surprising circumstance that with closely-allied birds, some species should regularly undergo a double annual moult, and others only a single one. The ptarmigan, for instance, moults twice or even thrice in the year, and the black-cock only once: some of the splendidly-coloured honey-suckers (Nectariniæ) of India and some sub-genera of obscurely-coloured pipits (Anthus) have a double, whilst others have only a single annual moult.[76] But the gradations in the manner of moulting, which are known to occur with various birds, shew us how species, or whole groups of species, might have originally acquired their double annual moult, or having once gained the habit, have again lost it. With certain bustards and plovers the vernal moult is far from complete, some feathers being renewed, and some changed in colour. There is also reason to believe that with certain bustards and rail-like birds, which properly undergo a double moult, some of the older males retain their nuptial plumage throughout the year. A few highly modified feathers may alone be added during the spring to the plumage, as occurs with the disc-formed tail-feathers of certain drongos (*Bhringa*) in India, and with the elongated feathers on the back, neck, and crest of certain herons. By such steps as these, the vernal moult might be ren-

[76] On the moulting of the ptarmigan, see Gould's 'Birds of Great Britain.' On the honey-suckers, Jerdon, 'Birds of India,' vol. i. p. 359, 365, 369. On the moulting of Anthus, see Blyth, in 'Ibis,' 1867, p. 32.

dered more and more complete, until a perfect double moult was acquired. A gradation can also be shewn to exist in the length of time during which either annual plumage is retained; so that the one might come to be retained for the whole year, the other being completely lost. Thus the *Machetes pugnax* retains his ruff in the spring for barely two months. The male widow-bird (*Chera progne*) acquires in Natal his fine plumage and long tail-feathers in December or January and loses them in March; so that they are retained during only about three months. Most species which undergo a double moult keep their ornamental feathers for about six months. The male, however, of the wild *Gallus bankiva* retains his neck-hackles for nine or ten months; and when these are cast off, the underlying black feathers on the neck are fully exposed to view. But with the domesticated descendant of this species, the neck-hackles of the male are immediately replaced by new ones; so that we here see, with respect to part of the plumage, a double moult changed under domestication into a single moult.[77]

The common drake (*Anas boschas*) is well known after the breeding-season to lose his male plumage for a period of three months, during which time he assumes that of the female. The male pintail-duck (*Anas acuta*) loses his plumage for the shorter period of six weeks or two months; and Montagu remarks that

---

77 For the foregoing statements in regard to partial moults, and on old males retaining their nuptial plumage, see Jerdon, on bustards and plovers, in 'Birds of India,' vol. iii. p. 617, 637, 709, 711. Also Blyth in 'Land and Water,' 1867, p. 84. On the Vidua, 'Ibis,' vol. iii. 1861, p. 133. On the Drongo shrikes, Jerdon, ibid. vol. i. p. 435. On the vernal moult of the *Herodias bubulcus*, Mr. S. S. Allen, in 'Ibis,' 1863, p. 33. On *Gallus bankiva*, Blyth, in 'Annals and Mag. of Nat. Hist.' vol. i. 1848, p. 455; see, also, on this subject, my 'Variation of Animals under Domestication,' vol. i. p. 236.

"this double moult within so short a time is a most "extraordinary circumstance, that seems to bid defiance "to all human reasoning." But he who believes in the gradual modification of species will be far from feeling surprise at finding gradations of all kinds. If the male pintail were to acquire his new plumage within a still shorter period, the new male feathers would almost necessarily be mingled with the old, and both with some proper to the female; and this apparently is the case with the male of a not distantly-allied bird, namely the *Merganser serrator*, for the males are said to "undergo a change of plumage, which assimilates them "in some measure to the female." By a little further acceleration in the process, the double moult would be completely lost.[78]

Some male birds, as before stated, become more brightly coloured in the spring, not by a vernal moult, but either by an actual change of colour in the feathers, or by their obscurely-coloured deciduary margins being shed. Changes of colour thus caused may last for a longer or shorter time. With the *Pelecanus onocrotalus* a beautiful rosy tint, with lemon-coloured marks on the breast, overspreads the whole plumage in the spring; but these tints, as Mr. Sclater states, "do not last long, dis-"appearing generally in about six weeks or two months "after they have been attained." Certain finches shed the margins of their feathers in the spring, and then become brighter-coloured, while other finches undergo no such change. Thus the *Fringilla tristis* of the United States (as well as many other American species), exhibits its bright colours only when the winter is past, whilst our goldfinch, which exactly represents this bird

---

[78] See Macgillivray, 'Hist. British Birds' (vol. v. p. 34, 70, and 223), on the moulting of the Anatidæ, with quotations from Waterton and Montagu. Also Yarrell, 'Hist. of British Birds,' vol. iii. p. 243.

in habits, and our siskin, which represents it still more closely in structure, undergo no such annual change. But a difference of this kind in the plumage of allied species is not surprising, for with the common linnet, which belongs to the same family, the crimson forehead and breast are displayed only during the summer in England, whilst in Madeira these colours are retained throughout the year.[79]

Display by Male Birds of their Plumage.—Ornaments of all kinds, whether permanently or temporarily gained, are sedulously displayed by the males, and apparently serve to excite, or attract, or charm the females. But the males will sometimes display their ornaments, when not in the presence of the females, as occasionally occurs with grouse at their balz-places, and as may be noticed with the peacock; this latter bird, however, evidently wishes for a spectator of some kind, and will shew off his finery, as I have often seen, before poultry or even pigs.[80] All naturalists who have closely attended to the habits of birds, whether in a state of nature or under confinement, are unanimously of opinion that the males delight to display their beauty. Audubon frequently speaks of the male as endeavouring in various ways to charm the female. Mr. Gould, after describing some peculiarities in a male humming-bird, says he has no doubt that it has the power of displaying them to the greatest advantage before the female. Dr. Jerdon[81]

---

79 On the pelican, see Sclater, in 'Proc. Zool. Soc.' 1868, p. 265. On the American finches, see Audubon, 'Ornith. Biography,' vol. i. p. 174, 221, and Jerdon, 'Birds of India,' vol. ii. p. 383. On the *Fringilla cannabina* of Madeira, Mr. E. Vernon Harcourt, 'Ibis,' vol. v., 1863, p. 230.

80 See also 'Ornamental Poultry,' by Rev. E. S. Dixon, 1848, p. 8.

81 'Birds of India,' introduct. vol. i. p. xxiv.; on the peacock, vol. iii. p. 507. See Gould's 'Introduction to the Trochilidæ,' 1861, p. 15 and 111.

insists that the beautiful plumage of the male serves "to fascinate and attract the female." Mr. Bartlett, at the Zoological Gardens, expressed himself to me in the strongest terms to the same effect.

It must be a grand sight in the forests of India "to "come suddenly on twenty or thirty pea-fowl, the males "displaying their gorgeous trains, and strutting about "in all the pomp of pride before the gratified females." The wild turkey-cock erects his glittering plumage, expands his finely-zoned tail and barred wing-feathers, and altogether, with his gorged crimson and blue wattles, makes a superb, though, to our eyes, grotesque appearance. Similar facts have already been given with respect to grouse of various kinds. Turning to another Order. The male *Rupicola crocea* (fig. 50) is one of the most beautiful birds in the world, being of a splendid orange, with some of the feathers curiously truncated and plumose. The female is brownish-green, shaded with red, and has a much smaller crest. Sir R. Schomburgk has described their courtship; he found one of their meeting-places where ten males and two females were present. The space was from four to five feet in diameter, and appeared to have been cleared of every blade of grass and smoothed as if by human hands. A male "was capering to the "apparent delight of several others. Now spreading "its wings, throwing up its head, or opening its tail "like a fan; now strutting about with a hopping gait "until tired, when it gabbled some kind of note, and "was relieved by another. Thus three of them suc-"cessively took the field, and then, with self-appro-"bation, withdrew to rest." The Indians, in order to obtain their skins, wait at one of the meeting-places till the birds are eagerly engaged in dancing, and then are able to kill, with their poisoned arrows, four or five males, one after the other.[82] With Birds of Paradise a dozen or more full-plumaged males congregate in a tree to hold a dancing-party, as it is called by the natives; and here flying about, raising their wings,

Fig. 50. Rupicola crocea, male (from Brehm).

elevating their exquisite plumes, and making them vibrate, the whole tree seems, as Mr. Wallace re-

[82] 'Journal of R. Geograph, Soc.' vol. x. 1840, p. 236.

marks, to be filled with waving plumes. When thus engaged, they become so absorbed that a skilful archer may shoot nearly the whole party. These birds, when kept in confinement in the Malay Archipelago, are said to take much care in keeping their feathers clean; often spreading them out, examining them, and removing every speck of dirt. One observer, who kept several pairs alive, did not doubt that the display of the male was intended to please the female.[83]

The gold-pheasant (*Thaumalea picta*) during his courtship not only expands and raises his splendid frill, but turns it, as I have myself seen, obliquely towards the female on whichever side she may be standing, obviously in order that a large surface may be displayed before her.[84] Mr. Bartlett has observed a male Polyplectron (fig. 51) in the act of courtship, and has shewn me a specimen stuffed in the attitude then assumed. The tail and wing-feathers of this bird are ornamented with beautiful ocelli, like those on the peacock's train. Now when the peacock displays himself, he expands and erects his tail transversely to his body, for he stands in front of the female, and has to shew off, at the same time, his rich blue throat and breast. But the breast of the Polyplectron is obscurely coloured, and the ocelli are not confined to the tail-feathers. Consequently the Polyplectron does not stand in front of the female; but he erects and expands his tail-feathers a little obliquely,

---

[83] 'Annals and Mag. of Nat. Hist.' vol. xiii. 1854, p. 157; also Wallace, ibid. vol. xx. 1857, p. 412, and 'The Malay Archipelago,' vol. ii. 1869, p. 252. Also Dr. Bennett, as quoted by Brehm, 'Thierleben,' B. iii. s. 326.

[84] Mr. T. W. Wood has given ('The Student,' April, 1870, p. 115) a full account of this manner of display, which he calls the lateral or one-sided, by the gold pheasant and by the Japanese pheasant, *Ph. versicolor*.

lowering the expanded wing on the same side, and raising that on the opposite side. In this attitude the ocelli over the whole body are exposed before the eyes of the admiring female in one grand bespangled ex-

Fig. 51. Polyplectron chinquis, male (from Brehm).

panse. To whichever side she may turn, the expanded wings and the obliquely-held tail are turned towards her. The male Tragopan pheasant acts in nearly the same manner, for he raises the feathers of the body, though not the wing itself, on the side which is opposite to the female, and which would otherwise be concealed, so that nearly all the beautifully-spotted feathers are exhibited at the same time.

The case of the Argus pheasant is still more striking. The immensely developed secondary wing-feathers, which are confined to the male, are ornamented with a row of from twenty to twenty-three ocelli, each above an inch in diameter. The feathers are also elegantly marked with oblique dark stripes and rows of spots, like those on the skin of a tiger and leopard combined. The ocelli are so beautifully shaded that, as the Duke of Argyll remarks,[85] they stand out like a ball lying loosely within a socket. But when I looked at the specimen in the British Museum, which is mounted with the wings expanded and trailing downwards, I was greatly disappointed, for the ocelli appeared flat or even concave. Mr. Gould, however, soon made the case clear to me, for he had made a drawing of a male whilst he was displaying himself. At such times the long secondary feathers in both wings are vertically erected and expanded; and these, together with the enormously elongated tail-feathers, make a grand semicircular upright fan. Now as soon as the wing-feathers are held in this position, and the light shines on them from above, the full effect of the shading comes out, and each ocellus at once resembles the ornament called a ball and socket. These feathers have been shewn to several artists, and all have expressed their admiration at the perfect shading.

[85] 'The Reign of Law,' 1867, p. 203.

It may well be asked, could such artistically-shaded ornaments have been formed by means of sexual selection? But it will be convenient to defer giving an answer to this question until we treat in the next chapter of the principle of gradation.

The primary wing-feathers, which in most gallinaceous birds are uniformly coloured, are in the Argus pheasant not less wonderful objects than the secondary wing-feathers. They are of a soft brown tint with numerous dark spots, each of which consists of two or three black dots with a surrounding dark zone. But the chief ornament is a space parallel to the dark-blue shaft, which in outline forms a perfect second feather lying within the true feather. This inner part is coloured of a lighter chesnut, and is thickly dotted with minute white points. I have shewn this feather to several persons, and many have admired it even more than the ball-and-socket feathers, and have declared that it was more like a work of art than of nature. Now these feathers are quite hidden on all ordinary occasions, but are fully displayed when the long secondary feathers are erected, though in a widely different manner; for they are expanded in front like two little fans or shields, one on each side of the breast near the ground.

The case of the male Argus pheasant is eminently interesting, because it affords good evidence that the most refined beauty may serve as a charm for the female, and for no other purpose. We must conclude that this is the case, as the primary wing-feathers are never displayed, and the ball-and-socket ornaments are not exhibited in full perfection, except when the male assumes the attitude of courtship. The Argus pheasant does not possess brilliant colours, so that his success in courtship appears to have depended on the great size of his plumes, and on the elaboration of the most elegant patterns. Many will declare that it is utterly incredible that a female bird should be able to appreciate fine shading and exquisite patterns. It is undoubtedly a marvellous fact that she should possess this almost human degree of taste, though perhaps she admires the general effect rather than each separate detail. He who thinks that he can safely gauge the discrimination and taste of the lower animals, may deny that the female Argus pheasant can appreciate such refined beauty; but he will then be compelled to admit that the extraordinary attitudes assumed by the male during the act of courtship, by which the wonderful beauty of his plumage is fully displayed, are purposeless; and this is a conclusion which I for one will never admit.

Although so many pheasants and allied gallinaceous birds carefully display their beautiful plumage before the females, it is remarkable, as Mr. Bartlett informs me, that this is not the case with the dull-coloured Eared and Cheer pheasants (*Crossoptilon auritum* and *Phasianus Wallichii*); so that these birds seem conscious that they have little beauty to display. Mr. Bartlett has never seen the males of either of these species fighting together, though he has not had such good opportunities for observing the Cheer as the Eared pheasant. Mr. Jenner Weir, also, finds that all male birds with rich or strongly-characterised plumage are more quarrelsome than the dull-coloured species belonging to the same groups. The goldfinch, for instance, is far more pugnacious than the linnet, and the black-bird than the thrush. Those birds which undergo a seasonal change of plumage likewise become much more pugnacious at the period when they are most gaily ornamented. No doubt the males of some obscurely-coloured birds fight desperately together, but it appears that when sexual selection has been highly influential, and has given bright colours to the males of any species, it has also very often given a strong tendency to pugnacity. We shall meet with nearly analogous cases when we treat of mammals. On the other hand, with birds the power of song and brilliant colours have rarely been both acquired by the males of the same species; but in this case, the advantage gained would have been identically the same, namely success in charming the female. Nevertheless it must be owned that the males of several brilliantly-coloured birds have had their feathers specially modified for the sake of producing instrumental music, though the beauty of this cannot be compared, at least according to our taste, with that of the vocal music of many songsters.

We will now turn to male birds which are not ornamented in any very high degree, but which nevertheless display, during their courtship, whatever attractions they may possess. These cases are in some respects more curious than the foregoing, and have been but little noticed. I owe the following facts, selected from a large body of valuable notes, sent to me by Mr. Jenner Weir, who has long kept birds of many kinds, including all the British Fringillidæ and Emberizidæ. The bullfinch makes his advances in front of the female, and then puffs out his breast, so that many more of the crimson feathers are seen at once than otherwise would be the case. At the same time he twists and bows his black tail from side to side in a ludicrous manner. The male chaffinch also stands in front of the female, thus shewing his red breast, and "blue bell," as the fanciers call his head; the wings at the same time being slightly expanded, with the pure white bands on the shoulders thus rendered conspicuous. The common linnet distends his rosy breast, slightly expands his brown wings and tail, so as to make the best of them by exhibiting their white edgings. We must, however, be cautious in concluding that the wings are spread out solely for display, as some birds act thus whose wings are not beautiful. This is the case with the domestic cock, but it is always the wing on the side opposite to the female which is expanded, and at the same time scraped on the ground. The male goldfinch behaves differently from all other finches: his wings are beautiful, the shoulders being black, with the dark-tipped wing-feathers spotted with white and edged with golden yellow. When he courts the female, he sways his body from side to side, and quickly turns his slightly expanded wings first to one side then to the other, with a golden flashing effect. No other British finch, as Mr. Weir informs me, turns during his courtship from side to side in this manner; not even the closely-allied male siskin, for he would not thus add to his beauty.

Most of the British Buntings are plain-coloured birds; but in the spring the feathers on the head of the male reed-bunting (*Emberiza schœniculus*) acquire a fine black colour by the abrasion of the dusky tips; and these are erected during the act of courtship. Mr. Weir has kept two species of Amadina from Australia: the *A. castanotis* is a very small and chastely-coloured finch, with a dark tail, white rump, and jet-black upper tail-coverts, each of the latter being marked with three large conspicuous oval spots of white.[86] This species, when courting the female, slightly spreads out and vibrates these parti-coloured tail-coverts in a very peculiar manner. The male *Amadina Lathami* behaves very

[86] For the description of these birds, see Gould's 'Handbook to the Birds of Australia,' vol. i. 1865, p. 417.

differently, exhibiting before the female his brilliantly-spotted breast and scarlet rump and scarlet upper tail-coverts. I may here add from Dr. Jerdon, that the Indian Bulbul (*Pycnonotus hæmorrhous*) has crimson *under* tail-coverts, and the beauty of these feathers, it might be thought, could never be well exhibited; but the bird "when excited often spreads them out laterally, "so that they can be seen even from above."[87] The common pigeon has iridescent feathers on the breast, and every one must have seen how the male inflates his breast whilst courting the female, thus showing off these feathers to the best advantage. One of the beautiful bronze-winged pigeons of Australia (*Ocyphaps lophotes*) behaves, as described to me by Mr. Weir, very differently: the male, whilst standing before the female, lowers his head almost to the ground, spreads out and raises perpendicularly his tail, and half expands his wings. He then alternately and slowly raises and depresses his body, so that the iridescent metallic feathers are all seen at once, and glitter in the sun.

Sufficient facts have now been given to shew with what care male birds display their various charms, and this they do with the utmost skill. Whilst preening their feathers, they have frequent opportunities for admiring themselves and of studying how best to exhibit their beauty. But as all the males of the same species display themselves in exactly the same manner, it appears that actions, at first perhaps intentional, have become instinctive. If so, we ought not to accuse birds of conscious vanity; yet when we see a peacock strutting about, with expanded and quivering tail-feathers, he seems the very emblem of pride and vanity.

The various ornaments possessed by the males are

---

87 'Birds of India,' vol. ii. p. 96.

certainly of the highest importance to them, for they have been acquired in some cases at the expense of greatly impeded powers of flight or of running. The African night-jar (*Cosmetornis*), which during the pairing-season has one of its primary wing-feathers developed into a streamer of extreme length, is thus much retarded in its flight, although at other times remarkable for its swiftness. The "unwieldy size" of the secondary wing-feathers of the male Argus pheasant are said "almost entirely to deprive the bird of flight." The fine plumes of male Birds of Paradise trouble them during a high wind. The extremely long tail-feathers of the male widow-birds (Vidua) of Southern Africa render "their flight heavy;" but as soon as these are cast off they fly as well as the females. As birds always breed when food is abundant, the males probably do not suffer much inconvenience in searching for food from their impeded powers of movement; but there can hardly be a doubt that they must be much more liable to be struck down by birds of prey. Nor can we doubt that the long train of the peacock and the long tail and wing-feathers of the Argus pheasant must render them a more easy prey to any prowling tiger-cat than would otherwise be the case. Even the bright colours of many male birds cannot fail to make them conspicuous to their enemies of all kinds. Hence it probably is, as Mr. Gould has remarked, that such birds are generally of a shy disposition, as if conscious that their beauty was a source of danger, and are much more difficult to discover or approach, than the sombre-coloured and comparatively tame females, or than the young and as yet unadorned males.[88]

---

[88] On the Cosmetornis, see Livingstone's 'Expedition to the Zambesi,' 1865, p. 66. On the Argus pheasant, Jardine's 'Nat. Hist. Lib.:

It is a more curious fact that the males of some birds which are provided with special weapons for battle, and which in a state of nature are so pugnacious that they often kill each other, suffer from possessing certain ornaments. Cock-fighters trim the hackles and cut off the comb and gills of their cocks; and the birds are then said to be dubbed. An undubbed, bird, as Mr. Tegetmeier insists, "is at a fearful disadvantage: the "comb and gills offer an easy hold to his adversary's "beak, and as a cock always strikes where he holds, "when once he has seized his foe, he has him entirely "in his power. Even supposing that the bird is not "killed, the loss of blood suffered by an undubbed cock "is much greater than that sustained by one that has "been trimmed."[89] Young turkey-cocks in fighting always seize hold of each other's wattles; and I presume that the old birds fight in the same manner. It may perhaps be objected that the comb and wattles are not ornamental, and cannot be of service to the birds in this way; but even to our eyes, the beauty of the glossy black Spanish cock is much enhanced by his white face and crimson comb; and no one who has ever seen the splendid blue wattles of the male Tragopan pheasant, when distended during the act of courtship, can for a moment doubt that beauty is the object gained. From the foregoing facts we clearly see that the plumes and other ornaments of the male must be of the highest importance to him; and we further see that beauty in some cases is even more important than success in battle.

Birds,' vol. xiv. p. 167. On Birds of Paradise, Lesson, quoted by Brehm, 'Thierleben,' B. iii. s. 325. On the widow-bird, Barrow's 'Travels in Africa,' vol. i. p. 243, and 'Ibis,' vol. iii. 1861, p. 133. Mr. Gould, on the shyness of male birds, 'Handbook to Birds of Australia,' vol. i. 1865, p. 210, 457.

89 Tegetmeier, 'The Poultry Book,' 1866, p. 139.

CHAPTER XIV.

BIRDS—*continued.*

Choice exerted by the female — Length of courtship — Unpaired birds — Mental qualities and taste for the beautiful — Preference or antipathy shewn by the female for particular males — Variability of birds — Variations sometimes abrupt — Laws of variation — Formation of ocelli — Gradations of character — Case of Peacock, Argus pheasant, and Urosticte.

WHEN the sexes differ in beauty, in the power of singing, or in producing what I have called instrumental music, it is almost invariably the male which excels the female. These qualities, as we have just seen, are evidently of high importance to the male. When they are gained for only a part of the year, this is always shortly before the breeding-season. It is the male alone who elaborately displays his varied attractions, and often performs strange antics on the ground or in the air, in the presence of the female. Each male drives away or, if he can, kills all his rivals. Hence we may conclude, that it is the object of the male to induce the female to pair with him, and for this purpose he tries to excite or charm her in various ways; and this is the opinion of all those who have carefully studied the habits of living birds. But there remains a question which has an all important bearing on sexual selection, namely, does every male of the same species equally excite and attract the female? or does she exert a choice, and prefer certain males? This question can be answered in the affirmative by much direct and indirect evidence. It is much more difficult to decide what qualities determine the choice of the females; but here again we have some direct and indirect evidence that it is to a large extent the external attractions of the male, though no doubt his vigour, courage, and other mental qualities come into play. We will begin with the indirect evidence.

Length of Courtship.—The lengthened period during which both sexes of certain birds meet day after day at an appointed place, probably depends partly on the courtship being a prolonged affair, and partly on the reiteration of the act of pairing. Thus in Germany and Scandinavia the balzens or leks of the Black-cocks, last from the middle of March, all through April into May. As many as forty or fifty, or even more birds congregate at the leks; and the same place is often frequented during successive years. The lek of the Capercailzie lasts from the end of March to the middle or even end of May. In North America "the partridge dances" of the *Tetrao phasianellus* "last for a month or more." Other kinds of grouse both in North America and Eastern Siberia[1] follow nearly the same habits. The fowlers discover the hillocks where the Ruffs congregate by the grass being trampled bare, and this shews that the same spot is long frequented. The Indians of Guiana are well acquainted with the cleared arenas, where they expect to find the beautiful Cocks of the Rock; and the natives of New Guinea know the trees where from ten to twenty full-plumaged male Birds of

[1] Nordmann describes ('Bull. Soc. Imp. des Nat. Moscow,' 1861, tom. xxxiv. p. 264) the balzen of *Tetrao urogalloides* in Amur Land. He estimated the number of assembled males at above a hundred, the females, which lie hid in the surrounding bushes, not being counted. The noises uttered differ from those of the *T. urogallus* or the capercailzie.

Paradise congregate. In this latter case it is not expressly stated that the females meet on the same trees, but the hunters, if not specially asked, would not probably mention their presence, as their skins are valueless. Small parties of an African weaver (*Ploceus*) congregate, during the breeding-season, and perform for hours their graceful evolutions. Large numbers of the Solitary snipe (*Scolopax major*) assemble during the dusk in a morass; and the same place is frequented for the same purpose during successive years; here they may be seen running about "like so many large rats," puffing out their feathers, flapping their wings, and uttering the strangest cries.[2]

Some of the above-mentioned birds, namely, the black-cock, capercailzie, pheasant-grouse, the ruff, the Solitary snipe, and perhaps some others, are, as it is believed, polygamists. With such birds it might have been thought that the stronger males would simply have driven away the weaker, and then at once have taken possession of as many females as possible; but if it be indispensable for the male to excite or please the female, we can understand the length of the courtship and the congregation of so many individuals of both sexes at the same spot. Certain species which are strictly monogamous likewise hold nuptial assemblages; this seems to be the case in Scandinavia with one of the ptarmigans, and their leks last from the middle of March to the middle of May. In Australia the lyre-bird or *Menura superba* forms "small round hillocks,"

---

[2] With respect to the assemblages of the above named grouse see Brehm, 'Thierleben,' B. iv. s. 350; also L. Lloyd, 'Game Birds of Sweden,' 1867, p. 19, 78. Richardson, 'Fauna Bor. Americana,' Birds, p. 362. References in regard to the assemblages of other birds have previously been given. On Paradisea see Wallace, in 'Annals and Mag. of Nat. Hist.' vol. xx. 1857, p. 412. On the snipe, Lloyd, ibid. p. 221.

and the *M. Alberti* scratches for itself shallow holes, or, as they are called by the natives, *corroborying places*, where it is believed both sexes assemble. The meetings of the *M. superba* are sometimes very large; and an account has lately been published[3] by a traveller, who heard in a valley beneath him, thickly covered with scrub, "a din which completely astonished" him; on crawling onwards he beheld to his amazement about one hundred and fifty of the magnificent lyre-cocks, "ranged in order of battle, and fighting with indescribable fury." The bowers of the Bower-birds are the resort of both sexes during the breeding-season; and "here the males meet and contend with each other "for the favours of the female, and here the latter "assemble and coquet with the males." With two of the genera, the same bower is resorted to during many years.[4]

The common magpie (*Corvus pica*, Linn.), as I have been informed by the Rev. W. Darwin Fox, used to assemble from all parts of Delamere Forest, in order to celebrate the "great magpie marriage." Some years ago these birds abounded in extraordinary numbers, so that a gamekeeper killed in one morning nineteen males, and another killed by a single shot seven birds at roost together. Whilst they were so numerous, they had the habit very early in the spring of assembling at particular spots, where they could be seen in flocks, chattering, sometimes fighting, bustling and flying about the trees. The whole affair was evidently considered by the birds as of the highest importance. Shortly after the meeting they all separated, and were then observed by Mr. Fox and others

[3] Quoted by Mr. T. W. Wood in the 'Student,' April, 1870, p. 125.

[4] Gould, 'Handbook of Birds of Australia,' vol. i. p. 300, 308, 448-451. On the ptarmigan, above alluded to, see Lloyd, ibid. p. 129.

to be paired for the season. In any district in which a species does not exist in large numbers, great assemblages cannot, of course, be held, and the same species may have different habits in different countries. For instance, I have never met with any account of regular assemblages of black game in Scotland, yet these assemblages are so well known in Germany and Scandinavia that they have special names.

Unpaired Birds.—From the facts now given, we may conclude that with birds belonging to widely-different groups their courtship is often a prolonged, delicate, and troublesome affair. There is even reason to suspect, improbable as this will at first appear, that some males and females of the same species, inhabiting the same district, do not always please each other and in consequence do not pair. Many accounts have been published of either the male or female of a pair having been shot, and quickly replaced by another. This has been observed more frequently with the magpie than with any other bird, owing perhaps to its conspicuous appearance and nest. The illustrious Jenner states that in Wiltshire one of a pair was daily shot no less than seven times successively, "but all to no purpose, "for the remaining magpie soon found another mate;" and the last pair reared their young. A new partner is generally found on the succeeding day; but Mr. Thompson gives the case of one being replaced on the evening of the same day. Even after the eggs are hatched, if one of the old birds is destroyed a mate will often be found; this occurred after an interval of two days, in a case recently observed by one of Sir J. Lubbock's keepers.[5] The first and most obvious

[5] On magpies, Jenner, in 'Phil. Transact.' 1824, p. 21. Macgillivray, 'Hist. British Birds,' vol. i. p. 570. Thompson, in 'Annals and Mag. of Nat. Hist.' vol. viii. 1842, p. 494.

conjecture is that male magpies must be much more numerous than the females; and that in the above cases, as well in many others which could be given, the males alone had been killed. This apparently holds good in some instances, for the gamekeepers in Delamere Forest assured Mr. Fox that the magpies and carrion-crows which they formerly killed in succession in large numbers near their nests were all males; and they accounted for this fact by the males being easily killed whilst bringing food to the sitting females. Macgillivray, however, gives, on the authority of an excellent observer, an instance of three magpies successively killed on the same nest which were all females; and another case of six magpies successively killed whilst sitting on the same eggs, which renders it probable that most of them were females, though the male will sit on the eggs, as I hear from Mr. Fox, when the female is killed.

Sir J. Lubbock's gamekeeper has repeatedly shot, but how many times he could not say, one of a pair of jays (*Garrulus glandarius*), and has never failed shortly afterwards to find the survivor rematched. The Rev. W. D. Fox, Mr. F. Bond, and others, have shot one of a pair of carrion-crows (*Corvus corone*), but the nest was soon again tenanted by a pair. These birds are rather common; but the peregrine falcon (*Falco peregrinus*) is rare, yet Mr. Thompson states that in Ireland "if "either an old male or female be killed in the breed-"ing-season (not an uncommon circumstance), another "mate is found within a very few days, so that the "eyries, notwithstanding such casualties, are sure to "turn out their complement of young." Mr. Jenner Weir has known the same thing to occur with the peregrine falcons at Beachy Head. The same observer informs me that three kestrels, all males (*Falco tinnun-*

culus), were killed one after the other whilst attending the same nest; two of these were in mature plumage, and the third in the plumage of the previous year. Even with the rare golden eagle (*Aquila chrysaëtos*), Mr. Birkbeck was assured by a trustworthy gamekeeper in Scotland, that if one is killed, another is soon found. So with the white owl (*Strix flammea*), it has been observed that "the survivor readily found a mate, and "the mischief went on."

White of Selborne, who gives the case of the owl, adds that he knew a man, who from believing that partridges when paired were disturbed by the males fighting, used to shoot them; and though he had widowed the same female several times she was always soon provided with a fresh partner. This same naturalist ordered the sparrows, which deprived the house-martins of their nests, to be shot: but the one which was left, "be it cock or hen, presently procured a mate, "and so for several times following." I could add analogous cases relating to the chaffinch, nightingale, and redstart. With respect to the latter bird (*Phœnicura ruticilla*), the writer remarks that it was by no means common in the neighbourhood, and he expresses much surprise how the sitting female could so soon give effectual notice that she was a widow. Mr. Jenner Weir has mentioned to me a nearly similar case: at Blackheath he never sees or hears the note of the wild bullfinch, yet when one of his caged males has died, a wild one in the course of a few days has generally come and perched near the widowed female, whose call-note is far from loud. I will give only one other fact, on the authority of this same observer; one of a pair of starlings (*Sturnus vulgaris*) was shot in the morning; by noon a new mate was found; this was again shot, but before night the pair was complete; so that the disconsolate widow or widower was thrice consoled during the same day. Mr. Engleheart also informs me that he used during several years to shoot one of a pair of starlings which built in a hole in a house at Blackheath; but the loss was always immediately repaired. During one season he kept an account and found that he had shot thirty-five birds from the same nest; these consisted of both males and females, but in what proportion he could not say: nevertheless after all this destruction, a brood was reared.[6]

These facts are certainly remarkable. How is it that so many birds are ready immediately to replace a lost mate? Magpies, jays, carrion-crows, partridges, and some other birds, are never seen during the spring by themselves, and these offer at first sight the most perplexing case. But birds of the same sex, although of course not truly paired, sometimes live in pairs or in small parties, as is known to be the case with pigeons and partridges. Birds also sometimes live in triplets, as has been observed with starlings, carrion-crows, parrots, and partridges. With partridges two females have been known to live with one male, and two males with one female. In all such cases it is probable that the union would be easily broken. The males of certain birds may occasionally be heard pouring forth their love-song long after the proper time, shewing that they have either lost or never gained a mate. Death from accident or disease of either one of a pair, would leave the other bird free and single; and there is reason to believe that female birds during the breeding-season

---

[6] On the peregrine falcon see Thompson, 'Nat. Hist. of Ireland: Birds,' vol. i. 1849, p. 39. On owls, sparrows, and partridges, see White, 'Nat. Hist. of Selborne,' edit. of 1825, vol. i. p. 139. On the Phœnicura, see Loudon's 'Mag. of Nat. Hist.' vol. vii. 1834, p. 245. Brehm, ('Thierleben,' B. iv. s. 991) also alludes to cases of birds thrice mated during same day.

are especially liable to premature death. Again, birds which have had their nests destroyed, or barren pairs, or retarded individuals, would easily be induced to desert their mates, and would probably be glad to take what share they could of the pleasures and duties of rearing offspring, although not their own.[7] Such contingencies as these probably explain most of the foregoing cases.[8] Nevertheless it is a strange fact that within the same district, during the height of the breeding-season, there should be so many males and females always ready to repair the loss of a mated bird. Why do not such spare birds immediately pair together? Have we not some reason to suspect, and the suspicion has occurred to Mr. Jenner Weir, that inasmuch as the act of courtship appears to be with many birds a prolonged and tedious affair, so it occasionally happens that certain males and females do not succeed during the proper season, in exciting each other's love, and consequently do not pair? This suspicion will appear somewhat less improbable after we have seen what

---

[7] See White ('Nat. Hist. of Selborne,' 1825, vol. i. p. 140) on the existence, early in the season, of small coveys of male partridges, of which fact I have heard other instances. See Jenner, on the retarded state of the generative organs in certain birds, in 'Phil. Transact.' 1824. In regard to birds living in triplets, I owe to Mr. Jenner Weir the cases of the starling and parrots, and to Mr. Fox, of partridges; on carrion-crows, see the 'Field,' 1868, p. 415. On various male birds singing after the proper period, see Rev. L. Jenyns, 'Observations in Natural History,' 1846, p. 87.

[8] The following case has been given ('The Times,' Aug. 6th, 1868) by the Rev. F. O. Morris, on the authority of the Hon. and Rev. O. W. Forester. "The gamekeeper here found a hawk's nest this year, with "five young ones in it. He took four and killed them, but left one "with its wings clipped as a decoy to destroy the old ones by. They "were both shot next day, in the act of feeding the young one, and "the keeper thought it was done with. The next day he came again "and found two other charitable hawks, who had come with an adopted "feeling to succour the orphan. These two he killed, and then left "the nest. On returning afterwards he found two more charitable

strong antipathies and preferences female birds occasionally evince towards particular males.

Mental Qualities of Birds, and their taste for the beautiful.—Before we discuss any further the question whether the females select the more attractive males or accept the first whom they may encounter, it will be advisable briefly to consider the mental powers of birds. Their reason is generally, and perhaps justly, ranked as low; yet some facts could be given [9] leading to an opposite conclusion. Low powers of reasoning, however, are compatible, as we see with mankind, with strong affections, acute perception, and a taste for the beautiful; and it is with these latter qualities that we are here concerned. It has often been said that parrots become so deeply attached to each other that when one dies the other for a long time pines; but Mr. Jenner Weir thinks that with most birds the strength of their affection has been much exaggerated. Nevertheless when one of a pair in a state of nature has been shot, the survivor has been heard for days afterwards uttering a plaintive call; and Mr. St. John gives [10] various facts proving the attachment of mated birds. Starlings, however, as we have seen, may be consoled thrice in the same day for the loss of their mates. In the Zoological Gardens parrots have clearly

"individuals on the same errand of mercy. One of these he killed; "the other he also shot, but could not find. No more came on the like "fruitless errand."

9 For instance, Mr. Yarrell states ('Hist. British Birds,' vol. iii. 1845, p. 585) that a gull was not able to swallow a small bird which had been given to it. The gull "paused for a moment, and then, as if suddenly "recollecting himself, ran off at full speed to a pan of water, shook the "bird about in it until well soaked, and immediately gulped it down. "Since that time he invariably has had recourse to the same expedient "in similar cases."

10 'A Tour in Sutherlandshire,' vol. i. 1849, p. 185.

recognised their former masters after an interval of some months. Pigeons have such excellent local memories that they have been known to return to their former homes after an interval of nine months, yet, as I hear from Mr. Harrison Weir, if a pair which would naturally remain mated for life be separated for a few weeks during the winter and matched with other birds, the two, when brought together again, rarely, if ever, recognise each other.

Birds sometimes exhibit benevolent feelings; they will feed the deserted young even of distinct species, but this perhaps ought to be considered as a mistaken instinct. They will also feed, as shewn in an earlier part of this work, adult birds of their own species which have become blind. Mr. Buxton gives a curious account of a parrot which took care of a frost-bitten and crippled bird of a distinct species, cleansed her feathers and defended her from the attacks of the other parrots which roamed freely about his garden. It is a still more curious fact that these birds apparently evince some sympathy for the pleasures of their fellows. When a pair of cockatoos made a nest in an acacia tree, "it "was ridiculous to see the extravagant interest taken "in the matter by the others of the same species." These parrots, also, evinced unbounded curiosity, and clearly had "the idea of property and possession."[11]

Birds possess acute powers of observation. Every mated bird, of course, recognises its fellow. Audubon states that with the mocking-thrushes of the United States (*Mimus polyglottus*) a certain number remain all the year round in Louisiana, whilst the others migrate to the Eastern States; these latter, on their return,

---

[11] Acclimatization of Parrots,' by C. Buxton, M.P. 'Annals and Mag. of Nat. Hist.' Nov. 1868, p. 381.

are instantly recognised, and always attacked, by their Southern brethren. Birds under confinement distinguish different persons, as is proved by the strong and permanent antipathy or affection which they shew, without any apparent cause, towards certain individuals. I have heard of numerous instances with jays, partridges, canaries, and especially bullfinches. Mr. Hussey has described in how extraordinary a manner a tamed partridge recognised everybody; and its likes and dislikes were very strong. This bird seemed "fond "of gay colours, and no new gown or cap could be put "on without catching his attention."[12] Mr. Hewitt has carefully described the habits of some ducks (recently descended from wild birds), which, at the approach of a strange dog or cat, would rush headlong into the water, and exhaust themselves in their attempts to escape; but they knew so well Mr. Hewitt's own dogs and cats that they would lie down and bask in the sun close to them. They always moved away from a strange man, and so they would from the lady who attended them, if she made any great change in her dress. Audubon relates that he reared and tamed a wild turkey which always ran away from any strange dog; this bird escaped into the woods, and some days afterwards Audubon saw, as he thought, a wild turkey, and made his dog chase it; but to his astonishment, the bird did not run away, and the dog, when he came up, did not attack the bird, for they mutually recognised each other as old friends.[13]

Mr. Jenner Weir is convinced that birds pay particular attention to the colours of other birds, sometimes

---

[12] 'The Zoologist,' 1847-1848, p. 1602.

[13] Hewitt on wild ducks, 'Journal of Horticulture,' Jan. 13, 1863, p. 39. Audubon on the wild turkey, 'Ornith. Biography,' vol. i. p. 14. On the mocking thrush, ibid. vol. i. p. 110.

out of jealousy, and sometimes as a sign of kinship. Thus he turned a reed-bunting (*Emberiza schœniculus*), which had acquired its black head, into his aviary, and the new-comer was not noticed by any bird, except by a bullfinch, which is likewise black-headed. This bullfinch was a very quiet bird, and had never before quarrelled with any of its comrades, including another reed-bunting, which had not as yet become black-headed: but the reed-bunting with a black head was so unmercifully treated, that it had to be removed. Mr. Weir was also obliged to turn out a robin, as it fiercely attacked all birds with any red in their plumage, but no other kinds; it actually killed a red-breasted crossbill, and nearly killed a goldfinch. On the other hand, he has observed that some birds, when first introduced into his aviary, fly towards the species which resemble them most in colour, and settle by their sides.

As male birds display with so much care their fine plumage and other ornaments in the presence of the females, it is obviously probable that these appreciate the beauty of their suitors. It is, however, difficult to obtain direct evidence of their capacity to appreciate beauty. When birds gaze at themselves in a looking-glass (of which many instances have been recorded) we cannot feel sure that it is not from jealousy at a supposed rival, though this is not the conclusion of some observers. In other cases it is difficult to distinguish between mere curiosity and admiration. It is perhaps the former feeling which, as stated by Lord Lilford,[14] attracts the Ruff strongly towards any bright object, so that, in the Ionian Islands, it "will dart down to a "bright-coloured handkerchief, regardless of repeated

[14] The 'Ibis,' vol. ii. 1860, p. 344.

"shots." The common lark is drawn down from the sky, and is caught in large numbers, by a small mirror made to move and glitter in the sun. Is it admiration or curiosity which leads the magpie, raven, and some other birds to steal and secrete bright objects, such as silver articles or jewels?

Mr. Gould states that certain humming-birds decorate the outside of their nests, "with the utmost taste; "they instinctively fasten thereon beautiful pieces of "flat lichen, the larger pieces in the middle, and the "smaller on the part attached to the branch. Now "and then a pretty feather is intertwined or fastened "to the outer sides, the stem being always so placed, "that the feather stands out beyond the surface." The best evidence, however, of a taste for the beautiful is afforded by the three genera of Australian bower-birds already mentioned. Their bowers (see fig. 46, p. 70), where the sexes congregate and play strange antics, are differently constructed, but what most concerns us is, that they are decorated in a different manner by the several species. The Satin bower-bird collects gaily-coloured articles, such as the blue tail-feathers of parrakeets, bleached bones and shells, which it sticks between the twigs, or arranges at the entrance. Mr. Gould found in one bower a neatly-worked stone tomahawk and a slip of blue cotton, evidently procured from a native encampment. These objects are continually rearranged, and carried about by the birds whilst at play. The bower of the Spotted bower-bird "is beautifully lined "with tall grasses, so disposed that the heads nearly "meet, and the decorations are very profuse." Round stones are used to keep the grass-stems in their proper places, and to make divergent paths leading to the bower. The stones and shells are often brought from a great distance. The Regent bird, as described by

Mr. Ramsay, ornaments its short bower with bleached land-shells belonging to five or six species, and with "berries of various colours, blue, red, and black, which "give it when fresh a very pretty appearance. Besides "these there were several newly-picked leaves and "young shoots of a pinkish colour, the whole shewing a "decided taste for the beautiful." Well may Mr. Gould say "these highly decorated halls of assembly must be "regarded as the most wonderful instances of bird-archi-"tecture yet discovered;" and the taste, as we see, of the several species certainly differs.[15]

Preference for particular Males by the Females.—Having made these preliminary remarks on the discrimination and taste of birds, I will give all the facts known to me, which bear on the preference shewn by the female for particular males. It is certain that distinct species of birds occasionally pair in a state of nature and produce hybrids. Many instances could be given: thus Macgillivray relates how a male blackbird and female thrush "fell in love with each other," and produced offspring.[16] Several years ago eighteen cases had been recorded of the occurrence in Great Britain of hybrids between the black grouse and pheasant;[17] but most of these cases may perhaps be accounted for by solitary birds not finding one of their own species to pair with. With other birds, as Mr. Jenner Weir has reason to believe, hybrids are sometimes the result of the casual intercourse of birds building in close proximity. But these

[15] On the ornamented nests of humming-birds, Gould, 'Introduction to the Trochilidæ,' 1861, p. 19. On the bower-birds, Gould, 'Handbook to the Birds of Australia,' 1865, vol. i. p. 444-461. Mr. Ramsay in the 'Ibis,' 1867, p. 456.

[16] 'Hist. of British Birds,' vol. ii. p. 92.

[17] 'Zoologist,' 1853-1854, p. 3946.

remarks do not apply to the many recorded instances of tamed or domestic birds, belonging to distinct species, which have become absolutely fascinated with each other, although living with their own species. Thus Waterton[18] states that out of a flock of twenty-three Canada geese, a female paired with a solitary Bernicle gander, although so different in appearance and size; and they produced hybrid offspring. A male Wigeon (*Mareca penelope*), living with females of the same species, has been known to pair with a Pintail duck, *Querquedula acuta*. Lloyd describes the remarkable attachment between a shield-drake (*Tadorna vulpanser*) and a common duck. Many additional instances could be given; and the Rev. E. S. Dixon remarks that "Those "who have kept many different species of geese to-"gether, well know what unaccountable attachments "they are frequently forming, and that they are quite "as likely to pair and rear young with individuals of a "race (species) apparently the most alien to themselves, "as with their own stock."

The Rev. W. D. Fox informs me that he possessed at the same time a pair of Chinese geese (*Anser cygnoides*), and a common gander with three geese. The two lots kept quite separate, until the Chinese gander seduced one of the common geese to live with him. Moreover, of the young birds hatched from the eggs of the common geese, only four were pure, the other eighteen proving hybrids; so that the Chinese gander seems to have had prepotent charms over the common gander. I will

---

18 Waterton, 'Essays on Nat. Hist.' 2nd series, p. 42, 117. For the following statements, see on the wigeon, Loudon's 'Mag. of Nat. Hist.' vol. ix. p. 616; L. Lloyd, 'Scandinavian Adventures,' vol. i. 1854, p. 452; Dixon, 'Ornamental and Domestic Poultry,' p. 137; Hewitt, in 'Journal of Horticulture,' Jan. 13, 1863, p. 40; Bechstein, 'Stubenvögel,' 1840, s. 230.

give only one other case; Mr. Hewitt states that a wild duck, reared in captivity, "after breeding a couple of "seasons with her own mallard, at once shook him off "on my placing a mail Pintail on the water. It was "evidently a case of love at first sight, for she swam "about the new-comer caressingly, though he appeared "evidently alarmed and averse to her overtures of "affection. From that hour she forgot her old partner. "Winter passed by, and the next spring the Pintail "seemed to have become a convert to her blandish-"ments, for they nested and produced seven or eight "young ones."

What the charm may have been in these several cases, beyond mere novelty, we cannot even conjecture. Colour, however, sometimes comes into play; for in order to raise hybrids from the siskin (*Fringilla spinus*) and the canary, it is much the best plan, according to Bechstein, to place birds of the same tint together. Mr. Jenner Weir turned a female canary into his aviary, where there were male linnets, goldfinches, siskins, green-finches, chaffinches, and other birds, in order to see which she would choose; but there never was any doubt, and the greenfinch carried the day. They paired and produced hybrid offspring.

With the members of the same species the fact of the female preferring to pair with one male rather than with another is not so likely to excite attention, as when this occurs between distinct species. Such cases can best be observed with domesticated or confined birds; but these are often pampered by high feeding, and sometimes have their instincts vitiated to an extreme degree. Of this latter fact I could give sufficient proofs with pigeons, and especially with fowls, but they cannot be here related. Vitiated instincts may also account for some of the hybrid unions above referred to; but in many of these cases the birds were allowed to range freely over large ponds, and there is no reason to suppose that they were unnaturally stimulated by high feeding.

With respect to birds in a state of nature, the first and most obvious supposition which will occur to everyone is that the female at the proper season accepts the first male whom she may encounter; but she has at least the opportunity for exerting a choice, as she is almost invariably pursued by many males. Audubon —and we must remember that he spent a long life in prowling about the forests of the United States and observing the birds—does not doubt that the female deliberately chooses her mate; th s, speaking of a woodpecker, he says the hen is followed by half-a-dozen gay suitors, who continue performing strange antics, "until "a marked preference is shewn for one." The female of the red-winged starling (*Agelæus phœniceus*) is likewise pursued by several males, "until, becoming fatigued, "she alights, receives their addresses, and soon makes "a choice." He describes also how several male nightjars repeatedly plunge through the air with astonishing rapidity, suddenly turning, and thus making a singular noise; "but no sooner has the female made "her choice, than the other males are driven away." With one of the vultures (*Cathartes aura*) of the United States, parties of eight or ten or more males and females assemble on fallen logs, "exhibiting the strongest desire "to please mutually," and after many caresses, each male leads off his partner on the wing. Audubon likewise carefully observed the wild flocks of Canada geese (*Anser Canadensis*), and gives a graphic description of their love-antics; he says that the birds which had been previously mated "renewed their courtship as early as the "month of January, while the others would be contend-

"ing or coquetting for hours every day, until all seemed "satisfied with the choice they had made, after which, "although they remained together, any person could "easily perceive that they were careful to keep in pairs. "I have observed also that the older the birds, the "shorter were the preliminaries of their courtship. "The bachelors and old maids, whether in regret, or "not caring to be disturbed by the bustle, quietly "moved aside and lay down at some distance from the "rest."[19] Many similar statements with respect to other birds could be cited from this same observer.

Turning now to domesticated and confined birds, I will commence by giving what little I have learnt respecting the courtship of fowls. I have received long letters on this subject from Messrs. Hewitt and Tegetmeier, and almost an essay from the late Mr. Brent. It will be admitted by every one that these gentlemen, so well known from their published works, are careful and experienced observers. They do not believe that the females prefer certain males on account of the beauty of their plumage; but some allowance must be made for the artificial state under which they have long been kept. Mr. Tegetmeier is convinced that a game-cock, though disfigured by being dubbed with his hackles trimmed, would be accepted as readily as a male retaining all his natural ornaments. Mr. Brent, however, admits that the beauty of the male probably aids in exciting the female; and her acquiescence is necessary. Mr. Hewitt is convinced that the union is by no means left to mere chance, for the female almost invariably prefers the most vigorous, defiant, and mettlesome male; hence it is almost useless, as he remarks, "to attempt

[19] Audubon, 'Ornitholog. Biography,' vol. i. p. 191, 349; vol. ii. p. 42, 275; vol. iii. p. 2.

"true breeding if a game-cock in good health and con-"dition runs the locality, for almost every hen on leaving "the roosting-place will resort to the game-cock, even "though that bird may not actually drive away the male "of her own variety." Under ordinary circumstances the males and females of the fowl seem to come to a mutual understanding by means of certain gestures, described to me by Mr. Brent. But hens will often avoid the officious attentions of young males. Old hens, and hens of a pugnacious disposition, as the same writer informs me, dislike strange males, and will not yield until well beaten into compliance. Ferguson, however, describes how a quarrelsome hen was subdued by the gentle courtship of a Shanghai cock.[20]

There is reason to believe that pigeons of both sexes prefer pairing with birds of the same breed; and dove-cot-pigeons dislike all the highly improved breeds.[21] Mr. Harrison Weir has lately heard from a trustworthy observer, who keeps blue pigeons, that these drive away all other coloured varieties, such as white, red, and yellow; and from another observer, that a female dun carrier could not be matched, after repeated trials, with a black male, but immediately paired with a dun. Generally colour alone appears to have little influence on the pairing of pigeons. Mr. Tegetmeier, at my request, stained some of his birds with magenta, but they were not much noticed by the others.

Female pigeons occasionally feel a strong antipathy towards certain males, without any assignable cause. Thus MM. Boitard and Corbié, whose experience extended over forty-five years, state: "Quand une femelle

---

20 'Rare and Prize Poultry,' 1854, p. 27.

21 'The Variation of Animals and Plants under Domestication,' vol. ii. p. 103.

"éprouve de l'antipathie pour un mâle avec lequel on "veut l'accoupler, malgré tous les feux de l'amour, "malgré l'alpiste et le chènevis dont on la nourrit "pour augmenter son ardeur, malgré un emprisonne- "ment de six mois et même d'un an, elle refuse con- "stamment ses caresses; les avances empressées, les "agaceries, les tournoiemens, les tendres roucoulemens, "rien ne peut lui plaire ni l'émouvoir; gonflée, bou- "deuse, blottie dans un coin de sa prison, elle n'en sort "que pour boire et manger, ou pour repousser avec une "espèce de rage des caresses devenues trop pressantes."[22] On the other hand, Mr. Harrison Weir has himself observed, and has heard from several breeders, that a female pigeon will occasionally take a strong fancy for a particular male, and will desert her own mate for him. Some females, according to another experienced observer, Riedel,[23] are of a profligate disposition, and prefer almost any stranger to their own mate. Some amorous males, called by our English fanciers "gay birds," are so successful in their gallantries, that, as Mr. H. Weir informs me, they must be shut up, on account of the mischief which they cause.

Wild turkeys in the United States, according to Audubon, "sometimes pay their addresses to the domes- "ticated females, and are generally received by them "with great pleasure." So that these females apparently prefer the wild to their own males.[24]

Here is a more curious case. Sir R. Heron during many years kept an account of the habits of the pea-fowl, which he bred in large numbers. He states that

[22] Boitard and Corbié, 'Les Pigeons,' 1824, p. 12. Prosper Lucas ('Traité de l'Héréd. Nat.' tom. ii. 1850, p. 296) has himself observed nearly similar facts with pigeons.

[23] 'Die Taubenzucht,' 1824, s. 86.

[24] 'Ornithological Biography,' vol. i. p. 13.

"the hens have frequently great preference to a par-
"ticular peacock. They were all so fond of an old pied "cock, that one year, when he was confined though "still in view, they were constantly assembled close "to the trellice-walls of his prison, and would not suffer "a japanned peacock to touch them. On his being let "out in the autumn, the oldest of the hens instantly "courted him, and was successful in her courtship. "The next year he was shut up in a stable, and then "the hens all courted his rival."[25] This rival was a japanned or black-winged peacock, which to our eyes is a more beautiful bird than the common kind.

Lichtenstein, who was a good observer and had excellent opportunities of observation at the Cape of Good Hope, assured Rudolphi that the female widow-bird (*Chera progne*) disowns the male, when robbed of the long tail-feathers with which he is ornamented during the breeding-season. I presume that this observation must have been made on birds under confinement.[26] Here is another striking case; Dr. Jaeger,[27] director of the Zoological Gardens of Vienna, states that a male silver-pheasant, who had been triumphant over the other males and was the accepted lover of the females, had his ornamental plumage spoiled. He was then immediately superseded by a rival, who got the upper hand and afterwards led the flock.

Not only does the female exert a choice, but in some few cases she courts the male, or even fights for his possession. Sir R. Heron states that with peafowl, the

---

25 'Proc. Zool. Soc.' 1835, p. 54. The japanned peacock is considered by Mr. Sclater as a distinct species, and has been named *Pavo nigripennis*.

26 Rudolphi, 'Beyträge zur Anthropologie,' 1812, s. 184.

27 'Die Darwin'sche Theorie, und ihre Stellung zu Moral und Religion,' 1869, s. 59.

first advances are always made by the female; something of the same kind takes place, according to Audubon, with the older females of the wild turkey. With the capercailzie, the females flit round the male, whilst he is parading at one of the places of assemblage, and solicit his attention.[28] We have seen that a tame wild-duck seduced after a long courtship an unwilling Pintail drake. Mr. Bartlett believes that the Lophophorus, like many other gallinaceous birds, is naturally polygamous, but two females cannot be placed in the same cage with a male, as they fight so much together. The following instance of rivalry is more surprising as it relates to bullfinches, which usually pair for life. Mr. Jenner Weir introduced a dull-coloured and ugly female into his aviary, and she immediately attacked another mated female so unmercifully that the latter had to be separated. The new female did all the courtship, and was at last successful, for she paired with the male; but after a time she met with a just retribution, for, ceasing to be pugnacious, Mr. Weir replaced the old female, and the male then deserted his new and returned to his old love.

In all ordinary cases the male is so eager that he will accept any female, and does not, as far as we can judge, prefer one to the other; but exceptions to this rule, as we shall hereafter see, apparently occur in some few groups. With domesticated birds, I have heard of only one case in which the males shew any preference for particular females, namely, that of the domestic cock, who, according to the high authority of Mr. Hewitt, prefers the younger to the older hens. On the other

---

[28] In regard to peafowl, see Sir R. Heron, 'Proc. Zoolog. Soc.' 1835, p. 54, and the Rev. E. S. Dixon, 'Ornamental Poultry,' 1848, p. 8. For the turkey, Audubon, ibid. p. 4. For the capercailzie, Lloyd, 'Game Birds of Sweden,' 1867, p. 23.

hand, in effecting hybrid unions between the male pheasant and common hens, Mr. Hewitt is convinced that the pheasant invariably prefers the older birds. He does not appear to be in the least influenced by their colour, but "is most "capricious in his attachments."[29] From some inexplicable cause he shews the most determined aversion to certain hens, which no care on the part of the breeder can overcome. Some hens, as Mr. Hewitt informs me, are quite unattractive even to the males of their own species, so that they may be kept with several cocks during a whole season, and not one egg out of forty or fifty will prove fertile. On the other hand with the Long-tailed duck (*Harelda glacialis*), "it has been remarked," says M. Ekström, "that certain females are much more "courted than the rest. Frequently, indeed, one sees "an individual surrounded by six or eight amorous "males." Whether this statement is credible, I know not; but the native sportsmen shoot these females in order to stuff them as decoys.[30]

With respect to female birds feeling a preference for particular males, we must bear in mind that we can judge of choice being exerted, only by placing ourselves in imagination in the same position. If an inhabitant of another planet were to behold a number of young rustics at a fair, courting and quarrelling over a pretty girl, like birds at one of their places of assemblage, he would be able to infer that she had the power of choice only by observing the eagerness of the wooers to please her, and to display their finery. Now with birds, the evidence stands thus; they have acute powers of observation, and they seem to have some

---

29 Mr. Hewitt, quoted in 'Tegetmeier's Poultry Book,' 1866, p. 165.

30 Quoted in Lloyd's 'Game Birds of Sweden,' p. 345.

taste for the beautiful both in colour and sound. It is certain that the females occasionally exhibit, from unknown causes, the strongest antipathies and preferences for particular males. When the sexes differ in colour or in other ornaments, the males with rare exceptions are the most highly decorated, either permanently or temporarily during the breeding-season. They sedulously display their various ornaments, exert their voices, and perform strange antics in the presence of the females. Even well-armed males, who, it might have been thought, would have altogether depended for success on the law of battle, are in most cases highly ornamented; and their ornaments have been acquired at the expense of some loss of power. In other cases ornaments have been acquired, at the cost of increased risk from birds and beasts of prey. With various species many individuals of both sexes congregate at the same spot, and their courtship is a prolonged affair. There is even reason to suspect that the males and females within the same district do not always succeed in pleasing each other and pairing.

What then are we to conclude from these facts and considerations? Does the male parade his charms with so much pomp and rivalry for no purpose? Are we not justified in believing that the female exerts a choice, and that she receives the addresses of the male who pleases her most? It is not probable that she consciously deliberates; but she is most excited or attracted by the most beautiful, or melodious, or gallant males. Nor need it be supposed that the female studies each stripe or spot of colour; that the peahen, for instance, admires each detail in the gorgeous train of the peacock—she is probably struck only by the general effect. Nevertheless after hearing how carefully the male Argus pheasant displays his elegant primary wing-feathers, and erects his ocellated plumes in the right position for their full effect; or again, how the male goldfinch alternately displays his gold-bespangled wings, we ought not to feel too sure that the female does not attend to each detail of beauty. We can judge, as already remarked, of choice being exerted, only from the analogy of our own minds; and the mental powers of birds, if reason be excluded, do not fundamentally differ from ours. From these various considerations we may conclude that the pairing of birds is not left to chance; but that those males, which are best able by their various charms to please or excite the female, are under ordinary circumstances accepted. If this be admitted, there is not much difficulty in understanding how male birds have gradually acquired their ornamental characters. All animals present individual differences, and as man can modify his domesticated birds by selecting the individuals which appear to him the most beautiful, so the habitual or even occasional preference by the female of the more attractive males would almost certainly lead to their modification; and such modifications might in the course of time be augmented to almost any extent, compatible with the existence of the species.

Variability of Birds, and especially of their secondary Sexual Characters.—Variability and inheritance are the foundations for the work of selection. That domesticated birds have varied greatly, their variations being inherited, is certain. That birds in a state of nature present individual differences is admitted by every one; and that they have sometimes been modified into distinct races, is generally admitted.[31] Variations are

31 According to Dr. Blasius ('Ibis,' vol. ii. 1860, p. 297), there are 425 indubitable species of birds which breed in Europe, besides 60

of two kinds, which insensibly graduate into each other, namely, slight differences between all the members of the same species, and more strongly-marked deviations which occur only occasionally. These latter are rare with birds in a state of nature, and it is very doubtful whether they have often been preserved through selection, and then transmitted to succeeding generations.[32] Nevertheless, it may be worth while to give the few cases relating chiefly to colour (simple albinism and melanism being excluded), which I have been able to collect.

Mr. Gould is well known rarely to admit the existence of varieties, for he esteems very slight differences as specific; now he states[33] that near Bogota certain humming-birds belonging to the genus Cynanthus are divided into two or three races or varieties, which differ from each other in the colouring of the tail,—"some

---

forms, which are frequently regarded as distinct species. Of the latter, Blasius thinks that only ten are really doubtful, and that the other fifty ought to be united with their nearest allies; but this shews that there must be a considerable amount of variation with some of our European birds. It is also an unsettled point with naturalists, whether several North American birds ought to be ranked as specifically distinct from the corresponding European species.

[32] 'Origin of Species,' fifth edit. 1869, p. 104. I had always perceived, that rare and strongly-marked deviations of structure, deserving to be called monstrosities, could seldom be preserved through natural selection, and that the preservation of even highly-beneficial variations would depend to a certain extent on chance. I had also fully appreciated the importance of mere individual differences, and this led me to insist so strongly on the importance of that unconscious form of selection by man, which follows from the preservation of the most valued individuals of each breed, without any intention on his part to modify the characters of the breed. But until I read an able article in the 'North British Review' (March, 1867, p. 289, *et seq.*), which has been of more use to me than any other Review, I did not see how great the chances were against the preservation of variations, whether slight or strongly pronounced, occurring only in single individuals.

[33] 'Introduct. to the Trochilidæ,' p. 102.

"having the whole of the feathers blue, while others "have the eight central ones tipped with beautiful "green." It does not appear that intermediate gradations have been observed in this or the following cases. In the males alone of one of the Australian parrakeets "the thighs in some are scarlet, in others "grass-green." In another parrakeet of the same country "some individuals have the band across the "wing-coverts bright-yellow, while in others the same "part is tinged with red."[34] In the United States some few of the males of the Scarlet Tanager (*Tanagra rubra*) have "a beautiful transverse band of glowing "red on the smaller wing-coverts;"[35] but this variation seems to be somewhat rare, so that its preservation through sexual selection would follow only under unusually favourable circumstances. In Bengal the Honey buzzard (*Pernis cristata*) has either a small rudimental crest on its head, or none at all; so slight a difference however would not have been worth notice, had not this same species possessed in Southern India "a well-marked occipital crest formed of several gra-"duated feathers."[36]

The following case is in some respects more interesting. A pied variety of the raven, with the head, breast, abdomen, and parts of the wings and tail-feathers white, is confined to the Feroe Islands. It is not very rare there, for Graba saw during his visit from eight to ten living specimens. Although the characters of this variety are not quite constant, yet it has been named by several distinguished ornithologists as a distinct species. The fact of the pied birds being pursued and

---

34 Gould, 'Handbook of Birds of Australia,' vol. ii. p. 32 and 68.

35 Audubon, 'Ornitholog. Biography,' 1838, vol. iv. p. 389.

36 Jerdon, 'Birds of India,' vol. i. p. 108; and Mr. Blyth, in 'Land and Water,' 1868, p. 381.

persecuted with much clamour by the other ravens of the island was the chief cause which led Brünnich to conclude that it was specifically distinct; but this is now known to be an error.[37]

In various parts of the northern seas a remarkable variety of the common Guillemot (*Uria troile*) is found; and in Feroe, one out of every five birds, according to Graba's estimation, consists of this variety. It is characterised[38] by a pure white ring round the eye, with a curved narrow white line, an inch and a half in length, extending back from the ring. This conspicuous character has caused the bird to be ranked by several ornithologists as a distinct species under the name of *U. lacrymans*, but it is now known to be merely a variety. It often pairs with the common kind, yet intermediate gradations have never been seen; nor is this surprising, for variations which appear suddenly are often, as I have elsewhere shewn,[39] transmitted either unaltered or not at all. We thus see that two distinct forms of the same species may co-exist in the same district, and we cannot doubt that if the one had possessed any great advantage over the other, it would soon have been multiplied to the exclusion of the latter. If, for instance, the male pied ravens, instead of being persecuted and driven away by their comrades, had been highly attractive, like the pied peacock before mentioned, to the common black females, their numbers would have rapidly increased. And this would have been a case of sexual selection.

---

[37] Graba, 'Tagebuch, Reise nach Färo,' 1830, s. 51-54. Macgillivray, 'Hist. British Birds,' vol. iii. p. 745. 'Ibis,' vol. v. 1863, p. 469.

[38] Graba, ibid. s. 54. Macgillivray, ibid. vol. v. p. 327.

[39] 'Variation of Animals and Plants under Domestication,' vol. ii. p. 92.

With respect to the slight individual differences which are common, in a greater or less degree, to all the members of the same species, we have every reason to believe that they are by far the most important for the work of selection. Secondary sexual characters are eminently liable to vary, both with animals in a state of nature and under domestication.[40] There is also reason to believe, as we have seen in our eighth chapter, that variations are more apt to occur in the male than in the female sex. All these contingencies are highly favourable for sexual selection. Whether characters thus acquired are transmitted to one sex or to both sexes, depends exclusively in most cases, as I hope to shew in the following chapter, on the form of inheritance which prevails in the groups in question.

It is sometimes difficult to form any opinion whether certain slight differences between the sexes of birds are simply the result of variability with sexually-limited inheritance, without the aid of sexual selection, or whether they have been augmented through this latter process. I do not here refer to the innumerable instances in which the male displays splendid colours or other ornaments, of which the female partakes only to a slight degree; for these cases are almost certainly due to characters primarily acquired by the male, having been transferred, in a greater or less degree, to the female. But what are we to conclude with respect to certain birds in which, for instance, the eyes differ slightly in colour in the two sexes?[41] In some cases the eyes differ conspicuously; thus with the storks

---

40 On these points see also 'Variation of Animals and Plants under Domestication,' vol. i. p. 253; vol. ii. p. 73, 75.

41 See, for instance, on the irides of a Podica and Gallicrex in 'Ibis,' vol. ii. 1860, p. 206; and vol. v. 1863, p. 426.

of the genus *Xenorhynchus* those of the male are blackish-hazel, whilst those of the females are gamboge-yellow; with many hornbills (Buceros), as I hear from Mr. Blyth,[42] the males have intense crimson, and the females white eyes. In the *Buceros bicornis*, the hind margin of the casque and a stripe on the crest of the beak are black in the male, but not so in the female. Are we to suppose that these black marks and the crimson colour of the eyes have been preserved or augmented through sexual selection in the males? This is very doubtful; for Mr. Bartlett shewed me in the Zoological Gardens that the inside of the mouth of this Buceros is black in the male and flesh-coloured in the female; and their external appearance or beauty would not be thus affected. I observed in Chili[43] that the iris in the condor, when about a year old, is dark-brown, but changes at maturity into yellowish-brown in the male, and into bright red in the female. The male has also a small, longitudinal, leaden-coloured, fleshy crest or comb. With many gallinaceous birds the comb is highly ornamental, and assumes vivid colours during the act of courtship; but what are we to think of the dull-coloured comb of the condor, which does not appear to us in the least ornamental? The same question may be asked in regard to various other characters, such as the knob on the base of the beak of the Chinese goose (*Anser cygnoides*), which is much larger in the male than in the female. No certain answer can be given to these questions; but we ought to be cautious in assuming that knobs and various fleshy appendages cannot be attractive to the female, when we remember that with savage races of man

[42] See also Jerdon, 'Birds of India,' vol. i. p. 243-245.

[43] 'Zoology of the Voyage of H.M.S. Beagle,' 1841, p. 6.

various hideous deformities—deep scars on the face with the flesh raised into protuberances, the septum of the nose pierced by sticks or bones, holes in the ears and lips stretched widely open—are all admired as ornamental.

Whether or not unimportant differences between the sexes, such as those just specified, have been preserved through sexual selection, these differences, as well as all others, must primarily depend on the laws of variation. On the principle of correlated development, the plumage often varies on different parts of the body, or over the whole body, in the same manner. We see this well illustrated in certain breeds of the fowl. In all the breeds the feathers on the neck and loins of the males are elongated, and are called hackles; now when both sexes acquire a top-knot, which is a new character in the genus, the feathers on the head of the male become hackle-shaped, evidently on the principle of correlation; whilst those on the head of the female are of the ordinary shape. The colour also of the hackles forming the top-knot of the male, is often correlated with that of the hackles on the neck and loins, as may be seen by comparing these feathers in the Golden and Silver-spangled Polish, the Houdans, and Crève-cœur breeds. In some natural species we may observe exactly the same correlation in the colours of these same feathers, as in the males of the splendid Golden and Amherst pheasants.

The structure of each individual feather generally causes any change in its colouring to be symmetrical; we see this in the various laced, spangled, and pencilled breeds of the fowl; and on the principle of correlation the feathers over the whole body are often modified in the same manner. We are thus enabled without much trouble to rear breeds with their plumage marked and coloured almost as symmetrically as in natural species. In laced and spangled fowls the coloured margins of the feathers are abruptly defined; but in a mongrel raised by me from a black Spanish cock glossed with green and a white game hen, all the feathers were greenish-black, excepting towards their extremities, which were yellowish-white; but between the white extremities and the black bases, there was on each feather a symmetrical, curved zone of dark-brown. In some instances the shaft of the feather determines the distribution of the tints; thus with the body-feathers of a mongrel from the same black Spanish cock and a silver-spangled Polish hen, the shaft, together with a narrow space on each side, was greenish-black, and this was surrounded by a regular zone of dark-brown, edged with brownish-white. In these cases we see feathers becoming symmetrically shaded, like those which give so much elegance to the plumage of many natural species. I have also noticed a variety of the common pigeon with the wing-bars symmetrically zoned with three bright shades, instead of being simply black on a slaty-blue ground, as in the parent-species.

In many large groups of birds it may be observed that the plumage is differently coloured in each species, yet that certain spots, marks, or stripes, though likewise differently coloured, are retained by all the species. Analogous cases occur with the breeds of the pigeon, which usually retain the two wing-bars, though they may be coloured red, yellow, white, black, or blue, the rest of the plumage being of some wholly different tint. Here is a more curious case, in which certain marks are retained, though coloured in almost an exactly reversed manner to what is natural; the aboriginal pigeon has a blue tail, with the terminal halves of the outer webs of the two outer tail-feathers white; now there is a sub-variety having a white instead of a blue tail, with precisely that small part black which is white in the parent-species.[44]

Formation and variability of the Ocelli or eye-like Spots on the Plumage of Birds.—As no ornaments are more beautiful than the ocelli on the feathers of various birds, on the hairy coats of some mammals, on the scales of reptiles and fishes, on the skin of amphibians, on the wings of many Lepidoptera and other insects, they deserve to be especially noticed. An ocellus consists of a spot within a ring of another colour, like the pupil within the iris, but the central spot is often surrounded by additional concentric zones. The ocelli on the tail-coverts of the peacock offer a familiar example, as well as those on the wings of the peacock-butterfly (Vanessa). Mr. Trimen has given me a description of a S. African moth (*Gynanisa Isis*), allied to our Emperor moth, in which a magnificent ocellus occupies nearly the whole surface of each hinder wing; it consists of a black centre, including a semi-transparent crescent-shaped mark, surrounded by successive ochre-yellow, black, ochre-yellow, pink, white, pink, brown, and whitish zones. Although we do not know the steps by which these wonderfully-beautiful and complex ornaments have been developed, the process at least with insects has probably been a simple one; for, as Mr. Trimen writes to me, "no characters of mere marking or coloration are so "unstable in the Lepidoptera as the ocelli, both in "number and size." Mr. Wallace, who first called my attention to this subject, shewed me a series of specimens of our common meadow-brown butterfly (*Hip-*

44 Bechstein, 'Naturgeschichte Deutschlands,' B. iv. 1795, s. 31, on a sub-variety of the Monck pigeon.

parchia Janira) exhibiting numerous gradations from a simple minute black spot to an elegantly-shaded ocellus. In a S. African butterfly (*Cyllo Leda*, Linn.) belonging to the same family, the ocelli are even still more variable. In some specimens (A, fig. 52) large spaces on the upper surface of the wings are coloured black, and include irregular white marks; and from this state a complete gradation can be traced into a

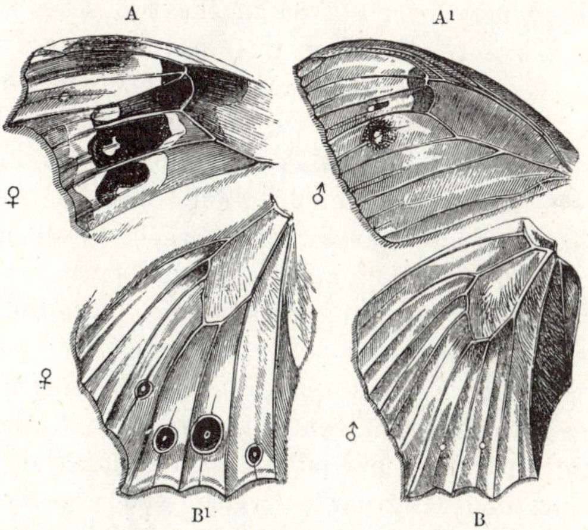

Fig. 52. Cyllo leda, Linn., from a drawing by Mr. Trimen, shewing the extreme range of variation in the ocelli.

A. Specimen, from Mauritius, upper surface of fore-wing.
A¹. Specimen, from Natal, ditto.
B. Specimen, from Java, upper surface of hind-wing.
B¹. Specimen, from Mauritius, ditto.

tolerably perfect (A^1) ocellus, and this results from the contraction of the irregular blotches of colour. In another series of specimens a gradation can be followed from excessively minute white dots, surrounded by a scarcely visible black line (B), into perfectly symmetrical and large ocelli (B^1).[45] In cases like these, the

[45] This woodcut has been engraved from a beautiful drawing, most kindly made for me by Mr. Trimen; see also his description of the

development of a perfect ocellus does not require a long course of variation and selection.

With birds and many other animals it seems, from the comparison of allied species, to follow, that circular spots are often generated by the breaking up and contraction of stripes. In the Tragopan pheasant faint white lines in the female represent the beautiful white spots in the male;[46] and something of the same kind may be observed in the two sexes of the Argus pheasant. However this may be, appearances strongly favour the belief that, on the one hand, a dark spot is often formed by the colouring-matter being drawn towards a central point from a surrounding zone, which is thus rendered lighter. And, on the other hand, that a white spot is often formed by the colour being driven away from a central point, so that it accumulates in a surrounding darker zone. In either case an ocellus is the result. The colouring matter seems to be a nearly constant quantity, but is redistributed, either centripetally or centrifugally. The feathers of the common guinea-fowl offer a good instance of white spots surrounded by darker zones; and wherever the white spots are large and stand near each other, the surrounding dark zones become confluent. In the same wing-feather of the Argus pheasant dark spots may be seen surrounded by a pale zone, and white spots by a dark zone. Thus the formation of an ocellus in its simplest state appears to be a simple affair. By what further steps the more complex ocelli, which

---

wonderful amount of variation in the coloration and shape of the wings of this butterfly, in his 'Rhopalocera Africæ Australis,' p. 186. See also an interesting paper by the Rev. H. H. Higgins, on the origin of the ocelli in the Lepidoptera in the 'Quarterly Journal of Science,' July, 1868, p. 325.

[46] Jerdon, 'Birds of India,' vol. iii. p. 517.

are surrounded by many successive zones of colour, have been generated, I will not pretend to say. But bearing in mind the zoned feathers of the mongrel offspring from differently-coloured fowls, and the extraordinary variability of the ocelli in many Lepidoptera, the formation of these beautiful ornaments can hardly be a highly complex process, and probably depends on some slight and graduated change in the nature of the tissues.

Gradation of Secondary Sexual Characters.—Cases of gradation are important for us, as they shew that it is at least possible that highly complex ornaments may have been acquired by small successive steps. In order to discover the actual steps by which the male of any existing bird has acquired his magnificent colours or other ornaments, we ought to behold the long line of his ancient and extinct progenitors; but this is obviously impossible. We may, however, generally gain a clue by comparing all the species of a group, if it be a large one; for some of them will probably retain, at least in a partial manner, traces of their former characters. Instead of entering on tedious details respecting various groups, in which striking instances of gradation could be given, it seems the best plan to take some one or two strongly-characterised cases, for instance that of the peacock, in order to discover if any light can thus be thrown on the steps by which this bird has become so splendidly decorated. The peacock is chiefly remarkable from the extraordinary length of his tail-coverts; the tail itself not being much elongated. The barbs along nearly the whole length of these feathers stand separate or are decomposed; but this is the case with the feathers of many species, and with some varieties of the domestic fowl and pigeon. The barbs coalesce towards the extremity of the shaft to form the oval disc or ocellus, which is certainly one of the most beautiful objects in the world. This consists of an iridescent, intensely blue, indented centre, surrounded by a rich green zone, and this by a broad coppery-brown zone, and this by five other narrow zones of slightly-different iridescent shades. A trifling character in the disc perhaps deserves notice; the barbs, for a space along one of the concentric zones are destitute, to a greater or less degree, of their barbules, so that a part of the disc is surrounded by an almost transparent zone, which gives to it a highly-finished aspect. But I have elsewhere described[47] an exactly analogous variation in the hackles of a sub-variety of the game-cock, in which the tips, having a metallic lustre, "are separated from "the lower part of the feather by a symmetrically-"shaped transparent zone, composed of the naked por-"tions of the barbs." The lower margin or base of the dark-blue centre of the ocellus is deeply indented on the line of the shaft. The surrounding zones likewise shew traces, as may be seen in the drawing (fig. 53), of indentations, or rather breaks. These indentations are common to the Indian and Javan peacocks (*Pavo cristatus* and *P. muticus*); and they seemed to me to deserve particular attention, as probably connected with the development of the ocellus; but for a long time I could not conjecture their meaning.

If we admit the principle of gradual evolution, there must formerly have existed many species which presented every successive step between the wonderfully elongated tail-coverts of the peacock and the short tail-

---

47 'Variation of Animals and Plants under Domestication,' vol. i. p. 254.

coverts of all ordinary birds; and again between the magnificent ocelli of the former, and the simpler ocelli or mere coloured spots of other birds; and so with all the other characters of the peacock. Let us look to the allied Gallinaceæ for any still-existing gradations. The species and sub-species of Polyplectron

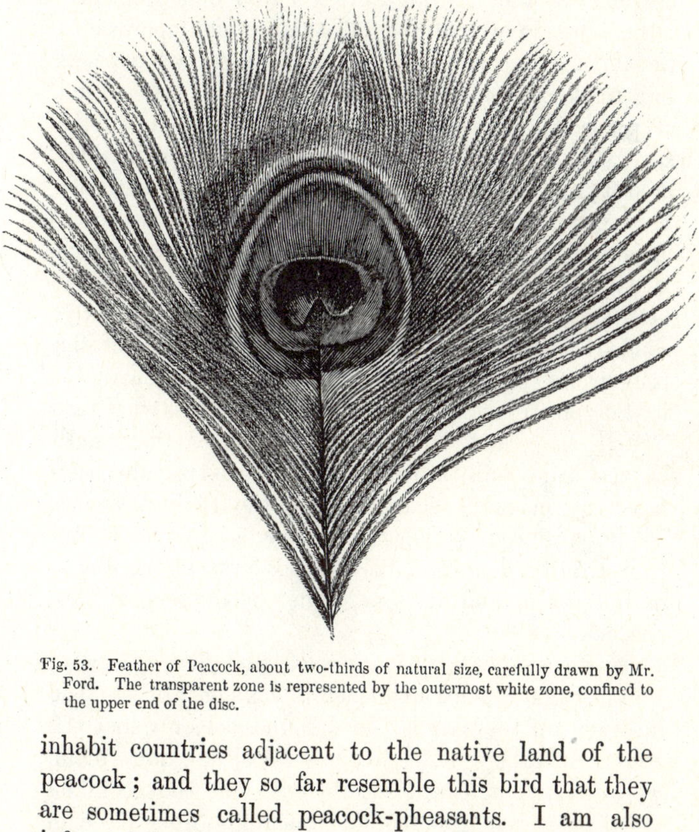

Fig. 53. Feather of Peacock, about two-thirds of natural size, carefully drawn by Mr. Ford. The transparent zone is represented by the outermost white zone, confined to the upper end of the disc.

inhabit countries adjacent to the native land of the peacock; and they so far resemble this bird that they are sometimes called peacock-pheasants. I am also informed by Mr. Bartlett that they resemble the peacock in their voice and in some of their habits. During the spring the males, as previously described, strut about before the comparatively plain-coloured females, expanding and erecting their tail and wing-feathers, which are ornamented with numerous ocelli. I request the reader to turn back to the drawing (fig. 51, p. 90) of a Polyplectron. In *P. Napoleonis* the ocelli are confined to the tail, and the back is of a rich metallic blue, in which respects this species approaches the Java peacock. *P. Hardwickii* possesses a peculiar top-knot, somewhat like that of this same kind of peacock. The ocelli on the wings and tail of the several species of Polyplectron are either circular or oval, and consist of a beautiful, iridescent, greenish-blue or greenish-purple disc, with a black border. This border in *P. chinquis* shades into brown which is edged with cream-colour, so that the ocellus is here surrounded with differently, though not brightly, shaded concentric zones. The unusual length of the tail-coverts is another highly remarkable character in Polyplectron; for in some of the species they are half as long, and in others two-thirds of the length of the true tail-feathers. The tail-coverts are ocellated, as in the peacock. Thus the several species of Polyplectron manifestly make a graduated approach in the length of their tail-coverts, in the zoning of the ocelli, and in some other characters, to the peacock.

Notwithstanding this approach, the first species of Polyplectron which I happened to examine almost made me give up the search; for I found not only that the true tail-feathers, which in the peacock are quite plain, were ornamented with ocelli, but that the ocelli on all the feathers differed fundamentally from those of the peacock, in there being two on the same feather, (fig. 54), one on each side of the shaft. Hence I concluded that the early progenitors of the peacock could not have resembled in any degree a Polyplectron. But on continuing my search, I observed that in some of the species the two ocelli stood very near each other; that in the tail-feathers of *P. Hardwickii* they touched each other; and, finally, that in the tail-coverts of this same species as well as of *P. malaccense* (fig. 55) they were actually confluent. As the central part alone is confluent, an indentation is left at both the upper and lower ends; and the surrounding coloured zones are likewise indented. A single ocellus is thus formed on each tail-covert, though still plainly betraying its double origin. These confluent ocelli differ from the single ocelli of the peacock in having an indentation at both ends, instead of at the lower or basal end alone. The explanation, however, of this difference is not difficult; in some species of Polyplectron the two oval ocelli on the same feather stand parallel to each other; in other species (as in *P. chinquis*) they converge towards one end; now the partial confluence of two convergent ocelli would manifestly leave a much

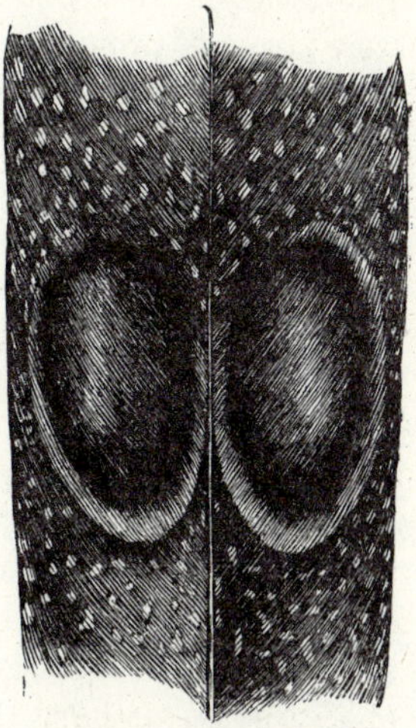

Fig. 54. Part of a tail-covert of Polyplectron chinquis, with the two ocelli of nat. size.

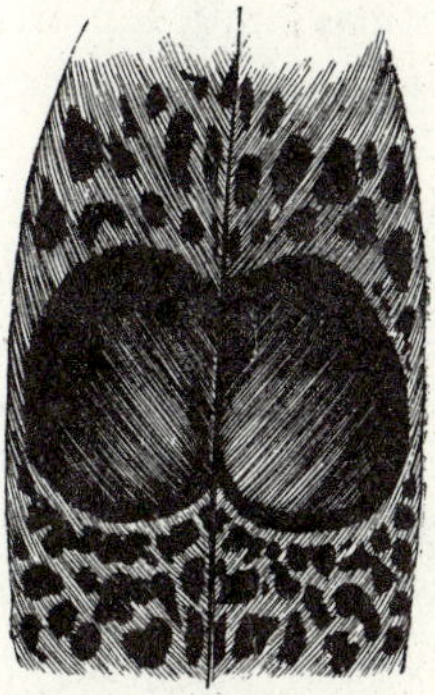

Fig. 55. Part of a tail-covert of Polyplectron malaccense, with the two ocelli, partially confluent, of nat. size.

deeper indentation at the divergent than at the convergent end. It is also manifest that if the convergence were strongly pronounced and the confluence complete, the indentation at the convergent end would tend to be quite obliterated.

The tail-feathers in both species of peacock are entirely destitute of ocelli, and this apparently is related to their being covered up and concealed by the long tail-coverts. In this respect they differ remarkably from the tail-feathers of Polyplectron, which in most of the species are ornamented with larger ocelli than those on the tail-coverts. Hence I was led carefully to examine the tail-feathers of the several species of Polyplectron in order to discover whether the ocelli in any of them shewed any tendency to disappear, and, to my great satisfaction, I was successful. The central tail-feathers of *P. Napoleonis* have the two ocelli on each side of the shaft perfectly developed; but the inner ocellus becomes less and less conspicuous on the more exterior tail-feathers, until a mere shadow or rudimentary vestige is left on the inner side of the outermost feather. Again, in *P. malaccense*, the ocelli on the tail-coverts are, as we have seen, confluent; and these feathers are of unusual length, being two-thirds of the length of the tail-feathers, so that in both these respects they resemble the tail-coverts of the peacock. Now in this species the two central tail-feathers alone are ornamented, each with two brightly-coloured ocelli, the ocelli having completely disappeared from the inner sides of all the other tail-feathers. Consequently the tail-coverts and tail-feathers of this species of Polyplectron make a near approach in structure and ornamentation to the corresponding feathers of the peacock.

As far, then, as the principle of gradation throws light on the steps by which the magnificent train of the peacock has been acquired, hardly anything more is needed. We may picture to ourselves a progenitor of the peacock in an almost exactly intermediate condition between the existing peacock, with his enormously elongated tail-coverts, ornamented with single ocelli, and an ordinary gallinaceous bird with short tail-coverts, merely spotted with some colour; and we shall then see in our mind's eye, a bird possessing tail-coverts, capable of erection and expansion, ornamented with two partially confluent ocelli, and long enough almost to conceal the tail-feathers,—the latter having already partially lost their ocelli; we shall see in short, a Polyplectron. The indentation of the central disc and surrounding zones of the ocellus in both species of peacock, seems to me to speak plainly in favour of this view; and this structure is otherwise inexplicable. The males of Polyplectron are no doubt very beautiful birds, but their beauty, when viewed from a little distance, cannot be compared, as I formerly saw in the Zoological Gardens, with that of the peacock. Many female progenitors of the peacock must, during a long line of descent, have appreciated this superiority; for they have unconsciously, by the continued preference of the most beautiful males, rendered the peacock the most splendid of living birds.

Argus pheasant.—Another excellent case for investigation is offered by the ocelli on the wing-feathers of the Argus pheasant, which are shaded in so wonderful a manner as to resemble balls lying within sockets, and which consequently differ from ordinary ocelli. No one, I presume, will attribute the shading, which has excited the admiration of many experienced artists, to chance —to the fortuitous concourse of atoms of colouring matter. That these ornaments should have been formed through the selection of many successive variations, not one of which was originally intended to produce the ball-and-socket effect, seems as incredible, as that one of Raphael's Madonnas should have been formed by the selection of chance daubs of paint made by a long succession of young artists, not one of whom intended at first to draw the human figure. In order to discover how the ocelli have been developed, we cannot look to a long line of progenitors, nor to various closely-allied forms, for such do not now exist. But fortunately the several feathers on the wing suffice to give us a clue to the problem, and they prove to demonstration that a gradation is at least possible from a mere spot to a finished ball-and-socket ocellus.

The wing-feathers, bearing the ocelli, are covered with dark stripes or rows of dark spots, each stripe or row running obliquely down the outer side of the shaft to an ocellus. The spots are generally elongated in a transverse line to the row in which they stand. They often become confluent, either in the line of the row—and then they form a longitudinal stripe—or transversely, that is, with the spots in the adjoining rows, and then they form transverse stripes. A spot sometimes breaks up into smaller spots, which still stand in their proper places.

It will be convenient first to describe a perfect ball-and-socket ocellus. This consists of an intensely black circular ring, surrounding a space shaded so as exactly to resemble a ball. The figure here given has been admirably drawn by Mr. Ford, and engraved, but a woodcut cannot exhibit the exquisite shading of the original. The ring is almost always slightly broken or interrupted (see fig. 56) at a point in the upper half, a little to the right of and above the white shade on the enclosed ball; it is also sometimes broken towards the base on the right hand. These little breaks have an important meaning. The ring is always much thickened, with the edges ill-defined towards the left-hand upper corner the feather being held erect, in the position in which it is here drawn. Beneath this thickened part there is on the surface of the ball an oblique almost pure-white mark, which shades off downwards into a pale-leaden hue, and this into yellowish and brown tints, which insensibly become darker and darker towards the lower part of the ball. It is this shading which gives so admirably the effect of light shining on a convex surface. If one of the balls be examined, it will be seen that the lower part is of a browner tint and is indistinctly separated by a curved oblique line from the upper part, which is yellower and more leaden; this oblique line runs at right angles to the longer axis of the white patch of light, and indeed of all the shading; but this difference in the tints, which cannot of course be shewn in the woodcut, does not in the least interfere with the perfect shading of the ball.[48] It should be particularly ob-

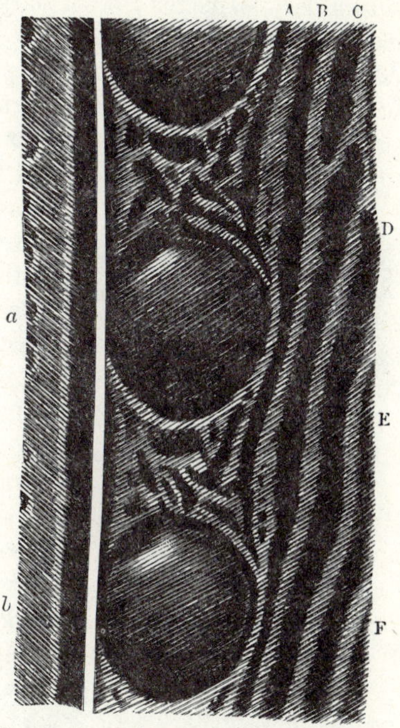

Fig. 56. Part of Secondary wing-feather of Argus pheasant, shewing two, *a* and *b*, perfect ocelli. A, B, C, &c., dark stripes running obliquely down, each to an ocellus.

[Much of the web on both sides, especially to the left of the shaft, has been cut off.]

---

48 When the Argus pheasant displays his wing-feathers like a great fan, those nearest to the body stand more upright than the outer ones,

served that each ocellus stands in obvious connection with a dark stripe, or row of dark spots, for both occur indifferently on the same feather. Thus in fig. 56 stripe A runs to ocellus *a;* B runs to ocellus *b;* stripe C is broken in the upper part, and runs down to the next succeeding ocellus, not represented in the woodcut; D to the next lower one, and so with the stripes E and F. Lastly, the several ocelli are separated from each other by a pale surface bearing irregular black marks.

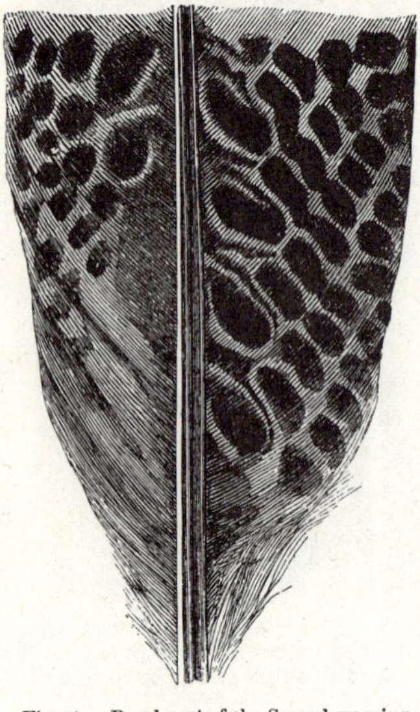

Fig. 57. Basal part of the Secondary wing-feather, nearest to the body.

I will next describe the other extreme of the series, namely the first trace of an ocellus. The short secondary wing-feather (fig. 57), nearest to the body, is marked like the other feathers, with oblique, longitudinal, rather irregular, rows of spots. The lowest spot, or that nearest the shaft, in the five lower rows (excluding the basal row) is a little larger than the other spots in the same row, and a little more elon-

so that the shading of the ball-and-socket ocelli ought to be slightly different on the different feathers, in order to bring out their full effect, relatively to the incidence of the light. Mr. T. W. Wood, who has the experienced eye of an artist, asserts ('Field,' Newspaper, May 28, 1870, p. 457) that this is the case; but after carefully examining two mounted specimens (the proper feathers from one having been given to me by Mr. Gould for more accurate comparison) I cannot perceive that this acme of perfection in the shading has been attained; nor can others to whom I have shewn these feathers recognise the fact.

gated in a transverse direction. It differs also from the other spots by being bordered on its upper side with some dull fulvous shading. But this spot is not in any way more remarkable than those on the plumage of many birds, and might easily be quite overlooked. The next higher spot in each row does not differ at all from the upper ones in the same row, although in the following series it becomes, as we shall see, greatly modified. The larger spots occupy exactly the same relative position on this feather as those occupied by the perfect ocelli on the longer wing-feathers.

By looking to the next two or three succeeding secondary wing-feathers, an absolutely insensible gradation can be traced from one of the above-described lower spots, together with the next higher one in the same row, to a curious ornament, which cannot be called an ocellus, and which I will name, from the want of a better term, an "elliptic ornament." These are shewn in the accompanying figure (fig. 58). We here see several oblique rows, A, B, C, D (see the lettered diagram), &c., of dark spots of the usual character. Each row of spots runs down to and is connected with one of the elliptic ornaments, in exactly the same manner as each stripe in fig. 56 runs down to, and is connected with, one of the ball-and-socket ocelli. Looking to any one row, for instance, B, the lowest spot or mark (*b*) is thicker and considerably longer than the upper spots, and has its left extremity pointed and curved upwards. This black mark is abruptly bordered on its upper side by a rather broad space of richly-shaded tints, beginning with a narrow brown zone, which passes into orange, and this into a pale leaden tint, with the end towards the shaft much paler. This mark corresponds in every respect with the larger, shaded spot, described in the last paragraph (fig. 57), but is more highly developed and more brightly coloured. To the right and above this spot (*b*), with its bright shading, there is a long, narrow, black mark (*c*), belonging to the same row, and which is arched a little downwards so as to face (*b*). It is also narrowly edged on the lower side with a fulvous tint. To the left of and above *c*, in the

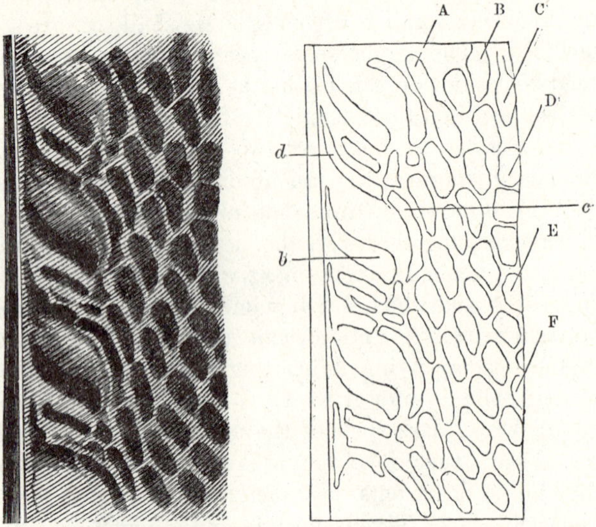

Fig. 58. Portion of one of the Secondary wing-feathers near to the body; shewing the so-called elliptic ornaments. The right-hand figure is given merely as a diagram for the sake of the letters of reference.

A, B, C, &c. Rows of spots running down to and forming the elliptic ornaments.
b. Lowest spot or mark in row B.
c. The next succeeding spot or mark in the same row.
d. Apparently a broken prolongation of the spot *c* in the same row B.

same oblique direction, but always more or less distinct from it, there is another black mark (*d*). This mark is generally sub-triangular and irregular in shape, but in the one lettered in the diagram is unusually narrow, elongated, and regular. It apparently consists of a lateral and broken prolongation of the mark (*c*), as I infer from traces of similar prolongations from the succeeding upper spots; but I do not feel sure of this. These three marks, *b*, *c*, and *d*, with the intervening bright shades, form together the so-called elliptic ornament. These ornaments stand in a line parallel to the shaft, and manifestly correspond in position with the ball-and-socket ocelli. Their extremely elegant appearance cannot be appreciated in the drawing, as the orange and leaden tints, contrasting so well with the black marks, cannot be shewn.

Between one of the elliptic ornaments and a perfect ball-and-socket ocellus, the gradation is so perfect that it is scarcely possible to decide when the latter term ought to be used. I regret that I have not given an additional drawing, besides fig. 58, which stands about half-way in the series between one of the simple spots and a perfect ocellus. The passage from the elliptic ornament into an ocellus is effected by the elongation and greater curvature in opposed directions of the lower black mark (*b*), and more especially of the upper one (*c*), together with the contraction of the irregular sub-triangular or narrow mark (*d*), so that at last these three marks become confluent, forming an irregular elliptic ring. This ring is gradually rendered more and more circular and regular, at the same time increasing in diameter. Traces of the junction of all three elongated spots or marks, especially of the two upper ones, can still be observed in many of the most perfect ocelli. The broken state of the black ring on the upper side of the ocellus in fig. 56 was pointed out. The irregular sub-triangular or narrow mark (*d*) manifestly forms, by its contraction and equalisation, the thickened portion of the ring on the left upper side of the perfect ball-and-socket ocellus. The lower part of the ring is invariably a little thicker than the other parts (see fig. 56), and this follows from the lower black mark of the elliptic ornament (*b*) having been originally thicker than the upper mark (*c*). Every step can be followed in the process of confluence and modification; and the black ring which surrounds the ball of the ocellus is unquestionably formed by the union and modification of the three black marks, *b*, *c*, *d*, of the elliptic ornament. The irregular zigzag black marks between the successive ocelli (see again fig. 56) are plainly due to the breaking up of the somewhat more regular but similar marks between the elliptic ornaments.

The successive steps in the shading of the ball-and-socket ocelli can be followed out with equal clearness. The brown, orange, and pale-leaden narrow zones which border the lower black mark of the elliptic ornament can be seen gradually to become more and more softened and shaded into each other, with the upper lighter part towards the left-hand corner rendered still lighter, so as to become almost white. But even in the most perfect ball-and-socket ocelli a slight difference in the tints, though not in the shading, between the upper and lower parts of the ball can be perceived (as was before especially noticed), the line of separation being oblique, in the same direction with the bright coloured shades of the elliptic ornaments. Thus almost every minute detail in the shape and colouring of the ball-and-socket ocelli can be shewn to follow from gradual changes in the elliptic ornaments; and the development of the latter can be traced by equally small steps from the union of two almost simple spots, the lower one (fig. 57) having some dull fulvous shading on the upper side.

The extremities of the longer secondary feathers which bear the perfect ball-and-socket ocelli are peculiarly ornamented. (Fig. 59.) The oblique longitudinal stripes suddenly cease upwards and become confused, and above this limit the whole upper end of the feather (*a*) is covered with white dots, surrounded by little black rings, standing on a dark ground. Even the oblique stripe belonging to the uppermost ocellus (*b*) is represented only by a very short irregular black mark with the usual, curved, transverse base. As this stripe is thus abruptly cut off above, we can understand, from what has gone before, how it is that the upper thickened part of the ring is absent in the uppermost ocellus; for, as before stated, this thickened part is apparently formed by a broken prolongation of the next higher spot in the same row. From the absence of the upper and thickened part of the ring, the uppermost ocellus, though perfect in all other respects, appears as if its top had been obliquely sliced off. It would, I think, perplex any one, who believes that the plumage of the Argus-pheasant was created as we now see it, to account for the imperfect condition of the uppermost ocelli. I should add that in the secondary wing-feather farthest from the body all the ocelli are

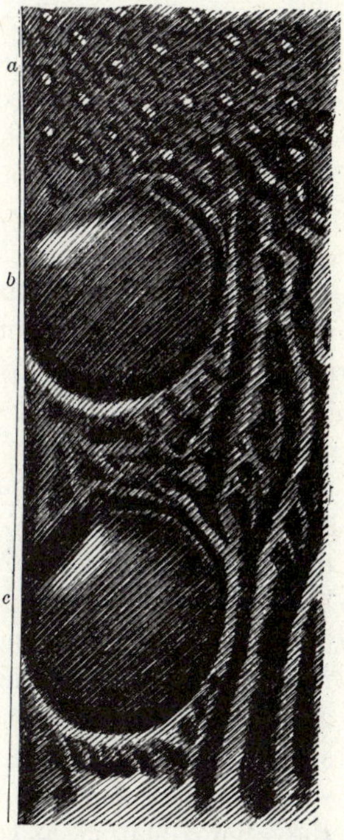

Fig. 59. Portion near summit of one of the Secondary wing-feathers, bearing perfect ball-and-socket ocelli.

a. Ornamented upper part.
b. Uppermost, imperfect ball-and-socket ocellus. (The shading above the white mark on the summit of the ocellus is here a little too dark.)
c. Perfect ocellus.

smaller and less perfect than on the other feathers, with the upper parts of the external black rings deficient, as in the case just mentioned. The imperfection here seems to be connected with the fact that the spots on this feather shew less tendency than usual to become confluent into stripes; on the contrary, they are often broken up into smaller spots, so that two or three rows run down to each ocellus.

We have now seen that a perfect series can be followed, from two almost simple spots, at first quite distinct from each other, to one of the wonderful ball-and-socket ornaments. Mr. Gould, who kindly gave me some of these feathers, fully agrees with me in the completeness of the gradation. It is obvious that the stages in development exhibited by the feathers on the same bird do not at all necessarily shew us the steps which have been passed through by the extinct progenitors of the species; but they probably give us the clue to the actual steps, and they at least prove to demonstration that a gradation is possible. Bearing in mind how carefully the male Argus pheasant displays his plumes before the female, as well as the many facts rendering it probable that female birds prefer the more attractive males, no one who admits the agency of sexual selection, will deny that a simple dark spot with some fulvous shading might be converted, through the approximation and modification of the adjoining spots, together with some slight increase of colour, into one of the so-called elliptic ornaments. These latter ornaments have been shewn to many persons, and all have admitted that they are extremely pretty, some thinking them even more beautiful than the ball-and-socket ocelli. As the secondary plumes became lengthened through sexual selection, and as the elliptic ornaments increased in diameter, their colours apparently became less bright; and then the ornamentation of the plumes had to be gained by improvements in the pattern and shading; and this process has been carried on until the wonderful ball-and-socket ocelli have been finally developed. Thus we can understand—and in no other way as it seems to me—the present condition and origin of the ornaments on the wing-feathers of the Argus pheasant.

From the light reflected by the principle of gradation; from what we know of the laws of variation; from the changes which have taken place in many of our domesticated birds; and, lastly, from the character (as we shall hereafter more clearly see) of the immature plumage of young birds—we can sometimes indicate with a certain amount of confidence, the probable steps by which the males have acquired their brilliant plumage and various ornaments; yet in many cases we are involved in darkness. Mr. Gould several years ago pointed out to me a humming-bird, the *Urosticte benjamini*, remarkable from the curious differences presented by the two sexes. The male, besides a splendid gorget, has greenish-black tail-feathers, with the four *central* ones tipped with white; in the female, as with most of the allied species, the three *outer* tail-feathers on each side are tipped with white, so that the male has the four central, whilst the female has the six exterior feathers ornamented with white tips. What makes the case curious is that, although the colouring of the tail differs remarkably in both sexes of many kinds of humming-birds, Mr. Gould does not know a single species, besides the Urosticte, in which the male has the four central feathers tipped with white.

The Duke of Argyll, in commenting on this case,[49]

---

49 'The Reign of Law,' 1867, p. 247.

passes over sexual selection, and asks, "What explanation "tion does the law of natural selection give of such "specific varieties as these?" He answers "none "whatever;" and I quite agree with him. But can this be so confidently said of sexual selection? Seeing in how many ways the tail-feathers of humming-birds differ, why should not the four central feathers have varied in this one species alone, so as to have acquired white tips? The variations may have been gradual, or somewhat abrupt as in the case recently given of the humming-birds near Bogota, in which certain individuals alone have the "central tail-feathers tipped "with beautiful green." In the female of the Urosticte I noticed extremely minute or rudimental white tips to the two outer of the four central black tail-feathers; so that here we have an indication of change of some kind in the plumage of this species. If we grant the possibility of the central tail-feathers of the male varying in whiteness, there is nothing strange in such variations having been sexually selected. The white tips, together with the small white ear-tufts, certainly add, as the Duke of Argyll admits, to the beauty of the male; and whiteness is apparently appreciated by other birds, as may be inferred from such cases as the snow-white male of the Bell-bird. The statement made by Sir R. Heron should not be forgotten, namely that his peahens, when debarred from access to the pied peacock, would not unite with any other male, and during that season produced no offspring. Nor is it strange that variations in the tail-feathers of the Urosticte should have been specially selected for the sake of ornament, for the next succeeding genus in the family takes its name of Metallura from the splendour of these feathers. Mr. Gould, after describing the peculiar plumage of the Urosticte, adds, "that ornament and variety is the sole "object, I have myself but little doubt."[50] If this be admitted, we can perceive that the males which were decked in the most elegant and novel manner would have gained an advantage, not in the ordinary struggle for life, but in rivalry with other males, and would consequently have left a larger number of offspring to inherit their newly-acquired beauty.

---

[50] 'Introduction to the Trochilidæ,' 1861, p. 110.

CHAPTER XV.

BIRDS—*continued.*

Discussion why the males alone of some species, and both sexes of other species, are brightly coloured—On sexually-limited inheritance, as applied to various structures and to brightly-coloured plumage—Nidification in relation to colour—Loss of nuptial plumage during the winter.

WE have in this chapter to consider, why with many kinds of birds the female has not received the same ornaments as the male; and why with many others, both sexes are equally, or almost equally, ornamented? In the following chapter we shall consider why in some few rare cases the female is more conspicuously coloured than the male.

In my 'Origin of Species'[1] I briefly suggested that the long tail of the peacock would be inconvenient, and the conspicuous black colour of the male capercailzie dangerous, to the female during the period of incubation; and consequently that the transmission of these characters from the male to the female offspring had been checked through natural selection. I still think that this may have occurred in some few instances: but after mature reflection on all the facts which I have been able to collect, I am now inclined to believe that when the sexes differ, the successive variations have generally been from the first limited in their transmission to the same sex in which they first appeared. Since my remarks appeared, the subject of sexual coloration

---

[1] Fourth edition, 1866, p. 241.

has been discussed in some very interesting papers by Mr. Wallace,[2] who believes that in almost all cases the successive variations tended at first to be transmitted equally to both sexes; but that the female was saved, through natural selection, from acquiring the conspicuous colours of the male, owing to the danger which she would thus have incurred during incubation.

This view necessitates a tedious discussion on a difficult point, namely whether the transmission of a character, which is at first inherited by both sexes, can be subsequently limited in its transmission, by means of selection, to one sex alone. We must bear in mind, as shewn in the preliminary chapter on sexual selection, that characters which are limited in their development to one sex are always latent in the other. An imaginary illustration will best aid us in seeing the difficulty of the case: we may suppose that a fancier wished to make a breed of pigeons, in which the males alone should be coloured of a pale blue, whilst the females retained their former slaty tint. As with pigeons characters of all kinds are usually transmitted to both sexes equally, the fancier would have to try to convert this latter form of inheritance into sexually-limited transmission. All that he could do would be to persevere in selecting every male pigeon which was in the least degree of a paler blue; and the natural result of this process, if steadily carried on for a long time, and if the pale variations were strongly inherited or often recurred, would be to make his whole stock of a lighter blue. But our fancier would be compelled to match, generation after generation, his pale blue males with slaty females, for he wishes to keep the

---

[2] 'Westminster Review,' July, 1867. 'Journal of Travel,' vol. i. 1868, p. 73.

latter of this colour. The result would generally be the production either of a mongrel piebald lot, or more probably the speedy and complete loss of the pale-blue colour, for the primordial slaty tint would be transmitted with prepotent force. Supposing, however, that some pale-blue males and slaty females were produced during each successive generation, and were always crossed together; then the slaty females would have, if I may use the expression, much blue blood in their veins, for their fathers, grandfathers, etc., will all have been blue birds. Under these circumstances it is conceivable (though I know of no distinct facts rendering it probable) that the slaty females might acquire so strong a latent tendency to pale-blueness, that they would not destroy this colour in their male offspring, their female offspring still inheriting the slaty tint. If so, the desired end of making a breed with the two sexes permanently different in colour might be gained.

The extreme importance, or rather necessity, of the desired character in the above case, namely, pale-blueness, being present though in a latent state in the female, so that the male offspring should not be deteriorated, will be best appreciated as follows: the male of Sœmmerring's pheasant has a tail thirty-seven inches in length, whilst that of the female is only eight inches; the tail of the male common pheasant is about twenty inches, and that of the female twelve inches long. Now if the female Sœmmerring pheasant with her *short* tail were crossed with the male common pheasant, there can be no doubt that the male hybrid offspring would have a much *longer* tail than that of the pure offspring of the common pheasant. On the other hand, if the female common pheasant, with her tail nearly *twice as long* as that of the female Sœmmerring pheasant, were crossed with the male of the latter, the male hybrid offspring would have a much *shorter* tail than that of the pure offspring of Sœmmerring's pheasant.[3]

Our fancier, in order to make his new breed with the males of a decided pale-blue tint, and the females unchanged, would have to continue selecting the males during many generations; and each stage of paleness would have to be fixed in the males, and rendered latent in the females. The task would be an extremely difficult one, and has never been tried, but might possibly succeed. The chief obstacle would be the early and complete loss of the pale-blue tint, from the necessity of reiterated crosses with the slaty female, the latter not having at first any *latent* tendency to produce pale-blue offspring.

On the other hand, if one or two males were to vary ever so slightly in paleness, and the variations were from the first limited in their transmission to the male sex, the task of making a new breed of the desired kind would be easy, for such males would simply have to be selected and matched with ordinary females. An analogous case has actually occurred, for there are breeds of the pigeon in Belgium[4] in which the males alone are marked with black striæ. In the case of the fowl, variations of colour limited in their transmission to the male sex habitually occur. Even when this form of inheritance prevails, it might well happen that some of the successive steps in the process of variation might be transferred to the female, who would then come to resemble in a slight degree the male, as occurs in some breeds of the fowl. Or again, the greater number, but

---

[3] Temminck says that the tail of the female *Phasianus Sœmmerringii* is only six inches long, 'Planches coloriées,' vol. v. 1838, p. 487 and 488: the measurements above given were made for me by Mr. Sclater. For the common pheasant, see Macgillivray, 'Hist. Brit. Birds,' vol. i. p. 118-121.

[4] Dr. Chapuis, 'Le Pigeon Voyageur Belge,' 1865, p. 87.

not all, of the successive steps might be transferred to both sexes, and the female would then closely resemble the male. There can hardly be a doubt that this is the cause of the male pouter pigeon having a somewhat larger crop, and of the male carrier pigeon having somewhat larger wattles, than their respective females; for fanciers have not selected one sex more than the other, and have had no wish that these characters should be more strongly displayed in the male than in the female, yet this is the case with both breeds.

The same process would have to be followed, and the same difficulties would be encountered, if it were desired to make a breed with the females alone of some new colour.

Lastly, our fancier might wish to make a breed with the two sexes differing from each other, and both from the parent-species. Here the difficulty would be extreme, unless the successive variations were from the first sexually limited on both sides, and then there would be no difficulty. We see this with the fowl; thus the two sexes of the pencilled Hamburghs differ greatly from each other, and from the two sexes of the aboriginal *Gallus bankiva;* and both are now kept constant to their standard of excellence by continued selection, which would be impossible unless the distinctive characters of both were limited in their transmission. The Spanish fowl offers a more curious case; the male has an immense comb, but some of the successive variations, by the accumulation of which it was acquired, appear to have been transferred to the female; for she has a comb many times larger than that of the females of the parent-species. But the comb of the female differs in one respect from that of the male, for it is apt to lop over; and within a recent period it has been ordered by the fancy that this should always be the case, and success has quickly followed the order. Now the lopping of the comb must be sexually limited in its transmission, otherwise it would prevent the comb of the male from being perfectly upright, which would be abhorrent to every fancier. On the other hand the uprightness of the comb in the male must likewise be a sexually-limited character, otherwise it would prevent the comb of the female from lopping over.

From the foregoing illustrations, we see that even with almost unlimited time at command, it would be an extremely difficult and complex process, though perhaps not impossible, to change through selection one form of transmission into the other. Therefore, without distinct evidence in each case, I am unwilling to admit that this has often been effected with natural species. On the other hand by means of successive variations, which were from the first sexually limited in their transmission, there would not be the least difficulty in rendering a male bird widely different in colour or in any other character from the female; the latter being left unaltered, or slightly altered, or specially modified for the sake of protection.

As bright colours are of service to the males in their rivalry with other males, such colours would be selected, whether or not they were transmitted exclusively to the same sex. Consequently the females might be expected often to partake of the brightness of the males to a greater or less degree; and this occurs with a host of species. If all the successive variations were transmitted equally to both sexes, the females would be undistinguishable from the males; and this likewise occurs with many birds. If, however, dull colours were of high importance for the safety of the female during incubation, as with many ground birds, the females which varied in brightness, or which received through inheritance from the males any marked accession of brightness, would sooner or later be destroyed. But the tendency in the males to continue for an indefinite period transmitting to their female offspring their own brightness, would have to be eliminated by a change in the form of inheritance; and this, as shewn by our previous illustration, would be extremely difficult. The more probable result of the long-continued destruction of the more brightly-coloured females, supposing the equal form of transmission to prevail, would be the lessening or annihilation of the bright colours of the males, owing to their continually crossing with the duller females. It would be tedious to follow out all the other possible results; but I may remind the reader, as shewn in the eighth chapter, that if sexually-limited variations in brightness occurred in the females, even if they were not in the least injurious to them and consequently were not eliminated, yet they would not be favoured or selected, for the male usually accepts any female, and does not select the more attractive individuals; consequently these variations would be liable to be lost, and would have little influence on the character of the race; and this will aid in accounting for the females being commonly less brightly-coloured than the males.

In the chapter just referred to, instances were given, and any number might have been added, of variations occurring at different ages, and inherited at the same age. It was also shewn that variations which occur late in life are commonly transmitted to the same sex in which they first appeared; whilst variations occurring early in life are apt to be transmitted to both sexes; not that all the cases of sexually-limited transmission can thus be accounted for. It was further shewn that if a male bird varied by becoming brighter whilst young, such variations would be of no service until the age for reproduction had arrived, and there was competition between rival males. If we suppose that three-fourths of the young males of any species are on an average destroyed by various enemies; then the chances would be as three to one against any one individual more brightly-coloured than usual surviving to propagate its kind. But in the case of birds which live on the ground and which commonly need the protection of dull colours, bright tints would be far more dangerous to the young and inexperienced than to the adult males. Consequently the males which varied in brightness whilst young would suffer much destruction and be eliminated through natural selection; on the other hand the males which varied in this manner when nearly mature, notwithstanding that they were exposed to some additional danger, might survive, and from being favoured through sexual selection, would procreate their kind. The brightly-coloured young males being destroyed and the mature ones being successful in their courtship, may account, on the principle of a relation existing between the period of variation and the form of transmission, for the males alone of many birds, having acquired and transmitted brilliant colours to their male offspring alone. But I by no means wish to maintain that the influence of age on the form of transmission is indirectly the sole cause of the great difference in brilliancy between the sexes of many birds.

As with all birds in which the sexes differ in colour, it is an interesting question whether the males alone have been modified through sexual selection, the females being left, as far as this agency is concerned, unchanged or only partially changed; or whether the females have been specially modified through natural selection for the sake of protection, I will discuss this question at considerable length, even at greater length than its intrinsic importance deserves; for various curious collateral points may thus be conveniently considered.

Before we enter on the subject of colour, more especially in reference to Mr. Wallace's conclusions, it may be useful to discuss under a similar point of view some other differences between the sexes. A breed of fowls formerly existed in Germany [5] in which the hens were furnished with spurs; they were good layers, but they so greatly disturbed their nests with their spurs that they could not be allowed to sit on their own eggs. Hence at one time it appeared to me probable that with the females of the wild Gallinaceæ the development of spurs had been checked through natural selection, from the injury thus caused to their nests. This seemed all the more probable as the wing-spurs, which could not be injurious during nidification, are often as well developed in the female as in the male; though in not a few cases they are rather larger in the male. When the male is furnished with leg-spurs the female almost always exhibits rudiments of them,—the rudiment sometimes consisting of a mere scale, as with the species of Gallus. Hence it might be argued that the females had aboriginally been furnished with well-developed spurs, but that these had subsequently been lost either through disuse or natural selection. But if this view be admitted, it would have to be extended to innumerable other cases; and it implies that the female progenitors of the existing spur-bearing species were once encumbered with an injurious appendage.

In some few genera and species, as in Galloperdix, Acomus, and the Javan peacock (*Pavo muticus*), the

---

[5] Bechstein, 'Naturgesch. Deutschlands,' 1793, B. iii. s. 339.

females, as well as the males, possess well-developed spurs. Are we to infer from this fact that they construct a different sort of nest, not liable to be injured by their spurs, from that made by their nearest allies, so that there has been no need for the removal of their spurs? Or are we to suppose that these females especially require spurs for their defence? It is a more probable conclusion that both the presence and absence of spurs in the females result from different laws of inheritance having prevailed, independently of natural selection. With the many females in which spurs appear as rudiments, we may conclude that some few of the successive variations, through which they were developed in the males, occurred very early in life, and were as a consequence transferred to the females. In the other and much rarer cases, in which the females possess fully developed spurs, we may conclude that all the successive variations were transferred to them; and that they gradually acquired the inherited habit of not disturbing their nests.

The vocal organs and the variously-modified feathers for producing sound, as well as the proper instincts for using them, often differ in the two sexes, but are sometimes the same in both. Can such differences be accounted for by the males having acquired these organs and instincts, whilst the females have been saved from inheriting them, on account of the danger to which they would have been exposed by attracting the attention of birds or beasts of prey? This does not seem to me probable, when we think of the multitude of birds which with impunity gladden the country with their voices during the spring.[6] It is a safer conclu-

[6] Daines Barrington, however, thought it probable ('Phil. Transact.' 1773, p. 164) that few female birds sing, because the talent would have

sion that as vocal and instrumental organs are of special service only to the males during their courtship, these organs were developed through sexual selection and continued use in this sex alone—the successive variations and the effects of use having been from the first limited in their transmission in a greater or less degree to the male offspring.

Many analogous cases could be advanced; for instance the plumes on the head, which are generally longer in the male than in the female, sometimes of equal length in both sexes, and occasionally absent in the female,—these several cases sometimes occurring in the same group of birds. It would be difficult to account for a difference of this kind between the sexes on the principle of the female having been benefited by possessing a slightly shorter crest than the male, and its consequent diminution or complete suppression through natural selection. But I will take a more favourable case, namely, the length of the tail. The long train of the peacock would have been not only inconvenient but dangerous to the peahen during the period of incubation and whilst accompanying her young. Hence there is not the least *à priori* improbability in the development of her tail having been checked through natural selection. But the females of various pheasants, which apparently are exposed on their open nests to as much danger as the peahen, have tails of considerable length. The females as well as the males of the *Menura superba* have long tails, and they build a domed nest, which is a great anomaly in so large a bird. Naturalists have wondered how the female Menura could manage her tail during incubation; but it

---

been dangerous to them during incubation. He adds, that a similar view may possibly account for the inferiority of the female to the male in plumage.

is now known[7] that she "enters the nest head first, "and then turns round with her tail sometimes over "her back, but more often bent round by her side. "Thus in time the tail becomes quite askew, and is a "tolerable guide to the length of time the bird has been sitting." Both sexes of an Australian kingfisher (*Tanysiptera sylvia*) have the middle tail-feathers greatly lengthened; and as the female makes her nest in a hole, these feathers become, as I am informed by Mr. R. B. Sharpe, much crumpled during nidification.

In these two cases the great length of the tail-feathers must be in some degree inconvenient to the female; and as in both species the tail-feathers of the female are somewhat shorter than those of the male, it might be argued that their full development had been prevented through natural selection. Judging from these cases, if with the peahen, the development of the tail had been checked only when it became inconveniently or dangerously long, she would have acquired a much longer tail than she actually possesses; for her tail is not nearly so long, relatively to the size of her body, as that of many female pheasants, nor longer than that of the female turkey. It must also be borne in mind, that in accordance with this view as soon as the tail of the peahen became dangerously long, and its development was consequently checked, she would have continually reacted on her male progeny, and thus have prevented the peacock from acquiring his present magnificent train. We may therefore infer that the length of the tail in the peacock and its shortness in the peahen are the result of the requisite variations in the male having been from the first transmitted to the male offspring alone.

---

[7] Mr. Ramsay, in 'Proc. Zoolog. Soc.' 1868, p. 50.

We are led to a nearly similar conclusion with respect to the length of the tail in the various species of pheasants. In the Eared pheasant (*Crossoptilon auritum*) the tail is of equal length in both sexes, namely, sixteen or seventeen inches; in the common pheasant it is about twenty inches long in the male, and twelve in the female; in Sœmmerring's pheasant, thirty-seven inches in the male, and only eight in the female; and lastly in Reeve's pheasant it is sometimes actually seventy-two inches long in the male and sixteen in the female. Thus in the several species, the tail of the female differs much in length, irrespectively of that of the male; and this can be accounted for as it seems to me, with much more probability, by the laws of inheritance,—that is by the successive variations having been from the first more or less closely limited in their transmission to the male sex,—than by the agency of natural selection, owing to the length of tail having been injurious in a greater or less degree to the females of the several species.

We may now consider Mr. Wallace's arguments in regard to the sexual coloration of birds. He believes that the bright tints originally acquired through sexual selection by the males, would in all or almost all cases have been transmitted to the females, unless the transference had been checked through natural selection. I may here remind the reader that various facts bearing on this view have already been given under reptiles, amphibians, fishes, and lepidoptera. Mr. Wallace rests his belief chiefly, but not exclusively, as we shall see in the next chapter, on the following statement,[8] that when both sexes are coloured in a strikingly-

8 'Journal of Travel,' edited by A. Murray, vol. i. 1868, p. 78.

conspicuous manner the nest is of such a nature as to conceal the sitting bird; but when there is a marked contrast of colour between the sexes, the male being gay and the female dull-coloured, the nest is open and exposes the sitting bird to view. This coincidence, as far as it goes, certainly supports the belief that the females which sit on open nests have been specially modified for the sake of protection. Mr. Wallace admits that there are, as might have been expected, some exceptions to his two rules, but it is a question whether the exceptions are not so numerous as seriously to invalidate them.

There is in the first place much truth in the Duke of Argyll's remark[9] that a large domed nest is more conspicuous to an enemy, especially to all tree-haunting carnivorous animals, than a smaller open nest. Nor must we forget that with many birds which build open nests the males sit on the eggs and aid in feeding the young as well as the females: this is the case, for instance, with *Pyranga æstiva*,[10] one of the most splendid birds in the United States, the male being vermilion, and the female light brownish-green. Now if brilliant colours had been extremely dangerous to birds whilst sitting on their open nests, the males in these cases would have suffered greatly. It might, however, be of such paramount importance to the male to be brilliantly coloured, in order to beat his rivals, that this would more than compensate for some additional danger.

Mr. Wallace admits that with the King-crows (Dicrurus), Orioles, and Pittidæ, the females are conspicuously coloured, yet they build open nests; but he urges that the birds of the first group are highly pug-

---

[9] 'Journal of Travel,' edited by A. Murray, vol. i. 1868, p. 281.
[10] Audubon, 'Ornithological Biography,' vol. i. p. 233.

nacious and could defend themselves; that those of the second group take extreme care in concealing their open nests, but this does not invariably hold good;[11] and that with the birds of the third group the females are brightly coloured chiefly on the under surface. Besides these cases the whole great family of pigeons, which are sometimes brightly, and almost always conspicuously coloured, and which are notoriously liable to the attacks of birds of prey, offers a serious exception to the rule, for pigeons almost always build open and exposed nests. In another large family, that of the Humming-birds, all the species build open nests, yet with some of the most gorgeous species the sexes are alike; and in the majority, the females, though less brilliant than the males, are very brightly coloured. Nor can it be maintained that all female humming-birds, which are brightly coloured, escape detection by their tints being green, for some display on their upper surfaces red, blue, and other colours.[12]

In regard to birds which build in holes or construct domed nests, other advantages, as Mr. Wallace remarks, besides concealment are gained, such as shelter from the rain, greater warmth, and in hot countries protection from the rays of the sun;[13] so that it is no valid

---

[11] Jerdon, 'Birds of India,' vol. ii. p. 108. Gould's 'Handbook of the Birds of Australia,' vol. i. p. 463.

[12] For instance, the female *Eupetomena macroura* has the head and tail dark blue with reddish loins; the female *Lampornis porphyrurus* is blackish-green on the upper surface, with the lores and sides of the throat crimson; the female *Eulampis jugularis* has the top of the head and back green, but the loins and the tail are crimson. Many other instances of highly conspicuous females could be given. See Mr. Gould's magnificent work on this family.

[13] Mr. Salvin noticed in Guatemala ('Ibis,' 1864, p. 375) that humming-birds were much more unwilling to leave their nests during very hot weather, when the sun was shining brightly, than during cool, cloudy, or rainy weather.

objection to his view that many birds having both sexes obscurely coloured build concealed nests.[14] The female Horn-bills (*Buceros*), for instance, of India and Africa are protected, during nidification, with extraordinary care, for the male plaisters up the hole in which the female sits on her eggs, and leaves only a small orifice through which he feeds her; she is thus kept a close prisoner during the whole period of incubation;[15] yet female hornbills are not more conspicuously coloured than many other birds of equal size which build open nests. It is a more serious objection to Mr. Wallace's view, as is admitted by him, that in some few groups the males are brilliantly coloured and the females obscure, and yet the latter hatch their eggs in domed nests. This is the case with the Grallinæ of Australia, the Superb Warblers (Maluridæ) of the same country, the Sun-birds (Nectariniæ), and with several of the Australian Honey-suckers or Meliphagidæ.[16]

If we look to the birds of England we shall see that there is no close and general relation between the colours of the female and the nature of the nest constructed by her. About forty of our British birds (excluding those of large size which could defend themselves) build in holes in banks, rocks, or trees, or construct domed nests. If we take the colours of the female goldfinch, bullfinch, or blackbird, as a standard of the degree of conspicuousness, which is not highly dangerous to the sitting female, then out of the above forty birds, the females of only twelve can be considered

---

[14] I may specify, as instances of obscurely-coloured birds building concealed nests, the species belonging to eight Australian genera, described in Gould's 'Handbook of the Birds of Australia,' vol. i. p. 340, 362, 365, 383, 387, 389, 391, 414.

[15] Jerdon, 'Birds of India,' vol. i. p. 244.

[16] On the nidification and colours of these latter species, see Gould's 'Handbook,' &c., vol. i. p. 504, 527.

as conspicuous to a dangerous degree, the remaining twenty-eight being inconspicuous.[17] Nor is there any close relation between a well-pronounced difference in colour between the two sexes, and the nature of the nest constructed. Thus the male house-sparrow (*Passer domesticus*) differs much from the female, the male tree-sparrow (*P. montanus*) differs hardly at all, and yet both build well-concealed nests. The two sexes of the common fly-catcher (*Muscicapa grisola*) can hardly be distinguished, whilst the sexes of the pied fly-catcher (*M. luctuosa*) differ considerably, and both build in holes. The female blackbird (*Turdus merula*) differs much, the female ring-ouzel (*T. torquatus*) differs less, and the female common thrush (*T. musicus*) hardly at all from their respective males; yet all build open nests. On the other hand, the not very distantly-allied water-ouzel (*Cinclus aquaticus*) builds a domed nest, and the sexes differ about as much as in the case of the ring-ouzel. The black and red grouse (*Tetrao tetrix* and *T. Scoticus*) build open nests, in equally well-concealed spots, but in the one species the sexes differ greatly, and in the other very little.

Notwithstanding the foregoing objections, I cannot doubt, after reading Mr. Wallace's excellent essay,

---

[17] I have consulted, on this subject, Macgillivray's 'British Birds,' and though doubts may be entertained in some cases in regard to the degree of concealment of the nest, and of the degree of conspicuousness of the female, yet the following birds, which all lay their eggs in holes or in domed nests, can hardly be considered, according to the above standard, as conspicuous: Passer, 2 species; Sturnus, of which the female is considerably less brilliant than the male; Cinclus; Motacilla boarula (?); Erithacus (?); Fruticola, 2 sp.; Saxicola; Ruticilla, 2 sp.; Sylvia, 3 sp.; Parus, 3 sp.; Mecistura; Anorthura; Certhia; Sitta; Yunx; Muscicapa, 2 sp.; Hirundo, 3 sp.; and Cypselus. The females of the following 12 birds may be considered as conspicuous according to the same standard, viz., Pastor, Motacilla alba, Parus major and P. cæruleus, Upupa, Picus, 4 sp., Coracias, Alcedo, and Merops.

that looking to the birds of the world, a large majority of the species in which the females are conspicuously coloured (and in this case the males with rare exceptions are equally conspicuous), build concealed nests for the sake of protection. Mr. Wallace enumerates[18] a long series of groups in which this rule holds good; but it will suffice here to give, as instances, the more familiar groups of kingfishers, toucans, trogons, puff-birds (Capitonidæ), plaintain-eaters (Musophagæ), woodpeckers, and parrots. Mr. Wallace believes that in these groups, as the males gradually acquired through sexual selection their brilliant colours, these were transferred to the females and were not eliminated by natural selection, owing to the protection which they already enjoyed from their manner of nidification. According to this view, their present manner of nesting was acquired before their present colours. But it seems to me much more probable that in most cases as the females were gradually rendered more and more brilliant from partaking of the colours of the male, they were gradually led to change their instincts (supposing that they originally built open nests), and to seek protection by building domed or concealed nests. No one who studies, for instance, Audubon's account of the differences in the nests of the same species in the Northern and Southern United States,[19] will feel any great difficulty in admitting that birds, either by a change (in the strict sense of the word) of their habits, or through the natural selection of so-called spontaneous variations of instinct, might readily be led to modify their manner of nesting.

[18] 'Journal of Travel,' edited by A. Murray, vol. i. p. 78.

[19] See many statements in the 'Ornithological Biography.' See, also, some curious observations on the nests of Italian birds by Eugenio Bettoni, in the 'Atti della Società Italiana,' vol. xi. 1869, p. 487.

This way of viewing the relation, as far as it holds good, between the bright colours of female birds and their manner of nesting, receives some support from certain analogous cases occurring in the Sahara Desert. Here, as in most other deserts, various birds, and many other animals, have had their colours adapted in a wonderful manner to the tints of the surrounding surface. Nevertheless there are, as I am informed by the Rev. Mr. Tristram, some curious exceptions to the rule; thus the male of the *Monticola cyanea* is conspicuous from his bright blue colour, and the female almost equally conspicuous from her mottled brown and white plumage; both sexes of two species of Dromolæa are of a lustrous black; so that these three birds are far from receiving protection from their colours, yet they are able to survive, for they have acquired the habit, when in danger, of taking refuge in holes or crevices in the rocks.

With respect to the above-specified groups of birds, in which the females are conspicuously coloured and build concealed nests, it is not necessary to suppose that each separate species had its nidifying instinct specially modified; but only that the early progenitors of each group were gradually led to build domed or concealed nests; and afterwards transmitted this instinct, together with their bright colours, to their modified descendants. This conclusion, as far as it can be trusted, is interesting, namely, that sexual selection, together with equal or nearly equal inheritance by both sexes, have indirectly determined the manner of nidification of whole groups of birds.

Even in the groups in which, according to Mr. Wallace, the females from being protected during nidification, have not had their bright colours eliminated through natural selection, the males often differ in a slight, and occasionally in a considerable degree, from the females. This is a significant fact, for such differences in colour must be accounted for on the principle of some of the variations in the males having been from the first limited in their transmission to the same sex; as it can hardly be maintained that these differences, especially when very slight, serve as a protection to the female. Thus all the species in the splendid group of the Trogons build in holes; and Mr. Gould gives figures [20] of both sexes of twenty-five species, in all of which, with one partial exception, the sexes differ sometimes slightly, sometimes conspicuously, in colour,—the males being always more beautiful than the females, though the latter are likewise beautiful. All the species of kingfisher build in holes, and with most of the species the sexes are equally brilliant, and thus far Mr. Wallace's rule holds good; but in some of the Australian species the colours of the females are rather less vivid than those of the male; and in one splendidly-coloured species, the sexes differ so much that they were at first thought to be specifically distinct.[21] Mr. R. B. Sharpe, who has especially studied this group, has shewn me some American species (Ceryle) in which the breast of the male is belted with black. Again, in Carcineutes, the difference between the sexes is conspicuous: in the male the upper surface is dull-blue banded with black, the lower surface being partly fawn-coloured, and there is much red about the head; in the female the upper surface is reddish-brown banded with black, and the lower surface white with black markings. It is an interesting fact, as shewing how the same peculiar style of sexual

[20] See his 'Monograph of the Trogonidæ,' first edition.

[21] Namely Cyanalcyon. Gould's 'Handbook of the Birds of Australia,' vol. i. p. 133; see, also, p. 130, 136.

colouring often characterises allied forms, that in three species of Dacelo the male differs from the female only in the tail being dull-blue banded with black, whilst that of the female is brown with blackish bars; so that here the tail differs in colour in the two sexes in exactly the same manner as the whole upper surface in the sexes of Carcineutes.

With parrots, which likewise build in holes, we find analogous cases: in most of the species both sexes are brilliantly coloured and undistinguishable, but in not a few species the males are coloured rather more vividly than the females, or even very differently from them. Thus, besides other strongly-marked differences, the whole under surface of the male King Lory (*Aprosmictus scapulatus*) is scarlet, whilst the throat and chest of the female is green tinged with red: in the *Euphema splendida* there is a similar difference, the face and wing-coverts moreover of the female being of a paler blue than in the male.[22] In the family of the tits (*Parinæ*), which build concealed nests, the female of our common blue tomtit (*Parus cæruleus*) is "much less brightly coloured" than the male; and in the magnificent Sultan yellow tit of India the difference is greater.[23]

Again in the great group of the woodpeckers,[24] the sexes are generally nearly alike, but in the *Megapicus validus* all those parts of the head, neck, and breast, which are crimson in the male are pale brown in the female. As in several woodpeckers the head of the male is bright crimson, whilst that of the female is

---

22 Every gradation of difference between the sexes may be followed in the parrots of Australia. See Gould's 'Handbook,' &c., vol. ii. p. 14-102.

23 Macgillivray's 'British Birds,' vol. ii. p. 433. Jerdon, 'Birds of India,' vol. ii. p. 282.

24 All the following facts are taken from M. Malherbe's magnificent 'Monographie des Picidées,' 1861.

plain, it occurred to me that this colour might possibly make the female dangerously conspicuous, whenever she put her head out of the hole containing her nest, and consequently that this colour, in accordance with Mr. Wallace's belief, had been eliminated. This view is strengthened by what Malherbe states with respect to *Indopicus carlotta;* namely, that the young females, like the young males, have some crimson about their heads, but that this colour disappears in the adult female, whilst it is intensified in the adult male. Nevertheless the following considerations render this view extremely doubtful: the male takes a fair share in incubation,[25] and would be thus far almost equally exposed to danger; both sexes of many species have their heads of an equally bright crimson; in other species the difference between the sexes in the amount of scarlet is so slight that it can hardly make any appreciable difference in the danger incurred; and lastly, the colouring of the head in the two sexes often differs slightly in other ways.

The cases, as yet given, of slight and graduated differences in colour between the males and females in the groups, in which as a general rule the sexes resemble each other, all relate to species which build domed or concealed nests. But similar gradations may likewise be observed in groups in which the sexes as a general rule resemble each other, but which build open nests. As I have before instanced the Australian parrots, so I may here instance, without giving any details, the Australian pigeons.[26] It deserves especial notice that in all these cases the slight differences in

---

[25] Audubon's 'Ornithological Biography,' vol. ii. p. 75; see also the 'Ibis,' vol. i. p. 268.

[26] Gould's 'Handbook of the Birds of Australia,' vol. ii. p. 109-149.

plumage between the sexes are of the same general nature as the occasionally greater differences. A good illustration of this fact has already been afforded by those kingfishers in which either the tail alone or the whole upper surface of the plumage differs in the same manner in the two sexes. Similar cases may be observed with parrots and pigeons. The differences in colour between the sexes of the same species are, also, of the same general nature as the differences in colour between the distinct species of the same group. For when in a group in which the sexes are usually alike, the male differs considerably from the female, he is not coloured in a quite new style. Hence we may infer that within the same group the special colours of both sexes when they are alike, and the colours of the male, when he differs slightly or even considerably from the female, have in most cases been determined by the same general cause; this being sexual selection.

It is not probable, as has already been remarked, that differences in colour between the sexes, when very slight, can be of service to the female as a protection. Assuming, however, that they are of service, they might be thought to be cases of transition; but we have no reason to believe that many species at any one time are undergoing change. Therefore we can hardly admit that the numerous females which differ very slightly in colour from their males are now all commencing to become obscure for the sake of protection. Even if we consider somewhat more marked sexual differences, is it probable, for instance, that the head of the female chaffinch, the crimson on the breast of the female bullfinch,—the green of the female greenfinch,—the crest of the female golden-crested wren, have all been rendered less bright by the slow process of selection for the sake of protection? I cannot think so; and still less with the slight differences between the sexes of those birds which build concealed nests. On the other hand, the differences in colour between the sexes, whether great or small, may to a large extent be explained on the principle of the successive variations, acquired by the males through sexual selection, having been from the first more or less limited in their transmission to the females. That the degree of limitation should differ in different species of the same group will not surprise any one who has studied the laws of inheritance, for they are so complex that they appear to us in our ignorance to be capricious in their action.[27]

As far as I can discover there are very few groups of birds containing a considerable number of species, in which all have both sexes brilliantly coloured and alike; but this appears to be the case, as I hear from Mr. Sclater, with the Musophagæ or plain-tain-eaters. Nor do I believe that any large group exists in which the sexes of all the species are widely dissimilar in colour: Mr. Wallace informs me that the chatterers of S. America (*Cotingidæ*) offer one of the best instances; but with some of the species, in which the male has a splendid red breast, the female exhibits some red on her breast; and the females of other species shew traces of the green and other colours of the males. Nevertheless we have a near approach to close sexual similarity or dissimilarity throughout several groups: and this, from what has just been said of the fluctuating nature of inheritance, is a somewhat surprising circumstance. But that the same laws should largely prevail with allied animals is not surprising. The domestic fowl has produced a

---

[27] See remarks to this effect in my work on 'Variation under Domestication,' vol. ii. chap. xii.

great number of breeds and sub-breeds, and in these the sexes generally differ in plumage; so that it has been noticed as a remarkable circumstance when in certain sub-breeds they resemble each other. On the other hand, the domestic pigeon has likewise produced a vast number of distinct breeds and sub-breeds, and in these, with rare exceptions, the two sexes are identically alike. Therefore if other species of Gallus and Columba were domesticated and varied, it would not be rash to predict that the same general rules of sexual similarity and dissimilarity, depending on the form of transmission, would, in both cases, hold good. In a similar maner the same form of transmission has generally prevailed throughout the same natural groups, although marked exceptions to this rule occur. Within the same family or even genus, the sexes may be identically alike or very different in colour. Instances have already been given relating to the same genus, as with sparrows, fly-catchers, thrushes and grouse. In the family of pheasants the males and females of almost all the species are wonderfully dissimilar, but are quite similar in the eared pheasant or *Crossoptilon auritum*. In two species of Chloehaga, a genus of geese, the males cannot be distinguished from the females, except by size; whilst in two others, the sexes are so unlike that they might easily be mistaken for distinct species.[28]

The laws of inheritance can alone account for the following cases, in which the female by acquiring at a late period of life certain characters proper to the male, ultimately comes to resemble him in a more or less complete manner. Here protection can hardly have come into play. Mr. Blyth informs me that the females of *Oriolus melanocephalus* and of some

[28] The 'Ibis,' vol. vi. 1864, p. 122.

allied species, when sufficiently mature to breed, differ considerably in plumage from the adult males; but after the second or third moults they differ only in their beaks having a slight greenish tinge. In the dwarf bitterns (Ardetta), according to the same authority, "the male acquires his final livery at the "first moult, the female not before the third or fourth "moult; in the meanwhile she presents an intermediate garb, which is ultimately exchanged for the "same livery as that of the male." So again the female *Falco peregrinus* acquires her blue plumage more slowly than the male. Mr. Swinhoe states that with one of the Drongo shrikes (*Dicrurus macrocercus*) the male whilst almost a nestling, moults his soft brown plumage and becomes of a uniform glossy greenish-black; but the female retains for a long time the white striæ and spots on the axillary feathers; and does not completely assume the uniform black colour of the male for the first three years. The same excellent observer remarks that in the spring of the second year the female spoonbill (Platalea) of China resembles the male of the first year, and that apparently it is not until the third spring that she acquires the same adult plumage as that possessed by the male at a much earlier age. The female *Bombycilla carolinensis* differs very little from the male, but the appendages, which like beads of red sealing-wax ornament the wing-feathers, are not developed in her so early in life as in the male. The upper mandible in the male of an Indian parrakeet (*Palæornis Javanicus*) is coral-red from his earliest youth, but in the female, as Mr. Blyth has observed with caged and wild birds, it is at first black and does not become red until the bird is at least a year old, at which age the sexes resemble each other in all respects. Both sexes of the wild turkey are ultimately furnished with a tuft of bristles on the breast, but in two-year-old birds the tuft is about four inches long in the male and hardly apparent in the female; when, however, the latter has reached her fourth year, it is from four to five inches in length.[29]

In these cases, the females follow a normal course of development in ultimately becoming like the males; and such cases must not be confounded with those in which diseased or old females assume masculine characters, or with those in which perfectly fertile females, whilst young, acquire through variation or some unknown cause the characters of the male.[30] But all these cases have so much in common that they depend, according to the hypothesis of pangenesis, on gemmules derived from each part of the male being present, though latent, in the female; their development following on some slight change in the elective affinities of her constituent tissues.

A few words must be added on changes of plumage in relation to the season of the year. From reasons formerly assigned there can be little doubt that the elegant plumes, long pendant feathers, crests, &c., of egrets, herons, and many other birds, which are developed and retained only during the summer, serve exclusively for ornamental or nuptial purposes, though

29 On Ardetta, Translation of Cuvier's 'Règne Animal,' by Mr. Blyth, footnote, p. 159. On the Peregrine Falcon, Mr. Blyth, in Charlesworth's 'Mag. of Nat. Hist.' vol. i. 1837, p. 304. On Dicrurus, 'Ibis,' 1863, p. 44. On the Platalea, 'Ibis,' vol. vi. 1864, p. 366. On the Bombycilla, Audubon's 'Ornitholog. Biography,' vol. i. p. 229. On the Palæornis, see, also, Jerdon, 'Birds of India,' vol. i. p. 263. On the wild turkey, Audubon, ibid. vol. i. p. 15: I hear from Judge Caton that in Illinois the female very rarely acquires a tuft.

30 Mr. Blyth has recorded (Translation of Cuvier's 'Règne Animal,' p. 158) various instances with Lanius, Ruticilla, Linaria, and Anas. Audubon has also recorded a similar case ('Ornith. Biog.' vol. v. p. 519) with *Tyranga æstiva*.

common to both sexes. The female is thus rendered more conspicuous during the period of incubation than during the winter; but such birds as herons and egrets would be able to defend themselves. As, however, plumes would probably be inconvenient and certainly of no use during the winter, it is possible that the habit of moulting twice in the year may have been gradually acquired through natural selection for the sake of casting off inconvenient ornaments during the winter. But this view cannot be extended to the many waders, in which the summer and winter plumages differ very little in colour. With defenceless species, in which either both sexes or the males alone become extremely conspicuous during the breeding-season,—or when the males acquire at this season such long wing or tail-feathers as to impede their flight, as with Cosmetornis and Vidua,—it certainly at first appears highly probable that the second moult has been gained for the special purpose of throwing off these ornaments. We must, however, remember that many birds, such as Birds of Paradise, the Argus pheasant and peacock, do not cast their plumes during the winter; and it can hardly be maintained that there is something in the constitution of these birds, at least of the Gallinaceæ, rendering a double moult impossible, for the ptarmigan moults thrice in the year.[31] Hence it must be considered as doubtful whether the many species which moult their ornamental plumes or lose their bright colours during the winter, have acquired this habit on account of the inconvenience or danger which they would otherwise have suffered.

I conclude, therefore, that the habit of moulting twice in the year was in most or all cases first acquired

---

31 See Gould's 'Birds of Great Britain.'

for some distinct purpose, perhaps for gaining a warmer winter covering; and that variations in the plumage occurring during the summer were accumulated through sexual selection, and transmitted to the offspring at the same season of the year. Such variations being inherited either by both sexes or by the males alone, according to the form of inheritance which prevailed. This appears more probable than that these species in all cases originally tended to retain their ornamental plumage during the winter, but were saved from this through natural selection, owing to the inconvenience or danger thus caused.

I have endeavoured in this chapter to shew that the arguments are not trustworthy in favour of the view that weapons, bright colours, and various ornaments, are now confined to the males owing to the conversion, by means of natural selection, of a tendency to the equal transmission of characters to both sexes into transmission to the male sex alone. It is also doubtful whether the colours of many female birds are due to the preservation, for the sake of protection, of variations which were from the first limited in their transmission to the female sex. But it will be convenient to defer any further discussion on this subject until I treat, in the following chapter, on the differences in plumage between the young and old.

CHAPTER XVI.

BIRDS—*concluded.*

The immature plumage in relation to the character of the plumage in both sexes when adult — Six classes of cases — Sexual differences between the males of closely-allied or representative species — The female assuming the characters of the male — Plumage of the young in relation to the summer and winter plumage of the adults — On the increase of beauty in the Birds of the World — Protective colouring — Conspicuously-coloured birds — Novelty appreciated — Summary of the four chapters on Birds.

WE must now consider the transmission of characters as limited by age in reference to sexual selection. The truth and importance of the principle of inheritance at corresponding ages need not here be discussed, as enough has already been said on the subject. Before giving the several rather complex rules or classes of cases, under which all the differences in plumage between the young and the old, as far as known to me, may be included, it will be well to make a few preliminary remarks.

With animals of all kinds when the young differ in colour from the adults, and the colours of the former are not, as far as we can see, of any special service, they may generally be attributed, like various embryological structures, to the retention by the young of the character of an early progenitor. But this view can be maintained with confidence, only when the young of several species closely resemble each other, and likewise resemble other adult species belonging to the same group; for the latter are the living proofs that such a state of things was formerly possible. Young lions and pumas are marked with feeble stripes or rows of spots, and as many allied species both young and old are similarly marked, no naturalist, who believes in the gradual evolution of species, will doubt that the progenitor of the lion and puma was a striped animal, the young having retained vestiges of the stripes, like the kittens of black cats, which when grown up are not in the least striped. Many species of deer, which when mature are not spotted, are whilst young covered with white spots, as are likewise some few species in their adult state. So again the young in the whole family of pigs (Suidæ), and in certain rather distantly-allied animals, such as the tapir, are marked with dark longitudinal stripes; but here we have a character apparently derived from an extinct progenitor, and now preserved by the young alone. In all such cases the old have had their colours changed in the course of time, whilst the young have remained but little altered, and this has been effected through the principle of inheritance at corresponding ages.

This same principle applies to many birds belonging to various groups, in which the young closely resemble each other, and differ much from their respective adult parents. The young of almost all the Gallinaceæ, and of some distantly-allied birds such as ostriches, are whilst covered with down longitudinally striped; but this character points back to a state of things so remote that it hardly concerns us. Young cross-bills (Loxia) have at first straight beaks like those of other finches, and in their immature striated plumage they resemble the mature redpole and female siskin, as well as the young of the goldfinch, greenfinch, and some other allied species. The young of many kinds of buntings (Emberiza) resemble each other, and likewise the adult state of the common bunting, *E. mili-*

aria. In almost the whole large group of thrushes the young have their breasts spotted—a character which is retained by many species throughout life, but is quite lost by others, as by the *Turdus migratorius.* So again with many thrushes, the feathers on the back are mottled before they are moulted for the first time, and this character is retained for life by certain eastern species. The young of many species of shrikes (Lanius), of some woodpeckers, and of an Indian pigeon (*Chalcophaps Indicus*), are transversely striped on the under surface; and certain allied species or genera when adult are similarly marked. In some closely-allied and resplendent Indian cuckoos (Chrysococcyx), the species when mature differ considerably from each other in colour, but the young cannot be distinguished. The young of an Indian goose (*Sarkidiornis melanonotus*) closely resemble in plumage an allied genus, Dendrocygna, when mature.[1] Similar facts will hereafter be given in regard to certain herons. Young black grouse (*Tetrao tetrix*) resemble the young as well as the old of certain other species, for instance the red grouse or *T. scoticus.* Finally, as Mr. Blyth, who has attended closely to this subject, has well remarked, the natural affinities of many species are best exhibited in their immature plumage; and as the true affinities of all organic beings depend on their descent from a common progenitor, this remark strongly confirms the belief that the immature plumage approximately shews us the former or ancestral condition of the species.

[1] In regard to thrushes, shrikes, and woodpeckers, see Mr. Blyth, in Charlesworth's 'Mag. of Nat. Hist.' vol. i. 1837, p. 304; also footnote to his translation of Cuvier's 'Règne Animal,' p. 159. I give the case of Loxia from Mr. Blyth's information. On thrushes, see also Audubon, 'Ornith. Biography,' vol. ii. p. 195. On Chrysococcyx and Chalcophaps, Blyth, as quoted in Jerdon's 'Birds of India,' vol. iii. p. 485. On Sarkidiornis, Blyth, in 'Ibis,' 1867, p. 175.

Although many young birds belonging to various orders thus give us a glimpse of the plumage of their remote progenitors, yet there are many other birds, both dull-coloured and bright-coloured, in which the young closely resemble their parents. With such species the young of the different species cannot resemble each other more closely than do the parents; nor can they present striking resemblances to allied forms in their adult state. They give us but little insight into the plumage of their progenitors, excepting in so far that when the young and the old are coloured in the same general manner throughout a whole group of species, it is probable that their progenitors were similarly coloured.

We may now consider the classes of cases or rules under which the differences and resemblances, between the plumage of the young and the old, of both sexes or of one sex alone, may be grouped. Rules of this kind were first enounced by Cuvier; but with the progress of knowledge they require some modification and amplification. This I have attempted to do, as far as the extreme complexity of the subject permits, from information derived from various sources; but a full essay on this subject by some competent ornithologist is much needed. In order to ascertain to what extent each rule prevails, I have tabulated the facts given in four great works, namely, by Macgillivray on the birds of Britain, Audubon on those of North America, Jerdon on those of India, and Gould on those of Australia. I may here premise, firstly, that the several cases or rules graduate into each other; and secondly, that when the young are said to resemble their parents, it is not meant that they are identically alike, for their colours are almost always rather less vivid, and the feathers are softer and often of a different shape.

RULES OR CLASSES OF CASES.

I. When the adult male is more beautiful or conspicuous than the adult female, the young of both sexes in their first plumage closely resemble the adult female, as with the common fowl and peacock; or, as occasionally occurs, they resemble her much more closely than they do the adult male.

II. When the adult female is more conspicuous than the adult male, as sometimes though rarely occurs, the young of both sexes in their first plumage resemble the adult male.

III. When the adult male resembles the adult female, the young of both sexes have a peculiar first plumage of their own, as with the robin.

IV. When the adult male resembles the adult female, the young of both sexes in their first plumage resemble the adults, as with the kingfisher, many parrots, crows, hedge-warblers.

V. When the adults of both sexes have a distinct winter and summer plumage, whether or not the male differs from the female, the young resemble the adults of both sexes in their winter dress, or much more rarely in their summer dress, or they resemble the females alone; or the young may have an intermediate character; or again they may differ greatly from the adults in both their seasonal plumages.

VI. In some few cases the young in their first plumage differ from each other according to sex; the young males resembling more or less closely the adult males, and the young females more or less closely the adult females.

CLASS I.—In this class, the young of both sexes resemble, more or less closely, the adult female, whilst the adult male differs, often in the most conspicuous manner, from the adult female. Innumerable instances in all Orders could be given; it will suffice to call to mind the common pheasant, duck, and house-sparrow. The cases under this class graduate into others. Thus the two sexes when adult may differ so slightly, and the young so slightly from the adults, that it is doubtful whether such cases ought to come under the present, or under the third or fourth classes. So again the young of both sexes, instead of being quite alike, may differ in a slight degree from each other, as in our sixth class. These transitional cases, however, are few in number, or at least are not strongly pronounced, in comparison with those which come strictly under the present class.

The force of the present law is well shewn in those groups, in which, as a general rule, the two sexes and the young are all alike; for when the male in these groups does differ from the female, as with certain parrots, kingfishers, pigeons, &c., the young of both sexes resemble the adult female.[2] We see the same fact exhibited still more clearly in certain anomalous cases; thus the male of *Heliothrix auriculata* (one of the humming-birds) differs conspicuously from the female in having a splendid gorget and fine ear-tufts, but the female is remarkable from having a much longer tail than that of the male; now the young of both sexes

---

[2] See, for instance, Mr. Gould's account ('Handbook of the Birds of Australia,' vol. i. p. 133) of Cyanalcyon (one of the Kingfishers) in which, however, the young male, though resembling the adult female, is less brilliantly coloured. In some species of Dacelo the males have blue tails, and the females brown ones; and Mr. R. B. Sharpe informs me that the tail of the young male of *D. Gaudichaudi* is at first brown. Mr. Gould has described (ibid. vol. ii. p. 14, 20, 37) the sexes and the young of certain Black Cockatoos and of the King Lory, with which the same rule prevails. Also Jerdon ('Birds of India,' vol. i. p. 260) on the *Palæornis rosa*, in which the young are more like the female than the male. See Audubon ('Ornith. Biograph.' vol. ii. p. 475) on the two sexes and the young of *Columba passerina*.

resemble (with the exception of the breast being spotted with bronze) the adult female in all respects including the length of her tail, so that the tail of the male actually becomes shorter as he reaches maturity, which is a most unusual circumstance.[3] Again, the plumage of the male goosander (*Mergus merganser*) is more conspicuously coloured, with the scapular and secondary wing-feathers much longer than in the female, but differently from what occurs, as far as I know, in any other bird, the crest of the adult male, though broader than that of the female, is considerably shorter, being only a little above an inch in length; the crest of the female being two and a half inches long. Now the young of both sexes resemble in all respects the adult female, so that their crests are actually of greater length though narrower than in the adult male.[4]

When the young and the females closely resemble each other and both differ from the male, the most obvious conclusion is that the male alone has been modified. Even in the anomalous cases of the Heliothrix and Mergus, it is probable that originally both adult sexes were furnished, the one species with a much elongated tail, and the other with a much elongated crest, these characters having since been partially lost by the adult males from some unexplained cause, and transmitted in their diminished state to their male offspring alone, when arrived at the corresponding age of maturity. The belief that in the present class the male alone has been modified, as far as the differences between the male and the female together with her young are concerned, is strongly supported by some

---

[3] I owe this information to Mr. Gould who shewed me the specimens; see also his 'Introduction to the Trochilidæ,' 1861, p. 120.

[4] Macgillivray, 'Hist. Brit. Birds,' vol. v. p. 207-214.

remarkable facts recorded by Mr. Blyth,[5] with respect to closely-allied species which represent each other in distinct countries. For with several of these representative species the adult males have undergone a certain amount of change and can be distinguished; the females and the young being undistinguishable, and therefore absolutely unchanged. This is the case with certain Indian chats (Thamnobia), with certain honey-suckers (Nectarinia), shrikes (Tephrodornis), certain kingfishers (Tanysiptera), Kallij pheasants (Gallophasis), and tree-partridges (Arboricola).

In some analogous cases, namely with birds having a distinct summer and winter plumage, but with the two sexes nearly alike, certain closely-allied species can easily be distinguished in their summer or nuptial plumage, yet are undistinguishable in their winter as well as in their immature plumage. This is the case with some of the closely-allied Indian wag-tails or Motacillæ. Mr. Swinhoe[6] informs me that three species of Ardeola, a genus of herons, which represent each other on separate continents, are "most strikingly different" when ornamented with their summer plumes, but are hardly, if at all, distinguishable during the winter. The young also of these three species in their immature plumage closely resemble the adults in their winter dress. This case is all the more interesting because with two other species of Ardeola both sexes retain, during the winter and summer, nearly the same plum-

---

[5] See his admirable paper in the 'Journal of the Asiatic Soc. of Bengal,' vol. xix. 1850, p. 223; see also Jerdon, 'Birds of India,' vol. i. introduction, p. xxix. In regard to Tanysiptera, Prof. Schlegel told Mr. Blyth that he could distinguish several distinct races, solely by comparing the adult males.

[6] See also Mr. Swinhoe, in 'Ibis,' July, 1863, p. 131; and a previous paper, with an extract from a note by Mr. Blyth, in 'Ibis,' Jan. 1861, p. 52.

age as that possessed by the three first species during the winter and in their immature state; and this plumage, which is common to several distinct species at different ages and seasons, probably shews us how the progenitor of the genus was coloured. In all these cases, the nuptial plumage which we may assume was originally acquired by the adult males during the breeding-season, and transmitted to the adults of both sexes at the corresponding season, has been modified, whilst the winter and immature plumages have been left unchanged.

The question naturally arises, how is it that in these latter cases the winter plumage of both sexes, and in the former cases the plumage of the adult females, as well as the immature plumage of the young, have not been at all affected? The species which represent each other in distinct countries will almost always have been exposed to somewhat different conditions, but we can hardly attribute the modification of the plumage in the males alone to this action, seeing that the females and the young, though similarly exposed, have not been affected. Hardly any fact in nature shews us more clearly how subordinate in importance is the direct action of the conditions of life, in comparison with the accumulation through selection of indefinite variations, than the surprising difference between the sexes of many birds; for both sexes must have consumed the same food and have been exposed to the same climate. Nevertheless we are not precluded from believing that in the course of time new conditions may produce some direct effect; we see only that this is subordinate in importance to the accumulated results of selection. When, however, a species migrates into a new country, and this must precede the formation of representative species, the changed conditions to which they will almost always have been exposed will cause them to undergo, judging from a widely-spread analogy, a certain amount of fluctuating variability. In this case sexual selection, which depends on an element eminently liable to change—namely the taste or admiration of the female—will have had new shades of colour or other differences to act on and accumulate; and as sexual selection is always at work, it would (judging from what we know of the results on domestic animals of man's unintentional selection), be a surprising fact if animals inhabiting separate districts, which can never cross and thus blend their newly-acquired characters, were not, after a sufficient lapse of time, differently modified. These remarks likewise apply to the nuptial or summer plumage, whether confined to the males or common to both sexes.

Although the females of the above closely-allied species, together with their young, differ hardly at all from each other, so that the males alone can be distinguished, yet in most cases the females of the species within the same genus obviously differ from each other. The differences, however, are rarely as great as between the males. We see this clearly in the whole family of the Gallinaceæ: the females, for instance, of the common and Japan pheasant, and especially of the gold and Amherst pheasant—of the silver pheasant and the wild fowl—resemble each other very closely in colour, whilst the males differ to an extraordinary degree. So it is with the females of most of the Cotingidæ, Fringillidæ, and many other families. There can indeed be no doubt that, as a general rule, the females have been modified to a less extent than the males. Some few birds, however, offer a singular and inexplicable exception; thus the females of *Paradisea apoda* and *P. papuana* differ from each other more than do their respective males;[7] the female of the latter species having the under surface pure white, whilst the female *P. apoda* is deep brown beneath. So, again, as I hear from Professor Newton, the males of two species of Oxynotus (shrikes), which represent each other in the islands of Mauritius and Bourbon,[8] differ but little in colour, whilst the females differ much. In the Bourbon species the female appears to have partially retained an immature condition of plumage, for at first sight she "might be taken for "the young of the Mauritian species." These differences may be compared with those which occur, independently of selection by man, and which we cannot explain, in certain sub-breeds of the game-fowl, in which the females are very different, whilst the males can hardly be distinguished.[9]

As I account so largely by sexual selection for the differences between the males of allied species, how can the differences between the females be accounted for in all ordinary cases? We need not here consider the species which belong to distinct genera; for with these, adaptation to different habits of life, and other agencies, will have come into play. In regard to the differences between the females within the same genus, it appears to me almost certain, after looking through various large groups, that the chief agent has been the transference, in a greater or less degree, to the female of the characters acquired by the males through sexual selection. In the several British finches, the two sexes differ either very slightly or considerably; and if we compare the females of the greenfinch, chaffinch, goldfinch, bullfinch, crossbill, sparrow, &c., we shall see that they

[7] Wallace, 'The Malay Archipelago,' vol. ii. 1869, p. 394.

[8] These species are described, with coloured figures, by M. F. Pollen, in 'Ibis,' 1866, p. 275.

[9] 'Variation of Animals, &c., under Domestication,' vol. i. p. 251.

differ from each other chiefly in the points in which they partially resemble their respective males; and the colours of the males may safely be attributed to sexual selection. With many gallinaceous species the sexes differ to an extreme degree, as with the peacock, pheasant, and fowl, whilst with other species there has been a partial or even complete transference of character from the male to the female. The females of the several species of Polyplectron exhibit in a dim condition, and chiefly on the tail, the splendid ocelli of their males. The female partridge differs from the male only in the red mark on her breast being smaller; and the female wild turkey only in her colours being much duller. In the guinea-fowl the two sexes are undistinguishable. There is no improbability in the plain, though peculiar spotted plumage of this latter bird having been acquired through sexual selection by the males, and then transmitted to both sexes; for it is not essentially different from the much more beautifully-spotted plumage, characteristic of the males alone of the Tragopan pheasants.

It should be observed that, in some instances, the transference of characters from the male to the female has been effected apparently at a remote period, the male having subsequently undergone great changes, without transferring to the female any of his later-gained characters. For instance, the female and the young of the black-grouse (*Tetrao tetrix*) resemble pretty closely both sexes and the young of the red-grouse *T. Scoticus;* and we may consequently infer that the black-grouse is descended from some ancient species, of which both sexes were coloured in nearly the same manner as the red-grouse. As both sexes of this latter species are more plainly barred during the breeding-season than at any other time, and as the male differs slightly from the female in his more strongly-pronounced red and brown tints,[10] we may conclude that his plumage has been, at least to a certain extent, influenced by sexual selection. If so, we may further infer that the nearly similar plumage of the female black-grouse was similarly produced at some former period. But since this period the male black-grouse has acquired his fine black plumage, with his forked and outwardly-curled tail-feathers; but of these characters there has hardly been any transference to the female, excepting that she shews in her tail a trace of the curved fork.

We may therefore conclude that the females of distinct though allied species have often had their plumage rendered more or less different by the transference in various degrees, of characters acquired, both during former and recent times, by the males through sexual selection. But it deserves especial attention that brilliant colours have been transferred much more rarely than other tints. For instance, the male of the red-throated bluebreast (*Cyanecula suecica*) has a rich blue breast, including a sub-triangular red mark; now marks of approximately the same shape have been transferred to the female, but the central space is fulvous instead of red, and is surrounded by mottled instead of blue feathers. The Gallinaceæ offer many analogous cases; for none of the species, such as partridges, quails, guinea-fowls, &c., in which the colours of the plumage have been largely transferred from the male to the female, are brilliantly coloured. This is well exemplified with the pheasants, in which the male is generally so much more brilliant than the female; but with the Eared and Cheer pheasants (*Crossoptilon*

---

[10] Macgillivray, 'Hist. British Birds,' vol. i. p. 172-174.

auritum and *Phasianus Wallichii*) the two sexes closely resemble each other and their colours are dull. We may go so far as to believe that if any part of the plumage in the males of these two pheasants had been brilliantly coloured, this would not have been transferred to the females. These facts strongly support Mr. Wallace's view that with birds which are exposed to much danger during nidification, the transference of bright colours from the male to the female has been checked through natural selection. We must not, however, forget that another explanation, before given, is possible; namely, that the males which varied and became bright, whilst they were young and inexperienced, would have been exposed to much danger, and would generally have been destroyed; the older and more cautious males, on the other hand, if they varied in a like manner, would not only have been able to survive, but would have been favoured in their rivalry with other males. Now variations occurring late in life tend to be transmitted exclusively to the same sex, so that in this case extremely bright tints would not have been transmitted to the females. On the other hand, ornaments of a less conspicuous kind, such as those possessed by the Eared and Cheer pheasants, would not have been dangerous, and if they appeared during early youth, would generally have been transmitted to both sexes.

In addition to the effects of the partial transference of characters from the males to the females, some of the differences between the females of closely-allied species may be attributed to the direct or definite action of the conditions of life.[11] With the males any such

[11] See, on this subject, chap. xxiii. in the 'Variation of Animals and Plants under Domestication.'

action would generally have been masked by the brilliant colours gained through sexual selection; but not so with the females. Each of the endless diversities in plumage, which we see in our domesticated birds is, of course, the result of some definite cause; and under natural and more uniform conditions, some one tint, assuming that it was in no way injurious, would almost certainly sooner or later prevail. The free intercrossing of the many individuals belonging to the same species would ultimately tend to make any change of colour, thus induced, uniform in character.

No one doubts that both sexes of many birds have had their colours adapted for the sake of protection; and it is possible that the females alone of some species may have been thus modified. Although it would be a difficult, perhaps an impossible process, as shewn in the last chapter, to convert through selection one form of transmission into another, there would not be the least difficulty in adapting the colours of the female, independently of those of the male, to surrounding objects, through the accumulation of variations which were from the first limited in their transmission to the female sex. If the variations were not thus limited, the bright tints of the male would be deteriorated or destroyed. Whether the females alone of many species have been thus specially modified, is at present very doubtful. I wish I could follow Mr. Wallace to the full extent; for the admission would remove some difficulties. Any variations which were of no service to the female as a protection would be at once obliterated, instead of being lost simply by not being selected, or from free intercrossing, or from being eliminated when transferred to the male and in any way injurious to him. Thus the plumage of the female would be kept constant in character. It would also be a relief if we could admit that the obscure tints of both sexes of many birds had been acquired and preserved for the sake of protection,—for example, of the hedge-warbler or kitty-wren (*Accentor modularis* and *Troglodytes vulgaris*), with respect to which we have no sufficient evidence of the action of sexual selection. We ought, however, to be cautious in concluding that colours which appear to us dull, are not attractive to the females of certain species; we should bear in mind such cases as that of the common house-sparrow, in which the male differs much from the female, but does not exhibit any bright tints. No one probably will dispute that many gallinaceous birds which live on the open ground have acquired their present colours, at least in part, for the sake of protection. We know how well they are thus concealed; we know that ptarmigans, whilst changing from their winter to their summer plumage, both of which are protective, suffer greatly from birds of prey. But can we believe that the very slight differences in tints and markings between, for instance, the female black and red-grouse serve as a protection? Are partridges, as they are now coloured, better protected than if they had resembled quails? Do the slight differences between the females of the common pheasant, the Japan and golden pheasants, serve as a protection, or might not their plumages have been interchanged with impunity? From what Mr. Wallace has observed of the habits of certain gallinaceous birds in the East he thinks that such slight differences are beneficial. For myself, I will only say that I am not convinced.

Formerly when I was inclined to lay much stress on the principle of protection, as accounting for the less bright colours of female birds, it occurred to me that possibly both sexes and the young might aboriginally have been brightly coloured in an equal degree; but that subsequently, the females from the danger incurred during incubation, and the young from being inexperienced, had been rendered dull as a protection. But this view is not supported by any evidence, and is not probable; for we thus in imagination expose during past times the females and the young to danger, from which it has subsequently been necessary to shield their modified descendants. We have, also, to reduce, through a gradual process of selection, the females and the young to almost exactly the same tints and markings, and to transmit them to the corresponding sex and period of life. It is also a somewhat strange fact, on the supposition that the females and the young have partaken during each stage of the process of modification of a tendency to be as brightly coloured as the males, that the females have never been rendered dull-coloured without the young participating in the same change; for there are no instances, as far as I can discover, of species with the females dull-coloured and the young bright-coloured. A partial exception, however, is offered by the young of certain woodpeckers, for they have "the whole upper part "of the head tinged with red," which afterwards either decreases into a mere circular red line in the adults of both sexes, or quite disappears in the adult females.[12]

Finally, with respect to our present class of cases, the most probable view appears to be that successive variations in brightness or in other ornamental characters, occurring in the males at a rather late period of life have alone been preserved; and that most or all of these variations owing to the late period of life at which they appeared, have been from the first transmitted only to the adult male offspring. Any varia-

[12] Audubon, 'Ornith. Biography,' vol. i. p. 193. Macgillivray, 'Hist. Brit. Birds,' vol. iii. p. 85. See also the case before given of *Indopicus carlotta*.

tions in brightness which occurred in the females or in the young would have been of no service to them, and would not have been selected; moreover, if dangerous, would have been eliminated. Thus the females and the young will either have been left unmodified, or, and this has much more commonly occurred, will have been partially modified by receiving through transference from the males some of the successive variations. Both sexes have perhaps been directly acted on by the conditions of life to which they have long been exposed; but the females from not being otherwise much modified will best exhibit any such effects. These changes and all others will have been kept uniform by the free intercrossing of many individuals. In some cases, especially with ground birds, the females and the young may possibly have been modified, independently of the males, for the sake of protection, so as to have acquired the same dull-coloured plumage.

CLASS II. *When the adult female is more conspicuous than the adult male, the young of both sexes in their first plumage resemble the adult male.*—This class is exactly the reverse of the last, for the females are here more brightly coloured or more conspicuous than the males; and the young, as far as they are known, resemble the adult males instead of the adult females. But the difference between the sexes is never nearly so great as occurs with many birds in the first class, and the cases are comparatively rare. Mr. Wallace who first called attention to the singular relation which exists between the less bright colours of the males and their performing the duties of incubation, lays great stress on this point,[13] as a crucial test that obscure colours have

---

[13] 'Westminster Review,' July, 1867, and A. Murray, 'Journal of Travel,' 1868, p. 83.

been acquired for the sake of protection during the period of nesting. A different view seems to me more probable. As the cases are curious and not numerous, I will briefly give all that I have been able to find.

In one section of the genus Turnix, quail-like birds, the female is invariably larger than the male (being nearly twice as large in one of the Australian species) and this is an unusual circumstance with the Gallinaceæ. In most of the species the female is more distinctly coloured and brighter than the male,[14] but in some few species the sexes are alike. In *Turnix taigoor* of India the male "wants the black on the throat and neck, "and the whole tone of the plumage is lighter and less "pronounced than that of the female." The female appears to be more vociferous, and is certainly much more pugnacious than the male; so that the females and not the males are often kept by the natives for fighting, like game-cocks. As male birds are exposed by the English bird-catchers for a decoy near a trap, in order to catch other males by exciting their rivalry, so the females of this Turnix are employed in India. When thus exposed the females soon begin their "loud "purring call, which can be heard a long way off, "and any females within ear-shot run rapidly to the "spot, and commence fighting with the caged bird." In this way from twelve to twenty birds, all breeding-females, may be caught in the course of a single day. The natives assert that the females after laying their eggs associate in flocks, and leave the males to sit on them. There is no reason to doubt the truth of this assertion, which is supported by some observa-

[14] For the Australian species, see Gould's 'Handbook,' &c., vol. ii. p. 178, 180, 186, and 188. In the British Museum specimens of the Australian Plain-wanderer (*Pedionomus torquatus*) may be seen, shewing similar sexual differences.

tions made in China by Mr. Swinhoe.[15] Mr. Blyth believes, that the young of both sexes resemble the adult male.

The females of the three species of Painted Snipes (Rhynchæa) "are not only larger, but much more richly

Fig. 60. Rhynchæa capensis (from Brehm).

"coloured than the males."[16] With all other birds, in which the trachea differs in structure in the two sexes

[15] Jerdon, 'Birds of India,' vol. iii. p. 596. Mr. Swinhoe, in 'Ibis,' 1865, p. 542; 1866, p. 131, 405.

[16] Jerdon, 'Birds of India,' vol. iii. p. 677.

it is more developed and complex in the male than in the female; but in the *Rhynchæa Australis* it is simple in the male, whilst in the female it makes four distinct convolutions before entering the lungs.[17] The female therefore of this species has acquired an eminently masculine character. Mr. Blyth ascertained, by examining many specimens, that the trachea is not convoluted in either sex of *R. Bengalensis*, which species so closely resembles *R. Australis* that it can hardly be distinguished except by its shorter toes. This fact is another striking instance of the law that secondary sexual characters are often widely different in closely-allied forms; though it is a very rare circumstance when such differences relate to the female sex. The young of both sexes of *R. Bengalensis* in their first plumage are said to resemble the mature male.[18] There is also reason to believe that the male undertakes the duty of incubation, for Mr. Swinhoe[19] found the females before the close of the summer associated in flocks, as occurs with the females of the Turnix.

The females of *Phalaropus fulicarius* and *P. hyperboreus* are larger, and in their summer plumage "more gaily "attired than the males." But the difference in colour between the sexes is far from conspicuous. The male alone of *P. fulicarius* undertakes, according to Professor Steenstrup, the duty of incubation, as is likewise shewn by the state of his breast-feathers during the breeding-season. The female of the dotterel plover (*Eudromias morinellus*) is larger than the male, and has the red and black tints on the lower surface, the white crescent on the breast, and the stripes over the eyes, more strongly pronounced. The male also takes at least a

17 Gould's 'Handbook of the Birds of Australia,' vol. ii. p. 275.

18 'The Indian Field,' Sept. 1858, p. 3.

19 'Ibis,' 1866, p. 298.

share in hatching the eggs; but the female likewise attends to the young.[20] I have not been able to discover whether with these species the young resemble the adult males more closely than the adult females; for the comparison is somewhat difficult to make on account of the double moult.

Turning now to the Ostrich order: the male of the common cassowary (*Casuarius galeatus*) would be thought by any one to be the female, from his smaller size and from the appendages and naked skin about his head being much less brightly coloured; and I am informed by Mr. Bartlett that in the Zoological Gardens it is certainly the male alone who sits on the eggs and takes care of the young.[21] The female is said by Mr. T. W. Wood[22] to exhibit during the breeding-season a most pugnacious disposition; and her wattles then become enlarged and more brilliantly coloured. So again the female of one of the emus (*Dromœus irroratus*) is considerably larger than the male, and she possesses a slight top-knot, but is otherwise undistinguishable in plumage. She appears, however, "to have greater power, when angry "or otherwise excited, of erecting, "like a turkey-cock, the feathers of her neck and

---

20 For these several statements, see Mr. Gould's 'Birds of Great Britain.' Prof. Newton informs me that he has long been convinced, from his own observations and from those of others, that the males of the above-named species take either the whole or a large share of the duties of incubation, and that they "shew much greater devotion "towards their young, when in danger, than do the females." So it is, as he informs me, with *Limosa lapponica* and some few other Waders, in which the females are larger and have more strongly contrasted colours than the males.

21 The natives of Ceram (Wallace, 'Malay Archipelago,' vol. ii. p. 150) assert that the male and female sit alternately on the eggs; but this assertion, as Mr. Bartlett thinks, may be accounted for by the female visiting the nest to lay her eggs.

22 'The Student,' April, 1870, p. 124.

"breast. She is usually the more courageous and "pugilistic. She makes a deep hollow guttural boom, "especially at night, sounding like a small gong. The "male has a slenderer frame and is more docile, with "no voice beyond a suppressed hiss when angry, or a "croak." He not only performs the whole duty of incubation, but has to defend the young from their mother; "for as soon as she catches sight of her pro-"geny she becomes violently agitated, and notwith-"standing the resistance of the father appears to use "her utmost endeavours to destroy them. For months "afterwards it is unsafe to put the parents together, "violent quarrels being the inevitable result, in which "the female generally comes off conqueror."[23] So that with this emu we have a complete reversal not only of the parental and incubating instincts, but of the usual moral qualities of the two sexes; the females being savage, quarrelsome and noisy, the males gentle and good. The case is very different with the African ostrich, for the male is somewhat larger than the female and has finer plumes with more strongly contrasted colours; nevertheless he undertakes the whole duty of incubation.[24]

I will specify the few other cases known to me, in which the female is more conspicuously coloured than the male, although nothing is known about their manner of incubation. With the carrion-hawk of the Falkland Islands (*Milvago leucurus*) I was much surprised to find by dissection that the individuals, which had all their tints strongly pronounced, with the cere and legs orange-coloured, were the adult females; whilst

[23] See the excellent account of the habits of this bird under confinement, by Mr. A. W. Bennett, in 'Land and Water,' May, 1868, p. 233.

[24] Mr. Sclater, on the incubation of the Struthiones, 'Proc. Zoo. Soc.,' June 9, 1863.

those with duller plumage and grey legs were the males or the young. In an Australian tree-creeper (*Climacteris erythrops*) the female differs from the male in "being adorned with beautiful, radiated, rufous mark-"ings on the throat, the male having this part quite "plain." Lastly in an Australian night-jar "the female "always exceeds the male in size and in the brilliance "of her tints; the males, on the other hand, have two "white spots on the primaries more conspicuous than "in the female."[25]

We thus see that the cases in which female birds are more conspicuously coloured than the males, with the young in their immature plumage resembling the adult males instead of the adult females, as in the previous class, are not numerous, though they are distributed in various Orders. The amount of difference, also, between the sexes is incomparably less than that which frequently occurs in the last class; so that the cause of the difference, whatever it may have been, has acted on the females in the present class either less energetically or less persistently than on the males in the last class. Mr. Wallace believes that the males have had their colours

---

[25] For the Milvago, see 'Zoology of the Voyage of the "Beagle,"' Birds, 1841, p. 16. For the Climacteris and night-jar (Eurostopodus), see Gould's 'Handbook of the Birds of Australia,' vol. i. p. 602 and 97. The New Zealand shieldrake (*Tadorna variegata*) offers a quite anomalous case: the head of the female is pure white, and her back is redder than that of the male; the head of the male is of a rich dark bronzed colour, and his back is clothed with finely pencilled slate-coloured feathers, so that he may altogether be considered as the more beautiful of the two. He is larger and more pugnacious than the female, and does not sit on the eggs. So that in all these respects this species comes under our first class of cases; but Mr. Sclater ('Proc. Zool. Soc.' 1866, p. 150) was much surprised to observe that the young of both sexes, when about three months old, resembled in their dark heads and necks the adult males, instead of the adult females; so that it would appear in this case that the females have been modified, whilst the males and the young have retained a former state of plumage.

rendered less conspicuous for the sake of protection during the period of incubation; but the difference between the sexes in hardly any of the foregoing cases appears sufficiently great for this view to be safely accepted. In some of the cases the brighter tints of the female are almost confined to the lower surface, and the males, if thus coloured, would not have been exposed to danger whilst sitting on the eggs. It should also be borne in mind that the males are not only in a slight degree less conspicuously coloured than the females, but are of less size, and have less strength. They have, moreover, not only acquired the maternal instinct of incubation, but are less pugnacious and vociferous than the females, and in one instance have simpler vocal organs. Thus an almost complete transposition of the instincts, habits, disposition, colour, size, and of some points of structure, has been effected between the two sexes.

Now if we might assume that the males in the present class have lost some of that ardour which is usual to their sex, so that they no longer search eagerly for the females; or, if we might assume that the females have become much more numerous than the males—and in the case of one Indian Turnix the females are said to be "much more commonly met with than the males"[26]—then it is not improbable that the females would have been led to court the males, instead of being courted by them. This indeed is the case to a certain extent, with some birds, as we have seen with the peahen, wild turkey, and certain kinds of grouse. Taking as our guide the habits of most male birds, the greater size and strength and the extraordinary pugnacity of the females of the Turnix and Emu, must mean that they endeavour to drive away rival females, in order to gain possession of

[26] Jerdon, 'Birds of India,' vol. iii. p. 598.

the male; and on this view, all the facts become clear; for the males would probably be most charmed or excited by the females which were the most attractive to them by their brighter colours, other ornaments, or vocal powers. Sexual selection would then soon do its work, steadily adding to the attractions of the females; the males and the young being left not at all, or but little modified.

CLASS III. *When the adult male resembles the adult female, the young of both sexes have a peculiar first plumage of their own.*—In this class both sexes when adult resemble each other, and differ from the young. This occurs with many birds of many kinds. The male robin can hardly be distinguished from the female, but the young are widely different with their mottled dusky-olive and brown plumage. The male and female of the splendid scarlet Ibis are alike, whilst the young are brown; and the scarlet-colour, though common to both sexes, is apparently a sexual character, for it is not well developed with birds under confinement, in the same manner as often occurs in the case of brilliantly coloured male birds. With many species of herons the young differ greatly from the adults, and their summer plumage, though common to both sexes, clearly has a nuptial character. Young swans are slate-coloured, whilst the mature birds are pure white; but it would be superfluous to give additional instances. These differences between the young and the old apparently depend, as in the two last classes, on the young having retained a former or ancient state of plumage, which has been exchanged for a new plumage by the old of both sexes. When the adults are brightly coloured, we may conclude from the remarks just made in relation to the scarlet ibis and to many herons, and from the analogy of the species in the first class, that such colours have been acquired through sexual selection by the nearly mature males; but that, differently from what occurs in the two first classes, the transmission, though limited to the same age, has not been limited to the same sex. Consequently both sexes when mature resemble each other and differ from the young.

CLASS IV. *When the adult male resembles the adult female, the young of both sexes in their first plumage resemble the adults.*—In this class the young and the adults of both sexes, whether brilliantly or obscurely coloured, resemble each other. Such cases are, I think, more common than those in the last class. We have in England instances in the kingfisher, some woodpeckers, the jay, magpie, crow, and many small dull-coloured birds, such as the hedge-warbler or kitty-wren. But the similarity in plumage between the young and the old is never absolutely complete, and graduates away into dissimilarity. Thus the young of some members of the kingfisher family are not only less vividly coloured than the adults, but many of the feathers on the lower surface are edged with brown,[27]—a vestige probably of a former state of the plumage. Frequently in the same group of birds, even within the same genus, for instance in an Australian genus of parrokeets (Platycercus), the young of some species closely resemble, whilst the young of other species differ considerably from their parents of both sexes, which are alike.[28] Both sexes and the young of the common jay are closely similar; but in the Canada jay (*Perisoreus canadensis*) the young differ so much from their parents that they were formerly described as distinct species.[29]

---

[27] Jerdon, 'Birds of India,' vol. i. p. 222, 228. Gould's 'Handbook of the Birds of Australia,' vol. i. 124, 130.

[28] Gould, Ibid. vol. ii. p. 37, 46, 56.

[29] Audubon, 'Ornith. Biography,' vol. ii. p. 55.

Before proceeding, I may remark that under the present and two next classes of cases the facts are so complex, and the conclusions so doubtful, that any one who feels no especial interest in the subject had better pass them over.

The brilliant or conspicuous colours which characterise many birds in the present class, can rarely or never be of service to them as a protection; so that they have probably been gained by the males through sexual selection, and then transferred to the females and the young. It is, however, possible that the males may have selected the more attractive females; and if these transmitted their characters to their offspring of both sexes, the same results would follow as from the selection of the more attractive males by the females. But there is some evidence that this contingency has rarely, if ever, occurred in any of those groups of birds, in which the sexes are generally alike; for if even a few of the successive variations had failed to be transmitted to both sexes, the females would have exceeded to a slight degree the males in beauty. Exactly the reverse occurs under nature; for in almost every large group, in which the sexes generally resemble each other, the males of some few species are in a slight degree more brightly coloured than the females. It is again possible that the females may have selected the more beautiful males, these males having reciprocally selected the more beautiful females; but it is doubtful whether this double process of selection would be likely to occur, owing to the greater eagerness of one sex than the other, and whether it would be more efficient than selection on one side alone. It is, therefore, the most probable view that sexual selection has acted, in the present class, as far as ornamental characters are concerned, in accordance with the general rule throughout the animal kingdom, that is, on the males; and that these have transmitted their gradually-acquired colours, either equally or almost equally, to their offspring of both sexes.

Another point is more doubtful, namely, whether the successive variations first appeared in the males after they had become nearly mature, or whilst quite young. In either case sexual selection must have acted on the male when he had to compete with rivals for the possession of the female; and in both cases the characters thus acquired have been transmitted to both sexes and all ages. But these characters, if acquired by the males when adult, may have been transmitted at first to the adults alone, and at some subsequent period transferred to the young. For it is known that when the law of inheritance at corresponding ages fails, the offspring often inherit characters at an earlier age than that at which they first appeared in their parents.[30] Cases apparently of this kind have been observed with birds in a state of nature. For instance Mr. Blyth has seen specimens of *Lanius rufus* and of *Colymbus glacialis* which had assumed. whilst young, in a quite anomalous manner, the adult plumage of their parents.[31] Again, the young of the common swan (*Cygnus olor*) do not cast off their dark feathers and become white until eighteen months or two years old; but Dr. F. Forel has described the case of three vigorous young birds, out of a brood of four, which were born pure white. These young birds were not albinoes, as shewn by the colour of their beaks

---

30 'Variation of Animals and Plants under Domestication,' vol. ii. p. 79.

31 Charlesworth, 'Mag. of Nat. Hist.' vol. i. 1837, p. 305, 306.

and legs, which nearly resembled the same parts in the adults.[32]

It may be worth while to illustrate the above three modes by which, in the present class, the two sexes and the young may have come to resemble each other, by the curious case of the genus Passer.[33] In the house-sparrow (*P. domesticus*) the male differs much from the female and from the young. These resemble each other, and likewise to a large extent both sexes and the young of the sparrow of Palestine (*P. brachydactylus*), as well as of some allied species. We may therefore assume that the female and young of the house-sparrow approximately shew us the plumage of the progenitor of the genus. Now with the tree-sparrow (*P. montanus*) both sexes and the young closely resemble the male of the house-sparrow; so that they have all been modified in the same manner, and all depart from the typical colouring of their early progenitor. This may have been effected by a male ancestor of the tree-sparrow having varied, firstly, when nearly mature, or, secondly, whilst quite young, having in either case transmitted his modified plumage to the females and the young; or, thirdly, he may have varied when adult and transmitted his plumage to both adult sexes, and, owing to the failure of the law of inheritance at corresponding ages, at some subsequent period to his young.

It is impossible to decide which of these three modes has generally prevailed throughout the present class of cases. The belief that the males varied whilst young, and transmitted their variations to their offspring of

---

32 'Bulletin de la Soc. Vaudoise des Sc. Nat.' vol. x. 1869, p. 132. The young of the Polish swan, *Cygnus immutabilis* of Yarrell, are always white; but this species, as Mr. Sclater informs me, is believed to be nothing more than a variety of the Domestic Swan (*Cygnus olor*).

33 I am indebted to Mr. Blyth for information in regard to this genus. The sparrow of Palestine belongs to the sub-genus Petronia.

both sexes is perhaps the most probable. I may here add that I have endeavoured, with little success, by consulting various works, to decide how far with birds the period of variation has generally determined the transmission of characters to one sex or to both. The two rules, often referred to (namely, that variations occurring late in life are transmitted to one and the same sex, whilst those which occur early in life are transmitted to both sexes), apparently hold good in the first,[34] second, and fourth classes of cases; but they fail in an equal number, namely, in the third, often in the fifth,[35] and in the sixth small class. They hold good, however, as far as I can judge, with a considerable majority of the species of birds. Whether or not this be so, we may conclude from the facts given in the eighth chapter that the period of variation has been one important element in determining the form of transmission.

With birds it is difficult to decide by what standard we ought to judge of the earliness or lateness of the period of variation, whether by the age in reference to the duration of life, or to the power of reproduction, or to the number of moults through which the species passes. The moulting of birds, even within the same family, sometimes differs much without any assignable

---

[34] For instance, the males of *Tanagra æstiva* and *Fringilla cyanea* require three years, the male of *Fringilla ciris* four years, to complete their beautiful plumage. (See Audubon, 'Ornith. Biography,' vol. i. p. 233, 280, 378.) The Harlequin duck takes three years (ibid. vol. iii. p. 614). The male of the Gold pheasant, as I hear from Mr. J. Jenner Weir, can be distinguished from the female when about three months old, but he does not acquire his full splendour until the end of the September in the following year.

[35] Thus the *Ibis tantalus* and *Grus Americanus* take four years, the Flamingo several years, and the *Ardea Ludovicana* two years, before they acquire their perfect plumage. See Audubon, ibid. vol. i. p. 221; vol. iii. p. 133, 139, 211.

cause. Some birds moult so early, that nearly all the body-feathers are cast off before the first wing-feathers are fully grown; and we cannot believe that this was the primordial state of things. When the period of moulting has been accelerated, the age at which the colours of the adult plumage were first developed would falsely appear to us to have been earlier than it really was. This may be illustrated by the practice followed by some bird-fanciers, who pull out a few feathers from the breast of nestling bullfinches, and from the head or neck of young gold-pheasants, in order to ascertain their sex; for in the males these feathers are immediately replaced by coloured ones.[36] The actual duration of life is known in but few birds, so that we can hardly judge by this standard. And with reference to the period at which the powers of reproduction are gained, it is a remarkable fact that various birds occasionally breed whilst retaining their immature plumage.[37]

The fact of birds breeding in their immature plumage seems opposed to the belief that sexual selection has

---

36 Mr. Blyth, in Charlesworth's 'Mag. of Nat. Hist.' vol. i. 1837, p. 300. Mr. Bartlett has informed me in regard to gold-pheasants.

37 I have noticed the following cases in Audubon's 'Ornith. Biography. The Redstart of America' (*Muscicapa ruticilla*, vol. i. p. 203). The *Ibis tantalus* takes four years to come to full maturity, but sometimes breeds in the second year (vol. iii. p. 133). The *Grus Americanus* takes the same time, but breeds before acquiring its full plumage (vol. iii. p. 211). The adults of *Ardea cærulea* are blue and the young white; and white, mottled, and mature blue birds may all be seen breeding together (vol. iv. p. 58): but Mr. Blyth informs me that certain herons apparently are dimorphic, for white and coloured individuals of the same age may be observed. The Harlequin duck (*Anas histrionica*, Linn.) takes three years to acquire its full plumage, though many birds breed in the second year (vol. iii. p. 614). The White-headed Eagle (*Falco leucocephalus*, vol. iii. p. 210) is likewise known to breed in its immature state. Some species of Oriolus (according to Mr. Blyth and Mr. Swinhoe, in 'Ibis,' July, 1863, p. 68) likewise breed before they attain their full plumage.

played as important a part, as I believe it has, in giving ornamental colours, plumes, &c., to the males, and, by means of equal transmission, to the females of many species. The objection would be a valid one, if the younger and less ornamented males were as successful in winning females and propagating their kind, as the older and more beautiful males. But we have no reason to suppose that this is the case. Audubon speaks of the breeding of the immature males of *Ibis tantalus* as a rare event, as does Mr. Swinhoe, in regard to the immature males of Oriolus.[38] If the young of any species in their immature plumage were more successful in winning partners than the adults, the adult plumage would probably soon be lost, as the males which retained their immature dress for the longest period would prevail, and thus the character of the species would ultimately be modified.[39] If, on the other hand, the young never succeeded in obtaining a female, the habit of early reproduction would perhaps be sooner or later quite eliminated, from being superfluous and entailing waste of power.

The plumage of certain birds goes on increasing in

---

[38] See the last foot-note.

[39] Other animals, belonging to quite distinct classes, are either habitually or occasionally capable of breeding before they have fully acquired their adult characters. This is the case with the young males of the salmon. Several amphibians have been known to breed whilst retaining their larval structure. Fritz Müller has shewn ('Facts and Arguments for Darwin,' Eng. trans. 1869, p. 79) that the males of several amphipod crustaceans become sexually mature whilst young; and I infer that this is a case of premature breeding, because they have not as yet acquired their fully-developed claspers. All such facts are highly interesting, as bearing on one means by which species may undergo great modifications of character, in accordance with Mr. Cope's views, expressed under the terms of the "retardation and acceleration of generic characters;" but I cannot follow the views of this eminent naturalist to their full extent. See Mr. Cope, "On the Origin of Genera," from the 'Proc. of Acad. Nat. Sc. of Philadelphia,' Oct. 1868.

beauty during many years after they are fully mature; this is the case with the train of the peacock, and with the crest and plumes of certain herons; for instance, the *Ardea Ludovicana*;[40] but it is very doubtful whether the continued development of such feathers is the result of the selection of successive beneficial variations, or merely of continuous growth. Most fishes continue increasing in size, as long as they are in good health and have plenty of food; and a somewhat similar law may prevail with the plumes of birds.

CLASS V. *When the adults of both sexes have a distinct winter and summer plumage, whether or not the male differs from the female, the young resemble the adults of both sexes in their winter dress, or much more rarely in their summer dress, or they resemble the females alone; or the young may have an intermediate character; or again, they may differ greatly from the adults in both their seasonal plumages.*—The cases in this class are singularly complex; nor is this surprising, as they depend on inheritance, limited in a greater or less degree in three different ways, namely by sex, age, and the season of the year. In some cases the individuals of the same species pass through at least five distinct states of plumage. With the species, in which the male differs from the female during the summer season alone, or, which is rarer, during both seasons,[41] the young generally resemble the females,—as with the so-called goldfinch of North America, and apparently with the splendid Maluri of Australia.[42] With

---

40 Jerdon, 'Birds of India,' vol. iii. p. 507, on the peacock. Audubon, ibid. vol. iii. p. 139, on the Ardea.

41 For illustrative cases see vol. iv. of Macgillivray's 'Hist. Brit. Birds;' on Tringa, &c., p. 229, 271; on the Machetes, p. 172; on the *Charadrius hiaticula*, p. 118; on the *Charadrius pluvialis*, p. 94.

42 For the goldfinch of N. America, *Fringilla tristis*, Linn., see

the species, the sexes of which are alike during both the summer and winter, the young may resemble the adults, firstly, in their winter dress; secondly, which occurs much more rarely, in their summer dress; thirdly, they may be intermediate between these two states; and, fourthly, they may differ greatly from the adults at all seasons. We have an instance of the first of these four cases in one of the egrets of India (*Buphus coromandus*), in which the young and the adults of both sexes are white during the winter, the adults becoming golden-buff during the summer. With the Gaper (*Anastomus oscitans*) of India we have a similar case, but the colours are reversed; for the young and the adults of both sexes are grey and black during the winter, the adults becoming white during the summer.[43] As an instance of the second case, the young of the razor-bill (*Alca torda*, Linn.), in an early state of plumage, are coloured like the adults during the summer; and the young of the white-crowned sparrow of North America (*Fringilla leucophrys*), as soon as fledged, have elegant white stripes on their heads, which are lost by the young and the old during the winter.[44] With respect to the third case, namely, that of the young having an intermediate character between the summer and winter adult plumages, Yarrell[45] insists that this occurs with many

---

Audubon, 'Ornith. Biography,' vol. i. p. 172. For the Maluri, Gould's 'Handbook of the Birds of Australia,' vol. i. p. 318.

[43] I am indebted to Mr. Blyth for information in regard to the Buphus; see also Jerdon, 'Birds of India,' vol. iii. p. 749. On the Anastomus, see Blyth, in 'Ibis,' 1867, p. 173.

[44] On the Alca, see Macgillivray, 'Hist. Brit. Birds,' vol. v. p. 347. On the *Fringilla leucophrys*, Audubon, ibid. vol. ii. p. 89. I shall have hereafter to refer to the young of certain herons and egrets being white.

[45] 'History of British Birds,' vol. i. 1839, p. 159.

waders. Lastly, in regard to the young differing greatly from both sexes in their adult summer and winter plumages, this occurs with some herons and egrets of North America and India,—the young alone being white.

I will make only a few remarks on these complicated cases. When the young resemble the female in her summer dress, or the adults of both sexes in their winter dress, the cases differ from those given under Classes I. and III. only in the characters originally acquired by the males during the breeding-season, having been lmited in their transmission to the corresponding season. When the adults have a distinct summer and winter plumage, and the young differ from both, the case is more difficult to understand. We may admit as probable that the young have retained an ancient state of plumage; we can account through sexual selection for the summer or nuptial plumage of the adults, but how are we to account for their distinct winter plumage? If we could admit that this plumage serves in all cases as a protection, its acquirement would be a simple affair; but there seems no good reason for this admission. It may be suggested that the widely different conditions of life during the winter and summer have acted in a direct manner on the plumage; this may have had some effect, but I have not much confidence in so great a difference, as we sometimes see, between the two plumages having been thus caused. A more probable explanation is, that an ancient style of plumage, partially modified through the transference of some characters from the summer plumage, has been retained by the adults during the winter. Finally, all the cases in our present class apparently depend on characters acquired by the adult males, having been variously limited in their transmission according to age, season, and sex; but it would not be worth while to attempt to follow out these complex relations.

CLASS VI. *The young in their first plumage differ from each other according to sex; the young males resembling more or less closely the adult males, and the young females more or less closely the adult females.*—The cases in the present class, though occurring in various groups, are not numerous; yet, if experience had not taught us to the contrary, it seems the most natural thing that the young should at first always resemble to a certain extent, and gradually become more and more like, the adults of the same sex. The adult male blackcap (*Sylvia atricapilla*) has a black head, that of the female being reddish-brown; and I am informed by Mr. Blyth, that the young of both sexes can be distinguished by this character even as nestlings. In the family of thrushes an unusual number of similar cases have been noticed; the male blackbird (*Turdus merula*) can be distinguished in the nest from the female, as the main wing-feathers, which are not moulted so soon as the body-feathers, retain a brownish tint until the second general moult.[46] The two sexes of the mocking bird (*Turdus polyglottus*, Linn.) differ very little from each other, yet the males can easily be distinguished at a very early age from the females by shewing more pure white.[47] The males of a forest-thrush and of a rock-thrush (viz. *Orocetes erythrogastra* and *Petrocincla cyanea*) have much of their plumage of a fine blue, whilst the females are brown; and the nestling males of both species have their main wing and tail-feathers edged with blue, whilst those of the female are

---

[46] Blyth, in Charlesworth's 'Mag. of Nat. Hist.' vol. i. 1837, p. 362; and from information given to me by him.

[47] Audubon, 'Ornith. Biography,' vol. i. p. 113.

edged with brown.[48] So that the very same feathers which in the young blackbird assume their mature character and become black after the others, in these two species assume this character and become blue before the others. The most probable view with reference to these cases is that the males, differently from what occurs in Class I., have transmitted their colours to their male offspring at an earlier age than that at which they themselves first acquired them; for if they had varied whilst quite young, they would probably have transmitted all their characters to their offspring of both sexes.[49]

In *Aïthurus polytmus* (one of the humming-birds) the male is splendidly coloured black and green, and two of the tail-feathers are immensely lengthened; the female has an ordinary tail and inconspicuous colours; now the young males, instead of resembling the adult female, in accordance with the common rule, begin from the first to assume the colours proper to their sex, and their tail-feathers soon become elongated. I owe this information to Mr. Gould, who has given me the following more striking and as yet unpublished case. Two humming-birds belonging to the genus Eustephanus, both beautifully coloured, inhabit the small island of Juan Fernandez, and have always been ranked as specifically distinct. But it has lately been ascertained that the one, which is of a rich ches-

48 Mr. C. A. Wright, in 'Ibis,' vol. vi. 1864, p. 65. Jerdon, 'Birds of India,' vol. i. p. 515.

49 The following additional cases may be mentioned: the young males of *Tanagra rubra* can be distinguished from the young females (Audubon, 'Ornith. Biography,' vol. iv. p. 392), and so it is with the nestlings of a blue nuthatch, *Dendrophila frontalis* of India (Jerdon, 'Birds of India,' vol. i. p. 389). Mr. Blyth also informs me that the sexes of the stonechat, *Saxicola rubicola*, are distinguishable at a very early age.

nut-brown colour with a golden-red head, is the male, whilst the other, which is elegantly variegated with green and white with a metallic-green head, is the female. Now the young from the first resemble to a certain extent the adults of the corresponding sex, the resemblance gradually becoming more and more complete.

In considering this last case, if as before we take the plumage of the young as our guide, it would appear that both sexes have been independently rendered beautiful; and not that the one sex has partially transferred its beauty to the other. The male apparently has acquired his bright colours through sexual selection in the same manner as, for instance, the peacock or pheasant in our first class of cases; and the female in the same manner as the female Rhynchæa or Turnix in our second class of cases. But there is much difficulty in understanding how this could have been effected at the same time with the two sexes of the same species. Mr. Salvin states, as we have seen in the eighth chapter, that with certain humming-birds the males greatly exceed in number the females, whilst with other species inhabiting the same country the females greatly exceed the males. If, then, we might assume that during some former lengthened period the males of the Juan Fernandez species had greatly exceeded the females in number, but that during another lengthened period the females had greatly exceeded the males, we could understand how the males at one time, and the females at another time, might have been rendered beautiful by the selection of the brighter-coloured individuals of either sex; both sexes transmitting their characters to their young at a rather earlier age than usual. Whether this is the true explanation I will not pretend to say; but the case is too remarkable to be passed over without notice.

We have now seen in numerous instances under all six classes, that an intimate relation exists between the plumage of the young and that of the adults, either of one sex or both sexes. These relations are fairly well explained on the principle that one sex—this being in the great majority of cases the male—first acquired through variation and sexual selection bright colours or other ornaments, and transmitted them in various ways, in accordance with the recognised laws of inheritance. Why variations have occurred at different periods of life, even sometimes with the species of the same group, we do not know; but with respect to the form of transmission, one important determining cause seems to have been the age at which the variations first appeared.

From the principle of inheritance at corresponding ages, and from any variations in colour which occurred in the males at an early age not being then selected, on the contrary being often eliminated as dangerous, whilst similar variations occurring at or near the period of reproduction have been preserved, it follows that the plumage of the young will often have been left unmodified, or but little modified. We thus get some insight into the colouring of the progenitors of our existing species. In a vast number of species in five out of our six classes of cases, the adults of one sex or both are brightly coloured, at least during the breeding-season, whilst the young are invariably less brightly coloured than the adults, or are quite dull-coloured; for no instance is known, as far as I can discover, of the young of dull-coloured species displaying bright colours, or of the young of brightly-coloured species being more brilliantly coloured than their parents. In the fourth class, however, in which the young and the old resemble each other, there are many species (though by no means all) brightly-coloured, and as these form whole groups, we may infer that their early progenitors were likewise brightly-coloured. With this exception, if we look to the birds of the world, it appears that their beauty has been greatly increased since that period, of which we have a partial record in their immature plumage.

On the Colour of the Plumage in relation to Protection.—It will have been seen that I cannot follow Mr. Wallace in the belief that dull colours when confined to the females have been in most cases specially gained for the sake of protection. There can, however, be no doubt, as formerly remarked, that both sexes of many birds have had their colours modified for this purpose, so as to escape the notice of their enemies; or, in some instances, so as to approach their prey unobserved, in the same manner as owls have had their plumage rendered soft, that their flight may not be overheard. Mr. Wallace remarks[50] that "it is only "in the tropics, among forests which never lose their "foliage, that we find whole groups of birds, whose "chief colour is green." It will be admitted by every one, who has ever tried, how difficult it is to distinguish parrots in a leaf-covered tree. Nevertheless, we must remember that many parrots are ornamented with crimson, blue, and orange tints, which can hardly be protective. Woodpeckers are eminently arboreal, but, besides green species, there are many black, and black-and-white kinds—all the species being apparently exposed to

---

[50] 'Westminster Review,' July, 1867, p. 5.

nearly the same dangers. It is therefore probable that strongly-pronounced colours have been acquired by tree-haunting birds through sexual selection, but that green tints have had an advantage through natural selection over other colours for the sake of protection.

In regard to birds which live on the ground, every-one admits that they are coloured so as to imitate the surrounding surface. How difficult it is to see a partridge, snipe, woodcock, certain plovers, larks, and night-jars when crouched on the ground. Animals inhabiting deserts offer the most striking instances, for the bare surface affords no concealment, and all the smaller quadrupeds, reptiles, and birds depend for safety on their colours. As Mr. Tristram has remarked,[51] in regard to the inhabitants of the Sahara, all are protected by their "isabelline or sand-colour." Calling to my recollection the desert-birds which I had seen in South America, as well as most of the ground-birds in Great Britain, it appeared to me that both sexes in such cases are generally coloured nearly alike. Accordingly I applied to Mr. Tristram, with respect to the birds of the Sahara, and he has kindly given me the following information. There are twenty-six species, belonging to fifteen genera, which manifestly have had their plumage coloured in a protective manner; and this colouring is all the more striking, as with most of these birds it is different from that of their congeners. Both sexes of thirteen out of the twenty-six species are coloured in the same manner; but these belong to genera in which this rule commonly prevails, so that they tell us nothing about the protective colours being the same in both sexes of desert-birds. Of

---

[51] 'Ibis,' 1859, vol. i. p. 429, *et seq.*

the other thirteen species, three belong to genera in which the sexes usually differ from each other, yet they have the sexes alike. In the remaining ten species, the male differs from the female; but the difference is confined chiefly to the under surface of the plumage, which is concealed when the bird crouches on the ground; the head and back being of the same sand-coloured hue in both sexes. So that in these ten species the upper surfaces of both sexes have been acted on and rendered alike, through natural selection, for the sake of protection; whilst the lower surfaces of the males alone have been diversified through sexual selection, for the sake of ornament. Here, as both sexes are equally well protected, we clearly see that the females have not been prevented through natural selection from inheriting the colours of their male parents: we must look to the law of sexually limited transmission, as before explained.

In all parts of the world both sexes of many soft-billed birds, especially those which frequent reeds or sedges, are obscurely coloured. No doubt if their colours had been brilliant, they would have been much more conspicuous to their enemies; but whether their dull tints have been specially gained for the sake of protection seems, as far as I can judge, rather doubtful. It is still more doubtful whether such dull tints can have been gained for the sake of ornament. We must, however, bear in mind that male birds, though dull-coloured, often differ much from their females, as with the common sparrow, and this leads to the belief that such colours have been gained through sexual selection, from being attractive. Many of the soft-billed birds are songsters; and a discussion in a former chapter should not be forgotten, in which it was shewn that the best songsters are rarely ornamented with bright tints. It would appear that female birds, as a general rule, have selected their mates either for their sweet voices or gay colours, but not for both charms combined. Some species which are manifestly coloured for the sake of protection, such as the jack-snipe, woodcock, and night-jar, are likewise marked and shaded, according to our standard of taste, with extreme elegance. In such cases we may conclude that both natural and sexual selection have acted conjointly for protection and ornament. Whether any bird exists which does not possess some special attraction, by which to charm the opposite sex, may be doubted. When both sexes are so obscurely coloured, that it would be rash to assume the agency of sexual selection, and when no direct evidence can be advanced shewing that such colours serve as a protection, it is best to own complete ignorance of the cause, or, which comes to nearly the same thing, to attribute the result to the direct action of the conditions of life.

There are many birds both sexes of which are conspicuously, though not brilliantly coloured, such as the numerous black, white, or piebald species; and these colours, are probably the result of sexual selection. With the common blackbird, capercailzie, blackcock, black Scoter-duck (Oidemia), and even with one of the Birds of Paradise (*Lophorina atra*), the males alone are black, whilst the females are brown or mottled; and there can hardly be a doubt that blackness in these cases has been a sexually selected character. Therefore it is in some degree probable that the complete or partial blackness of both sexes in such birds as crows, certain cockatoos, storks, and swans, and many marine birds, is likewise the result of sexual selection, accompanied by equal transmission to both sexes;

for blackness can hardly serve in any case as a protection. With several birds, in which the male alone is black, and in others in which both sexes are black, the beak or skin about the head is brightly coloured, and the contrast thus afforded adds greatly to their beauty; we see this in the bright yellow beak of the male blackbird, in the crimson skin over the eyes of the black-cock and capercailzie, in the variously and brightly-coloured beak of the Scoter-drake (Oidemia), in the red beak of the chough (*Corvus graculus*, Linn.), of the black swan, and black stork. This leads me to remark that it is not at all incredible that toucans may owe the enormous size of their beaks to sexual selection, for the sake of displaying the diversified and vivid stripes of colour, with which these organs are ornamented.[52] The naked skin at the base of the beak and round the eyes is likewise often brilliantly coloured; and Mr. Gould, in speaking of one species,[53] says that the colours of the beak "are doubtless in the finest "and most brilliant state during the time of pairing." There is no greater improbability in toucans being encumbered with immense beaks, though rendered as light as possible by their cancellated structure, for an object falsely appearing to us unimportant, namely, the display of fine colours, than that the male Argus

[52] No satisfactory explanation has ever been offered of the immense size, and still less of the bright colours, of the toucan's beak. Mr. Bates ('The Naturalist on the Amazons,' vol. ii. 1863, p. 341) states that they use their beak for reaching fruit at the extreme tips of the branches; and likewise, as stated by other authors, for extracting eggs and young birds from the nests of other birds. But as Mr. Bates admits, the beak "can scarcely be considered a very perfectly-formed instru-"ment for the end to which it is applied." The great bulk of the beak, as shewn by its breadth, depth, as well as length, is not intelligible on the view, that it serves merely as an organ of prehension.

[53] Ramphaston carinatus, Gould's 'Monograph of Ramphastidæ.'

pheasant and some other birds should be encumbered with plumes so long as to impede their flight.

In the same manner, as the males alone of various species are black, the females being dull-coloured; so in a few cases the males alone are either wholly or partially white, as with the several Bell-birds of South America (Chasmorhynchus), the Antarctic goose (*Bernicla antarctica*), the silver-pheasant, &c., whilst the females are brown or obscurely mottled. Therefore, on the same principle as before, it is probable that both sexes of many birds, such as white cockatoos, several egrets with their beautiful plumes, certain ibises, gulls, terns, &c., have acquired their more or less completely white plumage through sexual selection. The species which inhabit snowy regions of course come under a different head. The white plumage of some of the above-named birds appears in both sexes only when they are mature. This is likewise the case with certain gannets, tropic-birds, &c., and with the snow-goose (*Anser hyperboreus*). As the latter breeds on the "barren grounds," when not covered with snow, and as it migrates southward during the winter, there is no reason to suppose that its snow-white adult plumage serves as a protection. In the case of the *Anastomus oscitans* previously alluded to, we have still better evidence that the white plumage is a nuptial character, for it is developed only during the summer; the young in their immature state, and the adults in their winter dress, being grey and black. With many kinds of gulls (Larus), the head and neck become pure white during the summer, being grey or mottled during the winter and in the young state. On the other hand, with the smaller gulls, or sea-mews (Gavia), and with some terns (Sterna), exactly the reverse occurs; for the heads of the young birds during the first year, and of the adults during the winter, are either pure white, or much paler-coloured than during the breeding-season. These latter cases offer another instance of the capricious manner in which sexual selection appears often to have acted.[54]

The cause of aquatic birds having acquired a white plumage so much more frequently than terrestrial birds, probably depends on their large size and strong powers of flight, so that they can easily defend themselves or escape from birds of prey, to which moreover they are not much exposed. Consequently sexual selection has not here been interfered with or guided for the sake of protection. No doubt, with birds which roam over the open ocean, the males and females could find each other much more easily when made conspicuous either by being perfectly white, or intensely black; so that these colours may possibly serve the same end as the call-notes of many land-birds. A white or black bird, when it discovers and flies down to a carcase floating on the sea or cast up on the beach, will be seen from a great distance, and will guide other birds of the same and of distinct species, to the prey; but as this would be a disadvantage to the first finders, the individuals which were the whitest or blackest would not thus have procured more food than the less strongly coloured individuals. Hence conspicuous colours cannot have been gradually acquired for this purpose through natural selection.[55]

---

[54] On Larus, Gavia, and Sterna, see Macgillivray, 'Hist. Brit. Birds,' vol. v. p. 515, 584, 626. On the Anser hyperboreus, Audubon, 'Ornith. Biography,' vol. iv. p. 562. On the Anastomus, Mr. Blyth, in 'Ibis,' 1867, p. 173.

[55] It may be noticed that with vultures, which roam far and wide through the higher regions of the atmosphere, like marine birds over the ocean, three or four species are almost wholly or largely white, and

As sexual selection depends on so fluctuating an element as taste, we can understand how it is that within the same group of birds, with habits of life nearly the same, there should exist white or nearly white, as well as black, or nearly black species,—for instance, white and black cockatoos, storks, ibises, swans, terns, and petrels. Piebald birds likewise sometimes occur in the same groups, for instance, the black-necked swan, certain terns, and the common magpie. That a strong contrast in colour is agreeable to birds, we may conclude, by looking through any large collection of specimens or series of coloured plates, for the sexes frequently differ from each other in the male having the pale parts of a purer white, and the variously coloured dark parts of still darker tints than in the female.

It would even appear that mere novelty, or change for the sake of change, has sometimes acted like a charm on female birds, in the same manner as changes of fashion with us. The Duke of Argyll says,[56]—and I am glad to have the unusual satisfaction of following for even a short distance in his footsteps—"I am more "and more convinced that variety, mere variety, must "be admitted to be an object and an aim in Nature." I wish the Duke had explained what he here means by Nature. Is it meant that the Creator of the universe ordained diversified results for His own satisfaction, or for that of man? The former notion seems to me as much wanting in due reverence as the latter in probability. Capriciousness of taste in the birds themselves appears a more fitting explanation. For example; the males

many other species are black. This fact supports the conjecture that these conspicuous colours may aid the sexes in finding each other during the breeding-season.

56 'The Journal of Travel,' edited by A. Murray, vol. i. 1868, p. 286.

of some parrots can hardly be said to be more beautiful, at least according to our taste, than the females, but they differ from them in such points, as the male having a rose-coloured collar instead of, as in the female, "a bright emeraldine narrow green collar;" or in the male having a black collar instead of "a yellow demi-collar in front," with a pale roseate instead of a plum-blue head.[57] As so many male birds have for their chief ornament elongated tail-feathers or elongated crests, the shortened tail, formerly described in the male of a humming-bird, and the shortened crest of the male goosander almost seem like one of the many opposite changes of fashion which we admire in our own dresses.

Some members of the heron family offer a still more curious case of novelty in colouring having apparently been appreciated for the sake of novelty. The young of the *Ardea asha* are white, the adults being dark slate-coloured; and not only the young, but the adults of the allied *Buphus coromandus* in their winter plumage are white, this colour changing into a rich golden-buff during the breeding-season. It is incredible that the young of these two species, as well as of some other members of the same family,[58] should have been specially rendered pure white and thus made conspicuous to their enemies; or that the adults of one of these two species should have been specially rendered white during the winter in a country which is never

[57] See Jerdon on the genus Palæornis, 'Birds of India,' vol. i. p. 258-260.

[58] The young of *Ardea rufescens* and *A. cœrulea* of the U. States are likewise white, the adults being coloured in accordance with their specific names. Audubon ('Ornith. Biography,' vol. iii. p. 416; vol. iv. p. 58) seems rather pleased at the thought that this remarkable change of plumage will greatly "disconcert the systematists."

covered with snow. On the other hand we have reason to believe that whiteness has been gained by many birds as a sexual ornament. We may therefore conclude that an early progenitor of the *Ardea asha* and the *Buphus* acquired a white plumage for nuptial purposes, and transmitted this colour to their young; so that the young and the old became white like certain existing egrets; the whiteness having afterwards been retained by the young whilst exchanged by the adults for more strongly pronounced tints. But if we could look still further backwards in time to the still earlier progenitors of these two species, we should probably see the adults dark-coloured. I infer that this would be the case, from the analogy of many other birds, which are dark whilst young, and when adult are white; and more especially from the case of the *Ardea gularis*, the colours of which are the reverse of those of *A. asha*, for the young are dark-coloured and the adults white, the young having retained a former state of plumage. It appears therefore that the progenitors in their adult condition of the *Ardea asha*, the *Buphus*, and of some allies, have undergone, during a long line of descent, the following changes of colour: firstly a dark shade, secondly pure white, and thirdly, owing to another change of fashion (if I may so express myself), their present slaty, reddish, or golden-buff tints. These successive changes are intelligible only on the principle of novelty having been admired by birds for the sake of novelty.

Summary of the Four Chapters on Birds.—Most male birds are highly pugnacious during the breeding-season, and some possess weapons especially adapted for fighting with their rivals. But the most pugnacious and the best-armed males rarely or never depend for success solely on their power to drive away or kill their rivals, but have special means for charming the female. With some it is the power of song, or of emitting strange cries, or of producing instrumental music, and the males in consequence differ from the females in their vocal organs, or in the structure of certain feathers. From the curiously diversified means for producing various sounds we gain a high idea of the importance of this means of courtship. Many birds endeavour to charm the females by love-dances or antics, performed on the ground or in the air, and sometimes at prepared places. But ornaments of many kinds, the most brilliant tints, combs and wattles, beautiful plumes, elongated feathers, top-knots, and so forth, are by far the commonest means. In some cases mere novelty appears to have acted as a charm. The ornaments of the males must be highly important to them, for they have been acquired in not a few cases at the cost of increased danger from enemies, and even at some loss of power in fighting with their rivals. The males of very many species do not assume their ornamental dress until they arrive at maturity, or they assume it only during the breeding-season, or the tints then become more vivid. Certain ornamental appendages become enlarged, turgid, and brightly-coloured during the very act of courtship. The males display their charms with elaborate care and to the best effect; and this is done in the presence of the females. The courtship is sometimes a prolonged affair, and many males and females congregate at an appointed place. To suppose that the females do not appreciate the beauty of the males is to admit that their splendid decorations, all their pomp and display, are useless; and this is incredible. Birds have fine powers of discrimination, and in some few instances it can be shewn that they have a taste for the beautiful. The females, moreover, are known occasionally to exhibit a marked preference or antipathy for certain individual males.

If it be admitted that the females prefer, or are unconsciously excited by the more beautiful males, then the males would slowly but surely be rendered more and more attractive through sexual selection. That it is this sex which has been chiefly modified we may infer from the fact that in almost every genus in which the sexes differ, the males differ much more from each other than do the females; this is well shewn in certain closely-allied representative species in which the females can hardly be distinguished, whilst the males are quite distinct. Birds in a state of nature offer individual differences which would amply suffice for the work of sexual selection; but we have seen that they occasionally present more strongly-marked variations which recur so frequently that they would immediately be fixed, if they served to allure the female. The laws of variation will have determined the nature of the initial changes, and largely influenced the final result. The gradations, which may be observed between the males of allied species, indicate the nature of the steps which have been passed through, and explain in the most interesting manner certain characters, such as the indented ocelli of the tail-feathers of the peacock, and the wonderfully-shaded ocelli of the wing-feathers of the Argus pheasant. It is evident that the brilliant colours, top-knots, fine plumes, &c., of many male birds cannot have been acquired as a protection; indeed they sometimes lead to danger. That they are not due to the direct and definite action of the conditions of life, we may feel assured, because the females have been exposed to the same conditions, and yet often differ from the males to an extreme degree. Although it is probable that changed conditions acting during a lengthened period have produced some definite effect on both sexes, the more important result will have been an increased tendency to fluctuating variability or to augmented individual differences; and such differences will have afforded an excellent groundwork for the action of sexual selection.

The laws of inheritance, irrespectively of selection, appear to have determined whether the characters acquired by the males for the sake of ornament, for producing various sounds, and for fighting together, have been transmitted to the males alone or to both sexes, either permanently or periodically during certain seasons of the year. Why various characters should sometimes have been transmitted in one way and sometimes in another is, in most cases, not known; but the period of variability seems often to have been the determining cause. When the two sexes have inherited all characters in common they necessarily resemble each other; but as the successive variations may be differently transmitted, every possible gradation may be found, even within the same genus, from the closest similarity to the widest dissimilarity between the sexes. With many closely-allied species, following nearly the same habits of life, the males have come to differ from each other chiefly through the action of sexual selection; whilst the females have come to differ chiefly from partaking in a greater or lesser degree of the characters thus acquired by the males. The effects, moreover, of the definite action of the conditions of life, will not have been masked in the females, as in the case of the males, by the accumulation through sexual selection of strongly-pronounced colours and other ornaments. The individuals of both sexes, however affected, will have been kept at each successive period nearly uniform by the free intercrossing of many individuals.

With the species, in which the sexes differ in colour, it is possible that at first there existed a tendency to transmit the successive variations equally to both sexes; and that the females were prevented from acquiring the bright colours of the males, on account of the danger to which they would have been exposed during incubation. But it would be, as far as I can see, an extremely difficult process to convert, by means of natural selection, one form of transmission into another. On the other hand there would not be the least difficulty in rendering a female dull-coloured, the male being still kept bright-coloured, by the selection of successive variations, which were from the first limited in their transmission to the same sex. Whether the females of many species have actually been thus modified, must at present remain doubtful. When, through the law of the equal transmission of characters to both sexes, the females have been rendered as conspicuously coloured as the males, their instincts have often been modified, and they have been led to build domed or concealed nests.

In one small and curious class of cases the characters and habits of the two sexes have been completely transposed, for the females are larger, stronger, more vociferous and brightly-coloured than their males. They have, also, become so quarrelsome that they often fight together like the males of the most pugnacious species. If, as seems probable, they habitually drive away rival females, and by the display of their bright colours or other charms endeavour to attract the males, we can understand how it is that they have gradually been rendered, by means of sexual selection and sexually-limited transmission, more beautiful than the males—the latter being left unmodified or only slightly modified.

Whenever the law of inheritance at corresponding ages prevails, but not that of sexually-limited transmission, then if the parents vary late in life—and we know that this constantly occurs with our poultry, and occasionally with other birds—the young will be left unaffected, whilst the adults of both sexes will be modified. If both these laws of inheritance prevail and either sex varies late in life, that sex alone will be modified, the other sex and the young being left unaffected. When variations in brightness or in other conspicuous characters occur early in life, as no doubt often happens, they will not be acted on through sexual selection until the period of reproduction arrives; consequently they will be liable to be lost by the accidental deaths of the young, and if dangerous will be eliminated through natural selection. Thus we can understand how it is that variations arising late in life have chiefly been preserved for the ornamentation and arming of the males, the females and the young being left almost unaffected, and therefore like each other. With species having a distinct summer and winter plumage, the males of which either resemble or differ from the females during both seasons or during the summer alone, the degrees and kinds of resemblance between the young and the old are exceedingly complex; and this complexity apparently depends on characters, first acquired by the males, being transmitted in various ways and degrees, as limited by age, sex, and season.

As the young of so many species have been but little modified in colour and in other ornaments, we are enabled to form some judgment with respect to the plumage of their early progenitors; and we may infer that the beauty of our existing species, if we look to the whole class, has been largely increased since that period of which the immature plumage gives us an indirect record. Many birds, especially those which live much on the ground, have undoubtedly been obscurely coloured for the sake of protection. In some instances the upper exposed surface of the plumage has been thus coloured in both sexes, whilst the lower surface in the males alone has been variously ornamented through sexual selection. Finally, from the facts given in these four chapters, we may conclude that weapons for battle, organs for producing sound, ornaments of many kinds, bright and conspicuous colours, have generally been acquired by the males through variation and sexual selection, and have been transmitted in various ways according to the several laws of inheritance—the females and the young being left comparatively but little modified.[59]

---

[59] I am greatly indebted to the kindness of Mr. Sclater for having looked over these four chapters on birds, and the two following ones on mammals. By this means I have been saved from making mistakes about the names of the species, and from giving any facts which are actually known to this distinguished naturalist to be erroneous. But of course he is not at all answerable for the accuracy of the statements quoted by me from various authorities.

CHAPTER XVII.

SECONDARY SEXUAL CHARACTERS OF MAMMALS.

The law of battle — Special weapons, confined to the males — Cause of absence of weapons in the female — Weapons common to both sexes, yet primarily acquired by the male — Other uses of such weapons — Their high importance — Greater size of the male — Means of defence — On the preference shewn by either sex in the pairing of quadrupeds.

WITH mammals the male appears to win the female much more through the law of battle than through the display of his charms. The most timid animals, not provided with any special weapons for fighting, engage in desperate conflicts during the season of love. Two male hares have been seen to fight together until one was killed; male moles often fight, and sometimes with fatal results; male squirrels "engage in frequent contests, and often wound each other severely;" as do male beavers, so that "hardly a skin is without scars."[1] I observed the same fact with the hides of the guanacoes in Patagonia; and on one occasion several were so absorbed in fighting that they fearlessly rushed close by me. Livingstone speaks of the males of the many animals in Southern Africa as almost invariably shewing the scars received in former contests.

The law of battle prevails with aquatic as with ter-

[1] See Waterton's account of two hares fighting, 'Zoologist,' vol. i. 1843, p. 211. On moles, Bell, 'Hist. of British Quadrupeds,' 1st edit. p. 100. On squirrels, Audubon and Bachman, 'Viviparous Quadrupeds of N. America,' 1846, p. 269. On beavers, Mr. A. H. Green, in 'Journal of Lin. Soc. Zoolog.' vol. x. 1869, p. 362.

restrial mammals. It is notorious how desperately male seals fight, both with their teeth and claws, during the breeding-season; and their hides are likewise often covered with scars. Male sperm-whales are very jealous at this season; and in their battles "they often "lock their jaws together, and turn on their sides and "twist about;" so that it is believed by some naturalists that the frequently deformed state of their lower jaws is caused by these struggles.[2]

All male animals which are furnished with special weapons for fighting, are well known to engage in fierce battles. The courage and the desperate conflicts of stags have often been described; their skeletons have been found in various parts of the world, with the horns inextricably locked together, shewing how miserably the victor and vanquished had perished.[3] No animal in the world is so dangerous as an elephant in must. Lord Tankerville has given me a graphic description of the battles between the wild bulls in Chillingham Park, the descendants, degenerated in size but not in courage, of the gigantic *Bos primigenius*. In 1861 several contended for mastery; and it was observed that two of the younger bulls attacked in concert the old leader of the herd, overthrew and disabled him, so that he was believed by the keepers to be lying mortally wounded in a neighbouring wood. But a few days afterwards one of the young bulls singly approached the wood; and

---

[2] On the battles of seals, see Capt. C. Abbott in 'Proc. Zool. Soc.' 1868, p. 191; also Mr. H. Brown, ibid. 1869, p. 436; also L. Lloyd, 'Game Birds of Sweden,' 1867, p. 412; also Pennant. On the sperm-whale, see Mr. J. H. Thompson, in 'Proc. Zool. Soc.' 1867, p. 246.

[3] See Scrope ('Art of Deer-stalking,' p. 17) on the locking of the horns with the Cervus elephus. Richardson, in 'Fauna Bor. Americana,' 1829, p. 252, says that the wapiti, moose, and rein-deer have been found thus locked together. Sir A. Smith found at the Cape of Good Hope the skeletons of two gnus in the same condition.

then the "monarch of the chase," who had been lashing himself up for vengeance, came out and, in a short time killed his antagonist. He then quietly joined the herd, and long held undisputed sway. Admiral Sir B. J. Sulivan informs me that when he resided in the Falkland Islands he imported a young English stallion, which, with eight mares, frequented the hills near Port William. On these hills there were two wild stallions, each with a small troop of mares; "and it is certain "that these stallions would never have approached each "other without fighting. Both had tried singly to fight "the English horse and drive away his mares, but had "failed. One day they came in *together* and attacked "him. This was seen by the capitan who had charge of "the horses, and who, on riding to the spot, found one "of the two stallions engaged with the English horse, "whilst the other was driving away the mares, and had "already separated four from the rest. The capitan "settled the matter by driving the whole party into the "corral, for the wild stallions would not leave the "mares."

Male animals already provided with efficient cutting or tearing teeth for the ordinary purposes of life, as in the carnivora, insectivora, and rodents, are seldom furnished with weapons especially adapted for fighting with their rivals. The case is very different with the males of many other animals. We see this in the horns of stags and of certain kinds of antelopes in which the females are hornless. With many animals the canine teeth in the upper or lower jaw, or in both, are much larger in the males than in the females; or are absent in the latter, with the exception sometimes of a hidden rudiment. Certain antelopes, the musk-deer, camel, horse, boar, various apes, seals, and the walrus, offer instances of these several cases. In the females of the walrus the tusks are sometimes quite absent.[4] In the male elephant of India and in the male dugong[5] the upper incisors form offensive weapons. In the male narwhal one alone of the upper teeth is developed into the well-known, spirally-twisted, so called horn, which is sometimes from nine to ten feet in length. It is believed that the males use these horns for fighting together; for "an unbroken one can rarely be got, and occasionally "one may be found with the point of another jammed "into the broken place."[6] The tooth on the opposite side of the head in the male consists of a rudiment about ten inches in length, which is embedded in the jaw. It is not, however, very uncommon to find double-horned male walruses in which both teeth are well developed. In the females both teeth are rudimentary. The male cachalot has a larger head than that of the female, and it no doubt aids these animals in their aquatic battles. Lastly, the adult male ornithorhynchus is provided with a remarkable apparatus, namely a spur on the fore-leg, closely resembling the poison-fang of a venomous snake; its use is not known, but we may suspect that it serves as a weapon of offence.[7] It is represented by a mere rudiment in the female.

When the males are provided with weapons which the females do not possess, there can hardly be a doubt that they are used for fighting with other males, and that they have been acquired through sexual selection.

---

[4] Mr. Lamont ('Seasons with the Sea-Horses,' 1861, p. 143) says that a good tusk of the male walrus weighs 4 pounds, and is longer than that of the female, which weighs about 3 pounds. The males are described as fighting ferociously. On the occasional absence of the tusks in the female, see Mr. R. Brown, 'Proc. Zool. Soc.' 1868, p. 429.

[5] Owen, 'Anatomy of Vertebrates,' vol. iii. p. 283.

[6] Mr. R. Brown, in 'Proc. Zool. Soc.' 1869, p. 553.

[7] Owen on the Cachalot and Ornithorhynchus, ibid. vol. iii. p. 638, 641.

It is not probable, at least in most cases, that the females have actually been saved from acquiring such weapons, owing to their being useless and superfluous, or in some way injurious. On the contrary, as they are often used by the males of many animals for various purposes, more especially as a defence against their enemies, it is a surprising fact that they are so poorly developed or quite absent in the females. No doubt with female deer the development during each recurrent season of great branching horns, and with female elephants the development of immense tusks, would have been a great waste of vital power, on the admission that they were of no use to the females. Consequently variations in the size of these organs, leading to their suppression, would have come under the control of natural selection, and if limited in their transmission to the female offspring would not have interfered with their development through sexual selection in the males. But how on this view can we explain the presence of horns in the females of certain antelopes, and of tusks in the females of many animals, which are only of slightly less size than in the males? The explanation in almost all cases must, I believe, be sought in the laws of transmission.

As the reindeer is the single species in the whole family of Deer in which the female is furnished with horns, though somewhat smaller, thinner, and less branched than in the male, it might naturally be thought that they must be of some special use to her. There is, however, some evidence opposed to this view. The female retains her horns from the time when they are fully developed, namely in September, throughout the winter, until May, when she brings forth her young; whilst the male casts his horns much earlier, towards the end of November. As both sexes have the same requirements and follow the same habits of life, and as the male sheds his horns during the winter, it is very improbable that they can be of any special service to the female at this season, which includes the larger proportion of the time during which she bears horns. Nor is it probable that she can have inherited horns from some ancient progenitor of the whole family of deer, for, from the fact of the males alone of so many species in all quarters of the globe possessing horns, we may conclude that this was the primordial character of the group. Hence it appears that horns must have been transferred from the male to the female at a period subsequent to the divergence of the various species from a common stock; but that this was not effected for the sake of giving her any special advantage.[8]

We know that the horns are developed at a most unusually early age in the reindeer; but what the cause of this may have been is not known. The effect, however, has apparently been the transference of the horns to both sexes. It is intelligible on the hypothesis of pangenesis, that a very slight change in the constitution of the male, either in the tissues of the forehead or in the gemmules of the horns, might lead to their early development; and as the young of both sexes have nearly the same constitution before the period of reproduction, the horns, if developed at an early age in the male, would tend to be developed equally in both sexes. In support of this view, we should bear in mind that the horns are always transmitted through the female, and that she has a latent capacity for their development, as we see in old or diseased females.[9] Moreover the females

---

8 On the structure and shedding of the horns of the reindeer, Hoffberg, 'Amœnitates Acad.' vol. iv. 1788, p. 149. See Richardson, 'Fauna Bor. Americana,' p. 241, in regard to the American variety or species; also Major W. Ross King, 'The Sportsman in Canada,' 1866, p. 80.

9 Isidore Geoffroy St.-Hilaire, 'Essais de Zoolog. Généralc,' 1841 p. 513. Other masculine characters, besides the horns, are sometimes

of some other species of deer either normally or occasionally exhibit rudiments of horns; thus the female of *Cervulus moschatus* has "bristly tufts, ending in a knob, "instead of a horn;" and "in most specimens of the "female Wapiti (*Cervus Canadensis*) there is a sharp "bony protuberance in the place of the horn."[10] From these several considerations we may conclude that the possession of fairly well-developed horns by the female reindeer, is due to the males having first acquired them as weapons for fighting with other males; and secondarily to their development from some unknown cause at an unusually early age in the males, and their consequent transmission to both sexes.

Turning to the sheath-horned ruminants: with antelopes a graduated series can be formed, beginning with the species, the females of which are completely destitute of horns—passing to those which have horns so small as to be almost rudimentary, as in *Antilocapra Americana*—to those which have fairly well-developed horns, but manifestly smaller and thinner than in the male, and sometimes of a different shape,[11] and ending with those in which both sexes have horns of equal size. As with the reindeer, so with antelopes there exists a relation between the period of the development of the horns and their transmission to one or both sexes; it

similarly transferred to the female; thus Mr. Boner, in speaking of an old female chamois ('Chamois Hunting in the Mountains of Bavaria,' 1860, 2nd edit. p. 363), says, "not only was the head very male-look-"ing, but along the back there was a ridge of long hair, usually to be "found only in bucks."

[10] On the Cervulus, Dr. Gray, 'Catalogue of the Mammalia in British Museum,' part iii. p. 220. On the *Cervus Canadensis* or Wapiti see Hon. J. D. Caton, 'Ottawa Acad. of Nat. Sciences,' May, 1868, p. 9.

[11] For instance the horns of the female *Ant. Euchore* resemble those of a distinct species, viz. the *Ant. Dorcas* var. *Corine*, see Desmarest, 'Mammalogie,' p. 455.

is therefore probable that their presence or absence in the females of some species, and their more or less perfect condition in the females of other species, depend, not on their being of some special use, but simply on the form of inheritance which has prevailed. It accords with this view that even in the same restricted genus both sexes of some species, and the males alone of other species, are thus provided. It is a remarkable fact that, although the females of *Antilope bezoartica* are normally destitute of horns, Mr. Blyth has seen no less than three females thus furnished; and there was no reason to suppose that they were old or diseased. The males of this species have long straight spirated horns, nearly parallel to each other, and directed backwards. Those of the female, when present, are very different in shape, for they are not spirated, and spreading widely bend round, so that their points are directed forwards. It is a still more remarkable fact that in the castrated male, as Mr. Blyth informs me, the horns are of the same peculiar shape as in the female, but longer and thicker. In all cases the differences between the horns of the males and females, and of castrated and entire males, probably depend on various causes,—on the more or less complete transference of male characters to the females,—on the former state of the progenitors of the species,—and partly perhaps on the horns being differently nourished, in nearly the same manner as the spurs of the domestic cock, when inserted into the comb or other parts of the body, assume various abnormal forms from being differently nourished.

In all the wild species of goats and sheep the horns are larger in the male than in the female, and are sometimes quite absent in the latter.[12] In several domestic

---

[12] Gray, 'Catalogue Mamm. Brit. Mus.' part iii. 1852, p. 160.

breeds of the sheep and goat, the males alone are furnished with horns; and it is a significant fact, that in one such breed of sheep on the Guinea coast, the horns are not developed, as Mr. Winwood Reade informs me, in the castrated male; so that they are affected in this respect like the horns of stags. In some breeds, as in that of N. Wales, in which both sexes are properly horned, the ewes are very liable to be hornless. In these same sheep, as I have been informed by a trustworthy witness who purposely inspected a flock during the lambing-season, the horns at birth are generally more fully developed in the male than in the female. With the adult musk-ox (*Ovibos moschatus*) the horns of the male are larger than those of the female, and in the latter the bases do not touch.[13] In regard to ordinary cattle Mr. Blyth remarks: "In most of the wild bovine "animals the horns are both longer and thicker in the "bull than in the cow, and in the cow-banteng (*Bos* "*sondaicus*) the horns are remarkably small, and in-"clined much backwards. In the domestic races of "cattle, both of the humped and humpless types, the "horns are short and thick in the bull, longer and "more slender in the cow and ox; and in the Indian "buffalo, they are shorter and thicker in the bull, longer "and more slender in the cow. In the wild gaour "(*B. gaurus*) the horns are mostly both longer and "thicker in the bull than in the cow."[14] Hence with most sheath-horned ruminants the horns of the male are either longer or stronger than those of the female. With the *Rhinoceros simus*, as I may here add, the horns of the female are generally longer but less powerful than in the male; and in some other species of

---

[13] Richardson, 'Fauna Bor. Americana,' p. 278.

[14] 'Land and Water' 1867, p. 346.

rhinoceros they are said to be shorter in the female.[15] From these various facts we may conclude that horns of all kinds, even when they are equally developed in both sexes, were primarily acquired by the males in order to conquer other males, and have been transferred more or less completely to the female, in relation to the force of the equal form of inheritance.

The tusks of the elephant, in the different species or races, differ according to sex, in nearly the same manner as the horns of ruminants. In India and Malacca the males alone are provided with well-developed tusks. The elephant of Ceylon is considered by most naturalists as a distinct race, but by some as a distinct species, and here "not one in a hundred is found with "tusks, the few that possess them being exclusively "males."[16] The African elephant is undoubtedly distinct, and the female has large, well-developed tusks, though not so large as those of the male. These differences in the tusks of the several races and species of elephants—the great variability of the horns of deer, as notably in the wild reindeer—the occasional presence of horns in the female *Antilope bezoartica*—the presence of two tusks in some few male narwhals—the complete absence of tusks in some female walruses;—are all instances of the extreme variability of secondary sexual characters, and of their extreme liability to differ in closely-allied forms.

Although tusks and horns appear in all cases to have been primarily developed as sexual weapons, they often serve for other purposes. The elephant uses his tusks

---

15 Sir Andrew Smith, 'Zoology of S. Africa,' pl. xix. Owen, 'Anatomy of Vertebrates,' vol. iii. p. 624.

16 Sir J. Emerson Tennent, 'Ceylon,' 1859, vol. ii. p. 274. For Malacca, 'Journal of Indian Archipelago,' vol. iv. p. 357.

in attacking the tiger; according to Bruce, he scores the trunks of trees until they can be easily thrown down, and he likewise thus extracts the farinaceous cores of palms; in Africa he often uses one tusk, this being always the same, to probe the ground and thus to ascertain whether it will bear his weight. The common bull defends the herd with his horns; and the elk in Sweden has been known, according to Lloyd, to strike a wolf dead with a single blow of his great horns. Many similar facts could be given. One of the most curious secondary uses to which the horns of any animal are occasionally put, is that observed by Captain Hutton [17] with the wild goat (*Capra ægagrus*) of the Himalayas, and as it is said with the ibex, namely, that when the male accidentally falls from a height he bends inwards his head, and, by alighting on his massive horns, breaks the shock. The female cannot thus use her horns, which are smaller, but from her more quiet disposition she does not so much need this strange kind of shield.

Each male animal uses his weapons in his own peculiar fashion. The common ram makes a charge and butts with such force with the bases of his horns, that I have seen a powerful man knocked over as easily as a child. Goats and certain species of sheep, for instance the *Ovis cycloceros* of Afghanistan,[18] rear on their hind legs, and then not only butt, but "make a cut down "and a jerk up, with the ribbed front of their scimitar-"shaped horn, as with a sabre. When the *O. cycloceros* "attacked a large domestic ram, who was a noted "bruiser, he conquered him by the sheer novelty of his

---

[17] 'Calcutta Journal of Nat. Hist.' vol. ii. 1843, p. 526.

[18] Mr. Blyth, in 'Land and Water,' March, 1867, p. 134, on the authority of Capt. Hutton and others. For the wild Pembrokeshire goats see the 'Field,' 1869, p. 150.

"mode of fighting, always closing at once with his "adversary, and catching him across the face and nose "with a sharp drawing jerk of his head, and then "bounding out of the way before the blow could be "returned." In Pembrokeshire a male goat, the master of a flock which during several generations had run wild, was known to have killed several other males in single combat; this goat possessed enormous horns, measuring 39 inches in a straight line from tip to tip. The common bull, as every one knows, gores and tosses his opponent; but the Italian buffalo is said never to use his horns, he gives a tremendous blow with his convex forehead, and then tramples on his fallen enemy with his knees—an instinct which the common bull does not possess.[19] Hence a dog who pins a buffalo by the nose is immediately crushed. We must, however, remember that the Italian buffalo has long been domesticated, and it is by no means certain that the wild parent-form had similarly shaped horns. Mr. Bartlett informs me that when a female Cape buffalo (*Bubalus caffer*) was turned into an enclosure with a bull of the same species, she attacked him, and he in return pushed her about with great violence. But it was manifest to Mr. Bartlett that had not the bull shewn dignified forbearance, he could easily have killed her by a single lateral thrust with his immense horns. The giraffe uses his short hair-covered horns, which are rather longer in the male than in the female, in a curious manner; for with his long neck he swings his head to either side, almost upside down, with such force, that I have seen a hard plank deeply indented by a single blow.

---

19 M. E. M. Bailly, "sur l'usage des Cornes," &c., 'Annal. des Sc. Nat.' tom. ii. 1824, p. 369.

With antelopes it is sometimes difficult to imagine how they can possibly use their curiously-shaped horns; thus the spring-boc (*Ant. euchore*) has rather short upright horns, with the sharp points bent inwards almost at a right angle, so as to face each other; Mr. Bartlett does not know how they are used, but suggests that they would inflict a fearful wound down each side of the face of an antagonist. The slightly-curved horns of the *Oryx leucoryx* (fig. 61) are directed backwards, and are of such length that their points reach beyond the

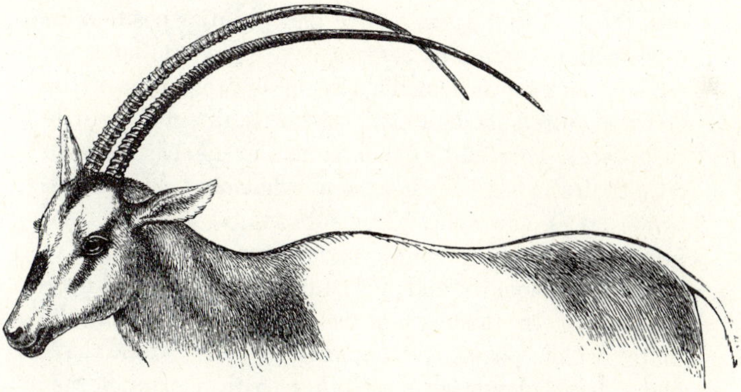

Fig. 61. Oryx leucoryx, male (from the Knowsley Menagerie).

middle of the back, over which they stand in an almost parallel line. Thus they seem singularly ill-fitted for fighting; but Mr. Bartlett informs me that when two of these animals prepare for battle, they kneel down, with their heads between their front legs, and in this attitude the horns stand nearly parallel and close to the ground, with the points directed forwards and a little upwards. The combatants then gradually approach each other and endeavour to get the upturned points under each other's bodies; if one succeeds in doing this, he suddenly springs up, throwing up his head at the same time, and can thus wound or perhaps even transfix his antagonist. Both animals always kneel down so as to guard as far as possible against this manœuvre. It has been recorded that one of these antelopes has used his horns with effect even against a lion; yet from being forced to place his head between the fore-legs in order to bring the points of the horns forward, he would generally be under a great disadvantage when attacked by any other animal. It is, therefore, not probable that the horns have been modified into their present great length and peculiar position, as a protection against beasts of prey. We can, however, see that as soon as some ancient male progenitor of the Oryx acquired moderately long horns, directed a little backwards, he would be compelled in his battles with rival males to bend his head somewhat inwards or downwards, as is now done by certain stags; and it is not improbable that he might have acquired the habit of at first occasionally and afterwards of regularly kneeling down. In this case it is almost certain that the males which possessed the longest horns would have had a great advantage over others with shorter horns; and then the horns would gradually have been rendered longer and longer, through sexual selection, until they acquired their present extraordinary length and position.

With stags of many kinds the branching of the horns offers a curious case of difficulty; for certainly a single straight point would inflict a much more serious wound than several diverging points. In Sir Philip Egerton's museum there is a horn of the red-deer (*Cervus elaphus*) thirty inches in length, with "not fewer than "fifteen snags or branches;" and at Moritzburg there is still preserved a pair of antlers of a red-deer, shot in

1699 by Frederick I., each of which bears the astonishing number of thirty-three branches. Richardson figures a pair of antlers of the wild reindeer with twenty-nine points.[20] From the manner in which the horns are branched, and more especially from deer being known occasionally to fight together by kicking with their fore-feet,[21] M. Bailly actually came to the conclusion that their horns were more injurious than useful to them! But this author overlooks the pitched battles between rival males. As I felt much perplexed about the use or advantage of the branches, I applied to Mr. McNeill of Colinsay, who has long and carefully observed the habits of red-deer, and he informs me that he has never seen some of the branches brought into action, but that the brow-antlers, from inclining downwards, are a great protection to the forehead, and their points are likewise used in attack. Sir Philip Egerton also informs me in regard both to red-deer and fallow-deer, that when they fight they suddenly dash together, and getting their horns fixed against each other's bodies a desperate struggle ensues. When one is at last forced to yield and turn round, the victor endeavours to plunge his brow-antlers into his defeated foe. It thus appears that the upper branches are used chiefly or exclusively for pushing and fencing. Nevertheless with some species the upper branches are used as weapons of offence; when a man was attacked by a

---

[20] Owen, on the Horns of Red-deer, 'British Fossil Mammals,' 1846, p. 478; 'Forest Creatures,' by Charles Boner, 1861, p. 76, 62. Richardson on the Horns of the Reindeer, 'Fauna Bor. Americana,' 1829, p. 240.

[21] Hon. J. D. Caton ('Ottawa Acad. of Nat. Science,' May, 1868, p. 9), says that the American deer fight with their fore-feet, after "the question of superiority has been once settled and acknowledged in the herd." Bailly, "Sur l'usage des Cornes," 'Annales des Sc. Nat.' tom. ii. 1824, p. 371.

Wapiti deer (*Cervus Canadensis*) in Judge Caton's park in Ottawa, and several men tried to rescue him, the stag "never raised his head from the ground; in fact he kept "his face almost flat on the ground, with his nose nearly "between his fore-feet, except when he rolled his head "to one side to take a new observation preparatory to "a plunge." In this position the terminal points of the horns were directed against his adversaries. "In "rolling his head he necessarily raised it somewhat, "because his antlers were so long that he could not "roll his head without raising them on one side, while "on the other side they touched the ground." The stag by this procedure gradually drove the party of rescuers backwards, to a distance of 150 or 200 feet; and the attacked man was killed.[22]

Although the horns of stags are efficient weapons, there can, I think, be no doubt that a single point would have been much more dangerous than a branched antler; and Judge Caton, who has had large experience with deer, fully concurs in this conclusion. Nor do the branching horns, though highly important as a means of defence against rival stags, appear perfectly well adapted for this purpose, as they are liable to become interlocked. The suspicion has therefore crossed my mind that they may serve partly as ornaments. That the branched antlers of stags, as well as the elegant lyrated horns of certain antelopes, with their graceful double curvature, (fig. 62), are ornamental in our eyes, no one will dispute. If, then, the horns, like the splendid accoutrements of the knights of old, add to the noble appearance of stags and antelopes, they may have been partly modified for this purpose,

---

22 See a most interesting account in the Appendix to Hon. J. D. Caton's paper, as above quoted.

though mainly for actual service in battle; but I have no evidence in favour of this belief.

Fig. 62. Strepsiceros Kudu (from Andrew Smith's 'Zoology of South Africa')

An interesting case has lately been published, from which it appears that the horns of a deer in one district in the United States are now being modified through sexual and natural selection. A writer in an excellent

American Journal[23] says, that he has hunted for the last twenty-one years in the Adirondacks, where the *Cervus Virginianus* abounds. About fourteen years ago he first heard of *spike-horn bucks.* These became from year to year more common; about five years ago he shot one, and subsequently another, and now they are frequently killed. "The spike-horn differs greatly "from the common antler of the *C. Virginianus.* It "consists of a single spike, more slender than the antler, "and scarcely half so long, projecting forward from the "brow, and terminating in a very sharp point. It gives "a considerable advantage to its possessor over the "common buck. Besides enabling him to run more "swiftly through the thick woods and underbrush "(every hunter knows that does and yearling bucks "run much more rapidly than the large bucks when "armed with their cumbrous antlers), the spike-horn "is a more effective weapon than the common antler. "With this advantage the spike-horn bucks are gaining "upon the common bucks, and may, in time, entirely "supersede them in the Adirondacks. Undoubtedly "the first spike-horn buck was merely an accidental "freak of nature. But his spike-horns gave him an "advantage, and enabled him to propagate his pecu-"liarity. His descendants, having a like advantage, "have propagated the peculiarity in a constantly "increasing ratio, till they are slowly crowding the "antlered deer from the region they inhabit."

Male quadrupeds which are furnished with tusks use them in various ways, as in the case of horns. The boar strikes laterally and upwards; the musk-deer with serious effect downwards.[24] The walrus,

23 'The American Naturalist,' Dec. 1869, p. 552.
24 Pallas, 'Spicilegia Zoologica,' fasc. xiii. 1779, p. 18.

though having so short a neck and so unwieldy a body, "can strike either upwards, or downwards, or side-"ways, with equal dexterity."[25] The Indian elephant fights, as I was informed by the late Dr. Falconer, in a different manner according to the position and curvature of his tusks. When they are directed forwards and upwards he is able to fling a tiger to a great distance—it is said to even thirty feet; when they are short and turned downwards he endeavours suddenly to pin the tiger to the ground, and in consequence is dangerous to the rider, who is liable to be jerked off the hoodah.[26]

Very few male quadrupeds possess weapons of two distinct kinds specially adapted for fighting with rival males. The male muntjac-deer (*Cervulus*), however, offers an exception, as he is provided with horns and exserted canine teeth. But one form of weapon, has often been replaced in the course of ages by another form, as we may infer from what follows. With ruminants the development of horns generally stands in an inverse relation with that of even moderately well-developed canine teeth. Thus camels, guanacoes, chevrotains and musk-deer, are hornless, and they have efficient canines; these teeth being "always of "smaller size in the females than in the males." The Camelidæ have in their upper jaws, in addition to their true canines, a pair of canine-shaped incisors.[27] Male deer and antelopes, on the other hand, possess horns, and they rarely have canine teeth; and these when present are always of small size, so that it is

[25] Lamont, 'Seasons with the Sea-Horses,' 1861, p. 141.

[26] See also Corse ('Philosoph. Transact.' 1799, p. 212) on the manner in which the short-tusked Mooknah variety of the elephant attacks other elephants.

[27] Owen, 'Anatomy of Vertebrates,' vol. iii. p. 349.

doubtful whether they are of any service in their battles. With *Antilope montana* they exist only as rudiments in the young male, disappearing as he grows old; and they are absent in the female at all ages; but the females of certain other antelopes and deer have been known occasionally to exhibit rudiments of these teeth.[28] Stallions have small canine teeth, which are either quite absent or rudimentary in the mare; but they do not appear to be used in fighting, for stallions bite with their incisors, and do not open their mouths widely like camels and guanacoes. Whenever the adult male possesses canines now in an inefficient state, whilst the female has either none or mere rudiments, we may conclude that the early male progenitor of the species was provided with efficient canines, which had been partially transferred to the females. The reduction of these teeth in the males seems to have followed from some change in their manner of fighting, often caused (but not in the case of the horse) by the development of new weapons.

Tusks and horns are manifestly of high importance to their possessors, for their development consumes much organised matter. A single tusk of the Asiatic elephant,—one of the extinct woolly species,—and of the African elephant, have been known to weigh respectively 150, 160, and 180 pounds; and even greater weights have been assigned by some authors.[29] With deer, in

---

28 See Rüppell (in 'Proc. Zoolog. Soc.' Jan. 12, 1836, p. 3) on the canines in deer and antelopes, with a note by Mr. Martin on a female American deer. See also Falconer ('Palæont. Memoirs and Notes,' vol. i. 1868, p. 576) on canines in an adult female deer. In old males of the musk-deer the canines (Pallas, 'Spic. Zoolog.' fasc. xiii. 1779, p. 18) sometimes grow to the length of three inches, whilst in old females a rudiment projects scarcely half an inch above the gums.

29 Emerson Tennent, 'Ceylon,' 1859, vol. ii. p. 275; Owen, 'British Fossil Mammals,' 1846, p. 245.

which the horns are periodically renewed, the drain on the constitution must be greater; the horns, for instance, of the moose weigh from fifty to sixty pounds, and those of the extinct Irish elk from sixty to seventy pounds,—the skull of the latter weighing on an average only five and a quarter pounds. With sheep, although the horns are not periodically renewed, yet their development, in the opinion of many agriculturists, entails a sensible loss to the breeder. Stags, moreover, in escaping from beasts of prey are loaded with an additional weight for the race, and are greatly retarded in passing through a woody country. The moose, for instance, with horns extending five and a half feet from tip to tip, although so skilful in their use that he will not touch or break a dead twig when walking quietly, cannot act so dexterously whilst rushing away from a pack of wolves. "During his "progress he holds his nose up, so as to lay the "horns horizontally back; and in this attitude cannot "see the ground distinctly."[30] The tips of the horns of the great Irish elk were actually eight feet apart! Whilst the horns are covered with velvet, which lasts with the red-deer for about twelve weeks, they are extremely sensitive to a blow; so that in Germany the stags at this time change their habits to a certain extent, and avoid dense forests, frequenting young woods and low thickets.[31] These facts remind us, that male birds have acquired ornamental plumes at the cost of retarded flight, and other ornaments at the cost of some loss of power in their battles with rival males.

---

[30] Richardson, 'Fauna Bor. Americana,' on the moose, *Alces palmata*, p. 236, 237; also on the expanse of the horns 'Land and Water,' 1869, p. 143 See also Owen, 'British Fossil Mammals,' on the Irish elk, p. 447, 455.

[31] 'Forest Creatures,' by C. Boner, 1861, p. 60.

With quadrupeds, when, as is often the case, the sexes differ in size, the males are, I believe, always larger and stronger. This holds good in a marked manner, as I am informed by Mr. Gould, with the marsupials of Australia, the males of which appear to continue growing until an unusually late age. But the most extraordinary case is that of one of the seals (*Callorhinus ursinus*), a full-grown female weighing less than one-sixth of a full-grown male.[32] The greater strength of the male is invariably displayed, as Hunter long ago remarked,[33] in those parts of the body which are brought into action in fighting with rival males,—for instance, in the massive neck of the bull. Male quadrupeds are also more courageous and pugnacious than the females. There can be little doubt that these characters have been gained, partly through sexual selection, owing to a long series of victories by the stronger and more courageous males over the weaker, and partly through the inherited effects of use. It is probable that the successive variations in strength, size, and courage, whether due to so-called spontaneous variability or to the effects of use, by the accumulation of which male quadrupeds have acquired these characteristic qualities, occurred rather late in life, and were consequently to a large extent limited in their transmission to the same sex.

Under this point of view I was anxious to obtain information in regard to the Scotch deer-hound, the sexes of which differ more in size than those of any other breed (though blood-hounds differ considerably), or than in any wild canine species known to me.

32 See the very interesting paper by Mr. J. A. Allen in 'Bull. Mus. Comp. Zoolog. of Cambridge; United States,' vol. ii. No. 1, p. 82. The weights were ascertained by a careful observer, Capt. Bryant.

33 'Animal Economy, p. 45.

Accordingly, I applied to Mr. Cupples, a well-known breeder of these dogs, who has weighed and measured many of his own dogs, and who, with great kindness, has collected for me the following facts from various sources. Superior male dogs, measured at the shoulder, range from twenty-eight inches, which is low, to thirty-three, or even thirty-four inches in height; and in weight from eighty pounds, which is low, to 120, or even more pounds. The females range in height from twenty-three to twenty-seven, or even to twenty-eight inches; and in weight from fifty to seventy, or even eighty pounds.[34] Mr. Cupples concludes that from ninety-five to one hundred pounds for the male, and seventy for the female, would be a safe average; but there is reason to believe that formerly both sexes attained a greater weight. Mr. Cupples has weighed puppies when a fortnight old; in one litter the average weight of four males exceeded that of two females by six and a half ounces; in another litter the average weight of four males exceeded that of one female by less than one ounce; the same males, when three weeks old, exceeded the female by seven and a half ounces, and at the age of six weeks by nearly fourteen ounces. Mr. Wright of Yeldersley House, in a letter to Mr. Cupples, says: "I have taken notes on the sizes and weights of puppies "of many litters, and as far as my experience goes, "dog-puppies as a rule differ very little from bitches "till they arrive at about five or six months old; and "then the dogs begin to increase, gaining upon the

[34] See also Richardson's 'Manual on the Dog,' p. 59. Much valuable information on the Scottish deer-hound is given by Mr. McNeill, who first called attention to the inequality in size between the sexes, in Scrope's 'Art of Deer Stalking.' I hope that Mr. Cupples will keep to his intention of publishing a full account and history of this famous breed.

"bitches both in weight and size. At birth, and for "several weeks afterwards, a bitch-puppy will occa- "sionally be larger than any of the dogs, but they are "invariably beaten by them later." Mr. McNeill, of Colinsay, concludes that "the males do not attain "their full growth till over two years old, though "the females attain it sooner." According to Mr. Cupples' experience, male dogs go on growing in stature till they are from twelve to eighteen months old, and in weight till from eighteen to twenty-four months old; whilst the females cease increasing in stature at the age of from nine to fourteen or fifteen months, and in weight at the age of from twelve to fifteen months. From these various statements it is clear that the full difference in size between the male and female Scotch deer-hound is not acquired until rather late in life. The males are almost exclusively used for coursing, for, as Mr. McNeill informs me, the females have not sufficient strength and weight to pull down a full-grown deer. From the names used in old legends, it appears, as I hear from Mr. Cupples, that at a very ancient period the males were the most celebrated, the females being mentioned only as the mothers of famous dogs. Hence during many generations, it is the male which has been chiefly tested for strength, size, speed, and courage, and the best will have been bred from. As, however, the males do not attain their full dimensions until a rather late period in life, they will have tended, in accordance with the law often indicated, to transmit their characters to their male offspring alone; and thus the great inequality in size between the sexes of the Scotch deer-hound may probably be accounted for.

The males of some few quadrupeds possess organs or

parts developed solely as a means of defence against the attacks of other males. Some kinds of deer use, as we have seen, the upper branches of their horns chiefly or exclusively for defending themselves; and the Oryx antelope, as I am informed by Mr. Bartlett, fences most skilfully with his long, gently curved horns; but these are likewise used as organs of offence. Rhinoceroses, as the same observer remarks, in fighting parry each other's sidelong blows with their horns, which loudly clatter together, as do the tusks of boars. Although wild boars fight desperately together, they seldom, according to Brehm, receive fatal blows, as these fall on each other's tusks, or on the layer of gristly skin covering the shoulder, which the German hunters call the shield; and here we have a part specially modified for defence. With boars in the prime of life (see fig. 63) the tusks in the lower jaw are used for fighting but they become in old age, as Brehm states, so much curved inwards and upwards, over the snout, that they can no longer be thus used. They may, however, still continue to serve, and even in a still more effective manner, as a means of defence. In compensation for the loss of the lower tusks as weapons of offence, those in the upper jaw, which always project a little laterally, increase so much in length during old age, and curve so much upwards, that they can be used as a means of attack. Nevertheless an old boar is not so

Fig. 63. Head of common wild boar, in prime of life (from Brehm).

dangerous to man as one at the age of six or seven years.[35]

In the full-grown male Babirusa pig of Celebes (fig. 64), the lower tusks are formidable weapons, like those of the European boar in the prime of life, whilst the upper tusks are so long and have their points so much curled inwards, sometimes even touching the

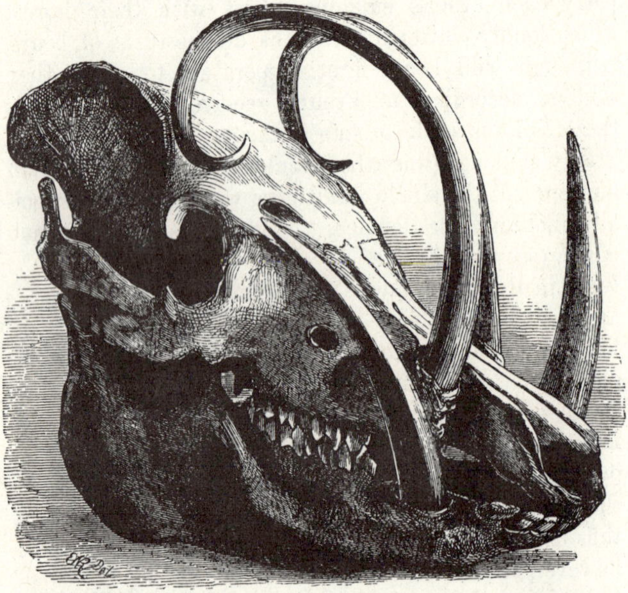

Fig. 64. Skull of the Babirusa Pig (from Wallace's 'Malay Archipelago')

forehead, that they are utterly useless as weapons of attack. They more nearly resemble horns than teeth, and are so manifestly useless as teeth that the animal was formerly supposed to rest his head by hooking them on to a branch. Their convex surfaces would, however,

35 Brehm, 'Thierleben,' B. ii. s. 729 732.

if the head were held a little laterally, serve as an excellent guard; and hence, perhaps it is that in old animals they "are generally broken off, as if by "fighting."[36] Here, then, we have the curious case of the upper tusks of the Babirusa regularly assuming during the prime of life, a structure which apparently renders them fitted only for defence; whilst in the European boar the lower and opposite tusks assume in a less degree and only during old age nearly the same form, and then serve in like manner solely for defence.

Fig. 65. Head of Æthiopian Wart-hog, from 'Proc. Zool. Soc.' 1869. (I now find that this drawing represents the head of a female, but it serves to shew, on a reduced scale, the characters of the male.)

In the wart-hog (*Phacochoerus æthiopicus*, fig. 65) the tusks in the upper jaw of the male curve upwards during the prime of life, and from being pointed, serve as formidable weapons. The tusks in the lower jaw are sharper than those in the upper, but from their shortness it seems hardly possible that they can be used as weapons of attack. They must, however, greatly

[36] See Mr. Wallace's interesting account of this animal, 'The Malay Archipelago,' 1869, vol. i. p. 435.

strengthen those in the upper jaw, from being ground so as to fit closely against their bases. Neither the upper nor the lower tusks appear to have been specially modified to act as guards, though, no doubt, they are thus used to a certain extent. But the wart-hog is not destitute of other special means of protection, for there exists, on each side of the face, beneath the eyes, a rather stiff, yet flexible, cartilaginous, oblong pad (fig. 65), which projects two or three inches outwards; and it appeared to Mr. Bartlett and myself, when viewing the living animal, that these pads, when struck from beneath by the tusks of an opponent, would be turned upwards, and would thus protect in an admirable manner the somewhat prominent eyes. These boars, as I may add on the authority of Mr. Bartlett, when fighting together, stand directly face to face.

Lastly, the African river-hog (*Potamochoerus penicillatus*) has a hard cartilaginous knob on each side of the face beneath the eyes, which answers to the flexible pad of the wart-hog; it has also two bony prominences on the upper jaw above the nostrils. A boar of this species in the Zoological Gardens recently broke into the cage of the wart-hog. They fought all night-long, and were found in the morning much exhausted, but not seriously wounded. It is a significant fact, as shewing the purpose of the above-described projections and excrescences, that these were covered with blood, and were scored and abraded in an extraordinary manner.

The mane of the lion forms a good defence against the one danger to which he is liable, namely the attacks of rival lions: for the males, as Sir. A. Smith informs me, engage in terrible battles, and a young lion dares not approach an old one. In 1857 a tiger at Bromwich broke into the cage of a lion, and a fearful scene ensued; "the lion's mane saved his neck "and head from being much injured, but the tiger at "last succeeded in ripping up his belly, and in a few "minutes he was dead."[37] The broad ruff round the throat and chin of the Canadian lynx (*Felis Canadensis*) is much longer in the male than in the female; but whether it serves as a defence I do not know. Male seals are well known to fight desperately together, and the males of certain kinds (*Otaria jubata*)[38] have great manes, whilst the females have small ones or none. The male baboon of the Cape of Good Hope (*Cynocephalus porcarius*) has a much longer mane and larger canine teeth than the female; and the mane probably serves as a protection, for on asking the keepers in the Zoological Gardens, without giving them any clue to my object, whether any of the monkeys especially attacked each other by the nape of the neck, I was answered that this was not the case, excepting with the above baboon. In the Hamadryas baboon, Ehrenberg compares the mane of the adult male to that of a young lion, whilst in the young of both sexes and in the female the mane is almost absent.

It appeared to me probable that the immense woolly mane of the male American bison, which reaches almost to the ground, and is much more developed in the males than in the females, served as a protection to them in their terrible battles; but an experienced hunter told Judge Caton that he had never observed anything which favoured this belief. The

---

37 'The Times,' Nov. 10th, 1857. In regard to the Canada lynx, see Audubon and Bachman, 'Quadrupeds of N. America,' 1846, p. 139.

38 Dr. Murie, on Otaria, 'Proc. Zoolog. Soc.' 1869, p. 109. Mr. J. A. Allen, in the paper above quoted (p. 75), doubts whether the hair, which is longer on the neck in the male than in the female, deserves to be called a mane.

stallion has a thicker and fuller mane than the mare; and I have made particular inquiries of two great trainers and breeders who have had charge of many entire horses, and am assured that they "invariably "endeavour to seize one another by the neck." It does not, however, follow from the foregoing statements, that when the hair on the neck serves as a defence, that it was originally developed for this purpose, though this is probable in some cases, as in that of the lion. I am informed by Mr. McNeill that the long hairs on the throat of the stag (*Cervus elephas*) serve as a great protection to him when hunted, for the dogs generally endeavour to seize him by the throat; but it is not probable that these hairs were specially developed for this purpose; otherwise the young and the females would, as we may feel assured, have been equally protected.

On Preference or Choice in Pairing, as shewn by either sex of Quadrupeds.—Before describing, in the next chapter, the differences between the sexes in voice, odour emitted, and ornamentation, it will be convenient here to consider whether the sexes exert any choice in their unions. Does the female prefer any particular male, either before or after the males may have fought together for supremacy; or does the male, when not a polygamist, select any particular female? The general impression amongst breeders seems to be that the male accepts any female; and this, owing to his eagerness, is, in most cases, probably the truth. Whether the female as a general rule indifferently accepts any male is much more doubtful. In the fourteenth chapter, on Birds, a considerable body of direct and indirect evidence was advanced, shewing that the female selects her partner; and it would be a strange anomaly if female quadrupeds, which stand higher in the scale of organisation and have higher mental powers, did not generally, or at least often, exert some choice. The female could in most cases escape, if wooed by a male that did not please or excite her; and when pursued, as so incessantly occurs, by several males, she would often have the opportunity, whilst they were fighting together, of escaping with, or at least of temporarily pairing with, some one male. This latter contingency has often been observed in Scotland with female red-deer, as I have been informed by Sir Philip Egerton.[39]

It is scarcely possible that much should be known about female quadrupeds exerting in a state of nature any choice in their marriage unions. The following very curious details on the courtship of one of the eared seals, *Callorhinus ursinus*, are given [40] on the authority of Capt. Bryant, who had ample opportunities for observation. He says, "Many of the females on "their arrival at the island where they breed appear "desirous of returning to some particular male, and "frequently climb the outlying rocks to overlook the "rookeries, calling out and listening as if for a familiar "voice. Then changing to another place they do the "same again . . . . As soon as a female reaches the "shore, the nearest male goes down to meet her, making "meanwhile a noise like the clucking of a hen to her "chickens. He bows to her and coaxes her until he "gets between her and the water so that she cannot "escape him. Then his manner changes, and with a

[39] Mr. Boner in his excellent description of the habits of the red-deer in Germany ('Forest Creatures,' 1861, p. 81) says, "while the "stag is defending his rights against one intruder, another invades the "sanctuary of his harem, and carries off trophy after trophy." Exactly the same thing occurs with seals, see Mr. J. A. Allen, ibid. p. 100.

[40] Mr. J. A. Allen in 'Bull. Mus. Comp. Zoolog. of Cambridge, United States,' vol. ii. No. 1, p. 99.

"harsh growl he drives her to a place in his harem. "This continues until the lower row of harems is "nearly full. Then the males higher up select the "time when their more fortunate neighbours are off "their guard to steal their wives. This they do by "taking them in their mouths and lifting them over "the heads of the other females, and carefully placing "them in their own harem, carrying them as cats do "their kittens. Those still higher up pursue the same "method until the whole space is occupied. Frequently "a struggle ensues between two males for the possession "of the same female, and both seizing her at once pull "her in two or terribly lacerate her with their teeth. "When the space is all filled, the old male walks around "complacently reviewing his family, scolding those "who crowd or disturb the others, and fiercely driving "off all intruders. This surveillance always keeps him "actively occupied."

As so little is known about the courtship of animals in a state of nature, I have endeavoured to discover how far our domesticated quadrupeds evince any choice in their unions. Dogs offer the best opportunity for observation, as they are carefully attended to and well understood. Many breeders have expressed a strong opinion on this head. Thus Mr. Mayhew remarks, "The females are "able to bestow their affections; and tender recollec- "tions are as potent over them as they are known to "be in other cases, where higher animals are con- "cerned. Bitches are not always prudent in their "loves, but are apt to fling themselves away on curs "of low degree. If reared with a companion of vulgar "appearance, there often springs up between the pair a "devotion which no time can afterwards subdue. The "passion, for such it really is, becomes of a more than "romantic endurance." Mr. Mayhew, who attended chiefly to the smaller breeds, is convinced that the females are strongly attracted by males of large size.[41] The well-known veterinary Blaine states[42] that his own female pug became so attached to a spaniel, and a female setter to a cur, that in neither case would they pair with a dog of their own breed until several weeks had elapsed. Two similar and trustworthy accounts have been given me in regard to a female retriever and a spaniel, both of which became enamoured with terrier-dogs.

Mr. Cupples informs me that he can personally vouch for the accuracy of the following more remarkable case, in which a valuable and wonderfully-intelligent female terrier loved a retriever, belonging to a neighbour, to such a degree that she had often to be dragged away from him. After their permanent separation, although repeatedly shewing milk in her teats, she would never acknowledge the courtship of any other dog, and to the regret of her owner, never bore puppies. Mr. Cupples also states that a female deerhound now (1868) in his kennel has thrice produced puppies, and on each occasion shewed a marked preference for one of the largest and handsomest, but not the most eager, of four deer-hounds living with her, all in the prime of life. Mr. Cupples has observed that the female generally favours a dog whom she has associated with and knows; her shyness and timidity at first incline her against a strange dog. The male, on the contrary, seems rather inclined towards strange females. It appears to be rare when the male refuses any particular female, but Mr. Wright, of Yeldersley House,

---

[41] 'Dogs: their Management,' by E. Mayhew, M.R.C.V.S., 2nd edit. 1864, p. 187-192.

[42] Quoted by Alex. Walker 'On Intermarriage,' 1838, p. 276; see also p. 244.

a great breeder of dogs, informs me that he has known some instances; he cites the case of one of his own deer-hounds, who would not take any notice of a particular female mastiff, so that another deer-hound had to be employed. It would be superfluous to give other cases, and I will only add that Mr. Barr, who has carefully bred many blood-hounds, states that in almost every instance particular individuals of the opposite sex shew a decided preference for each other. Finally Mr. Cupples, after attending to this subject for another year, has recently written to me, "I have had full con- "firmation of my former statement, that dogs in breed- "ing form decided preferences for each other, being "often influenced by size, bright colour, and individual "character, as well as by the degree of their previous "familiarity."

In regard to horses, Mr. Blenkiron, the greatest breeder of race-horses in the world, informs me that stallions are so frequently capricious in their choice, rejecting one mare and without any apparent cause taking to another, that various artifices have to be habitually used. The famous Monarque, for instance, would never consciously look at the dam of Gladiateur, and a trick had to be practised. We can partly see the reason why valuable race-horse stallions, which are in such demand, should be so particular in their choice. Mr. Blenkiron has never known a mare to reject a horse; but this has occurred in Mr. Wright's stable, so that the mare had to be cheated. Prosper Lucas[43] quotes various statements from French authorities, and remarks, "On voit des étalons qui s'éprennent d'une "jument, et négligent toutes les autres." He gives, on the authority of Baëlen, similar facts in regard to bulls.

---

[43] 'Traité de l'Héréd. Nat.' tom. ii. 1850, p. 296.

Hoffberg, in describing the domesticated reindeer of Lapland, says, "Fœmina majores et fortiores mares "præ cæteris admittunt, ad eos confugiunt, a juniori-"bus agitatæ, qui hos in fugam conjiciunt."[44] A clergyman, who has bred many pigs, assures me that sows often reject one boar and immediately accept another.

From these facts there can be no doubt that with most of our domesticated quadrupeds strong individual antipathies and preferences are frequently exhibited, and much more commonly by the female than by the male. This being the case, it is improbable that the unions of quadrupeds in a state of nature should be left to mere chance. It is much more probable that the females are allured or excited by particular males, who possess certain characters in a higher degree than other males; but what these characters are, we can seldom or never discover with certainty.

---

[44] 'Amœnitates Acad.' vol. iv. 1788, p. 160.

CHAPTER XVIII.

SECONDARY SEXUAL CHARACTERS OF MAMMALS—*continued.*

Voice—Remarkable sexual peculiarities in seals—Odour—Development of the hair—Colour of the hair and skin—Anomalous case of the female being more ornamented than the male—Colour and ornaments due to sexual selection—Colour acquired for the sake of protection—Colour, though common to both sexes, often due to sexual selection—On the disappearance of spots and stripes in adult quadrupeds—On the colours and ornaments of the Quadrumana—Summary.

QUADRUPEDS use their voices for various purposes, as a signal of danger, as a call from one member of a troop to another, or from the mother to her lost offspring, or from the latter for protection to their mother; but such uses need not here be considered. We are concerned only with the difference between the voices of the two sexes, for instance between that of the lion and lioness, or of the bull and cow. Almost all male animals use their voices much more during the rutting-season than at any other time; and some, as the giraffe and porcupine,[1] are said to be completely mute excepting at this season. As the throats (*i.e.* the larnyx and thyroid bodies[2]) of stags become periodically enlarged at the commencement of the breeding-season, it might be thought that their powerful voices must be then in some way of high importance to them; but this is very doubtful. From information given to me by two experienced observers, Mr. McNeill and Sir

---

1 Owen, 'Anatomy of Vertebrates,' vol. iii. p. 585. 2 Ibid. p. 595.

P. Egerton, it seems that young stags under three years old do not roar or bellow; and that the old ones begin bellowing at the commencement of the breeding-season, at first only occasionally and moderately, whilst they restlessly wander about in search of the females. Their battles are prefaced by loud and prolonged bellowing, but during the actual conflict they are silent. Animals of all kinds which habitually use their voices, utter various noises under any strong emotion, as when enraged and preparing to fight; but this may merely be the result of their nervous excitement, which leads to the spasmodic contraction of almost all the muscles of the body, as when a man grinds his teeth and clenches his hands in rage or agony. No doubt stags challenge each other to mortal combat by bellowing; but it is not likely that this habit could have led through sexual selection, that is by the loudest-voiced males having been the most successful in their conflicts, to the periodical enlargement of the vocal organs; for the stags with the most powerful voices, unless at the same time the strongest, best-armed, and most courageous, would not have gained any advantage over their rivals with weaker voices. The stags, moreover, which had weaker voices, though not so well able to challenge other stags, would have been drawn to the place of combat as certainly as those with stronger voices.

It is possible that the roaring of the lion may be of some actual service to him in striking terror into his adversary; for when enraged he likewise erects his mane and thus instinctively tries to make himself appear as terrible as possible. But it can hardly be supposed that the bellowing of the stag, even if it be of any service to him in this way, can have been important enough to have led to the periodical enlargement of the throat. Some writers suggest that the bellowing serves as a call to the female; but the experienced observers above quoted inform me that female deer do not search for the male, though the males search eagerly for the females, as indeed might be expected from what we know of the habits of other male quadrupeds. The voice of the female, on the other hand, quickly brings to her one or more stags,[3] as is well known to the hunters who in wild countries imitate her cry. If we could believe that the male had the power to excite or allure the female by his voice, the periodical enlargement of his vocal organs would be intelligible on the principle of sexual selection, together with inheritance limited to the same sex and season of the year; but we have no evidence in favour of this view. As the case stands, the loud voice of the stag during the breeding season does not seem to be of any special service to him, either during his courtship or battles, or in any other way. But may we not believe that the frequent use of the voice, under the strong excitement of love, jealousy, and rage, continued during many generations, may at last have produced an inherited effect on the vocal organs of the stag, as well as of other male animals? This appears to me, with our present state of knowledge, the most probable view.

The male gorilla has a tremendous voice, and when adult is furnished with a laryngeal sack, as is likewise the adult male orang.[4] The gibbons rank amongst the noisiest of monkeys, and the Sumatra species (*Hylobates syndactylus*) is also furnished with a laryngeal sack; but Mr. Blyth, who has had opportunities for observation,

---

3 See, for instance, Major W. Ross King ('The Sportsman in Canada,' 1866, p. 53, 131) on the habits of the moose and wild reindeer.

4 Owen, 'Anatomy of Vertebrates,' vol. iii. p. 600.

does not believe that the male is more noisy than the female. Hence, these latter monkeys probably use their voices as a mutual call; and this is certainly the case with some quadrupeds, for instance with the beaver.[5] Another gibbon, the *H. agilis*, is highly remarkable, from having the power of emitting a complete and correct octave of musical notes,[6] which we may reasonably suspect serves as a sexual charm; but I shall have to recur to this subject in the next chapter. The vocal organs of the American *Mycetes caraya* are one-third larger in the male than in the female, and are wonderfully powerful. These monkeys, when the weather is warm, make the forests resound during the morning and evening with their overwhelming voices. The males begin the dreadful concert, in which the females, with their less powerful voices, sometimes join, and which is often continued during many hours. An excellent observer, Rengger,[7] could not perceive that they were excited to begin their concert by any special cause; he thinks that like many birds, they delight in their own music, and try to excel each other. Whether most of the foregoing monkeys have acquired their powerful voices in order to beat their rivals and to charm the females—or whether the vocal organs have been strengthened and enlarged through the inherited effects of long-continued use without any particular good being gained—I will not pretend to say; but the former view, at least in the case of the *Hylobates agilis*, seems the most probable.

I may here mention two very curious sexual peculiarities occurring in seals, because they have been sup-

---

[5] Mr. Green, in 'Journal of Linn. Soc.' vol. x. Zoology, 1869, p. 362.

[6] C. L. Martin, 'General Introduction to the Nat. Hist. of Mamm. Animals,' 1841, p. 431.

[7] 'Naturgeschichte der Säugethiere von Paraguay,' 1830, s. 15, 21.

posed by some writers to affect the voice. The nose of the male sea-elephant (*Macrorhinus proboscideus*), when about three years old, is greatly elongated during the breeding-season, and can then be erected. In this state it is sometimes a foot in length. The female at no period of life is thus provided, and her voice is different. That of the male consists of a wild, hoarse, gurgling noise, which is audible at a great distance, and is believed to be strengthened by the proboscis. Lesson compares the erection of the proboscis, to the swelling of the wattles of male gallinaceous birds, whilst they court the females. In another allied kind of seal, namely, the bladder-nose (*Cystophora cristata*), the head is covered by a great hood or bladder. This is internally supported by the septum of the nose, which is produced far backwards and rises into a crest seven inches in height. The hood is clothed with short hair, and is muscular; it can be inflated until it more than equals the whole head in size! The males when rutting fight furiously on the ice, and their roaring "is "said to be sometimes so loud as to be heard four "miles off." When attacked by man they likewise roar or bellow; and whenever irritated the bladder is inflated. Some naturalists believe that the voice is thus strengthened, but various other uses have been assigned to this extraordinary structure. Mr. R. Brown thinks that it serves as a protection against accidents of all kinds. This latter view is not probable, if what the sealers have long maintained is correct, namely, that the hood or bladder is very poorly developed in the females and in the males whilst young.[8]

---

8 On the sea-elephant, see an article by Lesson, in 'Dict. Class. Hist. Nat.' tom. xiii. p. 418. For the Cystophora or Stemmatopus, see Dr. Dekay, 'Annals of Lyceum of Nat. Hist. New York,' vol. i. 1824, p. 94. Pennant has also collected information from the sealers on this

Odour.—With some animals, as with the notorious skunk of America, the overwhelming odour which they emit appears to serve exclusively as a means of defence. With shrew-mice (Sorex) both sexes possess abdominal scent-glands, and there can be little doubt, from the manner in which their bodies are rejected by birds and beasts of prey, that their odour is protective; nevertheless the glands become enlarged in the males during the breeding-season. In many quadrupeds the glands are of the same size in both sexes;[9] but their use is not known. In other species the glands are confined to the males, or are more developed in them than in the females; and they almost always become more active during the rutting-season. At this period the glands on the sides of the face of the male elephant enlarge and emit a secretion having a strong musky odour.

The rank effluvium of the male goat is well known, and that of certain male deer is wonderfully strong and persistent. On the banks of the Plata I have perceived the whole air tainted with the odour of the male *Cervus campestris*, at the distance of half a mile to leeward of a herd; and a silk handkerchief, in which I carried home a skin, though repeatedly used and washed, retained, when first unfolded, traces of the odour for one year and seven months. This animal does not emit its strong odour until more than a year old, and if cas-

---

animal. The fullest account is given by Mr. Brown, who doubts about the rudimentary condition of the bladder in the female, in 'Proc. Zoolog. Soc.' 1868, p. 435.

[9] As with the castoreum of the beaver, see Mr. L. H. Morgan's most interesting work, 'The American Beaver,' 1868, p. 300. Pallas ('Spic. Zoolog.' fasc. viii. 1779, p. 23) has well discussed the odoriferous glands of mammals. Owen ('Anat. of Vertebrates,' vol. iii. p. 634) also gives an account of these glands, including those of the elephant, and (p. 763) those of shrew-mice.

trated whilst young never emits it.[10] Besides the general odour, with which the whole body of certain ruminants seems to be permeated during the breeding-season, many deer, antelopes, sheep, and goats, possess odoriferous glands in various situations, more especially on their faces. The so-called tear-sacks or suborbital pits come under this head. These glands secrete a semi-fluid fetid matter, which is sometimes so copious as to stain the whole face, as I have seen in the case of an antelope. They are "usually larger in the male than in "the female, and their development is checked by cas- "tration."[11] According to Desmarest they are altogether absent in the female of *Antilope subgutturosa*. Hence, there can be no doubt that they stand in some close relation with the reproductive functions. They are also sometimes present, and sometimes absent, in nearly-allied forms. In the adult male musk-deer (*Moschus moschiferus*), a naked space round the tail is bedewed with an odoriferous fluid, whilst in the adult female, and in the male, until two years old, this space is covered with hair and is not odoriferous. The proper musk-sack, from its position, is necessarily confined to the male, and forms an additional scent-organ. It is a singular fact that the matter secreted by this latter gland does not, according to Pallas, change in consistence, or increase in quantity, during the rutting-season; nevertheless this naturalist admits that its presence is in some way connected with the act of repro-

---

10 Rengger, 'Naturgeschichte der Säugethiere von Paraguay,' 1830, s. 355. This observer also gives some curious particulars in regard to the odour emitted.

11 Owen, 'Anatomy of Vertebrates,' vol. iii. p. 632. See, also, Dr. Murie's observations on their glands in 'Proc. Zoolog. Soc.' 1870, p. 340. Desmarest, On the *Antilope subgutturosa*, 'Mammalogie,' 1820, p. 455.

duction. He gives, however, only a conjectural and unsatisfactory explanation of its use.[12]

In most cases, when during the breeding-season the male alone emits a strong odour, this probably serves to excite or allure the female. We must not judge on this head by our own taste, for it is well known that rats are enticed by certain essential oils, and cats by valerian, substances which are far from agreeable to us; and that dogs, though they will not eat carrion, sniff and roll in it. From the reasons given when discussing the voice of the stag, we may reject the idea that the odour serves to bring the females from a distance to the males. Active and long-continued use cannot here have come into play, as in the case of the vocal organs. The odour emitted must be of considerable importance to the male, inasmuch as large and complex glands, furnished with muscles for everting the sack, and for closing or opening the orifice, have in some cases been developed. The development of these organs is intelligible through sexual selection, if the more odoriferous males are the most successful in winning the females, and in leaving offspring to inherit their gradually-perfected glands and odours.

Development of the Hair.—We have seen that male quadrupeds often have the hair on their necks and shoulders much more developed than in the females; and many additional instances could be given. This sometimes serves as a defence to the male during his battles; but whether the hair in most cases has been specially developed for this purpose is very doubtful. We may feel almost certain that this is not the case,

---

[12] Pallas, 'Spicilegia Zoolog.' fasc. xiii. 1799, p. 24; Desmoulins, 'Dict. Class. d'Hist. Nat.' tom. iii. p. 586.

when a thin and narrow crest runs along the whole length of the back; for a crest of this kind would afford scarcely any protection, and the ridge of the back is not a likely place to be injured; nevertheless such crests are sometimes confined to the males, or are much more developed in them than in the females. Two antelopes, the *Tragelaphus scriptus*[13] (see fig. 68, p. 300) and *Portax picta*, may be given as instances. The crests of certain stags and of the male wild goat stand erect, when these animals are enraged or terrified;[14] but it can hardly be supposed that they have been acquired for the sake of exciting fear in their enemies. One of the above-named antelopes, the *Portax picta*, has a large well-defined brush of black hair on the throat, and this is much larger in the male than in the female. In the *Ammotragus tragelaphus* of North Africa, a member of the sheep-family, the front-legs are almost concealed by an extraordinary growth of hair, which depends from the neck and upper halves of the legs; but Mr. Bartlett does not believe that this mantle is of the least use to the male, in whom it is much more developed than in the female.

Male quadrupeds of many kinds differ from the females in having more hair, or hair of a different character, on certain parts of their faces. The bull alone has curled hair on the forehead.[15] In three closely-allied sub-genera of the goat family, the males alone possess beards, sometimes of large size; in two other sub-genera both sexes have a beard, but this

---

13 Dr. Gray, 'Gleanings from the Menagerie at Knowsley,' pl. 28.

14 Judge Caton on the wapiti, 'Transact. Ottawa Acad. Nat. Sciences,' 1868, p. 36, 40; Blyth, 'Land and Water,' on *Capra ægagrus*, 1867, p. 37.

15 'Hunter's Essays and Observations,' edited by Owen, 1861, vol. i. p. 236.

disappears in some of the domestic breeds of the common goat; and neither sex of the Hemitragus has a beard. In the ibex the beard is not developed during the summer, and is so small at other seasons that it may be called rudimentary.[16] With some monkeys the beard is confined to the male, as in the Orang, or is

Fig. 66. Pithecia Satanas, male (from Brehm).

much larger in the male than in the female, as in the *Mycetes caraya* and *Pithecia satanas* (fig. 66). So it is with the whiskers of some species of Macacus,[17] and, as we have seen, with the manes of some species of baboons.

[16] See Dr. Gray's 'Cat. of Mammalia in British Museum,' part iii. 1852, p. 144.

[17] Rengger, 'Säugethicre,' &c., s. 14; Desmarest, 'Mammalogie,' p. 66.

But with most kinds of monkeys the various tufts of hair about the face and head are alike in both sexes.

The males of various members of the Ox family (Bovidæ), and of certain antelopes, are furnished with a dewlap, or great fold of skin on the neck, which is much less developed in the female.

Now, what must we conclude with respect to such sexual differences as these? No one will pretend that the beards of certain male-goats, or the dewlap of the bull, or the crests of hair along the backs of certain male antelopes, are of any direct or ordinary use to them. It is possible that the immense beard of the male Pithecia, and the large beard of the male Orang, may protect their throats when fighting; for the keepers in the Zoological Gardens inform me that many monkeys attack each other by the throat: but it is not probable that the beard has been developed for a distinct purpose from that which the whiskers, moustache, and other tufts of hair on the face serve; and no one will suppose that these are useful as a protection. Must we attribute to mere purposeless variability in the male all these appendages of hair or skin? It cannot be denied that this is possible; for with many domesticated quadrupeds, certain characters, apparently not derived through reversion from any wild parent-form, have appeared in, and are confined to, the males, or are more largely developed in them than in the females,—for instance the hump in the male zebu-cattle of India, the tail in fat-tailed rams, the arched outline of the forehead in the males of several breeds of sheep, the mane in the ram of an African breed, and, lastly, the mane, long hairs on the hinder legs, and the dewlap in the male alone of the Berbura goat.[18] The mane which occurs in

[18] See the chapters on these several animals in vol. i. of my 'Variation of Animals under Domestication;' also vol. ii. p. 73; also chap. xx.

the rams alone of the above-mentioned African breed of sheep, is a true secondary sexual character, for it is not developed, as I hear from Mr. Winwood Reade, if the animal be castrated. Although we ought to be extremely cautious, as shewn in my work on 'Variation under Domestication,' in concluding that any character, even with animals kept by semi-civilised people, has not been subjected to selection by man, and thus augmented; yet in the cases just specified this is improbable, more especially as the characters are confined to the males, or are more strongly developed in them than in the females. If it were positively known that the African ram with a mane was descended from the same primitive stock with the other breeds of sheep, or the Berbura male-goat with his mane, dewlap, &c., from the same stock with other goats; and if selection has not been applied to these characters, then they must be due to simple variability, together with sexually-limited inheritance.

In this case it would appear reasonable to extend the same view to the many analogous characters occurring in animals under a state of nature. Nevertheless I cannot persuade myself that this view is applicable in many cases, as in that of the extraordinary development of hair on the throat and fore-legs of the male Ammotragus, or of the immense beard of the male Pithecia. With those antelopes in which the male when adult is more strongly-coloured than the female, and with those monkeys in which this is likewise the case, and in which the hair on the face is of a different colour from that on the rest of the head, being arranged in the most diversified and elegant manner, it seems probable that the crests and tufts of hair have

on the practice of selection by semi-civilised people. For the Berbura goat, see Dr. Gray, 'Catalogue,' ibid. p. 157.

been acquired as ornaments; and this I know is the opinion of some naturalists. If this view be correct, there can be little doubt that they have been acquired, or at least modified, through sexual selection.

Colour of the Hair and of the Naked Skin.—I will first give briefly all the cases known to me, of male quadrupeds differing in colour from the females. With Marsupials, as I am informed by Mr. Gould, the sexes rarely differ in this respect; but the great red kangaroo offers a striking exception, "delicate blue being "the prevailing tint in those parts of the female, "which in the male are red."[19] In the *Didelphis opossum* of Cayenne the female is said to be a little more red than the male. With Rodents Dr. Gray remarks: "African squirrels, especially those found in the tropi-"cal regions, have the fur much brighter and more "vivid at some seasons of the year than at others, and "the fur of the male is generally brighter than that "of the female."[20] Dr. Gray informs me that he specified the African squirrels, because, from their unusually bright colours, they best exhibit this difference. The female of the *Mus minutus* of Russia is of a paler and dirtier tint than the male. In some few bats the fur of the male is lighter and brighter than in the female.[21]

The terrestrial Carnivora and Insectivora rarely exhibit sexual differences of any kind, and their colours are almost always exactly the same in both sexes. The

---

19 *Osphranter rufus*, Gould, 'Mammals of Australia,' vol. ii. 1863. On the Didelphis, Desmarest, 'Mammalogie,' p. 256.

20 'Annals and Mag. of Nat. Hist.' Nov. 1867, p. 325. On the *Mus minutus*, Desmarest, 'Mammalogie,' p. 304.

21 J. A. Allen, in 'Bulletin of Mus. Comp. Zoolog. of Cambridge, United States,' 1869, p. 207.

ocelot (*Felis pardalis*), however, offers an exception, for the colours of the female, compared with those of the male, are "moins apparentes, le fauve étant plus terne, "le blanc moins pur, les raies ayant moins de largeur "et les taches moins de diamètre."[22] The sexes of the allied *Felis mitis* also differ, but even in a less degree, the general hues of the female being rather paler than in the male, with the spots less black. The marine Carnivora or Seals, on the other hand, sometimes differ considerably in colour, and they present, as we have already seen, other remarkable sexual differences. Thus the male of the *Otaria nigrescens* of the southern hemisphere is of a rich brown shade above; whilst the female, who acquires her adult tints earlier in life than the male, is dark-grey above, the young of both sexes being of a very deep chocolate colour. The male of the northern *Phoca groenlandica* is tawny grey, with a curious saddle-shaped dark mark on the back; the female is much smaller, and has a very different appearance, being "dull white or yellow-"ish straw-colour, with a tawny hue on the back;" the young at first are pure white, and can "hardly be dis-"tinguished among the icy hummocks and snow, their "colour thus acting as a protection."[23]

With Ruminants sexual differences of colour occur more commonly than in any other order. A difference of this kind is general with the Strepsicerene antelopes; thus the male nilghau (*Portax picta*) is bluish-grey and much darker than the female, with the square white patch on the throat, the white marks on the fetlocks,

---

[22] Desmarest, 'Mammalogie,' 1820, p. 223. On *Felis mitis*, Rengger, ibid. s. 194.

[23] Dr. Murie on the Otaria, 'Proc. Zool. Soc.' 1869, p. 108. Mr. R. Brown, on the *P. groenlandica*, ibid. 1868, p. 417. See also on the colours of seals, Desmarest, ibid. p. 243, 249.

and the black spots on the ears, all much more distinct. We have seen that in this species the crests and tufts of hair are likewise more developed in the male than in the hornless female. The male, as I am informed by Mr. Blyth, without shedding his hair, periodically becomes darker during the breeding-season. Young males cannot be distinguished from young females until above twelve months old; and if the male is emasculated before this period, he never, according to the same authority, changes colour. The importance of this latter fact, as distinctive of sexual colouring, becomes obvious, when we hear[24] that neither the red summer-coat nor the blue winter-coat of the Virginian deer is at all affected by emasculation. With most or all of the highly-ornamented species of Tragelaphus the males are darker than the hornless females, and their crests of hair are more fully developed. In the male of that magnificent antelope, the *Derbyan Eland*, the body is redder, the whole neck much blacker, and the white band which separates these colours, broader, than in the female. In the Cape Eland also, the male is slightly darker than the female.[25]

In the Indian Black-buck (*A. bezoartica*), which belongs to another tribe of antelopes, the male is very dark, almost black; whilst the hornless female is fawn-coloured. We have in this species, as Mr. Blyth informs me, an exactly parallel series of facts, as with the *Portax picta*, namely in the male periodically changing colour during the breed-

---

[24] Judge Caton, in 'Trans. Ottawa Acad. of Nat. Sciences,' 1868, p. 4.

[25] Dr. Gray, 'Cat. of Mamm. in Brit. Mus.' part iii. 1852, p. 134-142; also Dr. Gray, 'Gleanings from the Menagerie of Knowsley,' in which there is a splendid drawing of the Oreas derbianus: see the text on Tragelaphus. For the Cape Eland (*Oreas canna*), see Andrew Smith, 'Zoology of S. Africa,' pl. 41 and 42. There are also many of these antelopes in the Zoological Society's Gardens.

ing season, in the effects of emasculation on this change, and in the young of both sexes being undistinguishable from each other. In the *Antilope niger* the male is black, the female as well as the young being brown; in *A. sing-sing* the male is much brighter coloured than the hornless female, and his chest and belly are blacker; in the male *A. caama*, the marks and lines which occur on various parts of the body are black instead of as in the female brown; in the brindled gnu (*A. gorgon*) "the colours of the male are nearly the same as those "of the female, only deeper and of a brighter hue."[26] Other analogous cases could be added.

The Banteng bull (*Bos sondaicus*) of the Malayan archipelago is almost black, with white legs and buttocks; the cow is of a bright dun, as are the young males until about the age of three years, when they rapidly change colour. The emasculated bull reverts to the colour of the female. The female Kemas goat is paler, and the female *Capra ægagrus* is said to be more uniformly tinted than their respective males. Deer rarely present any sexual differences in colour. Judge Caton, however, informs me that with the males of the Wapiti deer (*Cervus Canadensis*) the neck, belly, and legs are much darker than the same parts in the female; but during the winter the darker tints gradually fade away and disappear. I may here mention that Judge Caton has in his park three races of the Virginian deer, which differ slightly in colour, but the differences are almost exclusively confined to the blue

[26] On the *Ant. niger*, see 'Proc. Zool. Soc.' 1850, p. 133. With respect to an allied species, in which there is an equal sexual difference in colour, see Sir S. Baker, 'The Albert Nyanza,' 1866, vol. ii. p. 327. For the *A. sing-sing*, Gray, 'Cat. B. Mus.' p. 100. Desmarest, 'Mammalogie,' p. 468, on the *A. caama*. Andrew Smith, 'Zoology of S. Africa,' on the Gnu.

winter or breeding coat; so that this case may be compared with those given in a previous chapter of closely-allied or representative species of birds which differ from each other only in their nuptial plumage.[27] The females of *Cervus paludosus* of S. America, as well as the young of both sexes, do not possess the black stripes on the nose, and the blackish-brown line on the breast which characterise the adult males.[28] Lastly, the mature male of the beautifully coloured and spotted Axis deer is considerably darker, as I am informed by Mr. Blyth, than the female; and this hue the castrated male never acquires.

The last Order which we have to consider—for I am not aware that sexual differences in colour occur in the other mammalian groups—is that of the Primates. The male of the *Lemur macaco* is coal-black, whilst the female is reddish-yellow, but highly variable in colour.[29] Of the Quadrumana of the New World, the females and young of *Mycetes caraya* are greyish-yellow and alike; in the second year the young male becomes reddish-brown, in the third year black, excepting the stomach, which, however, becomes quite black in the fourth or fifth year. There is also a strongly-marked difference in colour between the sexes in *Mycetes seniculus* and *Cebus capucinus;* the young of the former and I believe of the latter species resembling the females. With *Pithecia leucocephala* the young likewise resemble the females, which are brownish-

27 'Ottawa Academy of Sciences,' May 21, 1868, p. 3, 5.

28 S. Müller, on the Banteng, 'Zoog. Indischen Archipel.' 1839-1844, tab. 35: see also Raffles, as quoted by Mr. Blyth, in 'Land and Water,' 1867, p. 476. On goats, Dr. Gray, 'Cat. Brit. Mus.' p. 146; Desmarest, 'Mammalogie,' p. 482. On the *Cervus paludosus*, Rengger, ibid. s. 345.

29 Sclater, 'Proc. Zool. Soc.' 1866, p. 1. The same fact has also been fully ascertained by MM. Pollen and van Dam.

black above and light rusty-red beneath, the adult males being black. The ruff of hair round the face of *Ateles marginatus* is tinted yellow in the male and white in the female. Turning to the Old World, the males of *Hylobates hoolock* are always black, with the exception of a white band over the brows; the females vary from whity-brown to a dark tint mixed with black, but are never wholly black.[30] In the beautiful *Cercopithecus diana* the head of the adult male is of an intense black, whilst that of the female is dark grey; in the former the fur between the thighs is of an elegant fawn-colour, in the latter it is paler. In the equally beautiful and curious moustache monkey (*Cercopithecus cephus*) the only difference between the sexes is that the tail of the male is chesnut and that of the female grey; but Mr. Bartlett informs me that all the hues become more strongly pronounced in the male when adult, whilst in the female they remain as they were during youth. According to the coloured figures given by Solomon Müller, the male of *Semnopithecus chrysomelas* is nearly black, the female being pale brown. In the *Cercopithecus cynosurus* and *griseo-viridis* one part of the body which is confined to the male sex is of the most brilliant blue or green, and contrasts strikingly with the naked skin on the hinder part of the body, which is vivid red.

Lastly, in the Baboon family, the adult male of *Cynocephalus hamadryas* differs from the female not only by his immense mane, but slightly in the colour of the hair and of the naked callosities. In the drill (*Cynocephalus*

---

[30] On Mycetes, Rengger, ibid. s. 14; and Brehm, 'Illustrirtes Thierleben,' B. i. s. 96, 107. On Ateles, Desmarest, 'Mammalogie,' p. 75. On Hylobates, Blyth, 'Land and Water,' 1867, p. 135. On the Semnopithecus, S. Müller, 'Zoog. Indischen Archipel.' tab. x.

leucophœus) the females and young are much paler-coloured, with less green, than the adult males. No other member of the whole class of mammals is coloured in so extraordinary a manner as the adult male mandrill (*Cynocephalus mormon*). The face at this age becomes of a fine blue, with the ridge and tip of the nose of the most brilliant red. According to some authors the face is also marked with whitish stripes, and is shaded in parts

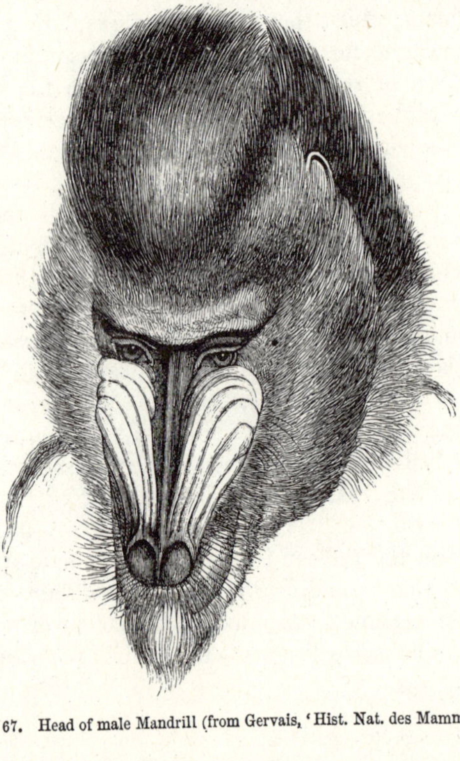

Fig. 67. Head of male Mandrill (from Gervais, 'Hist. Nat. des Mammifères').

with black, but the colours appear to be variable. On the forehead there is a crest of hair, and on the chin a yellow beard. "Toutes les parties supérieures de leurs "cuisses et le grand espace nu de leurs fesses sont "également colorés du rouge le plus vif, avec un "mélange de bleu qui ne manque réellement pas "d'élégance."[31] When the animal is excited all the naked parts become much more vividly tinted. Several authors have used the strongest expressions in describing these resplendent colours, which they compare with those of the most brilliant birds. Another most remarkable peculiarity is that when the great canine teeth are fully developed, immense protuberances of bone are formed on each cheek, which are deeply furrowed longitudinally, and the naked skin over them is brilliantly-coloured, as just described. (Fig. 67.) In the adult females and in the young of both sexes these protuberances are scarcely perceptible; and the naked parts are much less brightly coloured, the face being almost black, tinged with blue. In the adult female, however, the nose at certain regular intervals of time becomes tinted with red.

In all the cases hitherto given the male is more strongly or brightly coloured than the female, and differs in a greater degree from the young of both sexes. But as a reversed style of colouring is characteristic of the two sexes with some few birds, so with the Rhesus monkey (*Macacus rhesus*) the female has a large surface of naked skin round the tail, of a brilliant carmine red, which periodically becomes, as I was assured by the keepers in the Zoological Gardens, even more vivid, and her face is also pale red. On the other hand with

---

[31] Gérvais, 'Hist. Nat. des Mammifères,' 1854, p. 103. Figures are given of the skull of the male. Desmarest, 'Mammalogie,' p. 70. Geoffroy St.-Hilaire and F. Cuvier, 'Hist. Nat. des Mamm.' 1824, tom. i.

the adult male and with the young of both sexes, as I saw in the Gardens, neither the naked skin at the posterior end of the body, nor the face, shew a trace of red. It appears, however, from some published accounts, that the male does occasionally, or during certain seasons, exhibit some traces of the red. Although he is thus less ornamented than the female, yet in the larger size of his body, larger canine teeth, more developed whiskers, more prominent superciliary ridges, he follows the common rule of the male excelling the female.

I have now given all the cases known to me of a difference in colour between the sexes of mammals. The colours of the female either do not differ in a sufficient degree from those of the male, or are not of a suitable nature, to afford her protection, and therefore cannot be explained on this principle. In some, perhaps in many cases, the differences may be the result of variations confined to one sex and transmitted to the same sex, without any good having been thus gained, and therefore without the aid of selection. We have instances of this kind with our domesticated animals, as in the males of certain cats being rusty-red, whilst the females are tortoise-shell coloured. Analogous cases occur under nature; Mr. Bartlett has seen many black varieties of the jaguar, leopard, vulpine phalanger and wombat; and he is certain that all, or nearly all, were males. On the other hand, both sexes of wolves, foxes, and apparently of American squirrels, are occasionally born black. Hence it is quite possible that with some mammals the blackness of the males, especially when this colour is congenital, may simply be the result, without the aid of selection, of one or more variations having occurred, which from the first were sexually limited in their transmission. Nevertheless it can hardly be admitted that the diversified, vivid, and contrasted colours of certain quadrupeds, for instance of the above-mentioned monkeys and antelopes, can thus be accounted for. We should bear in mind that these colours do not appear in the male at birth, as in the case of most ordinary variations, but only at or near maturity; and that unlike ordinary variations, if the male be emasculated, they never appear or subsequently disappear. It is on the whole a much more probable conclusion that the strongly-marked colours and other ornamental characters of male quadrupeds are beneficial to them in their rivalry with other males, and have consequently been acquired through sexual selection. The probability of this view is strengthened by the differences in colour between the sexes occurring almost exclusively, as may be observed by going through the previous details, in those groups and subgroups of mammals, which present other and distinct secondary sexual characters; these being likewise due to the action of sexual selection.

Quadrupeds manifestly take notice of colour. Sir S. Baker repeatedly observed that the African elephant and rhinoceros attacked with special fury white or grey horses. I have elsewhere shewn[32] that half-wild horses apparently prefer pairing with those of the same colour, and that herds of fallow-deer of a different colour, though living together, have long kept distinct. It is a more significant fact that a female zebra would not admit the addresses of a male ass until he was painted so as to resemble a zebra, and then, as John Hunter remarks, "she received him very readily. In this curious fact,

---

[32] 'The Variation of Animals and Plants under Domestication,' 1868, vol. ii. p. 102, 103.

"we have instinct excited by mere colour, which had "so strong an effect as to get the better of every- "thing else. But the male did not require this, the "female being an animal somewhat similar to himself, "was sufficient to rouse him."[33]

In an early chapter we have seen that the mental powers of the higher animals do not differ in kind, though so greatly in degree, from the corresponding powers of man, especially of the lower and barbarous races; and it would appear that even their taste for the beautiful is not widely different from that of the Quadrumana. As the negro of Africa raises the flesh on his face into parallel ridges "or cicatrices, high above the "natural surface, which unsightly deformities, are con- "sidered great personal attractions;"[34]—as negroes, as well as savages in many parts of the world, paint their faces with red, blue, white, or black bars,—so the male mandrill of Africa appears to have acquired his deeply-furrowed and gaudily-coloured face from having been thus rendered attractive to the female. No doubt it is to us a most grotesque notion that the posterior end of the body should have been coloured for the sake of ornament even more brilliantly than the face; but this is really not more strange than that the tails of many birds should have been especially decorated.

With mammals we do not at present possess any evidence that the males take pains to display their charms before the female; and the elaborate manner in which this is performed by male birds, is the strongest argument in favour of the belief that the females admire,

---

[33] 'Essays and Observations by J. Hunter,' edited by Owen, 1861, vol. i. p. 194.

[34] Sir S. Baker, 'The Nile Tributaries of Abyssinia,' 1867.

or are excited by, the ornaments and colours displayed before them. There is, however, a striking parallelism between mammals and birds in all their secondary sexual characters, namely in their weapons for fighting with rival males, in their ornamental appendages, and in their colours. In both classes, when the male differs from the female, the young of both sexes almost always resemble each other, and in a large majority of cases resemble the adult female. In both classes the male assumes the characters proper to his sex shortly before the age for reproduction; if emasculated he either never acquires such characters or subsequently loses them. In both classes the change of colour is sometimes seasonal, and the tints of the naked parts sometimes become more vivid during the act of courtship. In both classes the male is almost always more vividly or strongly coloured than the female, and is ornamented with larger crests either of hair or feathers, or other appendages. In a few exceptional cases the female in both classes is more highly ornamented than the male. With many mammals, and at least in the case of one bird, the male is more odoriferous than the female. In both classes the voice of the male is more powerful than that of the female. Considering this parallelism there can be little doubt that the same cause, whatever it may be, has acted on mammals and birds; and the result, as far as ornamental characters are concerned, may safely be attributed, as it appears to me, to the long-continued preference of the individuals of one sex for certain individuals of the opposite sex, combined with their success in leaving a larger number of offspring to inherit their superior attractions.

Equal transmission of ornamental characters to both sexes.—With many birds, ornaments, which analogy leads us to believe were primarily acquired by the males, have been transmitted equally, or almost equally, to both sexes; and we may now enquire how far this view may be extended to mammals. With a considerable number of species, especially the smaller kinds, both sexes have been coloured, independently of sexual selection, for the sake of protection; but not, as far as I can judge, in so many cases, nor in nearly so striking a manner as in most of the lower classes. Audubon remarks that he often mistook the musk-rat,[35] whilst sitting on the banks of a muddy stream, for a clod of earth, so complete was the resemblance. The hare on her form is a familiar instance of concealment through colour; yet this principle partly fails in a closely-allied species, namely the rabbit, for as this animal runs to its burrow, it is made conspicuous to the sportsman and no doubt to all beasts of prey, by its upturned pure-white tail. No one has ever doubted that the quadrupeds which inhabit snow-clad regions, have been rendered white to protect them from their enemies, or to favour their stealing on their prey. In regions where snow never lies long on the ground a white coat would be injurious; consequently species thus coloured are extremely rare in the hotter parts of the world. It deserves notice that many quadrupeds, inhabiting moderately cold regions, although they do not assume a white winter dress, become paler during this season; and this apparently is the direct result of the conditions to which they have long been exposed. Pallas[36] states that in Siberia a change of this nature occurs with the wolf, two species of Mustela, the domestic horse, the *Equus he-*

---

35 *Fiber zibethicus*, Audubon and Bachman, 'The Quadrupeds of N. America,' 1846, p. 109.

36 'Novæ species Quadrupedum e Glirium ordine,' 1778, p. 7. What I have called the roe is the *Capreolus Sibiricus subecaudatus* of Pallas.

mionus, the domestic cow, two species of antelopes, the musk-deer, the roe, the elk, and reindeer. The roe, for instance, has a red summer and a greyish-white winter coat; and the latter may perhaps serve as a protection to the animal whilst wandering through the leafless thickets, sprinkled with snow and hoar-frost. If the above named animals were gradually to extend their range into regions perpetually covered with snow, their pale winter-coats would probably be rendered, through natural selection, whiter and whiter by degrees, until they became as white as snow.

Although we must admit that many quadrupeds have received their present tints as a protection, yet with a host of species, the colours are far too conspicuous and too singularly arranged to allow us to suppose that they serve for this purpose. We may take as an illustration certain antelopes: when we see that the square white patch on the throat, the white marks on the fetlocks, and the round black spots on the ears, are all more distinct in the male of the *Portax picta*, than in the female;—when we see that the colours are more vivid, that the narrow white lines on the flank and the broad white bar on the shoulder are more distinct in the male *Oreas Derbyanus* than in the female;—when we see a similar difference between the sexes of the curiously-ornamented *Tragelaphus scriptus* (fig. 68),—we may conclude that these colours and various marks have been at least intensified through sexual selection. It is inconceivable that such colours and marks can be of any direct or ordinary service to these animals; and as they have almost certainly been intensified through sexual selection, it is probable that they were originally gained through this same process, and then partially transferred to the females. If this view be admitted, there can be little doubt that the equally singular colours and marks of many other antelopes, though common to both sexes, have been gained and transmitted in a like manner. Both sexes, for instance, of the Koodoo (*Strepsiceros Kudu*, fig. 62) have nar-

Fig. 68. Tragelaphus scriptus, male (from the Knowsley Menagerie).

row white vertical lines on their hinder flanks, and an elegant angular white mark on their foreheads. Both sexes in the genus Damalis are very oddly coloured; in *D. pygarga* the back and neck are purplish-red, shading on the flanks into black, and abruptly separated from the white belly and a large white space on the buttocks; the head is still more oddly coloured, a large oblong white mask, narrowly-edged with black, covers the face up to the eyes (fig. 69); there are three white stripes on the forehead, and the ears are marked with white. The fawns of this species are of a uniform pale yellow-

Fig. 69. Damalis pygarga, male (from the Knowsley Menagerie).

ish-brown. In *Damalis albifrons* the colouring of the head differs from that in the last species in a single white stripe replacing the three stripes, and in the ears being almost wholly white.[37] After having studied to

[37] See the fine plates in A. Smith's 'Zoology of S. Africa,' and Dr. Gray's 'Gleanings from the Menagerie of Knowsley.'

the best of my ability the sexual differences of animals belonging to all classes, I cannot avoid the conclusion that the curiously-arranged colours of many antelopes, though common to both sexes, are the result of sexual selection primarily applied to the male.

The same conclusion may perhaps be extended to the tiger, one of the most beautiful animals in the world, the sexes of which cannot be distinguished by colour, even by the dealers in wild beasts. Mr. Wallace believes[38] that the striped coat of the tiger "so assi-"milates with the vertical stems of the bamboo, as to "assist greatly in concealing him from his approaching "prey." But this view does not appear to me satisfactory. We have some slight evidence that his beauty may be due to sexual selection, for in two species of Felis analogous marks and colours are rather brighter in the male than in the female. The zebra is conspicuously striped, and stripes on the open plains of South Africa cannot afford any protection. Burchell[39] in describing a herd says, "their sleek ribs glistened in the "sun, and the brightness and regularity of their striped "coats presented a picture of extraordinary beauty, in "which probably they are not surpassed by any other "quadruped." Here we have no evidence of sexual selection, as throughout the whole group of the Equidæ the sexes are identical in colour. Nevertheless he who attributes the white and dark vertical stripes on the flanks of various antelopes to sexual selection, will probably extend the same view to the Royal Tiger and beautiful Zebra.

We have seen in a former chapter that when young animals belonging to any class follow nearly the same

---

[38] 'Westminster Review,' July 1, 1867, p. 5.

[39] 'Travels in South Africa,' 1824, vol. ii. p. 315.

habits of life with their parents, and yet are coloured in a different manner, it may be inferred that they have retained the colouring of some ancient and extinct progenitor. In the family of pigs, and in the genus Tapir, the young are marked with longitudinal stripes, and thus differ from every existing adult species in these two groups. With many kinds of deer the young are marked with elegant white spots, of which their parents exhibit not a trace. A graduated series can be followed from the Axis deer, both sexes of which at all ages and during all seasons are beautifully spotted (the male being rather more strongly coloured than the female)—to species in which neither the old nor the young are spotted. I will specify some of the steps in this series. The Mantchurian deer (*Cervus Mantchuricus*) is spotted during the whole year, but the spots are much plainer, as I have seen in the Zoological Gardens, during the summer, when the general colour of the coat is lighter, than during the winter, when the general colour is darker and the horns are fully developed. In the hog-deer (*Hyelaphus porcinus*) the spots are extremely conspicuous during the summer when the coat is reddish-brown, but quite disappear during the winter when the coat is brown.[40] In both these species the young are spotted. In the Virginian deer the young are likewise spotted, and about five per cent. of the adult animals living in Judge Caton's park, as I am informed by him, temporarily exhibit at the period when the red summer coat is being replaced by the bluish winter coat, a row of spots on each flank, which are always the same in

---

[40] Dr. Gray, 'Gleanings from the Menagerie of Knowsley,' p. 64. Mr. Blyth, in speaking ('Land and Water,' 1869, p. 42) of the hog-deer of Ceylon, says it is more brightly spotted with white than the common hog-deer, at the season when it renews its horns.

number, though very variable in distinctness. From this condition there is but a very small step to the complete absence of spots at all seasons in the adults; and lastly, to their absence at all ages, as occurs with certain species. From the existence of this perfect series, and more especially from the fawns of so many species being spotted, we may conclude that the now living members of the deer family are the descendants of some ancient species which, like the Axis deer, was spotted at all ages and seasons. A still more ancient progenitor probably resembled to a certain extent the *Hyomoschus aquaticus*—for this animal is spotted, and the hornless males have large exserted canine teeth, of which some few true deer still retain rudiments. It offers, also, one of those interesting cases of a form linking together two groups, as it is intermediate in certain osteological characters between the pachyderms and ruminants, which were formerly thought to be quite distinct.[41]

A curious difficulty here arises. If we admit that coloured spots and stripes have been acquired as ornaments, how comes it that so many existing deer, the descendants of an aboriginally spotted animal, and all the species of pigs and tapirs, the descendants of an aboriginally striped animal, have lost in their adult state their former ornaments? I cannot satisfactorily answer this question. We may feel nearly sure that the spots and stripes disappeared in the progenitors of our existing species at or near maturity, so that they were retained by the young and, owing to the law of inheritance at corresponding ages, by the young of all succeeding generations. It may have been a great advantage to

[41] Falconer and Cautley, 'Proc. Geolog. Soc.' 1843; and Falconer's 'Pal. Memoirs,' vol. i. p. 196.

the lion and puma from the open nature of the localities which they commonly haunt, to have lost their stripes, and to have been thus rendered less conspicuous to their prey; and if the successive variations, by which this end was gained, occurred rather late in life, the young would have retained their stripes, as we know to be the case. In regard to deer, pigs, and tapirs, Fritz Müller has suggested to me that these animals by the removal through natural selection of their spots or stripes would have been less easily seen by their enemies; and they would have especially required this protection, as soon as the carnivora increased in size and number during the Tertiary periods. This may be the true explanation, but it is rather strange that the young should not have been equally well protected, and still more strange that with some species the adults should have retained their spots, either partially or completely, during part of the year. We know, though we cannot explain the cause, that when the domestic ass varies and becomes reddish-brown, grey or black, the stripes on the shoulders and even on the spine frequently disappear. Very few horses, except dun-coloured kinds, exhibit stripes on any part of their bodies, yet we have good reason to believe that the aboriginal horse was striped on the legs and spine, and probably on the shoulders.[42] Hence the disappearance of the spots and stripes in our adult existing deer, pigs, and tapirs, may be due to a change in the general colour of their coats; but whether this change was effected through sexual or natural selection, or was due to the direct action of the conditions of life, or some other unknown cause, it is impossible to decide. An observation made by Mr. Sclater well illustrates our ignorance of the laws which regulate the

[42] 'The Variation of Animals and Plants under Domestication,' 1868, vol. i. p. 61-64.

appearance and disappearance of stripes; the species of Asinus which inhabit the Asiatic continent are destitute of stripes, not having even the cross shoulder-stripe, whilst those which inhabit Africa are conspicuously striped, with the partial exception of *A. tæniopus*, which has only the cross shoulder-stripe and generally some faint bars on the legs; and this species inhabits the almost intermediate region of Upper Egypt and Abyssinia.[43]

Quadrumana.—Before we conclude, it will be advisable to add a few remarks to those already given on the

Fig. 70. Head of Semnopithecus rubicundus. This and the following figures (from Prof. Gervais) are given to shew the odd arrangement and development of the hair on the head.

[43] 'Proc. Zool. Soc.' 1862, p. 164. See, also, Dr. Hartmann, 'Ann. d. Landw.' Bd. xliii. s. 222.

ornamental characters of monkeys. In most of the species the sexes resemble each other in colour, but in some, as we have seen, the males differ from the females, especially in the colour of the naked parts of the skin, in the development of the beard, whiskers, and mane. Many species are coloured either in so ex-

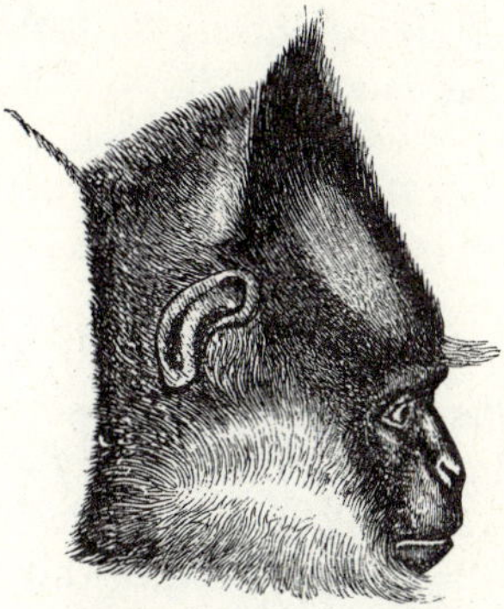

Fig. 71. Head of Semnopithecus comatus.

Fig. 72. Head of Cebus capucinus.

Fig. 73. Head of Ateles marginatus.

Fig. 74. Head of Cebus vellerosus.

traordinary or beautiful a manner, and are furnished with such curious and elegant crests of hair, that we can hardly avoid looking at these characters as having been gained for the sake of ornament. The accompanying figures (figs. 70 to 74) serve to shew the arrangement of the hair on the face and head in several species. It is scarcely conceivable that these crests of hair and the strongly-contrasted colours of the fur and skin can be the result of mere variability without the aid of selection; and it is inconceivable that they can be of any ordinary use to these animals. If so, they have probably been gained through sexual selection, though transmitted equally, or almost equally, to both sexes. With many of the Quadrumana, we have additional evidence of the action of sexual selection in the greater size and strength of the males, and in the greater development of their canine teeth, in comparison with the females.

With respect to the strange manner in which both sexes of some species are coloured, and of the beauty of others, a few instances will suffice. The face of the *Cercopithecus petaurista* (fig. 75) is black, the whiskers and beard being white, with a defined, round, white spot on the nose, covered with short white hair, which gives to the animal an almost ludicrous aspect. The *Semnopithecus frontatus* likewise, has a blackish face with a long black beard, and a large naked spot on the forehead of a bluish-white colour. The face of *Macacus lasiotus* is dirty flesh-coloured, with a defined red spot on each cheek. The appearance of *Cercocebus æthiops* is grotesque, with its black face, white whiskers and collar, chesnut head, and a large naked white spot over each eyelid. In very many species, the beard, whiskers, and crests of hair round the face are of a different colour from the rest of the head, and when different, are always of a lighter tint,[44] being often pure

[44] I observed this fact in the Zoological Gardens; and numerous cases may be seen in the coloured plates in Geoffroy St.-Hilaire and F. Cuvier, 'Hist. Nat. des Mammifères,' tom. i. 1824.

white, sometimes bright yellow, or reddish. The whole face of the South American *Brachyurus calvus* is of a "glowing scarlet hue;" but this colour does not appear

Fig. 75. Cercopithecus petaurista (from Brehm).

until the animal is nearly mature.[45] The naked skin of the face differs wonderfully in colour in the various species. It is often brown or flesh-colour, with parts perfectly white, and often as black as that of the most sooty negro. In the Brachyurus the scarlet tint is brighter than that of the most blushing Caucasian damsel. It is sometimes more distinctly orange than in any Mongolian, and in several species it is blue, passing into violet or grey. In all the species known to Mr. Bartlett, in which the adults of both sexes have strongly-coloured faces, the colours are dull or absent during early youth. This likewise holds good with the Mandrill and Rhesus, in which the face and the posterior parts of the body are brilliantly coloured in one sex alone. In these latter cases we have every reason to believe that the colours were acquired through sexual selection; and we are naturally led to extend the same view to the foregoing species, though both sexes when adult have their faces coloured in the same manner.

Although, according to our taste, many kinds of monkeys are far from beautiful, other species are universally admired for their elegant appearance and bright colours. The *Semnopithecus nemæus*, though peculiarly coloured, is described as extremely pretty; the orange-tinted face is surrounded by long whiskers of glossy whiteness, with a line of chesnut-red over the eyebrows; the fur on the back is of a delicate grey, with a square patch on the loins, the tail and the fore-arms all of a pure white; a gorget of chesnut surmounts the chest; the hind thighs are black, with the legs chesnut-red. I will mention only two other monkeys on account of their beauty; and I have selected these as they present slight sexual differences in colour, which renders it

[45] Bates, 'The Naturalist on the Amazons,' 1863, vol. ii. p. 310.

in some degree probable that both sexes owe their elegant appearance to sexual selection. In the moustache-monkey (*Cercopithecus cephus*) the general colour of the fur is mottled-greenish, with the throat white; in the male the end of the tail is chesnut; but the face is the most ornamented part, the skin being chiefly bluish-grey, shading into a blackish tint beneath the eyes, with the upper lip of a delicate blue, clothed on the lower edge with a thin black moustache; the whiskers

Fig. 76. Cercopithecus Diana (from Brehm).

are orange-coloured, with the upper partblack, forming a band which extends backwards to the ears, the latter being clothed with whitish hairs. In the Zoological Society's Gardens I have often overheard visitors admiring the beauty of another monkey, deservedly called *Cercopithecus Diana* (fig. 76); the general colour of the fur is grey; the chest and inner surface of the fore-legs are white; a large triangular defined space on the hinder part of the back is rich chesnut; in the male the inner sides of the thighs and the abdomen are delicate fawn-coloured, and the top of the head is black; the face and ears are intensely black, finely contrasted with a white transverse crest over the eye-brows and with a long white peaked beard, of which the basal portion is black.[46]

In these and many other monkeys, the beauty and singular arrangement of their colours, and still more the diversified and elegant arrangement of the crests and tufts of hair on their heads, force the conviction on my mind that these characters have been acquired through sexual selection exclusively as ornaments.

Summary.—The law of battle for the possession of the female appears to prevail throughout the whole great class of mammals. Most naturalists will admit that the greater size, strength, courage, and pugnacity of the male, his special weapons of offence, as well as his special means of defence, have all been acquired or modified through that form of selection which I have

[46] I have seen most of the above-named monkeys in the Zoological Society's Gardens. The description of the *Semnopithecus nemæus* is taken from Mr. W. C. Martin's 'Nat. Hist. of Mammalia,' 1841, p. 460; see also p. 475, 523.

called sexual selection. This does not depend on any superiority in the general struggle for life, but on certain individuals of one sex, generally the male sex, having been successful in conquering other males, and on their having left a larger number of offspring to inherit their superiority, than the less successful males.

There is another and more peaceful kind of contest, in which the males endeavour to excite or allure the females by various charms. This may be effected by the powerful odours emitted by the males during the breeding-season; the odoriferous glands having been acquired through sexual selection. Whether the same view can be extended to the voice is doubtful, for the vocal organs of the males may have been strengthened by use during maturity, under the powerful excitements of love, jealousy, or rage, and transmitted to the same sex. Various crests, tufts, and mantles of hair, which are either confined to the male, or are more developed in this sex than in the females, seem in most cases to be merely ornamental, though they sometimes serve as a defence against rival males. There is even reason to suspect that the branching horns of stags, and the elegant horns of certain antelopes, though properly serving as weapons of offence or of defence, have been partly modified for the sake of ornament.

When the male differs in colour from the female he generally exhibits darker and more strongly-contrasted tints. We do not in this class meet with the splendid red, blue, yellow, and green colours, so common with male birds and many other animals. The naked parts, however, of certain Quadrumana must be excepted; for such parts, often oddly situated, are coloured in some species in the most brilliant manner. The colours of the male in other cases may be due to simple variation, without the aid of selection. But when the colours are diversified and strongly pronounced, when they are not developed until near maturity, and when they are lost after emasculation, we can hardly avoid the conclusion that they have been acquired through sexual selection for the sake of ornament, and have been transmitted exclusively, or almost exclusively, to the same sex. When both sexes are coloured in the same manner, and the colours are conspicuous or curiously arranged, without being of the least apparent use as a protection, and especially when they are associated with various other ornamental appendages, we are led by analogy to the same conclusion, namely, that they have been acquired through sexual selection, although transmitted to both sexes. That conspicuous and diversified colours, whether confined to the males or common to both sexes, are as a general rule associated in the same groups and sub-groups with other secondary sexual characters, serving for war or for ornament, will be found to hold good if we look back to the various cases given in this and the last chapter.

The law of the equal transmission of characters to both sexes, as far as colour and other ornaments are concerned, has prevailed far more extensively with mammals than with birds; but in regard to weapons, such as horns and tusks, these have often been transmitted either exclusively, or in a much higher degree to the males than to the females. This is a surprising circumstance, for as the males generally use their weapons as a defence against enemies of all kinds, these weapons would have been of service to the female. Their absence in this sex can be accounted for, as far as we can see, only by the form of inheritance which has prevailed. Finally with quadrupeds the contest between the individuals of the same sex, whether peaceful or bloody, has with the rarest exceptions been confined to the males; so that these have been modified through sexual selection, either for fighting with each other or for alluring the opposite sex, far more commonly than the females.

CHAPTER XIX.

SECONDARY SEXUAL CHARACTERS OF MAN.

Differences between man and woman — Causes of such differences and of certain characters common to both sexes — Law of battle — Differences in mental powers — and voice — On the influence of beauty in determining the marriages of mankind — Attention paid by savages to ornaments — Their ideas of beauty in woman — The tendency to exaggerate each natural peculiarity.

WITH mankind the differences between the sexes are greater than in most species of Quadrumana, but not so great as in some, for instance, the mandrill. Man on an average is considerably taller, heavier, and stronger than woman, with squarer shoulders and more plainly-pronounced muscles. Owing to the relation which exists between muscular development and the projection of the brows,[1] the superciliary ridge is generally more strongly marked in man than in woman. His body, and especially his face, is more hairy, and his voice has a different and more powerful tone. In certain tribes the women are said, whether truly I know not, to differ slightly in tint from the men; and with Europeans, the women are perhaps the more brightly coloured of the two, as may be seen when both sexes have been equally exposed to the weather.

Man is more courageous, pugnacious, and energetic than woman, and has a more inventive genius. His

---

[1] Schaaffhausen, translation in 'Anthropological Review,' Oct. 1868, p. 419, 420, 427.

brain is absolutely larger, but whether relatively to the larger size of his body, in comparison with that of woman, has not, I believe been fully ascertained. In woman the face is rounder; the jaws and the base of the skull smaller; the outlines of her body rounder, in parts more prominent; and her pelvis is broader than in man;[2] but this latter character may perhaps be considered rather as a primary than a secondary sexual character. She comes to maturity at an earlier age than man.

As with animals of all classes, so with man, the distinctive characters of the male sex are not fully developed until he is nearly mature; and if emasculated they never appear. The beard, for instance, is a secondary sexual character, and male children are beardless, though at an early age they have abundant hair on their heads. It is probably due to the rather late appearance in life of the successive variations, by which man acquired his masculine characters, that they are transmitted to the male sex alone. Male and female children resemble each other closely, like the young of so many other animals in which the adult sexes differ; they likewise resemble the mature female much more closely, than the mature male. The female, however, ultimately assumes certain distinctive characters, and in the formation of her skull, is said to be intermediate between the child and the man.[3] Again, as the young of closely allied though distinct species do not differ nearly so much from each other as do the adults, so it is with the children of the different races o man. Some have even maintained that race-differences

---

[2] Ecker, translation in 'Anthropological Review,' Oct. 1868, p. 351-356. The comparison of the form of the skull in men and women has been followed out with much care by Welcker.

[3] Ecker and Welcker, ibid. p. 352, 355; Vogt, 'Lectures on Man,' Eng. translat. p. 81.

cannot be detected in the infantile skull.[4] In regard to colour, the new-born negro child is reddish nut-brown, which soon becomes slaty-grey; the black colour being fully developed within a year in the Sudan, but not until three years in Egypt. The eyes of the negro are at first blue, and the hair chesnut-brown rather than black, being curled only at the ends. The children of the Australians immediately after birth are yellowish-brown, and become dark at a later age. Those of the Guaranys of Paraguay are whitish-yellow, but they acquire in the course of a few weeks the yellowish-brown tint of their parents. Similar observations have been made in other parts of America.[5]

I have specified the foregoing familiar differences between the male and female sex in mankind, because they are curiously the same as in the Quadrumana. With these animals the female is mature at an earlier age than the male; at least this is certainly the case with the *Cebus azaræ*.[6] With most of the species the males are larger and much stronger than the females, of which fact the gorilla offers a well-known instance. Even in so trifling a character as the greater prominence of the superciliary ridge, the males of certain monkeys differ from the females,[7] and agree in this respect with mankind. In the gorilla and certain other monkeys, the

---

[4] Schaaffhausen, 'Anthropolog. Review,' ibid. p. 429.

[5] Pruner-Bey, on negro infants, as quoted by Vogt, 'Lectures on Man,' Eng. translat. 1864, p. 189: for further facts on negro infants, as quoted from Winterbottom and Camper, see Lawrence, 'Lectures on Physiology,' &c. 1822, p. 451. For the infants of the Guaranys, see Rengger, 'Säugethiere,' &c. s. 3. See also Godron, 'De l'Espèce,' tom. ii. 1859, p. 253. For the Australians, Waitz, 'Introduct. to Anthropology,' Eng. translat. 1863, p. 99.

[6] Rengger, 'Säugethiere,' &c. 1830, s. 49.

[7] As in *Macacus cynomolgus* (Desmarest, 'Mammalogie,' p. 65) and in *Hylobates agilis* (Geoffroy St.-Hilaire and F. Cuvier, 'Hist. Nat. des Mamm.' 1824, tom. i. p. 2).

cranium of the adult male presents a strongly-marked sagittal crest, which is absent in the female; and Ecker found a trace of a similar difference between the two sexes in the Australians.[8] With monkeys when there is any difference in the voice, that of the male is the more powerful. We have seen that certain male monkeys, have a well-developed beard, which is quite deficient, or much less developed in the female. No instance is known of the beard, whiskers, or moustache being larger in a female than in the male monkey. Even in the colour of the beard there is a curious parallelism between man and the Quadrumana, for when in man the beard differs in colour from the hair of the head, as is often the case, it is, I believe, invariably of a lighter tint, being often reddish. I have observed this fact in England, and Dr. Hooker, who attended to this little point for me in Russia, found no exception to the rule. In Calcutta, Mr. J. Scott, of the Botanic Gardens, was so kind as to observe with care the many races of men to be seen there, as well as in some other parts of India, namely, two races in Sikhim, the Bhoteas, Hindoos, Burmese, and Chinese. Although most of these races have very little hair on the face, yet he always found that when there was any difference in colour between the hair of the head and the beard, the latter was invariably of a lighter tint. Now with monkeys, as has already been stated, the beard frequently differs in a striking manner in colour from the hair of the head, and in such cases it is invariably of a lighter hue, being often pure white, sometimes yellow or reddish.[9]

---

[8] 'Anthropological Review,' Oct. 1868, p. 353.

[9] Mr. Blyth informs me that he has never seen more than one instance of the beard, whiskers, &c., in a monkey becoming white with old age, as is so commonly the case with us. This, however, occurred in an aged

In regard to the general hairyness of the body, the women in all races are less hairy than the men, and in some few Quadrumana the under side of the body of the female is less hairy than that of the male.[10] Lastly, male monkeys, like men, are bolder and fiercer than the females. They lead the troop, and when there is danger, come to the front. We thus see how close is the parallelism between the sexual differences of man and the Quadrumana. With some few species, however, as with certain baboons, the gorilla and orang, there is a considerably greater difference between the sexes, in the size of the canine teeth, in the development and colour of the hair, and especially in the colour of the naked parts of the skin, than in the case of mankind.

The secondary sexual characters of man are all highly variable, even within the limits of the same race or sub-species; and they differ much in the several races. These two rules generally hold good throughout the animal kingdom. In the excellent observations made on board the *Novara*,[11] the male Australians were found to exceed the females by only 65 millim. in height, whilst with the Javanese the average excess was 218 millim., so that in this latter race the difference in height

---

and confined *Macacus cynomolgus*, whose moustaches were "remarkably long and human-like." Altogether this old monkey presented a ludicrous resemblance to one of the reigning monarchs of Europe, after whom he was universally nick-named. In certain races of man the hair on the head hardly ever becomes grey; thus Mr. D. Forbes has never seen, as he informs me, an instance with the Aymaras and Quichuas of S. America.

[10] This is the case with the females of several species of Hylobates, see Geoffroy St.-Hilaire and F. Cuvier, 'Hist. Nat. des Mamm.' tom. i. See, also, on *H. lar*. 'Penny Encyclopedia,' vol. ii. p. 149, 150.

[11] The results were deduced by Dr. Weisbach from the measurements made by Drs. K. Scherzer and Schwarz, see 'Reise der *Novara*: Anthropolog. Theil,' 1867, s. 216, 231, 234, 236, 239, 269.

between the sexes is more than thrice as great as with the Australians. The numerous measurements of various other races, with respect to stature, the circumference of the neck and chest, and the length of the back-bone and arms, which were carefully made, nearly all shewed that the males differed much more from each other than did the females. This fact indicates that, as far as these characters are concerned, it is the male which has been chiefly modified, since the races diverged from their common and primeval source.

The development of the beard and the hairiness of the body differ remarkably in the men belonging to distinct races, and even to different families in the same race. We Europeans see this amongst ourselves. In the island of St. Kilda, according to Martin,[12] the men do not acquire beards, which are very thin, until the age of thirty or upwards. On the Europæo-Asiatic continent, beards prevail until we pass beyond India, though with the natives of Ceylon they are frequently absent, as was noticed in ancient times by Diodorus.[13] Beyond India beards disappear, as with the Siamese, Malays, Kalmucks, Chinese, and Japanese; nevertheless the Ainos,[14] who inhabit the northernmost islands of the Japan archipelago, are the most hairy men in the world. With negroes the beard is scanty or absent, and they have no whiskers; in both sexes the body is almost destitute of fine down.[15] On the other hand, the Pa-

---

[12] 'Voyage to St. Kilda' (3rd edit. 1753) p. 37.

[13] Sir J. E. Tennent, 'Ceylon,' vol. ii. 1859, p. 107.

[14] Quatrefages, 'Revue des Cours Scientifiques,' Aug. 29, 1868, p. 630; Vogt, 'Lectures on Man,' Eng. translat. p. 127.

[15] On the beards of negroes, Vogt, 'Lectures,' &c. ibid. p. 127; Waitz, 'Introduct. to Anthropology,' Engl. translat. 1863, vol. i. p. 96. It is remarkable that in the United States ('Investigations in Military and Anthropological Statistics of American Soldiers,' 1869, p. 569) the

puans of the Malay archipelago, who are nearly as black as negroes, possess well-developed beards.[16] In the Pacific Ocean the inhabitants of the Fiji archipelago have large bushy beards, whilst those of the not-distant archipelagoes of Tonga and Samoa are beardless; but these men belong to distinct races. In the Ellice group all the inhabitants belong to the same race; yet on one island alone, namely Nunemaya, "the men have splendid beards;" whilst on the other islands "they "have, as a rule, a dozen straggling hairs for a beard."[17]

Throughout the great American continent the men may be said to be beardless; but in almost all the tribes a few short hairs are apt to appear on the face, especially during old age. With the tribes of North America, Catlin estimates that eighteen out of twenty men are completely destitute by nature of a beard; but occasionally there may be seen a man, who has neglected to pluck out the hairs at puberty, with a soft beard an inch or two in length. The Guaranys of Paraguay differ from all the surrounding tribes in having a small beard, and even some hair on the body, but no whiskers.[18] I am informed by Mr. D. Forbes, who particularly attended to this subject, that the Aymaras and Quichuas of the Cordillera are remarkably hairless, yet in old age a few straggling hairs occasionally appear on the chin. The men of these two tribes have very little hair on the various parts of the body where hair grows abundantly

---

pure negroes and their crossed offspring seem to have bodies almost as hairy as those of Europeans.

16 Wallace, 'The Malay Arch.' vol. ii. 1869, p. 178.

17 Dr. J. Barnard Davis on Oceanic Races, in 'Anthropolog. Review,' April, 1870, p. 185, 191.

18 Catlin, 'North American Indians,' 3rd edit. 1842, vol. ii. p. 227. On the Guaranys, see Azara, 'Voyages dans l'Amérique Mérid.' tom. ii. 1809, p. 58; also Rengger, 'Säugethiere von Paraguay,' s. 3.

in Europeans, and the women have none on the corresponding parts. The hair on the head, however, attains an extraordinary length in both sexes, often reaching almost to the ground; and this is likewise the case with some of the N. American tribes. In the amount of hair, and in the general shape of the body, the sexes of the American aborigines do not differ from each other so much as with most other races of mankind.[19] This fact is analagous with what occurs with some allied monkeys; thus the sexes of the chimpanzee are not as different as those of the gorilla or orang.[20]

In the previous chapters we have seen that with mammals, birds, fishes, insects, &c., many characters, which there is every reason to believe were primarily gained through sexual selection by one sex alone, have been transferred to both sexes. As this same form of transmission has apparently prevailed to a large extent with mankind, it will save much useless repetition if we consider the characters peculiar to the male sex together with certain other characters common to both sexes.

Law of Battle.—With barbarous nations, for instance with the Australians, the women are the constant cause of war both between the individuals of the same tribe and between distinct tribes. So no doubt it was in ancient times; "nam fuit ante Helenam mulier teter- "rima belli causa." With the North American Indians, the contest is reduced to a system. That excellent ob-

---

[19] Prof. and Mrs. Agassiz ('Journey in Brazil,' p. 530) remark that the sexes of the American Indians differ less than those of the negroes and of the higher races. See also Rengger, ibid. p. 3, on the Guaranys.

[20] Rütimeyer, 'Die Grenzen der Thierwelt; eine Betrachtung zu Darwin's Lehre,' 1868, s. 54.

server, Hearne,[21] says:—"It has ever been the custom "among these people for the men to wrestle for any "woman to whom they are attached; and, of course, the "strongest party always carries off the prize. A weak "man, unless he be a good hunter, and well-beloved, "is seldom permitted to keep a wife that a stronger "man thinks worth his notice. This custom prevails "throughout all the tribes, and causes a great spirit "of emulation among their youth, who are upon all "occasions, from their childhood, trying their strength "and skill in wrestling." With the Guanas of South America, Azara states that the men rarely marry till twenty or more years old, as before that age they cannot conquer their rivals.

Other similar facts could be given; but even if we had no evidence on this head, we might feel almost sure, from the analogy of the higher Quadrumana,[22] that the law of battle had prevailed with man during the early stages of his development. The occasional appearance at the present day of canine teeth which project above the others, with traces of a diastema or open space for the reception of the opposite canines, is in all probability a case of reversion to a former state, when the progenitors of man were provided with these weapons, like so many existing male Quadrumana. It was remarked in a former chapter that as man gradually became erect, and continually used his hands and arms for fighting with sticks and stones, as well as for the other purposes of life, he would have used his

---

[21] 'A Journey from Prince of Wales Fort.' 8vo. edit. Dublin, 1796, p. 104. Sir J. Lubbock ('Origin of Civilisation,' 1870, p. 69) gives other and similar cases in North America. For the Guanas of S. America see Azara, 'Voyages,' &c. tom. ii. p. 94.

[22] On the fighting of the male gorillas, see Dr. Savage, in 'Boston Journal of Nat. Hist.' vol. v. 1847, p. 423. On *Presbytis entellus*, see the 'Indian Field,' 1859, p. 146.

jaws and teeth less and less. The jaws, together with their muscles, would then have become reduced through disuse, as would the teeth through the not well understood principles of correlation and the economy of growth; for we everywhere see that parts which are no longer of service are reduced in size. By such steps the original inequality between the jaws and teeth in the two sexes of mankind would ultimately have been quite obliterated. The case is almost parallel with that of many male Ruminants, in which the canine teeth have been reduced to mere rudiments, or have disappeared, apparently in consequence of the development of horns. As the prodigious difference between the skulls of the two sexes in the Gorilla and Orang, stands in close relation with the development of the immense canine teeth in the males, we may infer that the reduction of the jaws and teeth in the early male progenitors of man led to a most striking and favourable change in his appearance.

There can be little doubt that the greater size and strength of man, in comparison with woman, together with his broader shoulders, more developed muscles, rugged outline of body, his greater courage and pugnacity, are all due in chief part to inheritance from some early male progenitor, who, like the existing anthropoid apes, was thus characterised. These characters will, however, have been preserved or even augmented during the long ages whilst man was still in a barbarous condition, by the strongest and boldest men having succeeded best in the general struggle for life, as well as in securing wives, and thus having left a large number of offspring. It is not probable that the greater strength of man was primarily acquired through the inherited effects of his having worked harder than woman for his own subsistence and that of his family;

for the women in all barbarous nations are compelled to work at least as hard as the men. With civilised people the arbitrament of battle for the possession of the women has long ceased; on the other hand, the men, as a general rule, have to work harder than the women for their mutual subsistence; and thus their greater strength will have been kept up.

Difference in the Mental Powers of the two Sexes.—With respect to differences of this nature between man and woman, it is probable that sexual selection has played a very important part. I am aware that some writers doubt whether there is any inherent difference; but this is at least probable from the analogy of the lower animals which present other secondary sexual characters. No one will dispute that the bull differs in disposition from the cow, the wild-boar from the sow, the stallion from the mare, and, as is well known to the keepers of menageries, the males of the larger apes from the females. Woman seems to differ from man in mental disposition, chiefly in her greater tenderness and less selfishness; and this holds good even with savages, as shewn by a well-known passage in Mungo Park's Travels, and by statements made by many other travellers. Woman, owing to her maternal instincts, displays these qualities towards her infants in an eminent degree; therefore it is likely that she should often extend them towards her fellow-creatures. Man is the rival of other men; he delights in competition, and this leads to ambition which passes too easily into selfishness. These latter qualities seem to be his natural and unfortunate birthright. It is generally admitted that with woman the powers of intuition, of rapid perception, and perhaps of imitation, are more strongly marked than in man; but some, at least, of these faculties are characteristic of the lower races, and therefore of a past and lower state of civilisation.

The chief distinction in the intellectual powers of the two sexes is shewn by man attaining to a higher eminence, in whatever he takes up, than woman can attain—whether requiring deep thought, reason, or imagination, or merely the use of the senses and hands. If two lists were made of the most eminent men and women in poetry, painting, sculpture, music, —comprising composition and performance, history, science, and philosophy, with half-a-dozen names under each subject, the two lists would not bear comparison. We may also infer, from the law of the deviation of averages, so well illustrated by Mr. Galton, in his work on 'Hereditary Genius,' that if men are capable of decided eminence over women in many subjects, the average standard of mental power in man must be above that of woman.

The half-human male progenitors of man, and men in a savage state, have struggled together during many generations for the possession of the females. But mere bodily strength and size would do little for victory, unless associated with courage, perseverance, and determined energy. With social animals, the young males have to pass through many a contest before they win a female, and the older males have to retain their females by renewed battles. They have, also, in the case of man, to defend their females, as well as their young, from enemies of all kinds, and to hunt for their joint subsistence. But to avoid enemies, or to attack them with success, to capture wild animals, and to invent and fashion weapons, requires the aid of the higher mental faculties, namely, observation, reason, invention, or imagination. These various faculties will thus have been continually put to the test, and selected during manhood; they will, moreover, have been strengthened by use during this same period of life. Consequently, in accordance with the principle often alluded to, we might expect that they would at least tend to be transmitted chiefly to the male offspring at the corresponding period of manhood.

Now, when two men are put into competition, or a man with a woman, who possess every mental quality in the same perfection, with the exception that the one has higher energy, perseverance, and courage, this one will generally become more eminent, whatever the object may be, and will gain the victory.[23] He may be said to possess genius—for genius has been declared by a great authority to be patience; and patience, in this sense, means unflinching, undaunted perseverance. But this view of genius is perhaps deficient; for without the higher powers of the imagination and reason, no eminent success in many subjects can be gained. But these latter as well as the former faculties will have been developed in man, partly through sexual selection,—that is, through the contest of rival males, and partly through natural selection,—that is, from success in the general struggle for life; and as in both cases the struggle will have been during maturity, the characters thus gained will have been transmitted more fully to the male than to the female offspring. Thus man has ultimately become superior to woman. It is, indeed, fortunate that the law of the equal transmission of characters to both sexes has commonly prevailed throughout the whole class of mammals; otherwise it is probable that man would have

---

[23] J. Stuart Mill remarks ('The Subjection of Women,' 1869, p. 122), "the things in which man most excels woman are those which require "most plodding, and long hammering at single thoughts." What is this but energy and perseverance?

become as superior in mental endowment to woman, as the peacock is in ornamental plumage to the peahen.

It must be borne in mind that the tendency in characters acquired at a late period of life by either sex, to be transmitted to the same sex at the same age, and of characters acquired at an early age to be transmitted to both sexes, are rules which, though general, do not always hold good. If they always held good, we might conclude (but I am here wandering beyond my proper bounds) that the inherited effects of the early education of boys and girls would be transmitted equally to both sexes; so that the present inequality between the sexes in mental power could not be effaced by a similar course of early training; nor can it have been caused by their dissimilar early training. In order that woman should reach the same standard as man, she ought, when nearly adult, to be trained to energy and perseverance, and to have her reason and imagination exercised to the highest point; and then she would probably transmit these qualities chiefly to her adult daughters. The whole body of women, however, could not be thus raised, unless during many generations the women who excelled in the above robust virtues were married, and produced offspring in larger numbers than other women. As before remarked with respect to bodily strength, although men do not now fight for the sake of obtaining wives, and this form of selection has passed away, yet they generally have to undergo, during manhood, a severe struggle in order to maintain themselves and their families; and this will tend to keep up or even increase their mental powers, and, as a consequence, the present inequality between the sexes.[24]

---

[24] An observation by Vogt bears on this subject: he says, it is a "remarkable circumstance, that the difference between the sexes, as

Voice and Musical Powers.—In some species of Quadrumana there is a great difference between the adult sexes, in the power of the voice and in the development of the vocal organs; and man appears to have inherited this difference from his early progenitors. His vocal cords are about one-third longer than in woman, or than in boys; and emasculation produces the same effect on him as on the lower animals, for it "arrests that pro-"minent growth of the thyroid, &c., which accompanies "the elongation of the cords."[25] With respect to the cause of this difference between the sexes, I have nothing to add to the remarks made in the last chapter on the probable effects of the long-continued use of the vocal organs by the male under the excitement of love, rage, and jealousy. According to Sir Duncan Gibb,[26] the voice differs in the different races of mankind; and with the natives of Tartary, China, &c., the voice of the male is said not to differ so much from that of the female, as in most other races.

The capacity and love for singing or music, though not a sexual character in man, must not here be passed over. Although the sounds emitted by animals of all kinds serve many purposes, a strong case can be made out, that the vocal organs were primarily used and perfected in relation to the propagation of the species. Insects and some few spiders are the lowest animals which voluntarily produce any sound; and this is generally effected by the aid of beautifully constructed

"regards the cranial cavity, increases with the development of the "race, so that the male European excels much more the female, than "the negro the negress. Welcker confirms this statement of Huschke "from his measurements of negro and German skulls." But Vogt admits ('Lectures on Man,' Eng. translat. 1864, p. 81) that more observations are requisite on this point.

25 Owen, 'Anatomy of Vertebrates,' vol. iii. p. 603.

26 'Journal of the Anthropolog. Soc.' April, 1869, p. lvii. and lxvi.

stridulating organs, which are often confined to the males alone. The sounds thus produced consist, I believe in all cases, of the same note, repeated rhythmically;[27] and this is sometimes pleasing even to the ears of man. Their chief, and in some cases exclusive use appears to be either to call or to charm the opposite sex.

The sounds produced by fishes are said in some cases to be made only by the males during the breeding season. All the air-breathing Vertebrata necessarily possess an apparatus for inhaling and expelling air, with a pipe capable of being closed at one end. Hence when the primeval members of this class were strongly excited and their muscles violently contracted, purposeless sounds would almost certainly have been produced; and these, if they proved in any way serviceable, might readily have been modified or intensified by the preservation of properly adapted variations. The Amphibians are the lowest Vertebrates which breathe air; and many of these animals, namely, frogs and toads, possess vocal organs, which are incessantly used during the breeding-season, and which are often more highly developed in the male than in the female. The male alone of the tortoise utters a noise, and this only during the season of love. Male alligators roar or bellow during the same season. Every one knows how largely birds use their vocal organs as a means of courtship; and some species likewise perform what may be called instrumental music.

In the class of Mammals, with which we are here more particularly concerned, the males of almost all the species use their voices during the breeding-season much more than at any other time; and some are abso-

---

[27] Dr. Scudder, "Notes on Stridulation," in 'Proc. Boston Soc. of Nat. Hist.' vol. xi. April, 1868.

lutely mute excepting at this season. Both sexes of other species, or the females alone, use their voices as a love-call. Considering these facts, and that the vocal organs of some quadrupeds are much more largely developed in the male than in the female, either permanently or temporarily during the breeding season; and considering that in most of the lower classes the sounds produced by the males, serve not only to call but to excite or allure the female, it is a surprising fact that we have not as yet any good evidence that these organs are used by male mammals to charm the females. The American *Mycetes caraya* perhaps forms an exception, as does more probably one of those apes which come nearer to man, namely, the *Hylobates agilis.* This gibbon has an extremely loud but musical voice. Mr. Waterhouse states,[28] "It appeared to me that in ascending and "descending the scale, the intervals were always exactly "half-tones; and I am sure that the highest note was "the exact octave to the lowest. The quality of the "notes is very musical; and I do not doubt that a good "violinist would be able to give a correct idea of the "gibbon's composition, excepting as regards its loud-"ness." Mr. Waterhouse then gives the notes. Professor Owen, who is likewise a musician, confirms the foregoing statement, and remarks that this gibbon "alone of brute mammals may be said to sing." It appears to be much excited after its performance. Unfortunately its habits have never been closely observed in a state of nature; but from the analogy of almost all other animals, it is highly probable that it utters its musical notes especially during the season of courtship.

---

[28] Given in W. C. L. Martin's 'General Introduct. to Nat. Hist. of Mamm. Animals,' 1841, p. 432; Owen, 'Anatomy of Vertebrates,' vol. iii. p. 600.

The perception, if not the enjoyment, of musical cadences and of rhythm is probably common to all animals, and no doubt depends on the common physiological nature of their nervous systems. Even Crustaceans, which are not capable of producing any voluntary sound, possess certain auditory hairs, which have been seen to vibrate when the proper musical notes are struck.[29] It is well known that some dogs howl when hearing particular tones. Seals apparently appreciate music, and their fondness for it "was well "known to the ancients, and is often taken advantage of by the hunters at the present day."[30] With all those animals, namely insects, amphibians, and birds, the males of which during the season of courtship incessantly produce musical notes or mere rhythmical sounds, we must believe that the females are able to appreciate them, and are thus excited or charmed; otherwise the incessant efforts of the males and the complex structures often possessed exclusively by them would be useless.

With man song is generally admitted to be the basis or origin of instrumental music. As neither the enjoyment nor the capacity of producing musical notes are faculties of the least direct use to man in reference to his ordinary habits of life, they must be ranked amongst the most mysterious with which he is endowed. They are present, though in a very rude and as it appears almost latent condition, in men of all races, even the most savage; but so different is the taste of the different races, that our music gives not the least pleasure to savages, and their music is to us hideous and unmeaning. Dr. Seemann, in some interesting

---

[29] Helmholtz, 'Théorie Phys. de la Musique,' 1868, p. 187.
[30] Mr. R. Brown, in 'Proc. Zoo. Soc.' 1868, p. 410.

remarks on this subject,[31] "doubts whether even amongst "the nations of Western Europe, intimately connected "as they are by close and frequent intercourse, the "music of the one is interpreted in the same sense "by the others. By travelling eastwards we find that "there is certainly a different language of music. "Songs of joy and dance-accompaniments are no longer, "as with us, in the major keys, but always in the minor." Whether or not the half-human progenitors of man possessed, like the before-mentioned gibbon, the capacity of producing, and no doubt of appreciating, musical notes, we have every reason to believe that man possessed these faculties at a very remote period, for singing and music are extremely ancient arts. Poetry, which may be considered as the offspring of song, is likewise so ancient that many persons have felt astonishment that it should have arisen during the earliest ages of which we have any record.

The musical faculties, which are not wholly deficient in any race, are capable of prompt and high development, as we see with Hottentots and Negroes, who have readily become excellent musicians, although they do not practise in their native countries anything that we should esteem as music. But there is nothing anomalous in this circumstance: some species of birds which never naturally sing, can without much difficulty be taught to perform; thus the house-sparrow has learnt the song of a linnet. As these two species are closely allied, and belong to the order of Insessores, which includes nearly all the singing-birds in the world, it is quite possible or probable that a progenitor of the spar-

---

31 'Journal of Anthropolog. Soc.' Oct. 1870, p. clv. See also the several later chapters in Sir John Lubbock's 'Prehistoric Times,' second edition, 1869, which contain an admirable account of the habits of savages.

row may have been a songster. It is a much more remarkable fact that parrots, which belong to a group distinct from the Insessores, and have differently-constructed vocal organs, can be taught not only to speak, but to pipe or whistle tunes invented by man, so that they must have some musical capacity. Nevertheless it would be extremely rash to assume that parrots are descended from some ancient progenitor which was a songster. Many analogous cases could be advanced of organs and instincts originally adapted for one purpose, having been utilised for some quite distinct purpose.[32] Hence the capacity for high musical development, which the savage races of man possess, may be due either to our semi-human progenitors having practised some rude form of music, or simply to their having acquired for some distinct purposes the proper vocal organs. But in this latter case we must assume that they already possessed, as in the above instance of the parrots, and as seems to occur with many animals, some sense of melody.

Music affects every emotion, but does not by itself excite in us the more terrible emotions of horror, rage, &c. It awakens the gentler feelings of tenderness and love, which readily pass into devotion. It likewise stirs up in us the sensation of triumph and the glorious ardour for war. These powerful and mingled feelings may well give rise to the sense of sublimity. We can concentrate, as

---

[32] Since this chapter has been printed I have seen a valuable article by Mr. Chauncey Wright ('North Amer. Review,' Oct. 1870, page 293), who, in discussing the above subject, remarks, "There are many con-" sequences of the ultimate laws or uniformities of nature through " which the acquisition of one useful power will bring with it many " resulting advantages as well as limiting disadvantages, actual or " possible, which the principle of utility may not have comprehended " in its action." This principle has an important bearing, as I have attempted to shew in the second chapter of this work, on the acquisition by man of some of his mental characteristics.

Dr. Seemann observes, greater intensity of feeling in a single musical note than in pages of writing. Nearly the same emotions, but much weaker and less complex, are probably felt by birds when the male pours forth his full volume of song, in rivalry with other males, for the sake of captivating the female. Love is still the commonest theme of our own songs. As Herbert Spencer remarks, music "arouses dormant sentiments of which we had not "conceived the possibility, and do not know the meaning; "or, as Richter says, tells us of things we have not seen "and shall not see." [33] Conversely, when vivid emotions are felt and expressed by the orator or even in common speech, musical cadences and rhythm are instinctively used. Monkeys also express strong feelings in different tones—anger and impatience by low,—fear and pain by high notes.[34] The sensations and ideas excited in us by music, or by the cadences of impassioned cratory, appear from their vagueness, yet depth, like mental reversions to the emotions and thoughts of a long-past age.

All these facts with respect to music become to a certain extent intelligible if we may assume that musical tones and rhythm were used by the half-

---

[33] See the very interesting discussion on the Origin and Function of Music, by Mr. Herbert Spencer, in his collected 'Essays,' 1858, p. 359. Mr. Spencer comes to an exactly opposite conclusion to that at which I have arrived. He concludes that the cadences used in emotional speech afford the foundation from which music has been developed; whilst I conclude that musical notes and rhythm were first acquired by the male or female progenitors of mankind for the sake of charming the opposite sex. Thus musical tones became firmly associated with some of the strongest passions an animal is capable of feeling, and are consequently used instinctively, or through association, when strong emotions are expressed in speech. Mr. Spencer does not offer any satisfactory explanation, nor can I, why high or deep notes should be expressive, both with man and the lower animals, of certain emotions. Mr. Spencer gives also an interesting discussion on the relations between poetry, recitative, and song.

[34] Rengger, 'Säugethiere von Paraguay,' s. 49.

human progenitors of man, during the season of courtship, when animals of all kinds are excited by the strongest passions. In this case, from the deeply-laid principle of inherited associations, musical tones would be likely to excite in us, in a vague and indefinite manner, the strong emotions of a long-past age. Bearing in mind that the males of some quadrumanous animals have their vocal organs much more developed than in the females, and that one anthropomorphous species pours forth a whole octave of musical notes and may be said to sing, the suspicion does not appear improbable that the progenitors of man, either the males or females, or both sexes, before they had acquired the power of expressing their mutual love in articulate language, endeavoured to charm each other with musical notes and rhythm. So little is known about the use of the voice by the Quadrumana during the season of love, that we have hardly any means of judging whether the habit of singing was first acquired by the male or female progenitors of mankind. Women are generally thought to possess sweeter voices than men, and as far as this serves as any guide we may infer that they first acquired musical powers in order to attract the other sex.[35] But if so, this must have occurred long ago, before the progenitors of man had become sufficiently human to treat and value their women merely as useful slaves. The impassioned orator, bard, or musician, when with his varied tones and cadences he excites the strongest emotions in his hearers, little suspects that he uses the same means by which, at an extremely remote period, his half-human ancestors aroused each other's ardent passions, during their mutual courtship and rivalry.

---

[35] See an interesting discussion on this subject by Häckel, 'Generelle Morph.' B. ii. 1866, s. 246.

On the influence of beauty in determining the marriages of mankind.—In civilised life man is largely, but by no means exclusively, influenced in the choice of his wife by external appearance; but we are chiefly concerned with primeval times, and our only means of forming a judgment on this subject is to study the habits of existing semi-civilised and savage nations. If it can be shewn that the men of different races prefer women having certain characteristics, or conversely that the women prefer certain men, we have then to enquire whether such choice, continued during many generations, would produce any sensible effect on the race, either on one sex or both sexes; this latter circumstance depending on the form of inheritance which prevails.

It will be well first to shew in some detail that savages pay the greatest attention to their personal appearance.[36] That they have a passion for ornament is notorious; and an English philosopher goes so far as to maintain that clothes were first made for ornament and not for warmth. As Professor Waitz remarks, "however poor "and miserable man is, he finds a pleasure in adorning "himself." The extravagance of the naked Indians of South America in decorating themselves is shewn "by "a man of large stature gaining with difficulty enough "by the labour of a fortnight to procure in exchange

---

[36] A full and excellent account of the manner in which savages in all parts of the world ornament themselves is given by the Italian traveller, Prof. Mantegazza, 'Rio de la Plata, Viaggi e Studi,' 1867, p. 525-545; all the following statements, when other references are not given, are taken from this work. See, also, Waitz, 'Introduct. to Anthropolog.' Eng. transl. vol. i. 1863, p. 275, *et passim*. Lawrence also gives very full details in his 'Lectures on Physiology,' 1822. Since this chapter was written Sir J. Lubbock has published his 'Origin of Civilisation,' 1870, in which there is an interesting chapter on the present subject, and from which (p. 42, 48) I have taken some facts about savages dyeing their teeth and hair, and piercing their teeth.

"the *chica* necessary to paint himself red."[37] The ancient barbarians of Europe during the Reindeer period brought to their caves any brilliant or singular objects which they happened to find. Savages at the present day everywhere deck themselves with plumes, necklaces, armlets, earrings, &c. They paint themselves in the most diversified manner. "If painted nations," as Humboldt observes, "had been examined with the same "attention as clothed nations, it would have been per-"ceived that the most fertile imagination and the most "mutable caprice have created the fashions of painting, "as well as those of garments."

In one part of Africa the eyelids are coloured black; in another the nails are coloured yellow or purple. In many places the hair is dyed of various tints. In different countries the teeth are stained black, red, blue, &c., and in the Malay Archipelago it is thought shameful to have white teeth like those of a dog. Not one great country can be named, from the Polar regions in the north to New Zealand in the south, in which the aborigines do not tattoo themselves. This practice was followed by the Jews of old and by the ancient Britons. In Africa some of the natives tattoo themselves, but it is much more common to raise protuberances by rubbing salt into incisions made in various parts of the body; and these are considered by the inhabitants of Kordofan and Durfur "to be great personal attractions." In the Arab countries no beauty can be perfect until the cheeks "or temples have been gashed."[38] In South America, as Humboldt remarks, "a mother would be accused of

[37] Humboldt, 'Personal Narrative,' Eng. translat. vol. iv. p. 515; on the imagination shewn in painting the body, p. 522; on modifying the form of the calf of the leg, p. 466.

[38] 'The Nile Tributaries,' 1867; 'The Albert N'yanza,' 1866, vol. i. p. 218.

"culpable indifference towards her children, if she did "not employ artificial means to shape the calf of the leg "after the fashion of the country." In the Old and New World the shape of the skull was formerly modified during infancy in the most extraordinary manner, as is still the case in many places, and such deformities are considered ornamental. For instance, the savages of Colombia[39] deem a much flattened head "an essential "point of beauty."

The hair is treated with especial care in various countries; it is allowed to grow to full length, so as to reach to the ground, or is combed into "a compact "frizzled mop, which is the Papuan's pride and glory."[40] In Northern Africa "a man requires a period of from "eight to ten years to perfect his coiffure." With other nations the head is shaved, and in parts of South America and Africa even the eyebrows are eradicated. The natives of the Upper Nile knock out the four front teeth, saying that they do not wish to resemble brutes. Further south, the Batokas knock out the two upper incisors, which, as Livingstone[41] remarks, gives the face a hideous appearance, owing to the growth of the lower jaw; but these people think the presence of the incisors most unsightly, and on beholding some Europeans, cried out, "Look at the great teeth!" The great chief Sebituani tried in vain to alter this fashion. In various parts of Africa and in the Malay Archipelago the natives file the incisor teeth into points like those of a saw, or pierce them with holes, into which they insert studs.

---

39 Quoted by Prichard, 'Phys. Hist. of Mankind,' 4th edit. vol. i. 1851, p. 321.

40 On the Papuans, Wallace, 'The Malay Archipelago,' vol. ii. p. 445. On the coiffure of the Africans, Sir S. Baker, 'The Albert N'yanza,' vol. i. p. 210.

41 'Travels, p. 533.

As the face with us is chiefly admired for its beauty, so with savages it is the chief seat of mutilation. In all quarters of the world the septum, and more rarely the wings of the nose are pierced, with rings, sticks, feathers, and other ornaments inserted into the holes. The ears are everywhere pierced and similarly ornamented, and with the Botucudos and Lenguas of South America the hole is gradually so much enlarged that the lower edge touches the shoulder. In North and South America and in Africa either the upper or lower lip is pierced; and with the Botocudos the hole in the lower lip is so large that a disc of wood four inches in diameter is placed in it. Mantegazza gives a curious account of the shame felt by a South American native, and of the ridicule which he excited, when he sold his *tembeta*,—the large coloured piece of wood which is passed through the hole. In central Africa the women perforate the lower lip and wear a crystal, which, from the movement of the tongue, has "a wriggling motion "indescribably ludicrous during conversation." The wife of the chief of Latooka told Sir S. Baker[42] that his "wife would be much improved if she would extract "her four front teeth from the lower jaw, and wear the "long pointed polished crystal in her under lip." Further south with the Makalolo, the upper lip is perforated, and a large metal and bamboo ring, called a *pelelé*, is worn in the hole. "This caused the lip in one case to "project two inches beyond the tip of the nose; and "when the lady smiled the contraction of the muscles "elevated it over the eyes. 'Why do the women wear "'these things?' the venerable chief, Chinsurdi, was "asked. Evidently surprised at such a stupid question, "he replied, 'For beauty! They are the only beautiful

[42] 'The Albert N'yanza,' 1866, vol. i. p. 217.

"'things women have; men have beards, women have "'none. What kind of a person would she be without "'the pelelé? She would not be a woman at all with a "'mouth like a man, but no beard.'"[43]

Hardly any part of the body, which can be unnaturally modified, has escaped. The amount of suffering thus caused must have been wonderfully great, for many of the operations require several years for their completion, so that the idea of their necessity must be imperative. The motives are various; the men paint their bodies to make themselves appear terrible in battle; certain mutilations are connected with religious rites; or they mark the age of puberty, or the rank of the man, or they serve to distinguish the tribes. As with savages the same fashions prevail for long periods,[44] mutilations, from whatever cause first made, soon come to be valued as distinctive marks. But self-adornment, vanity, and the admiration of others, seem to be the commonest motives. In regard to tattooing, I was told by the missionaries in New Zealand, that when they tried to persuade some girls to give up the practice, they answered, "We must just have a few "lines on our lips; else when we grow old we shall be "so very ugly." With the men of New Zealand, a most capable judge[45] says, "to have fine tattooed faces was "the great ambition of the young, both to render them-"selves attractive to the ladies, and conspicuous in war." A star tattooed on the forehead and a spot on the chin

---

[43] Livingstone, 'British Association,' 1860; report given in the 'Athenæum,' July 7, 1860, p. 29.

[44] Sir S. Baker (ibid. vol. i. p. 210) speaking of the natives of Central Africa says, "every tribe has a distinct and unchanging fashion for "dressing the hair." See Agassiz ('Journey in Brazil,' 1868, p. 318) on the invariability of the tattooing of the Amazonian Indians.

[45] Rev. R. Taylor, 'New Zealand and its Inhabitants,' 1855, p. 152.

are thought by the women in one part of Africa to be irresistible attractions.[46] In most, but not all parts of the world, the men are more highly ornamented than the women, and often in a different manner; sometimes, though rarely, the women are hardly at all ornamented. As the women are made by savages to perform the greatest share of the work, and as they are not allowed to eat the best kinds of food, so it accords with the characteristic selfishness of man that they should not be allowed to obtain, or to use, the finest ornaments. Lastly it is a remarkable fact, as proved by the foregoing quotations, that the same fashions in modifying the shape of the head, in ornamenting the hair, in painting, tattooing, perforating the nose, lips, or ears, in removing or filing the teeth, &c., now prevail and have long prevailed in the most distant quarters of the world. It is extremely improbable that these practices which are followed by so many distinct nations are due to tradition from any common source. They rather indicate the close similarity of the mind of man, to whatever race he may belong, in the same manner as the almost universal habits of dancing, masquerading, and making rude pictures.

Having made these preliminary remarks on the admiration felt by savages for various ornaments, and for deformities most unsightly in our eyes, let us see how far the men are attracted by the appearance of their women, and what are their ideas of beauty. As I have heard it maintained that savages are quite indifferent about the beauty of their women, valuing them solely as slaves, it may be well to observe that this conclusion does not at all agree with the care which the women take in ornamenting themselves, or with

---

[46] Mantegazza, 'Viaggi e Studi,' p. 542.

their vanity. Burchell[47] gives an amusing account of a Bush-woman, who used so much grease, red ochre, and shining powder, "as would have ruined any but a very rich husband." She displayed also "much vanity and too evident a consciousness of her superiority." Mr. Winwood Reade informs me that the negroes of the West Coast often discuss the beauty of their women. Some competent observers have attributed the fearfully common practice of infanticide partly to the desire felt by the women to retain their good looks.[48] In several regions the women wear charms and love-philters to gain the affections of the men; and Mr. Brown enumerates four plants used for this purpose by the women of North-Western America.[49]

Hearne,[50] who lived many years with the American Indians, and who was an excellent observer, says, in speaking of the women, "Ask a Northern Indian what "is beauty, and he will answer, a broad flat face, small "eyes, high cheek-bones, three or four broad black lines "across each cheek, a low forehead, a large broad chin, "a clumsy hook nose, a tawny hide, and breasts hanging "down to the belt." Pallas, who visited the northern parts of the Chinese empire, says "those women are "preferred who have the Mandschú form; that is to say, "a broad face, high cheek-bones, very broad noses, and "enormous ears;"[51] and Vogt remarks that the obliquity of the eye, which is proper to the Chinese and Japanese,

---

[47] 'Travels in S. Africa,' 1824, vol. i. p. 414.

[48] See, for references, 'Gerland über das Aussterben der Naturvölker,' 1868, s. 51, 53, 55; also Azara, 'Voyages,' &c. tom. ii. p. 116.

[49] On the vegetable productions used by the North-Western American Indians, 'Pharmaceutical Journal,' vol. x.

[50] 'A Journey from Prince of Wales Fort,' 8vo. edit. 1796, p. 89.

[51] Quoted by Prichard, 'Phys. Hist. of Mankind,' 3rd edit. vol. iv. 1844, p. 519; Vogt, 'Lectures on Man,' Eng. translat. p. 129. On the opinion of the Chinese on the Cingalese, E. Tennent, 'Ceylon,' vol. ii. 1859, p. 107.

is exaggerated in their pictures for the purpose, as it "it seems, of exhibiting its beauty, as contrasted with "the eye of the red-haired barbarians." It is well known, as Huc repeatedly remarks, that the Chinese of the interior think Europeans hideous with their white skins and prominent noses. The nose is far from being too prominent, according to our ideas, in the natives of Ceylon; yet "the Chinese in the seventh century, ac-"customed to the flat features of the Mogul races, were "surprised at the prominent noses of the Cingalese; and "Thsang described them as having 'the beak of a bird, "with the body of a man.'"

Finlayson, after minutely describing the people of Cochin China, says that their rounded heads and faces are their chief characteristics; and he adds, "the "roundness of the whole countenance is more striking "in the women, who are reckoned beautiful in propor-"tion as they display this form of face." The Siamese have small noses with divergent nostrils, a wide mouth, rather thick lips, a remarkably large face, with very high and broad cheek-bones. It is, therefore, not wonderful that "beauty, according to our notion is a stranger "to them. Yet they consider their own females to be "much more beautiful than those of Europe."[52]

It is well known that with many Hottentot women the posterior part of the body projects in a wonderful manner; they are steatopygous; and Sir Andrew Smith is certain that this peculiarity is greatly admired by the men.[53] He once saw a woman who was considered a

---

[52] Prichard, as taken from Crawfurd and Finlayson, 'Phys. Hist. of Mankind,' vol. iv. p. 534, 535.

[53] Idem illustrissimus viator dixit mihi præcinctorium vel tabula fœminæ, quod nobis teterrimum est, quondam permagno æstimari ab hominibus in hac gente. Nunc res mutata est, et censet talem conformationem minime optandam est."

beauty, and she was so immensely developed behind, that when seated on level ground she could not rise, and had to push herself along until she came to a slope. Some of the women in various negro tribes are similarly characterised; and, according to Burton, the Somal men "are "said to choose their wives by ranging them in a line, "and by picking her out who projects farthest *a tergo*, "Nothing can be more hateful to a negro than the "opposite form."[54]

With respect to colour, the negroes rallied Mungo Park on the whiteness of his skin and the prominence of his nose, both of which they considered as "unsightly "and unnatural conformations." He in return praised the glossy jet of their skins and the lovely depression of their noses; this they said was "honey-mouth," nevertheless they gave him food. The African Moors, also, "knitted their brows and seemed to shudder" at the whiteness of his skin. On the eastern coast, the negro boys when they saw Burton, cried out "Look at the "white man; does he not look like a white ape?" On the western coast, as Mr. Winwood Reade informs me, the negroes admire a very black skin more than one of a lighter tint. But their horror of whiteness may be partly attributed, according to this same traveller, to the belief held by most negroes that demons and spirits are white.

The Banyai of the more southern part of the continent are negroes, but "a great many of them are of a light "coffee-and-milk colour, and, indeed, this colour is con- "sidered handsome throughout the whole country;" so that here we have a different standard of taste. With the

54 'The Anthropological Review,' November, 1864, p. 237. For additional references, see Waitz, 'Introduct. to Anthropology,' Eng. translat. 1863, vol. i. p. 105.

Kafirs, who differ much from negroes, "the skin, except "among the tribes near Delagoa Bay, is not usually "black, the prevailing colour being a mixture of black "and red, the most common shade being chocolate. "Dark complexions, as being most common are natu-"rally held in the highest esteem. To be told that he "is light-coloured, or like a white man, would be deemed "a very poor compliment by a Kafir. I have heard of "one unfortunate man who was so very fair that no "girl would marry him." One of the titles of the Zulu king is "You who are black."[55] Mr. Galton, in speaking to me about the natives of S. Africa, remarked that their ideas of beauty seem very different from ours; for in one tribe two slim, slight, and pretty girls were not admired by the natives.

Turning to other quarters of the world; in Java, a yellow, not a white girl, is considered, according to Madame Pfeiffer, a beauty. A man of Cochin-China "spoke with contempt of the wife of the English "Ambassador, that she had white teeth like a dog, "and a rosy colour like that of potato-flowers." We have seen that the Chinese dislike our white skin, and that the N. Americans admire "a tawny hide." In S. America, the Yura-caras, who inhabit the wooded, damp slopes of the eastern Cordillera, are remarkably pale-coloured, as their name in their own language expresses; nevertheless they consider European women as very inferior to their own.[56]

---

[55] 'Mungo Park's Travels in Africa,' 4to. 1816, p. 53, 131. Burton's statement is quoted by Schaaffhausen, 'Archiv für Anthropolog.' 1866, s. 163. On the Banyai, Livingstone, 'Travels,' p. 64. On the Kafirs, the Rev. J. Shooter, 'The Kafirs of Natal and the Zulu Country,' 1857 p. 1.

[56] For the Javanese and Cochin-Chinese,' see Waitz, 'Introduct. to Anthropology,' Eng. translat. vol. i. p. 305. On the Yura-caras, A.

In several of the tribes of North America the hair on the head grows to a wonderful length; and Catlin gives a curious proof how much this is esteemed, for the chief of the Crows was elected to this office from having the longest hair of any man in the tribe, namely ten feet and seven inches. The Aymaras and Quichuas of S. America, likewise have very long hair; and this, as Mr. D. Forbes informs me, is so much valued for the sake of beauty, that cutting it off was the severest punishment which he could inflict on them. In both halves of the continent the natives sometimes increase the apparent length of their hair by weaving into it fibrous substances. Although the hair on the head is thus cherished, that on the face is considered by the North American Indians "as very vulgar," and every hair is carefully eradicated. This practice prevails throughout the American continent from Vancouver's Island in the north to Tierra del Fuego in the south. When York Minster, a Fuegian on board the "Beagle" was taken back to his country, the natives told him he ought to pull out the few short hairs on his face. They also threatened a young missionary, who was left for a time with them, to strip him naked, and pluck the hairs from his face and body, yet he was far from a hairy man. This fashion is carried to such an extreme that the Indians of Paraguay eradicate their eyebrows and eyelashes, saying that they do not wish to be like horses.[57]

It is remarkable that throughout the world the races

---

d'Orbigny, as quoted in Prichard, 'Phys. Hist. of Mankind,' vol. v. 3rd edit. p. 476.

[57] 'North American Indians,' by G. Catlin, 3rd edit. 1842, vol. i. p. 49; vol. ii. p. 227. On the natives of Vancouver Island, see Sproat, 'Scenes and Studies of Savage Life,' 1868, p. 25. On the Indians of Paraguay, Azara, 'Voyages,' tom. ii. p. 105.

which are almost completely destitute of a beard dislike hairs on the face and body, and take pains to eradicate them. The Kalmucks are beardless, and they are well known, like the Americans, to pluck out all straggling hairs; and so it is with the Polynesians, some of the Malays, and the Siamese. Mr. Veitch states that the Japanese ladies "all objected to our whiskers, considering them very ugly, and told us to cut them off, and "be like Japanese men." The New Zealanders are beardless; they carefully pluck out the hairs on the face, and have a saying that "There is no woman for a "hairy man."[58]

On the other hand, bearded races admire and greatly value their beards; among the Anglo-Saxons every part of the body, according to their laws, had a recognised value; "the loss of the beard being estimated at twenty "shillings, while the breaking of a thigh was fixed at "only twelve."[59] In the East men swear solemnly by their beards. We have seen that Chinsurdi, the chief of the Makalolo in Africa, evidently thought that beards were a great ornament. With the Fijians in the Pacific the beard is "profuse and bushy, and is his "greatest pride;" whilst the inhabitants of the adjacent archipelagoes of Tonga and Samoa are "beardless, "and abhor a rough chin." In one island alone of the Ellice group "the men are heavily bearded, and not a "little proud thereof."[60]

---

[58] On the Siamese, Prichard, ibid. vol. iv. p. 533. On the Japanese, Veitch in 'Gardeners' Chronicle,' 1860, p. 1104. On the New Zealanders Mantegazza, 'Viaggi e Studi,' 1867, p. 526. For the other nations mentioned, see references in Lawrence, 'Lectures on Physiology,' &c. 1822, p. 272.

[59] Lubbock, 'Origin of Civilisation,' 1870, p. 321.

[60] Dr. Barnard Davis quotes Mr. Pritchard and others for these facts in regard to the Polynesians, in 'Anthropological Review,' April, 1870, .185, 191.

We thus see how widely the different races of man differ in their taste for the beautiful. In every nation sufficiently advanced to have made effigies of their gods or of their deified rulers, the sculptors no doubt have endeavoured to express their highest ideal of beauty and grandeur.[61] Under this point of view it is well to compare in our mind the Jupiter or Apollo of the Greeks with the Egyptian or Assyrian statues; and these with the hideous bas-reliefs on the ruined buildings of Central America.

I have met with very few statements opposed to the above conclusion. Mr. Winwood Reade, however, who has had ample opportunities for observation, not only with the negroes of the West Coast of Africa, but with those of the interior who have never associated with Europeans, is convinced that their ideas of beauty are *on the whole* the same as ours. He has repeatedly found that he agreed with negroes in their estimation of the beauty of the native girls; and that their appreciation of the beauty of European women corresponded with ours. They admire long hair, and use artificial means to make it appear abundant; they admire also a beard, though themselves very scantily provided. Mr. Reade feels doubtful what kind of nose is most appreciated: a girl has been heard to say, "I "do not want to marry him, he has got no nose;" and this shews that a very flat nose is not an object of admiration. We should, however, bear in mind that the depressed and very broad noses and projecting jaws of the negroes of the West Coast are exceptional types with the inhabitants of Africa. Notwithstanding the foregoing statements, Mr. Reade does not think it pro-

---

61 Ch. Comte has remarks to this effect in his 'Traité de Législation,' 3rd edit. 1837, p. 136.

bable that negroes would ever prefer the "most beautiful European woman, on the mere grounds of physical admiration, to a good-looking negress."[62]

The truth of the principle, long ago insisted on by Humboldt,[63] that man admires and often tries to exaggerate whatever characters nature may have given him, is shewn in many ways. The practice of beardless races extirpating every trace of a beard, and generally all the hairs on the body, offers one illustration. The skull has been greatly modified during ancient and modern times by many nations; and there can be little doubt that this has been practised, especially in N. and S. America, in order to exaggerate some natural and admired peculiarity. Many American Indians are known to admire a head flattened to such an extreme degree as to appear to us like that of an idiot. The natives on the north-western coast compress the head into a pointed cone; and it is their constant practice to gather the hair into a knot on the top of the head, for the sake, as Dr. Wilson remarks, "of increasing the apparent elevation of the favourite conoid form." The inhabitants of Arakhan "admire a broad, smooth forehead, and in order to produce it, they fasten a plate of lead on the heads of the new-born children." On the other hand,

---

[62] The Fuegians, as I have been informed by a missionary who long resided with them, consider European women as extremely beautiful; but from what we have seen of the judgment of the other aborigines of America, I cannot but think that this must be a mistake, unless indeed the statement refers to the few Fuegians who have lived for some time with Europeans, and who must consider us as superior beings. I should add that a most experienced observer, Capt. Burton, believes that a woman whom we consider beautiful is admired throughout the world, 'Anthropological Review,' March, 1864, p. 245.

[63] 'Personal Narrative,' Eng. translat. vol. iv. p. 518, and elsewhere. Mantegazza, in his 'Viaggi e Studi,' 1867, strongly insists on this same principle.

"a broad, well-rounded occiput is considered a great "beauty" by the natives of the Fiji islands.[64]

As with the skull, so with the nose; the ancient Huns during the age of Attila were accustomed to flatten the noses of their infants with bandages, "for the sake "of exaggerating a natural conformation." With the Tahitians, to be called *long-nose* is considered as an insult, and they compress the noses and foreheads of their children for the sake of beauty. So it is with the Malays of Sumatra, the Hottentots, certain Negroes, and the natives of Brazil.[65] The Chinese have by nature unusually small feet;[66] and it is well known that the women of the upper classes distort their feet to make them still smaller. Lastly, Humboldt thinks that the American Indians prefer colouring their bodies with red paint in order to exaggerate their natural tint; and until recently European women added to their naturally bright colours by rouge and white cosmetics; but I doubt whether many barbarous nations have had any such intention in painting themselves.

In the fashions of our own dress we see exactly the same principle and the same desire to carry every point to an extreme; we exhibit, also, the same spirit of emulation. But the fashions of savages are far more permanent than ours; and whenever their bodies are

---

[64] On the skulls of the American tribes, see Nott and Gliddon, 'Types of Mankind,' 1854, p. 440; Prichard, 'Phys. Hist. of Mankind,' vol. i. 3rd edit. p. 321; on the natives of Arakhan, ibid. vol. iv. p. 537. Wilson, 'Physical Ethnology,' Smithsonian Institution, 1863, p. 288; on the Fijians, p. 290. Sir J. Lubbock ('Prehistoric Times,' 2nd edit. 1869, p. 506) gives an excellent résumé on this subject.

[65] On the Huns, Godron, 'De l'Espèce,' tom. ii. 1859, p. 300. On the Tahitians, Waitz, 'Anthropolog.' Eng. translat. vol. i. p. 305. Marsden, quoted by Prichard, 'Phys. Hist. of Mankind,' 3rd edit. vol. v. p. 67. Lawrence, 'Lectures on Physiology,' p. 337.

[66] This fact was ascertained in the 'Reise der *Novara*: Anthropolog. Theil,' Dr. Weisbach, 1867, s. 265.

artificially modified this is necessarily the case. The Arab women of the Upper Nile occupy about three days in dressing their hair; they never imitate other tribes, "but simply vie with each other in the superlativeness "of their own style." Dr. Wilson, in speaking of the compressed skulls of various American races, adds, "such "usages are among the least eradicable, and long sur-"vive the shock of revolutions that change dynasties "and efface more important national peculiarities."[67] The same principle comes largely into play in the art of selection; and we can thus understand, as I have elsewhere explained,[68] the wonderful development of all the races of animals and plants which are kept merely for ornament. Fanciers always wish each character to be somewhat increased; they do not admire a medium standard; they certainly do not desire any great and abrupt change in the character of their breeds; they admire solely what they are accustomed to behold, but they ardently desire to see each characteristic feature a little more developed.

No doubt the perceptive powers of man and the lower animals are so constituted that brilliant colours and certain forms, as well as harmonious and rhythmical sounds, give pleasure and are called beautiful; but why this should be so, we know no more than why certain bodily sensations are agreeable and others disagreeable. It is certainly not true that there is in the mind of man any universal standard of beauty with respect to the human body. It is, however, possible that certain tastes may in the course of time become inherited, though I know of no evidence in favour of this belief;

---

[67] 'Smithsonian Institution, 1863, p. 289. On the fashions of Arab women, Sir S. Baker, 'The Nile Tributaries,' 1867, p. 121.

[68] 'The Variation of Animals and Plants under Domestication,' vol. i. p. 214; vol. ii. p. 240.

and if so, each race would possess its own innate ideal standard of beauty. It has been argued[69] that ugliness consists in an approach to the structure of the lower animals, and this no doubt is true with the more civilised nations, in which intellect is highly appreciated; but a nose twice as prominent, or eyes twice as large as usual, would not be an approach in structure to any of the lower animals, and yet would be utterly hideous. The men of each race prefer what they are accustomed to behold; they cannot endure any great change; but they like variety, and admire each characteristic point carried to a moderate extreme.[70] Men accustomed to a nearly oval face, to straight and regular features, and to bright colours, admire, as we Europeans know, these points when strongly developed. On the other hand, men accustomed to a broad face, with high cheek-bones, a depressed nose, and a black skin, admire these points strongly developed. No doubt characters of all kinds may easily be too much developed for beauty. Hence a perfect beauty, which implies many characters modified in a particular manner, will in every race be a prodigy. As the great anatomist Bichat long ago said, if every one were cast in the same mould, there would be no such thing as beauty. If all our women were to become as beautiful as the Venus de Medici, we should for a time be charmed; but we should soon wish for variety; and as soon as we had obtained variety, we should wish to see certain characters in our women a little exaggerated beyond the then existing common standard.

---

[69] Schaaffhausen, 'Archiv für Anthropologie,' 1866, s. 164.

[70] Mr. Bain has collected ('Mental and Moral Science,' 1868, p. 304-314) about a dozen more or less different theories of the idea of beauty; but none are quite the same with that here given.

CHAPTER XX.

SECONDARY SEXUAL CHARACTERS OF MAN—*continued.*

On the effects of the continued selection of women according to a different standard of beauty in each race — On the causes which interfere with sexual selection in civilised and savage nations — Conditions favourable to sexual selection during primeval times — On the manner of action of sexual selection with mankind — On the women in savage tribes having some power to choose their husbands — Absence of hair on the body, and development of the beard — Colour of the skin — Summary.

WE have seen in the last chapter that with all barbarous races ornaments, dress, and external appearance are highly valued; and that the men judge of the beauty of their women by widely different standards. We must next inquire whether this preference and the consequent selection during many generations of those women, which appear to the men of each race the most attractive, has altered the character either of the females alone or of both sexes. With mammals the general rule appears to be that characters of all kinds are inherited equally by the males and females; we might therefore expect that with mankind any characters gained through sexual selection by the females would commonly be transferred to the offspring of both sexes. If any change has thus been effected it is almost certain that the different races will have been differently modified, as each has its own standard of beauty.

With mankind, especially with savages, many causes interfere with the action of sexual selection as far as the bodily frame is concerned. Civilised men are largely attracted by the mental charms of women, by their wealth, and especially by their social position; for men rarely marry into a much lower rank of life. The men who succeed in obtaining the more beautiful women, will not have a better chance of leaving a long line of descendants than other men with plainer wives, with the exception of the few who bequeath their fortunes according to primogeniture. With respect to the opposite form of selection, namely of the more attractive men by the women, although in civilised nations women have free or almost free choice, which is not the case with barbarous races, yet their choice is largely influenced by the social position and wealth of the men; and the success of the latter in life largely depends on their intellectual powers and energy, or on the fruits of these same powers in their forefathers.

There is, however, reason to believe that sexual selection has effected something in certain civilised and semi-civilised nations. Many persons are convinced, as it appears to me with justice, that the members of our aristocracy, including under this term all wealthy families in which primogeniture has long prevailed, from having chosen during many generations from all classes the more beautiful women as their wives, have become handsomer, according to the European standard of beauty, than the middle classes; yet the middle classes are placed under equally favourable conditions of life for the perfect development of the body. Cook remarks that the superiority in personal appearance "which is observable in the erees or nobles in all the "other islands (of the Pacific) is found in the Sandwich "islands;" but this may be chiefly due to their better food and manner of life.

The old traveller Chardin, in describing the Persians, says their "blood is now highly refined by frequent "intermixtures with the Georgians and Circassians, "two nations which surpass all the world in personal "beauty. There is hardly a man of rank in Persia "who is not born of a Georgian or Circassian mother." He adds that they inherit their beauty, "not from their "ancestors, for without the above mixture, the men of "rank in Persia, who are descendants of the Tartars, "would be extremely ugly."[1] Here is a more curious case: the priestesses who attended the temple of Venus Erycina at San-Giuliano in Sicily, were selected for their beauty out of the whole of Greece; they were not vestal virgins, and Quatrefages,[2] who makes this statement, says that the women of San-Giuliano are famous at the present day as the most beautiful in the island, and are sought by artists as models. But it is obvious that the evidence in the above cases is doubtful.

The following case, though relating to savages, is well worth giving from its curiosity. Mr. Winwood Reade informs me that the Jollofs, a tribe of negroes on the west coast of Africa, "are remarkable for their uni-"formly fine appearance." A friend of his asked one of these men, "How is it that every one whom I meet is "so fine-looking, not only your men, but your women? The Jollof answered, "It is very easily explained: it "has always been our custom to pick out our worse-"looking slaves and to sell them." It need hardly be added that with all savages female slaves serve as concubines. That this negro should have attributed, whether rightly or wrongly, the fine appearance of his tribe, to the long-continued elimination of the ugly

[1] These quotations are taken from Lawrence ('Lectures on Physiology,' &c. 1822, p. 393), who attributes the beauty of the upper classes in England to the men having long selected the more beautiful women.

[2] "Anthropologie," 'Revue des Cours Scientifiques,' Oct. 1868, p. 721.

women, is not so surprising as it may at first appear; for I have elsewhere shewn[3] that negroes fully appreciate the importance of selection in the breeding of their domestic animals, and I could give from Mr. Reade additional evidence on this head.

On the Causes which prevent or check the Action of Sexual Selection with Savages.—The chief causes are, firstly, so-called communal marriages or promiscuous intercourse; secondly, infanticide, especially of female infants; thirdly, early betrothals; and lastly, the low estimation in which women are held, as mere slaves. These four points must be considered in some detail.

It is obvious that as long as the pairing of man, or of any other animal, is left to chance, with no choice exerted by either sex, there can be no sexual selection; and no effect will be produced on the offspring by certain individuals having had an advantage over others in their courtship. Now it is asserted that there exist at the present day tribes which practise what Sir J. Lubbock by courtesy calls communal marriages; that is, all the men and women in the tribe are husbands and wives to each other. The licentiousness of many savages is no doubt astonishingly great, but it seems to me that more evidence is requisite before we fully admit that their existing intercourse is absolutely promiscuous. Nevertheless all those who have most closely studied the subject,[4] and whose judgment is worth much more

---

[3] 'The Variation of Animals and Plants under Domestication,' vol. i. p. 207.

[4] Sir J. Lubbock, 'The Origin of Civilisation,' 1870, chap. iii. especially p. 60-67. Mr. M'Lennan, in his extremely valuable work on 'Primitive Marriage,' 1865, p. 163, speaks of the union of the sexes "in the earliest times as loose, transitory, and in some degree promis-"cuous." Mr. M'Lennan and Sir J. Lubbock have collected much evidence on the extreme licentiousness of savages at the present time. Mr. L. H. Morgan, in his interesting memoir on the classificatory system

than mine, believe that communal marriage was the original and universal form throughout the world, including the intermarriage of brothers and sisters. The indirect evidence in favour of this belief is extremely strong, and rests chiefly on the terms of relationship which are employed between the members of the same tribe, implying a connection with the tribe alone, and not with either parent. But the subject is too large and complex for even an abstract to be here given, and I will confine myself to a few remarks. It is evident in the case of communal marriages, or where the marriage-tie is very loose, that the relationship of the child to its father cannot be known. But it seems almost incredible that the relationship of the child to its mother should ever have been completely ignored, especially as the women in most savage tribes nurse their infants for a long time. Accordingly in many cases the lines of descent are traced through the mother alone, to the exclusion of the father. But in many other cases the terms employed express a connection with the tribe alone, to the exclusion even of the mother. It seems possible that the connection between the related members of the same barbarous tribe, exposed to all sorts of danger, might be so much more important, owing to the need of mutual protection and aid, than that between the mother and her child, as to lead to the sole use of terms expressive of the former relationships; but Mr. Morgan is convinced that this view of the case is by no means sufficient.

The terms of relationship used in different parts of

---

of relationship ('Proc. American Acad. of Sciences,' vol. vii. Feb. 1868, p. 475) concludes that polygamy and all forms of marriage during primeval times were essentially unknown. It appears, also, from Sir J. Lubbock's work, that Bachofen likewise believes that communal intercourse originally prevailed.

the world may be divided, according to the author just quoted, into two great classes, the classificatory and descriptive,—the latter being employed by us. It is the classificatory system which so strongly leads to the belief that communal and other extremely loose forms of marriage were originally universal. But as far as I can see, there is no necessity on this ground for believing in absolutely promiscuous intercourse. Men and women, like many of the lower animals, might formerly have entered into strict though temporary unions for each birth, and in this case nearly as much confusion would have arisen in the terms of relationship as in the case of promiscuous intercourse. As far as sexual selection is concerned, all that is required is that choice should be exerted before the parents unite, and it signifies little whether the unions last for life or only for a season.

Besides the evidence derived from the terms of relationship, other lines of reasoning indicate the former wide prevalence of communal marriage. Sir J. Lubbock ingeniously accounts[5] for the strange and widely-extended habit of exogamy,—that is, the men of one tribe always taking wives from a distinct tribe,—by communism having been the original form of marriage; so that a man never obtained a wife for himself unless he captured her from a neighbouring and hostile tribe, and then she would naturally have become his sole and valuable property. Thus the practice of capturing wives might have arisen; and from the honour so gained might ultimately have become the universal habit. We can also, according to Sir J. Lubbock,[5] thus understand "the necessity of expiation for mar-

[5] Address to British Association 'On the Social and Religious Condition of the Lower Races of Man,' 1870, p. 20.

"riage as an infringement of tribal rites, since, according to old ideas, a man had no right to appropriate to himself that which belonged to the whole tribe." Sir J. Lubbock further gives a most curious body of facts shewing that in old times high honour was bestowed on women who were utterly licentious; and this, as he explains, is intelligible, if we admit that promiscuous intercourse was the aboriginal and therefore long revered custom of the tribe.[6]

Although the manner of development of the marriage-tie is an obscure subject, as we may infer from the divergent opinions on several points between the three authors who have studied it most closely, namely, Mr. Morgan, Mr. M'Lennan, and Sir J. Lubbock, yet from the foregoing and several other lines of evidence it seems certain that the habit of marriage has been gradually developed, and that almost promiscuous intercourse was once extremely common throughout the world. Nevertheless from the analogy of the lower animals, more particularly of those which come nearest to man in the series, I cannot believe that this habit prevailed at an extremely remote period, when man had hardly attained to his present rank in the zoological scale. Man, as I have attempted to shew, is certainly descended from some ape-like creature. With the existing Quadrumana, as far as their habits are known, the males of some species are monogamous, but live during only a part of the year with the females, as seems to be the case with the Orang. Several kinds, as some of the Indian and American monkeys, are strictly monogamous, and associate all the year round with their wives. Others are polygamous, as the Gorilla and several

[6] 'Origin of Civilisation,' 1870, p. 86. In the several works above quoted there will be found copious evidence on relationship through the females alone, or with the tribe alone.

American species, and each family lives separate. Even when this occurs, the families inhabiting the same district are probably to a certain extent social: the Chimpanzee, for instance, is occasionally met with in large bands. Again, other species are polygamous, but several males, each with their own females, live associated in a body, as with several species of Baboons.[7] We may indeed conclude from what we know of the jealousy of all male quadrupeds, armed, as many of them are, with special weapons for battling with their rivals, that promiscuous intercourse in a state of nature is extremely improbable. The pairing may not last for life, but only for each birth; yet if the males which are the strongest and best able to defend or otherwise assist their females and young offspring, were to select the more attractive females, this would suffice for the work of sexual selection.

Therefore, if we look far enough back in the stream of time, it is extremely improbable that primeval men and women lived promiscuously together. Judging from the social habits of man as he now exists, and from most savages being polygamists, the most probable view is that primeval man aboriginally lived in small communities, each with as many wives as he could support and obtain, whom he would have jealously guarded against all other men. Or he may have lived with several wives by himself, like the Gorilla; for all the natives "agree that but one adult male is "seen in a band; when the young male grows up, a "contest takes place for mastery, and the strongest, b

[7] Brehm ('Illust. Thierleben,' B. i. p. 77) says *Cynocephalus hamadryas* lives in great troops containing twice as many adult females as adult males. See Rengger on American polygamous species, and Owen ('Anat. of Vertebrates,' vol. iii. p. 746) on American monogamous species. Other references might be added.

"killing and driving out the others, establishes himself "as the head of the community."[8] The younger males, being thus expelled and wandering about, would, when at last successful in finding a partner, prevent too close interbreeding within the limits of the same family.

Although savages are now extremely licentious, and although communal marriages may formerly have largely prevailed, yet many tribes practise some form of marriage, but of a far more lax nature than with civilised nations. Polygamy, as just stated, is almost universally followed by the leading men in every tribe. Nevertheless there are tribes, standing almost at the bottom of the scale, which are strictly monogamous. This is the case with the Veddahs of Ceylon: they have a saying, according to Sir J. Lubbock,[9] "that "death alone can separate husband and wife." An intelligent Kandyan chief, of course a polygamist, "was perfectly scandalized at the utter barbarism of "living with only one wife, and never parting until "separated by death." It was, he said, "just like the "Wanderoo monkeys." Whether savages who now enter into some form of marriage, either polygamous or monogamous, have retained this habit from primeval times, or whether they have returned to some form of marriage, after passing through a stage of promiscuous intercourse, I will not pretend to conjecture.

Infanticide.—This practice is now very common throughout the world, and there is reason to believe that it prevailed much more extensively during former times.[10] Barbarians find it difficult to support them-

---

8 Dr. Savage, in 'Boston Journal of Nat. Hist.' vol. v. 1845-47, p. 423.

9 'Prehistoric Times,' 1869, p. 424.

10 Mr. M'Lennan, 'Primitive Marriage,' 1865. See especially on exogamy and infanticide, p. 130, 138, 165.

selves and their children, and it is a simple plan to kill their infants. In South America some tribes, as Azara states, formerly destroyed so many infants of both sexes, that they were on the point of extinction. In the Polynesian Islands women have been known to kill from four or five to even ten of their children; and Ellis could not find a single woman who had not killed at least one. Wherever infanticide prevails the struggle for existence will be in so far less severe, and all the members of the tribe will have an almost equally good chance of rearing their few surviving children. In most cases a larger number of female than of male infants are destroyed, for it is obvious that the latter are of most value to the tribe, as they will when grown up aid in defending it, and can support themselves. But the trouble experienced by the women in rearing children, their consequent loss of beauty, the higher estimation set on them and their happier fate, when few in number, are assigned by the women themselves, and by various ob-observers, as additional motives for infanticide. In Australia, where female infanticide is still common, Sir G. Grey estimated the proportion of native women to men as one to three; but others say as two to three. In a village on the eastern frontier of India, Colonel Macculloch found not a single female child.[11]

When, owing to female infanticide, the women of a tribe are few in number, the habit of capturing wives from neighbouring tribes would naturally arise. Sir J. Lubbock, however, as we have seen, attributes the practice in chief part, to the former existence of communal marriage, and to the men having consequently captured

[11] Dr. Gerland ('Ueber das Aussterben der Naturvölker,' 1868) has collected much information on infanticide, see especially s. 27, 51, 54. Azara ('Voyages,' &c. tom. ii. p. 94, 116) enters in detail on the motives. See also M'Lennan (ibid. p. 139) for cases in India.

women from other tribes to hold as their sole property. Additional causes might be assigned, such as the communities being very small, in which case, marriageable women would often be deficient. That the habit of capture was most extensively practised during former times, even by the ancestors of civilised nations, is clearly shewn by the preservation of many curious customs and ceremonies, of which Mr. M'Lennan has given a most interesting account. In our own marriages the "best man" seems originally to have been the chief abettor of the bridegroom in the act of capture. Now as long as men habitually procured their wives through violence and craft, it is not probable that they would have selected the more attractive women; they would have been too glad to have seized on any woman. But as soon as the practice of procuring wives from a distinct tribe was effected through barter, as now occurs in many places, the more attractive women would generally have been purchased. The incessant crossing, however, between tribe and tribe, which necessarily follows from any form of this habit would have tended to keep all the people inhabiting the same country nearly uniform in character; and this would have greatly interfered with the power of sexual selection in differentiating the tribes.

The scarcity of women, consequent on female infanticide, leads, also, to another practice, namely polyandry, which is still common in several parts of the world, and which formerly, as Mr. M'Lennan believes, prevailed almost universally; but this latter conclusion is doubted by Mr. Morgan and Sir J. Lubbock.[12] Whenever two

---

[12] 'Primitive Marriage,' p. 208; Sir J. Lubbock, 'Origin of Civilisation,' p. 100. See also Mr. Morgan, loc. cit., on former prevalence of polyandry.

or more men are compelled to marry one woman, it is certain that all the women of the tribe will get married, and there will be no selection by the men of the more attractive women. But under these circumstances the women no doubt will have the power of choice, and will prefer the more attractive men. Azara, for instance, describes how carefully a Guana woman bargains for all sorts of privileges, before accepting some one or more husbands; and the men in consequence take unusual care of their personal appearance.[13] The very ugly men would perhaps altogether fail in getting a wife, or get one later in life, but the handsomer men, although the most successful in obtaining wives, would not, as far as we can see, leave more offspring to inherit their beauty than the less handsome husbands of the same women.

Early Betrothals and Slavery of Women.—With many savages it is the custom to betroth the females whilst mere infants; and this would effectually prevent preference being exerted on either side according to personal appearance. But it would not prevent the more attractive women from being afterwards stolen or taken by force from their husbands by the more powerful men; and this often happens in Australia, America, and other parts of the world. The same consequences with reference to sexual selection would to a certain extent follow when women are valued almost exclusively as slaves or beasts of burden, as is the case with most savages. The men, however, at all times would prefer the handsomest slaves according to their standard of beauty.

We thus see that several customs prevail with savages

[13] 'Voyages,' &c. tom. ii. p. 92-95.

which would greatly interfere with, or completely stop, the action of sexual selection. On the other hand, the conditions of life to which savages are exposed, and some of their habits, are favourable to natural selection; and this always comes into play together with sexual selection. Savages are known to suffer severely from recurrent famines; they do not increase their food by artificial means; they rarely refrain from marriage,[14] and generally marry young. Consequently they must be subjected to occasional hard struggles for existence, and the favoured individuals will alone survive.

Turning to primeval times when men had only doubtfully attained the rank of manhood, they would probably have lived, as already stated, either as polygamists or temporarily as monogamists. Their intercourse, judging from analogy, would not then have been promiscuous. They would, no doubt, have defended their females to the best of their power from enemies of all kinds, and would probably have hunted for their subsistence, as well as for that of their offspring. The most powerful and able males would have succeeded best in the struggle for life and in obtaining attractive females. At this early period the progenitors of man, from having only feeble powers of reason, would not have looked forward to distant contingencies. They would have been governed more by their instincts and even less by their reason than are savages at the present day. They would not at that period have partially lost one of the strongest of all instincts, common to all the lower animals, namely the love of their young offspring; and

---

[14] Burchell says ('Travels in S. Africa, vol. ii. 1824, p. 58), that among the wild nations of Southern Africa, neither men nor women ever pass their lives in a state of celibacy. Azara ('Voyages dans l'Amérique Merid.' tom. ii. 1809, p. 21) makes precisely the same remark in regard to the wild Indians of South America.

consequently they would not have practised infanticide. There would have been no artificial scarcity of women, and polyandry would not have been followed; there would have been no early betrothals; women would not have been valued as mere slaves; both sexes, if the females as well as the males were permitted to exert any choice, would have chosen their partners, not for mental charms, or property, or social position, but almost solely from external appearance. All the adults would have married or paired, and all the offspring, as far as that was possible, would have been reared; so that the struggle for existence would have been periodically severe to an extreme degree. Thus during these primordial times all the conditions for sexual selection would have been much more favourable than at a later period, when man had advanced in his intellectual powers, but had retrograded in his instincts. Therefore, whatever influence sexual selection may have had in producing the differences between the races of man, and between man and the higher Quadrumana, this influence would have been much more powerful at a very remote period than at the present day.

On the Manner of Action of Sexual Selection with mankind.—With primeval men under the favourable conditions just stated, and with those savages who at the present time enter into any marriage tie (but subject to greater or less interference according as the habits of female infanticide, early betrothals, &c., are more or less practised), sexual selection will probably have acted in the following manner. The strongest and most vigorous men,—those who could best defend and hunt for their families, and during later times the chiefs or head-men,—those who were provided with the best weapons and who possessed the most property, such as a larger number of dogs or other animals, would have succeeded in rearing a greater average number of offspring, than would the weaker, poorer and lower members of the same tribes. There can, also, be no doubt that such men would generally have been able to select the more attractive women. At present the chiefs of nearly every tribe throughout the world succeed in obtaining more than one wife. Until recently, as I hear from Mr. Mantell, almost every girl in New Zealand, who was pretty, or promised to be pretty, was *tapu* to some chief. With the Kafirs, as Mr. C. Hamilton states,[15] "the chiefs generally have the pick "of the women for many miles round, and are most "persevering in establishing or confirming their privi"lege." We have seen that each race has its own style of beauty, and we know that it is natural to man to admire each characteristic point in his domestic animals, dress, ornaments, and personal appearance, when carried a little beyond the common standard. If then the several foregoing propositions be admitted, and I cannot see that they are doubtful, it would be an inexplicable circumstance, if the selection of the more attractive women by the more powerful men of each tribe, who would rear on an average a greater number of children, did not after the lapse of many generations modify to a certain extent the character of the tribe.

With our domestic animals, when a foreign breed is introduced into a new country, or when a native breed is long and carefully attended to, either for use or ornament, it is found after several generations to have undergone, whenever the means of comparison exist, a greater or less amount of change. This follows from unconscious selection during a long series of generations

---

[15] 'Anthropological Review,' Jan. 1870, p. xvi.

—that is, the preservation of the most approved individuals—without any wish or expectation of such a result on the part of the breeder. So again, if two careful breeders rear during many years animals of the same family, and do not compare them together or with a common standard, the animals are found after a time to have become to the surprise of their owners slightly different.[16] Each breeder has impressed, as Von Nathusius well expresses it, the character of his own mind — his own taste and judgment — on his animals. What reason, then, can be assigned why similar results should not follow from the long-continued selection of the most admired women by those men of each tribe, who were able to rear to maturity the greater number of children? This would be unconscious selection, for an effect would be produced, independently of any wish or expectation on the part of the men who preferred certain women to others.

Let us suppose the members of a tribe, in which some form of marriage was practised, to spread over an unoccupied continent; they would soon split up into distinct hordes, which would be separated from each other by various barriers, and still more effectually by the incessant wars between all barbarous nations. The hordes would thus be exposed to slightly different conditions and habits of life, and would sooner or later come to differ in some small degree. As soon as this occurred, each isolated tribe would form for itself a slightly different standard of beauty;[17] and then un-

[16] 'The Variation of Animals and Plants under Domestication,' vol. ii. p. 210-217.

[17] An ingenious writer argues, from a comparison of the pictures of Raphael, Rubens, and modern French artists, that the idea of beauty is not absolutely the same even throughout Europe: see the 'Lives of Haydn and Mozart,' by M. Bombet, English translat. p. 278.

conscious selection would come into action through the more powerful and leading savages preferring certain women to others. Thus the differences between the tribes, at first very slight, would gradually and inevitably be increased to a greater and greater degree.

With animals in a state of nature, many characters proper to the males, such as size, strength, special weapons, courage and pugnacity, have been acquired through the law of battle. The semi-human progenitors of man, like their allies the Quadrumana, will almost certainly have been thus modified; and, as savages still fight for the possession of their women, a similar process of selection has probably gone on in a greater or less degree to the present day. Other characters proper to the males of the lower animals, such as bright colours and various ornaments, have been acquired by the more attractive males having been preferred by the females. There are, however, exceptional cases in which the males, instead of having been the selected, have been the selectors. We recognise such cases by the females having been rendered more highly ornamented than the males,—their ornamental characters having been transmitted exclusively or chiefly to their female offspring. One such case has been described in the order to which man belongs, namely, with the Rhesus monkey.

Man is more powerful in body and mind than woman, and in the savage state he keeps her in a far more abject state of bondage than does the male of any other animal; therefore it is not surprising that he should have gained the power of selection. Women are everywhere conscious of the value of their beauty; and when they have the means, they take more delight in decorating themselves with all sorts of ornaments than do men. They borrow the plumes of male birds, with which nature decked this sex in order to charm the females. As women have long been selected for beauty, it is not surprising that some of the successive variations should have been transmitted in a limited manner; and consequently that women should have transmitted their beauty in a somewhat higher degree to their female than to their male offspring. Hence women have become more beautiful, as most persons will admit, than men. Women, however, certainly transmit most of their characters, including beauty, to their offspring of both sexes; so that the continued preference by the men of each race of the more attractive women, according to their standard of taste, would tend to modify in the same manner all the individuals of both sexes belonging to the race.

With respect to the other form of sexual selection (which with the lower animals is much the most common), namely, when the females are the selectors, and accept only those males which excite or charm them most, we have reason to believe that it formerly acted on the progenitors of man. Man in all probability owes his beard, and perhaps some other characters, to inheritance from an ancient progenitor who gained in this manner his ornaments. But this form of selection may have occasionally acted during later times; for in utterly barbarous tribes the women have more power in choosing, rejecting, and tempting their lovers, or of afterwards changing their husbands, than might have been expected. As this is a point of some importance, I will give in detail such evidence as I have been able to collect.

Hearne describes how a woman in one of the tribes of Arctic America repeatedly ran away from her husband and joined a beloved man; and with the Charruas of S. America, as Azara states, the power of divorce is perfectly free. With the Abipones, when a man chooses a wife he bargains with the parents about the price. But "it frequently happens that the "girl rescinds what has been agreed upon between the "parents and the bridegroom, obstinately rejecting the "very mention of marriage." She often runs away, hides herself, and thus eludes the bridegroom. In the Fiji Islands the man seizes on the woman whom he wishes for his wife by actual or pretended force; but "on reaching the home of her abductor, should she not "approve of the match, she runs to some one who can "protect her; if, however, she is satisfied, the matter is "settled forthwith." In Tierra del Fuego a young man first obtains the consent of the parents by doing them some service, and then he attempts to carry off the girl; "but if she is unwilling, she hides herself "in the woods until her admirer is heartily tired of "looking for her, and gives up the pursuit; but this "seldom happens." With the Kalmucks there is a regular race between the bride and bridegroom, the former having a fair start; and Clarke "was assured "that no instance occurs of a girl being caught, unless "she has a partiality to the pursuer." So with the wild tribes of the Malay archipelago there is a similar racing match; and it appears from M. Bourien's account, as Sir J. Lubbock remarks, that "the race 'is not to "'the swift, nor the battle to the strong,' but to the "young man who has the good fortune to please his "intended bride."

Turning to Africa: the Kafirs buy their wives, and girls are severely beaten by their fathers if they will not accept a chosen husband; yet it is manifest from many facts given by the Rev. Mr. Shooter, that they have considerable power of choice. Thus very ugly, though rich men, have been known to fail in getting wives. The girls, before consenting to be betrothed, compel the men to shew themselves off, first in front and then behind, and "exhibit their paces." They have been known to propose to a man, and they not rarely run away with a favoured lover. With the degraded bush-women of S. Africa, "when a girl has "grown up to womanhood without having been be-"trothed, which, however, does not often happen, her "lover must gain her approbation, as well as that of "the parents."[18] Mr. Winwood Reade made inquiries for me with respect to the negroes of Western Africa, and he informs me that "the women, at least among "the more intelligent Pagan tribes, have no difficulty "in getting the husbands whom they may desire, al-"though it is considered unwomanly to ask a man to "marry them. They are quite capable of falling in "love, and of forming tender, passionate, and faithful "attachments."

We thus see that with savages the women are not in quite so abject a state in relation to marriage as has often been supposed. They can tempt the men whom they prefer, and can sometimes reject those whom they dislike, either before or after marriage. Preference on the part of the women, steadily acting in any one direction, would ultimately affect the character of the tribe; for the women would generally choose not merely the handsomer men, according to their standard of taste,

[18] Azara, 'Voyages,' &c. tom. ii. p. 23. Dobrizhoffer, 'An Account of the Abipones,' vol. ii. 1822, p. 207. Williams on the Fiji Islanders, as quoted by Lubbock, 'Origin of Civilisation,' 1870, p. 79. On the Fuegians, King and FitzRoy, 'Voyages of the *Adventure* and *Beagle*,' vol. ii. 1839, p. 182. On the Kalmucks, quoted by M'Lennan, 'Primitive Marriage,' 1865, p. 32. On the Malays, Lubbock, ibid. p. 76. The Rev. J. Shooter, 'On the Kafirs of Natal,' 1857, p. 52-60. On the Bush-women, Burchell, 'Travels in S. Africa,' vol. ii. 1824, p. 59.

but those who were at the same time best able to defend and support them. Such well-endowed pairs would commonly rear a larger number of offspring than the less well endowed. The same result would obviously follow in a still more marked manner if there was selection on both sides; that is if the more attractive, and at the same time more powerful men were to prefer, and were preferred by, the more attractive women. And these two forms of selection seem actually to have occurred, whether or not simultaneously, with mankind, especially during the earlier periods of our long history.

We will now consider in a little more detail, relatively to sexual selection, some of the characters which distinguish the several races of man from each other and from the lower animals, namely, the more or less complete absence of hair from the body and the colour of the skin. We need say nothing about the great diversity in the shape of the features and of the skull between the different races, as we have seen in the last chapter how different is the standard of beauty in these respects. These characters will therefore probably have been acted on through sexual selection; but we have no means of judging, as far as I can see, whether they have been acted on chiefly through the male or female side. The musical faculties of man have likewise been already discussed.

Absence of Hair on the Body, and its Development on the Face and Head.—From the presence of the woolly hair or lanugo on the human fœtus, and of rudimentary hairs scattered over the body during maturity, we may infer that man is descended from some animal which was born hairy and remained so during life. The loss of hair is an inconvenience and probably an injury to man even under a hot climate, for he is thus exposed to sudden chills, especially during wet weather. As Mr. Wallace remarks, the natives in all countries are glad to protect their naked backs and shoulders with some slight covering. No one supposes that the nakedness of the skin is any direct advantage to man, so that his body cannot have been divested of hair through natural selection.[19] Nor have we any grounds for believing, as shewn in a former chapter, that this can be due to the direct action of the conditions to which man has long been exposed, or that it is the result of correlated development.

The absence of hair on the body is to a certain extent a secondary sexual character; for in all parts of the world women are less hairy than men. Therefore we may reasonably suspect that this is a character which has been gained through sexual selection. We know that the faces of several species of monkeys, and large surfaces at the posterior end of the body in other species, have been denuded of hair; and this we may safely attribute to sexual selection, for these surfaces are not only vividly coloured, but sometimes, as with the male mandrill and female rhesus, much more

---

19 'Contributions to the Theory of Natural Selection,' 1870, p. 346. Mr. Wallace believes (p. 350) "that some intelligent power has guided "or determined the development of man;" and he considers the hairless condition of the skin as coming under this head. The Rev. T. R. Stebbing, in commenting on this view ('Transactions of Devonshire Assoc. for Science,' 1870) remarks, that had Mr. Wallace "employed "his usual ingenuity on the question of man's hairless skin, he might "have seen the possibility of its selection through its superior beauty "or the health attaching to superior cleanliness. At any rate it is "surprising that he should picture to himself a superior intelligence "plucking the hair from the backs of savage men (to whom, according "to his own account it would have been useful and beneficial), in order "that the descendants of the poor shorn wretches might after many "deaths from cold and damp in the course of many generations," have been forced to raise themselves in the scale of civilisation through the practice of various arts, in the manner indicated by Mr. Wallace.

vividly in the one sex than in the other. As these animals gradually reach maturity the naked surfaces, as I am informed by Mr. Bartlett, grow larger, relatively to the size of their bodies. The hair, however, appears to have been removed in these cases, not for the sake of nudity, but that the colour of the skin should be more fully displayed. So again with many birds the head and neck have been divested of feathers through sexual selection, for the sake of exhibiting the brightly-coloured skin.

As woman has a less hairy body than man, and as this character is common to all races, we may conclude that our female semi-human progenitors were probably first partially divested of hair; and that this occurred at an extremely remote period before the several races had diverged from a common stock. As our female progenitors gradually acquired this new character of nudity, they must have transmitted it in an almost equal degree to their young offspring of both sexes; so that its transmission, as in the case of many ornaments with mammals and birds, has not been limited either by age or sex. There is nothing surprising in a partial loss of hair having been esteemed as ornamental by the ape-like progenitors of man, for we have seen that with animals of all kinds innumerable strange characters have been thus esteemed, and have consequently been modified through sexual selection. Nor is it surprising that a character in a slight degree injurious should have been thus acquired; for we know that this is the case with the plumes of some birds, and with the horns of some stags.

The females of certain anthropoid apes, as stated in a former chapter, are somewhat less hairy on the under surface than are the males; and here we have what might have afforded a commencement for the process of denudation. With respect to the completion of the process through sexual selection, it is well to bear in mind the New Zealand proverb, "there is no woman "for a hairy man." All who have seen photographs of the Siamese hairy family will admit how ludicrously hideous is the opposite extreme of excessive hairiness. Hence the king of Siam had to bribe a man to marry the first hairy woman in the family, who transmitted this character to her young offspring of both sexes.[20]

Some races are much more hairy than others, especially on the male side; but it must not be assumed that the more hairy races, for instance Europeans, have retained a primordial condition more completely than have the naked races, such as the Kalmucks or Americans. It is a more probable view that the hairiness of the former is due to partial reversion, for characters which have long been inherited are always apt to return. It does not appear that a cold climate has been influential in leading to this kind of reversion; excepting perhaps with the negroes, who have been reared during several generations, in the United States,[21] and possibly with the Ainos, who inhabit the northern

[20] 'The Variation of Animals and Plants under Domestication,' vol. ii. 1868, p. 327.

[21] 'Investigations into Military and Anthropological Statistics of American Soldiers,' by B. A. Gould, 1869; p. 568:—Observations were carefully made on the pilosity of 2129 black and coloured soldiers, whilst they were bathing; and by looking to the published table, "it "is manifest at a glance that there is but little, if any, difference "between the white and the black races in this respect." It is, however, certain that negroes in their native and much hotter land of Africa, have remarkably smooth bodies. It should be particularly observed, that pure blacks and mulattoes were included in the above enumeration; and this is an unfortunate circumstance, as in accordance with the principle, the truth of which I have elsewhere proved, crossed races would be eminently liable to revert to the primordial hairy character of their early ape-like progenitors.

islands of the Japan archipelago. But the laws of inheritance are so complex than we can seldom understand their action. If the greater hairiness of certain races be the result of reversion, unchecked by any form of selection, the extreme variability of this character, even within the limits of the same race, ceases to be remarkable.

With respect to the beard, if we turn to our best guide, namely the Quadrumana, we find beards equally well developed in both sexes of many species, but in others, either confined to the males, or more developed in them than in the females. From this fact, and from the curious arrangement, as well as the bright colours, of the hair about the heads of many monkeys, it is highly probable, as before explained, that the males first acquired their beards as an ornament through sexual selection, transmitting them in most cases, in an equal or nearly equal degree, to their offspring of both sexes. We know from Eschricht[22] that with mankind, the female as well as the male fœtus is furnished with much hair on the face, especially round the mouth; and this indicates that we are descended from a progenitor, of which both sexes were bearded. It appears therefore at first sight probable that man has retained his beard from a very early period, whilst woman lost her beard at the same time when her body became almost completely divested of hair. Even the colour of the beard with mankind seems to have been inherited from an ape-like progenitor; for when there is any difference in tint between the hair of the head and the beard, the latter is lighter coloured in all monkeys and in man. There is less improbability in the men of the bearded

---

22 "Ueber die Richtung der Haare am Menschlichen Körper," in Müller's 'Archiv für Anat. und Phys.' 1837, s. 40.

races having retained their beards from primordial times, than in the case of the hair on the body; for with those Quadrumana, in which the male has a larger beard than that of the female, it is fully developed only at maturity, and the later stages of development may have been exclusively transmitted to mankind. We should then see what is actually the case, namely, our male children, before they arrive at maturity, as destitute of beards as are our female children. On the other hand the great variability of the beard within the limits of the same race and in different races indicates that reversion has come into action. However this may be, we must not overlook the part which sexual selection may have played even during later times; for we know that with savages, the men of the beardless races take infinite pains in eradicating every hair from their faces, as something odious, whilst the men of the bearded races feel the greatest pride in their beards. The women, no doubt, participate in these feelings, and if so sexual selection can hardly have failed to have effected something in the course of later times.[23]

It is rather difficult to form a judgment how the long

[23] Mr. Sproat ('Scenes and Studies of Savage Life,' 1868, p. 25) suggests, with reference to the beardless natives of Vancouver's Island, that the custom of plucking out the hairs on the face, "continued from "one generation to another, would perhaps at last produce a race "distinguishable by a thin and straggling growth of beard." But the custom would not have arisen until the beard had already become, from some independent cause, greatly reduced. Nor have we any direct evidence that the continued eradication of the hair would lead to any inherited effect. Owing to this cause of doubt, I have not hitherto alluded to the belief held by some distinguished ethnologists, for instance M. Gosse of Geneva, that artificial modifications of the skull tend to be inherited. I have no wish to dispute this conclusion; and we now know from Dr. Brown-Séquard's remarkable observations, especially those recently communicated (1870) to the British Association, that with guinea-pigs the effects of operations are inherited.

hair on our heads became developed. Eschricht[24] states that in the human fœtus the hair on the face during the fifth month is longer than that on the head; and this indicates that our semi-human progenitors were not furnished with long tresses, which consequently must have been a late acquisition. This is likewise indicated by the extraordinary difference in the length of the hair in the different races; in the negro the hair forms a mere curly mat; with us it is of great length, and with the American natives it not rarely reaches to the ground. Some species of Semnopithecus have their heads covered with moderately long hair, and this probably serves as an ornament and was acquired through sexual selection. The same view may be extended to mankind, for we know that long tresses are now and were formerly much admired, as may be observed in the works of almost every poet; St. Paul says, "if a woman "have long hair, it is a glory to her;" and we have seen that in North America a chief was elected solely from the length of his hair.

Colour of the Skin.—The best kind of evidence that the colour of the skin has been modified through sexual selection is wanting in the case of mankind; for the sexes do not differ in this respect, or only slightly and doubtfully. On the other hand we know from many facts already given that the colour of the skin is regarded by the men of all races as a highly important element in their beauty; so that it is a character which would be likely to be modified through selection, as has occurred in innumerable instances with the lower animals. It seems at first sight a monstrous supposition that the jet blackness of the negro has been gained

[24] 'Ueber die Richtung,' ibid. s. 40.

through sexual selection; but this view is supported by various analogies, and we know that negroes admire their own blackness. With mammals, when the sexes differ in colour, the male is often black or much darker than the female; and it depends merely on the form of inheritance whether this or any other tint shall be transmitted to both sexes or to one alone. The resemblance of *Pithecia satanas* with his jet black skin, white rolling eyeballs, and hair parted on the top of the head, to a negro in miniature, is almost ludicrous.

The colour of the face differs much more widely in the various kinds of monkeys than it does in the races of man; and we have good reason to believe that the red, blue, orange, almost white and black tints of their skin, even when common to both sexes, and the bright colours of their fur, as well as the ornamental tufts of hair about the head, have all been acquired through sexual selection. As the newly-born infants of the most distinct races do not differ nearly as much in colour as do the adults, although their bodies are completely destitute of hair, we have some slight indication that the tints of the different races were acquired subsequently to the removal of the hair, which, as before stated, must have occurred at a very early period.

Summary.—We may conclude that the greater size, strength, courage, pugnacity, and even energy of man, in comparison with the same qualities in woman, were acquired during primeval times, and have subsequently been augmented, chiefly through the contests of rival males for the possession of the females. The greater intellectual vigour and power of invention in man is probably due to natural selection combined with the inherited effects of habit, for the most able men will have succeeded best in defending and providing for themselves, their wives and offspring. As far as the extreme intricacy of the subject permits us to judge, it appears that our male ape-like progenitors acquired their beards as an ornament to charm or excite the opposite sex, and transmitted them to man as he now exists. The females apparently were first denuded of hair in like manner as a sexual ornament; but they transmitted this character almost equally to both sexes. It is not improbable that the females were modified in other respects for the same purpose and through the same means; so that women have acquired sweeter voices and become more beautiful than men.

It deserves particular attention that with mankind all the conditions for sexual selection were much more favourable, during a very early period, when man had only just attained to the rank of manhood, than during later times. For he would then, as we may safely conclude, have been guided more by his instinctive passions, and less by foresight or reason. He would not then have been so utterly licentious as many savages now are; and each male would have jealously guarded his wife or wives. He would not then have practised infanticide; nor valued his wives merely as useful slaves; nor have been betrothed to them during infancy. Hence we may infer that the races of men were differentiated, as far as sexual selection is concerned, in chief part during a very remote epoch; and this conclusion throws light on the remarkable fact that at the most ancient period, of which we have as yet obtained any record, the races of man had already come to differ nearly or quite as much as they do at the present day.

The views here advanced, on the part which sexual selection has played in the history of man, want scientific precision. He who does not admit this agency in the case of the lower animals, will properly disregard all that I have written in the later chapters on man. We cannot positively say that this character, but not that, has been thus modified; it has, however, been shewn that the races of man differ from each other and from their nearest allies amongst the lower animals, in certain characters which are of no service to them in their ordinary habits of life, and which it is extremely probable would have been modified through sexual selection. We have seen that with the lowest savages the people of each tribe admire their own characteristic qualities,—the shape of the head and face, the squareness of the cheek-bones, the prominence or depression of the nose, the colour of the skin, the length of the hair on the head, the absence of hair on the face and body, or the presence of a great beard, and so forth. Hence these and other such points could hardly fail to have been slowly and gradually exaggerated from the more powerful and able men in each tribe, who would succeed in rearing the largest number of offspring, having selected during many generations as their wives the most strongly characterised and therefore most attractive women. For my own part I conclude that of all the causes which have led to the differences in external appearance between the races of man, and to a certain extent between man and the lower animals, sexual selection has been by far the most efficient.

CHAPTER XXI.

GENERAL SUMMARY AND CONCLUSION.

Main conclusion that man is descended from some lower form—Manner of development—Genealogy of man—Intellectual and moral faculties—Sexual selection—Concluding remarks.

A BRIEF summary will here be sufficient to recall to the reader's mind the more salient points in this work. Many of the views which have been advanced are highly speculative, and some no doubt will prove erroneous; but I have in every case given the reasons which have led me to one view rather than to another. It seemed worth while to try how far the principle of evolution would throw light on some of the more complex problems in the natural history of man. False facts are highly injurious to the progress of science, for they often long endure; but false views, if supported by some evidence, do little harm, as every one takes a salutary pleasure in proving their falseness; and when this is done, one path towards error is closed and the road to truth is often at the same time opened.

The main conclusion arrived at in this work, and now held by many naturalists who are well competent to form a sound judgment, is that man is descended from some less highly organised form. The grounds upon which this conclusion rests will never be shaken, for the close similarity between man and the lower animals in embryonic development, as well as in innumerable points of structure and constitution, both of high and of the most trifling importance,—the rudiments which he retains, and the abnormal reversions to which he is occasionally liable,—are facts which cannot be disputed. They have long been known, but until recently they told us nothing with respect to the origin of man. Now when viewed by the light of our knowledge of the whole organic world, their meaning is unmistakeable. The great principle of evolution stands up clear and firm, when these groups of facts are considered in connection with others, such as the mutual affinities of the members of the same group, their geographical distribution in past and present times, and their geological succession. It is incredible that all these facts should speak falsely. He who is not content to look, like a savage, at the phenomena of nature as disconnected, cannot any longer believe that man is the work of a separate act of creation. He will be forced to admit that the close resemblance of the embryo of man to that, for instance, of a dog—the construction of his skull, limbs, and whole frame, independently of the uses to which the parts may be put, on the same plan with that of other mammals—the occasional reappearance of various structures, for instance of several distinct muscles, which man does not normally possess, but which are common to the Quadrumana—and a crowd of analogous facts—all point in the plainest manner to the conclusion that man is the co-descendant with other mammals of a common progenitor.

We have seen that man incessantly presents individual differences in all parts of his body and in his mental faculties. These differences or variations seem to be induced by the same general causes, and to obey the same laws as with the lower animals. In both cases similar laws of inheritance prevail. Man tends to increase at a greater rate than his means of subsistence;

consequently he is occasionally subjected to a severe struggle for existence, and natural selection will have effected whatever lies within its scope. A succession of strongly-marked variations of a similar nature are by no means requisite; slight fluctuating differences in the individual suffice for the work of natural selection. We may feel assured that the inherited effects of the long-continued use or disuse of parts will have done much in the same direction with natural selection. Modifications formerly of importance, though no longer of any special use, will be long inherited. When one part is modified, other parts will change through the principle of correlation, of which we have instances in many curious cases of correlated monstrosities. Something may be attributed to the direct and definite action of the surrounding conditions of life, such as abundant food, heat, or moisture; and lastly, many characters of slight physiological importance, some indeed of considerable importance, have been gained through sexual selection.

No doubt man, as well as every other animal, presents structures, which as far as we can judge with our little knowledge, are not now of any service to him, nor have been so during any former period of his existence, either in relation to his general conditions of life, or of one sex to the other. Such structures cannot be accounted for by any form of selection, or by the inherited effects of the use and disuse of parts. We know, however, that many strange and strongly-marked peculiarities of structure occasionally appear in our domesticated productions, and if the unknown causes which produce them were to act more uniformly, they would probably become common to all the individuals of the species. We may hope hereafter to understand something about the causes of such occasional modifications, especially through the study of monstrosities: hence the labours of experimentalists, such as those of M. Camille Dareste, are full of promise for the future. In the greater number of cases we can only say that the cause of each slight variation and of each monstrosity lies much more in the nature or constitution of the organism, than in the nature of the surrounding conditions; though new and changed conditions certainly play an important part in exciting organic changes of all kinds.

Through the means just specified, aided perhaps by others as yet undiscovered, man has been raised to his present state. But since he attained to the rank of manhood, he has diverged into distinct races, or as they may be more appropriately called sub-species. Some of these, for instance the Negro and European, are so distinct that, if specimens had been brought to a naturalist without any further information, they would undoubtedly have been considered by him as good and true species. Nevertheless all the races agree in so many unimportant details of structure and in so many mental peculiarities, that these can be accounted for only through inheritance from a common progenitor; and a progenitor thus characterised would probably have deserved to rank as man.

It must not be supposed that the divergence of each race from the other races, and of all the races from a common stock, can be traced back to any one pair of progenitors. On the contrary, at every stage in the process of modification, all the individuals which were in any way best fitted for their conditions of life, though in different degrees, would have survived in greater numbers than the less well fitted. The process would have been like that followed by man, when he does not intentionally select particular individuals, but breeds from all the superior and neglects all the inferior individuals. He thus slowly but surely modifies his stock, and unconsciously forms a new strain. So with respect to modifications, acquired independently of selection, and due to variations arising from the nature of the organism and the action of the surrounding conditions, or from changed habits of life, no single pair will have been modified in a much greater degree than the other pairs which inhabit the same country, for all will have been continually blended through free intercrossing.

By considering the embryological structure of man, —the homologies which he presents with the lower animals,—the rudiments which he retains,—and the reversions to which he is liable, we can partly recall in imagination the former condition of our early progenitors; and can approximately place them in their proper position in the zoological series. We thus learn that man is descended from a hairy quadruped, furnished with a tail and pointed ears, probably arboreal in its habits, and an inhabitant of the Old World. This creature, if its whole structure had been examined by a naturalist, would have been classed amongst the Quadrumana, as surely as would the common and still more ancient progenitor of the Old and New World monkeys. The Quadrumana and all the higher mammals are probably derived from an ancient marsupial animal, and this through a long line of diversified forms, either from some reptile-like or some amphibian-like creature, and this again from some fish-like animal. In the dim obscurity of the past we can see that the early progenitor of all the Vertebrata must have been an aquatic animal, provided with branchiæ, with the two sexes united in the same individual, and with the most important organs of the body (such as the brain and heart) imperfectly developed. This animal seems to have been more like the larvæ of our existing marine Ascidians than any other known form.

The greatest difficulty which presents itself, when we are driven to the above conclusion on the origin of man, is the high standard of intellectual power and of moral disposition which he has attained. But every one who admits the general principle of evolution, must see that the mental powers of the higher animals, which are the same in kind with those of mankind, though so different in degree, are capable of advancement. Thus the interval between the mental powers of one of the higher apes and of a fish, or between those of an ant and scale-insect, is immense. The development of these powers in animals does not offer any special difficulty; for with our domesticated animals, the mental faculties are certainly variable, and the variations are inherited. No one doubts that these faculties are of the utmost importance to animals in a state of nature. Therefore the conditions are favourable for their development through natural selection. The same conclusion may be extended to man; the intellect must have been all-important to him, even at a very remote period, enabling him to use language, to invent and make weapons, tools, traps, &c.; by which means, in combination with his social habits, he long ago became the most dominant of all living creatures.

A great stride in the development of the intellect will have followed, as soon as, through a previous considerable advance, the half-art and half-instinct of language came into use; for the continued use of language will have reacted on the brain, and produced an inherited effect; and this again will have reacted on the improvement of language. The large size of the brain in man, in comparison with that of the lower animals, relatively to the size of their bodies, may be attributed in chief part, as Mr. Chauncey Wright has well remarked,[1] to the early use of some simple form of language,—that wonderful engine which affixes signs to all sorts of objects and qualities, and excites trains of thought which would never arise from the mere impression of the senses, and if they did arise could not be followed out. The higher intellectual powers of man, such as those of ratiocination, abstraction, self-consciousness, &c., will have followed from the continued improvement of other mental faculties; but without considerable culture of the mind, both in the race and in the individual, it is doubtful whether these high powers would be exercised, and thus fully attained.

The development of the moral qualities is a more interesting and difficult problem. Their foundation lies in the social instincts, including in this term the family ties. These instincts are of a highly complex nature, and in the case of the lower animals give special tendencies towards certain definite actions; but the more important elements for us are love, and the distinct emotion of sympathy. Animals endowed with the social instincts take pleasure in each other's company, warn each other of danger, defend and aid each other in many ways. These instincts are not extended to all the individuals of the species, but only to those of the same community. As they are highly beneficial to the species, they have in all probability been acquired through natural selection.

A moral being is one who is capable of comparing

---

[1] On the "Limits of Natural Selection," in the 'North American Review,' Oct. 1870, p. 295.

his past and future actions and motives,—of approving of some and disapproving of others; and the fact that man is the one being who with certainty can be thus designated makes the greatest of all distinctions between him and the lower animals. But in our third chapter I have endeavoured to shew that the moral sense follows, firstly, from the enduring and always present nature of the social instincts, in which respect man agrees with the lower animals; and secondly, from his mental faculties being highly active and his impressions of past events extremely vivid, in which respects he differs from the lower animals. Owing to this condition of mind, man cannot avoid looking backwards and comparing the impressions of past events and actions. He also continually looks forward. Hence after some temporary desire or passion has mastered his social instincts, he will reflect and compare the now weakened impression of such past impulses, with the ever present social instinct; and he will then feel that sense of dissatisfaction which all unsatisfied instincts leave behind them. Consequently he resolves to act differently for the future—and this is conscience. Any instinct which is permanently stronger or more enduring than another, gives rise to a feeling which we express by saying that it ought to be obeyed. A pointer dog, if able to reflect on his past conduct, would say to himself, I ought (as indeed we say of him) to have pointed at that hare and not have yielded to the passing temptation of hunting it.

Social animals are partly impelled by a wish to aid the members of the same community in a general manner, but more commonly to perform certain definite actions. Man is impelled by the same general wish to aid his fellows, but has few or no special instincts. He differs also from the lower animals in being able to express his desires by words, which thus become the guide to the aid required and bestowed. The motive to give aid is likewise somewhat modified in man: it no longer consists solely of a blind instinctive impulse, but is largely influenced by the praise or blame of his fellow men. Both the appreciation and the bestowal of praise and blame rest on sympathy; and this emotion, as we have seen, is one of the most important elements of the social instincts. Sympathy, though gained as an instinct, is also much strengthened by exercise or habit. As all men desire their own happiness, praise or blame is bestowed on actions and motives, according as they lead to this end; and as happiness is an essential part of the general good, the greatest-happiness principle indirectly serves as a nearly safe standard of right and wrong. As the reasoning powers advance and experience is gained, the more remote effects of certain lines of conduct on the character of the individual, and on the general good, are perceived; and then the self-regarding virtues, from coming within the scope of public opinion, receive praise, and their opposites receive blame. But with the less civilised nations reason often errs, and many bad customs and base superstitions come within the same scope, and consequently are esteemed as high virtues, and their breach as heavy crimes.

The moral faculties are generally esteemed, and with justice, as of higher value than the intellectual powers. But we should always bear in mind that the activity of the mind in vividly recalling past impressions is one of the fundamental though secondary bases of conscience. This fact affords the strongest argument for educating and stimulating in all possible ways the intellectual faculties of every human being. No doubt a man with a torpid mind, if his social affections and sympathies are well developed, will be led to good actions, and may have a fairly sensitive conscience. But whatever renders the imagination of men more vivid and strengthens the habit of recalling and comparing past impressions, will make the conscience more sensitive, and may even compensate to a certain extent for weak social affections and sympathies.

The moral nature of man has reached the highest standard as yet attained, partly through the advancement of the reasoning powers and consequently of a just public opinion, but especially through the sympathies being rendered more tender and widely diffused through the effects of habit, example, instruction, and reflection. It is not improbable that virtuous tendencies may through long practice be inherited. With the more civilised races, the conviction of the existence of an all-seeing Deity has had a potent influence on the advancement of morality. Ultimately man no longer accepts the praise or blame of his fellows as his chief guide, though few escape this influence, but his habitual convictions controlled by reason afford him the safest rule. His conscience then becomes his supreme judge and monitor. Nevertheless the first foundation or origin of the moral sense lies in the social instincts, including sympathy; and these instincts no doubt were primarily gained, as in the case of the lower animals, through natural selection.

The belief in God has often been advanced as not only the greatest, but the most complete of all the distinctions between man and the lower animals. It is however impossible, as we have seen, to maintain that this belief is innate or instinctive in man. On the other hand a belief in all-pervading spiritual agencies seems to be universal; and apparently follows from a considerable advance in the reasoning powers of man, and from a still greater advance in his faculties of imagination, curiosity and wonder. I am aware that the assumed instinctive belief in God has been used by many persons as an argument for His existence. But this is a rash argument, as we should thus be compelled to believe in the existence of many cruel and malignant spirits, possessing only a little more power than man; for the belief in them is far more general than of a beneficent Deity. The idea of a universal and beneficent Creator of the universe does not seem to arise in the mind of man, until he has been elevated by long-continued culture.

He who believes in the advancement of man from some lowly-organised form, will naturally ask how does this bear on the belief in the immortality of the soul. The barbarous races of man, as Sir J. Lubbock has shewn, possess no clear belief of this kind; but arguments derived from the primeval beliefs of savages are, as we have just seen, of little or no avail. Few persons feel any anxiety from the impossibility of determining at what precise period in the development of the individual, from the first trace of the minute germinal vesicle to the child either before or after birth, man becomes an immortal being; and there is no greater cause for anxiety because the period in the gradually ascending organic scale cannot possibly be determined.[2]

I am aware that the conclusions arrived at in this work will be denounced by some as highly irreligious; but he who thus denounces them is bound to shew why it is more irreligious to explain the origin of man as a distinct species by descent from some lower form,

[2] The Rev. J. A. Picton gives a discussion to this effect in his 'New Theories and the Old Faith,' 1870.

through the laws of variation and natural selection, than to explain the birth of the individual through the laws of ordinary reproduction. The birth both of the species and of the individual are equally parts of that grand sequence of events, which our minds refuse to accept as the result of blind chance. The understanding revolts at such a conclusion, whether or not we are able to believe that every slight variation of structure,—the union of each pair in marriage,—the dissemination of each seed,—and other such events, have all been ordained for some special purpose.

Sexual selection has been treated at great length in these volumes; for, as I have attempted to shew, it has played an important part in the history of the organic world. As summaries have been given to each chapter, it would be superfluous here to add a detailed summary. I am aware that much remains doubtful, but I have endeavoured to give a fair view of the whole case. In the lower divisions of the animal kingdom, sexual selection seems to have done nothing: such animals are often affixed for life to the same spot, or have the two sexes combined in the same individual, or what is still more important, their perceptive and intellectual faculties are not sufficiently advanced to allow of the feelings of love and jealousy, or of the exertion of choice. When, however, we come to the Arthropoda and Vertebrata, even to the lowest classes in these two great Sub-Kingdoms, sexual selection has effected much; and it deserves notice that we here find the intellectual faculties developed, but in two very distinct lines, to the highest standard, namely in the Hymenoptera (ants, bees, &c.) amongst the Arthropoda, and in the Mammalia, including man, amongst the Vertebrata.

In the most distinct classes of the animal kingdom,

with mammals, birds, reptiles, fishes, insects, and even crustaceans, the differences between the sexes follow almost exactly the same rules. The males are almost always the wooers; and they alone are armed with special weapons for fighting with their rivals. They are generally stronger and larger than the females, and are endowed with the requisite qualities of courage and pugnacity. They are provided, either exclusively or in a much higher degree than the females, with organs for producing vocal or instrumental music, and with odoriferous glands. They are ornamented with infinitely diversified appendages, and with the most brilliant or conspicuous colours, often arranged in elegant patterns, whilst the females are left unadorned. When the sexes differ in more important structures, it is the male which is provided with special sense-organs for discovering the female, with locomotive organs for reaching her, and often with prehensile organs for holding her. These various structures for securing or charming the female are often developed in the male during only part of the year, namely the breeding season. They have in many cases been transferred in a greater or less degree to the females; and in the latter case they appear in her as mere rudiments. They are lost by the males after emasculation. Generally they are not developed in the male during early youth, but appear a short time before the age for reproduction. Hence in most cases the young of both sexes resemble each other; and the female resembles her young offspring throughout life. In almost every great class a few anomalous cases occur in which there has been an almost complete transposition of the characters proper to the two sexes; the females assuming characters which properly belong to the males. This surprising uniformity in the laws regulating the differences between the sexes in so many and such widely separated classes, is intelligible if we admit the action throughout all the higher divisions of the animal kingdom of one common cause, namely sexual selection.

Sexual selection depends on the success of certain individuals over others of the same sex in relation to the propagation of the species; whilst natural selection depends on the success of both sexes, at all ages, in relation to the general conditions of life. The sexual struggle is of two kinds; in the one it is between the individuals of the same sex, generally the male sex, in order to drive away or kill their rivals, the females remaining passive; whilst in the other, the struggle is likewise between the individuals of the same sex, in order to excite or charm those of the opposite sex, generally the females, which no longer remain passive, but select the more agreeable partners. This latter kind of selection is closely analogous to that which man unintentionally, yet effectually, brings to bear on his domesticated productions, when he continues for a long time choosing the most pleasing or useful individuals, without any wish to modify the breed.

The laws of inheritance determine whether characters gained through sexual selection by either sex shall be transmitted to the same sex, or to both sexes; as well as the age at which they shall be developed. It appears that variations which arise late in life are commonly transmitted to one and the same sex. Variability is the necessary basis for the action of selection, and is wholly independent of it. It follows from this, that variations of the same general nature have often been taken advantage of and accumulated through sexual selection in relation to the propagation of the species, and through natural selection in relation to the general purposes of life. Hence secondary sexual characters, when equally transmitted to both sexes can be distinguished from ordinary specific characters only by the light of analogy. The modifications acquired through sexual selection are often so strongly pronounced that the two sexes have frequently been ranked as distinct species, or even as distinct genera. Such strongly-marked differences must be in some manner highly important; and we know that they have been acquired in some instances at the cost not only of inconvenience, but of exposure to actual danger.

The belief in the power of sexual selection rests chiefly on the following considerations. The characters which we have the best reason for supposing to have been thus acquired are confined to one sex; and this alone renders it probable that they are in some way connected with the act of reproduction. These characters in innumerable instances are fully developed only at maturity; and often during only a part of the year, which is always the breeding-season. The males (passing over a few exceptional cases) are the most active in courtship; they are the best armed, and are rendered the most attractive in various ways. It is to be especially observed that the males display their attractions with elaborate care in the presence of the females; and that they rarely or never display them excepting during the season of love. It is incredible that all this display should be purposeless. Lastly we have distinct evidence with some quadrupeds and birds that the individuals of the one sex are capable of feeling a strong antipathy or preference for certain individuals of the opposite sex.

Bearing these facts in mind, and not forgetting the marked results of man's unconscious selection, it seems to me almost certain that if the individuals of one sex were during a long series of generations to prefer pairing with certain individuals of the other sex, characterised in some peculiar manner, the offspring would slowly but surely become modified in this same manner. I have not attempted to conceal that, excepting when the males are more numerous than the females, or when polygamy prevails, it is doubtful how the more attractive males succeed in leaving a larger number of offspring to inherit their superiority in ornaments or other charms than the less attractive males; but I have shewn that this would probably follow from the females,—especially the more vigorous females which would be the first to breed, preferring not only the more attractive but at the same time the more vigorous and victorious males.

Although we have some positive evidence that birds appreciate bright and beautiful objects, as with the Bower-birds of Australia, and although they certainly appreciate the power of song, yet I fully admit that it is an astonishing fact that the females of many birds and some mammals should be endowed with sufficient taste for what has apparently been effected through sexual selection; and this is even more astonishing in the case of reptiles, fish, and insects. But we really know very little about the minds of the lower animals. It cannot be supposed that male Birds of Paradise or Peacocks, for instance, should take so much pains in erecting, spreading, and vibrating their beautiful plumes before the females for no purpose. We should remember the fact given on excellent authority in a former chapter, namely that several peahens, when debarred from an admired male, remained widows during a whole season rather than pair with another bird.

Nevertheless I know of no fact in natural history more wonderful than that the female Argus pheasant should be able to appreciate the exquisite shading of the ball-and-socket ornaments and the elegant patterns on the wing-feathers of the male. He who thinks that the male was created as he now exists must admit that the great plumes, which prevent the wings from being used for flight, and which, as well as the primary feathers, are displayed in a manner quite peculiar to this one species during the act of courtship, and at no other time, were given to him as an ornament. If so, he must likewise admit that the female was created and endowed with the capacity of appreciating such ornaments. I differ only in the conviction that the male Argus pheasant acquired his beauty gradually, through the females having preferred during many generations the more highly ornamented males; the æsthetic capacity of the females having been advanced through exercise or habit in the same manner as our own taste is gradually improved. In the male, through the fortunate chance of a few feathers not having been modified, we can distinctly see how simple spots with a little fulvous shading on one side might have been developed by small and graduated steps into the wonderful ball-and-socket ornaments; and it is probable that they were actually thus developed.

Everyone who admits the principle of evolution, and yet feels great difficulty in admitting that female mammals, birds, reptiles, and fish, could have acquired the high standard of taste which is implied by the beauty of the males, and which generally coincides with our own standard, should reflect that in each member of the vertebrate series the nerve-cells of the brain are the direct offshoots of those possessed by the common progenitor of the whole group. It thus becomes intelligible that the brain and mental faculties should be capable under similar conditions of nearly the same course of development, and consequently of performing nearly the same functions.

The reader who has taken the trouble to go through the several chapters devoted to sexual selection, will be able to judge how far the conclusions at which I have arrived are supported by sufficient evidence. If he accepts these conclusions, he may, I think, safely extend them to mankind; but it would be superfluous here to repeat what I have so lately said on the manner in which sexual selection has apparently acted on both the male and female side, causing the two sexes of man to differ in body and mind, and the several races to differ from each other in various characters, as well as from their ancient and lowly-organised progenitors.

He who admits the principle of sexual selection will be led to the remarkable conclusion that the cerebral system not only regulates most of the existing functions of the body, but has indirectly influenced the progressive development of various bodily structures and of certain mental qualities. Courage, pugnacity, perseverance, strength and size of body, weapons of all kinds, musical organs, both vocal and instrumental, bright colours, stripes and marks, and ornamental appendages, have all been indirectly gained by the one sex or the other, through the influence of love and jealousy, through the appreciation of the beautiful in sound, colour or form, and through the exertion of a choice; and these powers of the mind manifestly depend on the development of the cerebral system.

Man scans with scrupulous care the character and pedigree of his horses, cattle, and dogs before he matches them; but when he comes to his own marriage he rarely, or never, takes any such care. He is impelled by nearly the same motives as are the lower animals when left to their own free choice, though he is in so far superior to them that he highly values mental charms and virtues. On the other hand he is strongly attracted by mere wealth or rank. Yet he might by selection do something not only for the bodily constitution and frame of his offspring, but for their intellectual and moral qualities. Both sexes ought to refrain from marriage if in any marked degree inferior in body or mind; but such hopes are Utopian and will never be even partially realised until the laws of inheritance are thoroughly known. All do good service who aid towards this end. When the principles of breeding and of inheritance are better understood, we shall not hear ignorant members of our legislature rejecting with scorn a plan for ascertaining by an easy method whether or not consanguineous marriages are injurious to man.

The advancement of the welfare of mankind is a most intricate problem: all ought to refrain from marriage who cannot avoid abject poverty for their children; for poverty is not only a great evil, but tends to its own increase by leading to recklessness in marriage. On the other hand, as Mr. Galton has remarked, if the prudent avoid marriage, whilst the reckless marry, the inferior members will tend to supplant the better members of society. Man, like every other animal, has no doubt advanced to his present high condition through a struggle for existence consequent on his rapid multiplication; and if he is to advance still higher he must remain subject to a severe struggle. Otherwise he would soon sink into indolence, and the more highly-gifted men would not be more successful in the battle of life than the less gifted. Hence our natural rate of increase, though leading to many and obvious evils, must not be greatly diminished by any means. There should be open competition for all men; and the most able should not be prevented by laws or customs from succeeding best and rearing the largest number of offspring. Important as the struggle for existence has been and even still is, yet as far as the highest part of man's nature is concerned there are other agencies more important. For the moral qualities are advanced, either directly or indirectly, much more through the effects of habit, the reasoning powers, instruction, religion, &c., than through natural selection; though to this latter agency the social instincts, which afforded the basis for the development of the moral sense, may be safely attributed.

The main conclusion arrived at in this work, namely that man is descended from some lowly-organised form, will, I regret to think, be highly distasteful to many persons. But there can hardly be a doubt that we are descended from barbarians. The astonishment which I felt on first seeing a party of Fuegians on a wild and broken shore will never be forgotten by me, for the reflection at once rushed into my mind—such were our ancestors. These men were absolutely naked and bedaubed with paint, their long hair was tangled, their mouths frothed with excitement, and their expression was wild, startled, and distrustful. They possessed hardly any arts, and like wild animals lived on what they could catch; they had no government, and were merciless to every one not of their own small tribe. He who has seen a savage in his native land will not feel much shame, if forced to acknowledge that the blood of some more humble creature flows in his veins. For my own part I would as soon be descended from that heroic little monkey, who braved his dreaded enemy in order to save the life of his keeper; or from that old baboon, who, descending from the mountains, carried away in triumph his young comrade from a crowd of astonished dogs—as from a savage who delights to torture his enemies, offers up bloody sacrifices, practises infanticide without remorse, treats his wives like slaves, knows no decency, and is haunted by the grossest superstitions.

Man may be excused for feeling some pride at having risen, though not through his own exertions, to the very summit of the organic scale; and the fact of his having thus risen, instead of having been aboriginally placed there, may give him hopes for a still higher destiny in the distant future. But we are not here concerned with hopes or fears, only with the truth as far as our reason allows us to discover it. I have given the evidence to the best of my ability; and we must acknowledge, as it seems to me, that man with all his noble qualities, with sympathy which feels for the most debased, with benevolence which extends not only to other men but to the humblest living creature, with his god-like intellect which has penetrated into the movements and constitution of the solar system—with all these exalted powers—Man still bears in his bodily frame the indelible stamp of his lowly origin.

INDEX.

B.

C.

D.

E.

G.

H.

I.

J.

K.

L.

N.

P.

Q.

S.

T.

U.

V.

W.

X.

Y.

Z.

THE END.

TABLE

OF THE

PRINCIPAL ADDITIONS AND CORRECTIONS TO THE EDITION OF 1874.

1st Ed., 1871. In 2 Vols. — Vol. I.	2nd Ed., 1874. In 1 Vol.	2nd Ed., 1888. In 2 Vols. — Vol. I.	—
Page 22	Page 15–17	Page 21–23	Discussion on the rudimentary points in the human ear revised.
26	19	27	Cases of men born with hairy bodies.
27	20, note	29, note	Mantegazza on the last molar tooth in man.
29	23.	33	The rudiments of a tail in man.
32	24, note	35, note	Bianconi on homologous structures, as explained by adaptation on mechanical principles.
40	70	105	Intelligence in a baboon.
42	71	108	Sense of humour in dogs.
44	72, 73	110, 111	Further facts on imitation in man and animals.
47	75	114–119	Reasoning power in the lower animals.
50	80	122	Acquisition of experience by animals.
53	83	126–128	Power of abstraction in animals.
58	88, 89	135, 136	Power of forming concepts in relation to language.
64	92	140, 141	Pleasure from certain sounds, colours, and forms.
78	104	158	Fidelity in the elephant.
79	104	159	Galton on gregariousness of cattle.
81	105, 106	161, 162	Parental affection.
90	112, 113, note	172, 173, note	Persistence of enmity and hatred.
91	114	174	Nature and strength of shame, regret, and remorse.
94	117, note	179, note	Suicide amongst savages.
97	120, note	184, note	The motives of conduct.
112	28	42	Selection, as applied to primeval man.
122	35, 36	53, 54	Resemblances between idiots and animals.
124, note	39, note	58, note	Division of the malar bone.

1st Ed., 1871. In 2 Vols. — Vol. I.	2nd Ed., 1874. In 1 Vol.	2nd Ed., 1888. In 2 Vols. — Vol. I.	———
Page	Page	Page	
125, note	36–38, note	54–56, note	Supernumerary mammæ and digits.
128, 129	41, 42	62, 63	Further cases of muscles proper to animals appearing in man.
146	55, note	82, note	Broca: average capacity of skull diminished by the preservation of the inferior members of society.
149	57	86	Belt on advantages to man from his hairlessness.
150	58, 59	87, 88	Disappearance of the tail in man and certain monkeys.
169	134, 135	207, 208	Injurious forms of selection in civilised nations.
180	143	220	Indolence of man, when free from a struggle for existence.
193	151	234	Gorilla protecting himself from rain with his hands.
208, note	161, note	250, note	Hermaphroditism in fish.
209	163	252	Rudimentary mammæ in male mammals.
239	188–190	292	Changed conditions lessen fertility and cause ill-health amongst savages.
245	195, 196	303	Darkness of skin a protection against the sun.
250	199–206	309–318	Note by Professor Huxley on the development of the brain in man and apes.
256	209, 210	323	Special organs of male parasitic worms for holding the female.
275, 276	224, 225	346, 347	Greater variability of male than female; direct action of the environment in causing differences between the sexes.
290	235	363	Period of development of protuberances on birds' heads determines their transmission to one or both sexes.
301	243, 244	376, 377	Causes of excess of male births.
314	254	391	Proportion of the sexes in the bee family.
315	255, 256	393, 394	Excess of males perhaps sometimes determined by selection.
327	264	406, 407	Bright colours of lowly organised animals.
338	272	418, 419	Sexual selection amongst spiders.
339	273	420	Cause of smallness of male spiders.
345	277	426	Use of phosphorescence of the glow-worm.
349	280	431	The humming noises of flies.
350	281	432	Use of bright colours to Hemiptera (bugs).

1st Ed., 1871. In 2 Vols. — Vol. I.	2nd Ed., 1874. In 1 Vol.	2nd Ed., 1888. In 2 Vols. — Vol. I.	—
Page 351	Page 282	Page 433	Musical apparatus of Homoptera.
354	284, 285	437–439	Development of stridulating apparatus in Orthoptera.
359	288, note	443, note	
366	292, 293	449, 450	Hermann Müller on sexual differences of bees.
387	308	471	Sounds produced by moths.
397	315	482	Display of beauty by butterflies.
401	319	488, 489	Female butterflies, taking the more active part in courtship, brighter than their males.
412	324, 325	497, 498	Further cases of mimicry in butterflies and moths.
417	326	500, 501	Cause of bright and diversified colours of caterpillars.
Vol. II.		Vol. II.	
2	331	2	Brush-like scales of male Mallotus.
14	341	15–17	Further facts on courtship of fishes, and the spawning of Macropus.
23	347	25	Dufossé on the sounds made by fishes.
26	349	29	Belt on a frog protected by bright colouring.
30	352	34	Further facts on mental powers of snakes.
32	353	35	Sounds produced by snakes; the rattlesnake.
36	357	41	Combats of Chameleons.
72	383	79	Marshall on protuberances on birds' heads.
91	398	99–101	Further facts on display by the Argus pheasant.
108	411	120	Attachment between paired birds.
118	417	131	Female pigeon rejecting certain males.
120	419	133	Albino birds not finding partners, in a state of nature.
124	423	139	Direct action of climate on birds' colours.
147–50	438–441	161–166	Further facts on the ocelli in the Argus pheasant.
152	443	169	Display by humming-birds in courtship.
157	446	174	Cases with pigeons of colour transmitted to one sex alone.
232	495, 496	249, 250	Taste for the beautiful permanent enough to allow of sexual selection with the lower animals.
247	505	265	Horns of sheep originally a masculine character.
248	506	266	Castration affecting horns of animals.
256	513, 514	277	Prong-horned variety of *Cervus virginianus*.

1st Ed., 1871. In 2 Vols. — Vol. I.	2nd Ed., 1874. In 1 Vol.	2nd Ed., 1888. In 2 Vols. — Vol. I.	—
Page 260	Page 516	Page 281	Relative sizes of male and female whales and seals.
266	521	288	Absence of tusks in male miocene pigs.
286	534	309	Dobson on sexual differences of bats.
299	542, 543	322, 323	Reeks on advantage from peculiar colouring.
316	556	341	Difference of complexion in men and women of an African tribe.
337	572	365	Speech subsequent to singing.
356	586	387	Schopenhauer on importance of courtship to mankind.
359 *et seq.*	588 *et seq.*	391 *et seq.*	Revision of discussion on communal marriages and promiscuity.
373	598, 599	406–409	Power of choice of woman in marriage, amongst savages.
380	603	414	Long-continued habit of plucking out hairs may produce an inherited effect.